Recent discoveries of diamond and coesite in the upper crustal rocks of the Earth have drastically changed scientists' ideas concerning the limits of crustal metamorphism. Previously, it was thought that such ultrahigh pressure minerals could form only in the Earth's deep mantle or as a result of high energy impacts of extraterrestrial objects on the Earth's surface.

In examining the geological aspects of diamond and coesite in the Earth's crust, this book attempts to define an entirely new field of metamorphism. In doing so, it provides unique insights into the formation of diamond and coesite at very high pressures and explores new ideas regarding the tectonic setting of this style of metamorphism. After presenting a general overview of the geology and tectonics of UHPM, the text relates experimental and petrogenetic studies of UHPM minerals to P-T stability fields. Then the principal mineralogical indicators of UHPM are discussed, and details relating them to possible new, yet undiscovered areas are outlined. Several chapters discuss the structural style of the deformation of these UHPM rocks, relating them to subduction and continental collision. Estimated thermal and kinetic parameters are modeled to produce constraints on the conditions leading to UHPM. Separate chapters provide petrologic and tectonic accounts of UHPM occurrences in the Western Alps, Norway, China and Russia, as well as of the UHPM peridotites in Ronda, Spain and Beni Bouchera, Morocco.

This book will be of particular interest to researchers and graduate students of metamorphic petrology and global tectonics.

ULTRAHIGH PRESSURE METAMORPHISM

Cambridge Topics in Petrology

Series editors
Paul C. Hess, Brown University
Alan B. Thompson, ETH, Zurich

Cambridge Topics in Petrology is a new series of monographs and advanced texts on important areas of current research within the field of petrology. The series will cover a range of topics that reflect the increasingly complex body of research that falls under the umbrella of igneous and metamorphic petrology. The series aims to become a resource that assimilates, synthesizes and analyzes the wealth of widely disseminated information on the subject, and makes it available to a wider readership in a form accessible to the advanced student and researcher.

ULTRAHIGH PRESSURE
METAMORPHISM

Edited by

ROBERT G. COLEMAN   XIAOMIN WANG

Stanford University

CAMBRIDGE UNIVERSITY PRESS
Cambridge, New York, Melbourne, Madrid, Cape Town, Singapore, São Paulo

Cambridge University Press
The Edinburgh Building, Cambridge CB2 2RU, UK

Published in the United States of America by Cambridge University Press, New York

www.cambridge.org
Information on this title: www.cambridge.org/9780521432146

First published 1995
This digitally printed first paperback version 2005

A catalogue record for this publication is available from the British Library

Library of Congress Cataloguing in Publication data
Ultrahigh pressure metamorphism / edited by Robert G. Coleman, Xiaomin Wang.
p. cm.
ISBN 0 521 43214 6 hardback
1. Metamorphism (Geology) 2. High pressure (Science)
I. Coleman, Robert Griffin, 1923– . II. Wang, Xiaomin, 1962–
QE475.A2U48   1995
552'.4– dc20   94–13874

ISBN-13 978-0-521-43214-6 hardback
ISBN-10 0-521-43214-6 hardback

ISBN-13 978-0-521-54799-4 paperback
ISBN-10 0-521-54799-7 paperback

Contents

List of contributors   vii

Preface   ix

1   Overview of the geology and tectonics of UHPM
    R. G. Coleman and X. Wang   1

2   Experimental and petrogenetic study of UHPM
    H.-J. Massonne   33

3   Principal mineralogic indicators of UHP in crustal rocks
    C. Chopin and N. V. Sobolev   96

4   Structures in UHPM rocks: A case study from the Alps
    A. Michard, C. Henry, and C. Chopin   132

5   Creation, preservation, and exhumation of UHPM
    rocks   *B. R. Hacker and S. M. Peacock*   159

6   The role of serpentinite melanges in the unroofing of UHPM
    rocks: An example from the Western Alps of Italy
    M. C. Blake, Jr., D. E. Moore, and A. S. Jayko   182

7   Ultra-high-pressure metamorphic rocks in the Western
    Alps   *R. Compagnoni, T. Hirajima, and C. Chopin*   206

8   HP and UHP eclogites and garnet peridotites in the
    Scandinavian Caledonides   *E. J. Krogh and D. A. Carswell*   244

9   Microcoesites and microdiamonds in Norway: An
    overview   *D. C. Smith*   299

10   UHPM terrane in east central China   *X. Wang, R. Zhang,*
    and J. G. Liou   356

11   A model for the tectonic history of HP and UHPM regions
    in east central China   *E. A. Eide*   391

12   Diamond-bearing metamorphic rocks of the Kokchetav
    massif (Northern Kazakhstan)   *V. S. Shatsky, N. V. Sobolev,*
    and M. A. Vavilov   427

13   Orogenic ultramafic rocks of UHP (diamond facies)
    origin   *D. G. Pearson, G. R. Davies, and P. H. Nixon*   456

Index   511

Contributors

M. C. Blake, Jr.
U.S. Geological Survey, Menlo Park, CA 94025–2115

D. A. Carswell
University of Sheffield, Department of Earth Sciences, Beaumont Building, Brookhill, Sheffield S3 7HF, United Kingdom

Christian Chopin
Laboratoire de Géologie, Ecole Normale Supérieure, URA 1316 du CNRS, 24 rue Lhomond, 75231 Paris, France

R. G. Coleman
Department of Geological and Environmental Sciences, Stanford University, Stanford, CA 94305–2115

Roberto Compagnoni
Department of Mineralogical and Petrological Sciences, University of Turin, Via Valperga Caluso 37, 10125 Torino, Italy

G. R. Davies
Faculteit der Aardwetenschappen, Vrije Universiteit, De Boelelaan 1085, 1081 HV Amsterdam, The Netherlands

Elizabeth A. Eide
Stanford University, Stanford, CA 94305–2115

B. R. Hacker
Department of Geological and Environmental Sciences, Stanford University, Stanford, CA 94305–2115

Caroline Henry
Institut de Géologie, 11 rue Emile Argand, CH 2007, Neuchatel, Suisse

Takao Hirajima
Department of Geology and Mineralogy, Faculty of Science, Kyoto University, Kyoto 606, Japan

A. S. Jayko
U.S. Geological Survey, Menlo Park, CA 94025

E. J. Krogh
Institut for Bioogi og Geologi, University i Tromsø, P. O. Box 3085, Guleng, N-9001, Tromsø, Norway

J. G. Liou
Department of Geological and Environmental Sciences, Stanford University, Stanford, CA 94305–2115

H.-J. Massonne
Institut für Mineralogie, Ruhr-Universität, Bochum, Germany

André Michard
Laboratoire de Géologie, Ecole Normale Supérieure, URA 1316 du CNRS, 24 rue Lhomond, 75231 Paris, France

D. E. Moore
U.S. Geological Survey, Menlo Park, CA 94025

P. H. Nixon
Department of Earth Sciences, The University of Leeds, Leeds LS2 9JT, United Kingdom

S. M. Peacock
Department of Geology, Arizona State University, Tempe, AZ 85287–1404

D. G. Pearson
Department of Earth Sciences, The Open University, Walton Hall, Milton Keynes, Bucks MK7 6AA, United Kingdom

V. S. Shatsky
Institute of Mineralogy and Petrography, Siberian Branch of Russian Academy of Sciences, 630090 Novosibirsk, Russia

David C. Smith
Laboratoire de Minéralogie, Muséum National d'Histoire Naturelle, 61 Rue Buffon, 75005 Paris, France

N. V. Sobolev
Institute of Mineralogy and Petrography, Siberian Branch of Russian Academy of Sciences, 630090 Novosibirsk, Russia

M. A. Vavilov
Institute of Mineralogy and Petrography, Siberian Branch of Russian Academy of Sciences, 630090 Novosibirsk, Russia

Xiaomin Wang
Department of Geological and Environmental Sciences, Stanford University, Stanford, CA 94305–2115

Ruyuan Zhang
Department of Geology, Stanford University, Stanford, CA 94305–2115

Preface

The new discoveries of diamond and coesite previously thought to form only in the mantle or under cataclysmic conditions have mobilized the petrologic community into an intense search throughout the world for new occurrences of these minerals in crustal rocks. The advancement in petrologic knowledge has taken shape in three separate and interconnected spheres that have allowed us to explore further into the depths of our planet. Continued high temperature and pressure laboratory experimental synthesis produced heretofore unrecognized naturally occurring species that initiated a search for these minerals in nature. Identification and chemical analysis of micron-size mineral inclusions in resistant minerals by the electron and ion probe have provided the ground truth for verifying the presence of such ultrahigh pressure minerals (UHPM) occurring naturally. At a much different scale, geophysical seismic profiles across young orogenic zones containing UHP minerals, such as the Alps, reveal that fragments of subducted crust indeed penetrate into the mantle and that plate movements are capable of moving crustal materials to such great depths. In all new discoveries, such as this one, there has been an enthusiasm developed for discovering yet another occurrence of diamond or coesite. Nearly a decade ago these discoveries were not predicted, and so the reality of these new finds is only now being considered by the earth science community; therefore, we can look forward to new paradigms explaining the origin of UHP minerals.

Our efforts to recruit experts in this field so as to collate the rapidly expanding field of UHPM into a book started three years ago. Since that time new discoveries have been made, and amazing progress in explaining the tectonic setting of such occurrences is developing. This book is designed to bring together our knowledge and to provide a forum to describe carefully certain occurrences. We have no illusions that at least some of the data given in this book will become anachronistic before it is published, but these collected chapters provide a starting point for anyone interested in this exciting new development in earth science.

Our chapter authors have done an excellent job in summarizing the data up to this time, and we would like to thank them for their contributions and insights. We would also like to acknowledge the efforts of reviewers who have given us insights and materially improved the expression and content of the chapters. The enthusiasm and counseling of Catherine Flack, our Cambridge editor, has been of great help to us during the preparation of this book.

Reviewers

Anderson, T. B.	Kornprobst, J.	Platt, J.
Austrheim, H.	Larsen, R.	Rubie, D.
Compagnoni, R.	Liou, J. G.	Schertl, H.-P.
Eide, E.	McWilliams, M.	Thompson, G.
Ernst, G.	Moore, D.	Wang, X.
Janse, A. J.	Patrick, B.	Wilshire, H.

Stanford University   R. G. Coleman and Xiaomin Wang
January 1994

1

Overview of the Geology and Tectonics of UHPM

R. G. COLEMAN AND X. WANG

Abstract

An historical account provides a background to the recent discoveries of coesite and diamond. Descriptions of the general geologic and tectonic setting for ultrahigh pressure metamorphism (UHPM) areas in the Western Gneiss Belt of Norway; Kockchetav area, Kazakhstan; Dabie Shan, Eastern China; Western Alps and the Bohemian massif, Eastern Europe, indicate a similarity in tectonic setting. It is shown that the size of these areas is large and that they consist of coherent lower crustal material previously consolidated as part of older continental fragments. The protoliths of these rocks is usually characterized as old, dry, and cold supracrustal rocks. These older crustal rocks often appear to have undergone anatexis prior to being metamorphosed as part of a continent–continent collision. Special conditions are required to exhume these rocks which have reached depths in excess of 100 km. Numerous tectonic schemes have been proposed, but no single mechanism can be applied to all of the known occurrences of UHPM. Future studies will enhance our sketchy knowledge of the tectonic situation that allows UHPM rocks to be formed and then rapidly exhumed so as to be preserved within crustal exposures of metamorphic rocks.

Introduction

The recent discoveries of coesite and diamond in metamorphic rocks that are part of the Earth's upper crust has drastically changed our ideas concerning the limits of metamorphism. Formation of these ultrahigh pressure minerals requires pressures in excess of that expected within the crust of the Earth. Therefore it appears that fairly large masses of lighter crustal rock have been subducted into the Earth's upper mantle where such pressures exist. The present preservation of these high pressure minerals within the crust requires that

1

large blocks of the subducted crust return to the Earth's surface, preserving the history of deep subduction into the mantle. This book provides details of the mineralogy and stability fields of the UHPM minerals. Descriptions of the tectonic setting and occurrence of these UPHM rocks provide a background for future studies in this new style of metamorphism.

Historical account

The term *ultrahigh pressure metamorphism (UHPM)* refers to a *metamorphic* process that occurs at pressure greater than ~28 kbar (the minimum pressure required for formation of coesite at ~700°C). These processes generate coesite- and/or diamond-bearing metamorphic rocks or rocks containing such equivalent high-pressure mineral assemblage.

Coesite and diamond are well known in kimberlite pipes and meteorite craters. However, they are increasingly recognized in metamorphic rocks in several collision zones (Fig. 1.1). The first alleged coesite-bearing metamorphic rocks were reported by two Russian geologists (Chesnokov and Popov, 1965). They described inclusions of quartz aggregate in eclogitic garnets from the south Ural Mountains. These quartz inclusions are surrounded by radiating fractures that were believed to be caused by the volume increase in transformation from coesite to quartz. This occurrence, unfortunately, has never been confirmed and was not publicized perhaps because the paper was published in Russian.

The first confirmed coesite occurrence was reported by Chopin (1984). He described coesite and quartz pseudomorphs after coesite as inclusions in almost pure pyrope in a pyrope-quartzite from the Dora-Maira area of the Western Alps of Italy. This discovery was the first to demonstrate clearly that there are some crustal rocks have been metamorphosed at pressures >28 kbar at a temperature of 700°C. These unusually high pressures and temperatures for the coesite-quartz equilibrium have been experimentally determined (Mirwald and Massonne, 1980; Bohlen and Boettcher, 1982). Soon after the initial discovery, coesite was reported in eclogites from the Western Gneiss Region of Norway by Smith (1984). Coesite was confirmed by Raman spectroscopy. The impact of such discoveries was so great that work intensified in the following few years for these areas, especially in the western Alps. These researches include experimental, mineralogical, petrological, kinetic, and tectonic modeling studies (Chopin, 1984,1986; Gillet et al., 1984; Schreyer, 1985,1988; Van der Molen and Van Roermund, 1986; Smith and Lappin, 1989; Chopin et al., 1991; Michard et al., 1991; Reinecke, 1991; Schertl et al., 1991). A review paper by Schreyer (1988) published in *Episodes* summarized all the evidence and the geological implications up to that time.

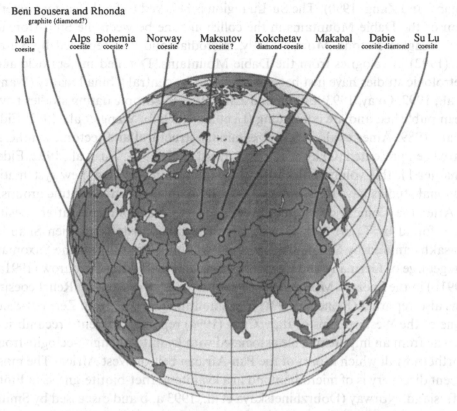

Figure 1.1. Global perspective showing the occurrence of UHPM localities in the Eastern Hemisphere.

No new localities of coesite were found during the next five years. In 1989, coesite was discovered in eclogites from central China (the Dabie Mountains) (Okay et al., 1989; Wang et al.,1989). Raman spectroscopy was also used by Wang et al. (1989) to confirm this coesite occurrence. For the first time, coesite was found within both garnet and omphacite at the same locality. Quartz pseudomorphs after coesite and calcite pseudomorphs after aragonite were also described by Wang and Liou (1991) for marble, metapelite and gneiss from the Dabie Mountains. Jadeite (Jd_{85}) was found in gneiss from the Dabie Mountains by Okay et al. (1989). About the same time, diamond-bearing gneiss from the Kokchetav massif of Russia was confirmed by Sobolev and Shatsky (1990). The recognition of diamond in gneissic terranes indicates that continental rocks might have been metamorphosed at even higher pressure (>40 kbar).

In the next few years, many other coesite localities were reported. Relict coesite was reported in eclogites from Jiangsu and Shandong (Su-Lu) region of eastern China (Hirajima et al., 1990; Zhang et al., 1990) following the earlier reports of quartz pseudomorphs after coesite (Yang and Smith, 1989;

Enami and Zang, 1990). The Su-Lu region is believed to be the eastern extension of the Dabie Mountains in the collision zone between the Sino-Korean
and Yangtze cratons. More recently, microdiamond was reported by Xu et
al. (1992) in eclogites from the Dabie Mountains. Detailed mineralogic and
petrologic studies have just become available for central China locality (Wang
et al., 1992; Okay, 1993; Enami and Zang, 1990). Isotopic dating studies have
been published and are continuing (Li et al., 1989a,b; Wang et al., 1992; Eide
et al., 1989; Ames et al., 1993). Preliminary structural and tectonic synthesis
have been presented (Okay et al., 1989; Ernst et al., 1991; Xu et al., 1992; Eide,
Chapter 11, this volume). Because of the interest in this area, new systematic
regional studies have been initiated by several independent scientific groups.

After the Chinese coesite discoveries, quartz pseudomorphs after coesite
were found by Tagiri and Bakirov (1990) in talc schists from Tien Shan in
Kasakhstan and by Schmadicke et al. (1991) in eclogites from the Saxonian
Erzgebirge of Germany and Czechoslovakia and by Bakun-Czubarow (1991a,
1991b) in the Sneznik Mountains of the East Sudeten of Poland. Relict coesite
was also reported (Reinecke,1991) in piemontite-quartzite from Zermatt-Saas
zone of the Western Alps of Italy. Caby (1994) reports apparent Precambrian
coesite from an impure marble associated with kyanite-phengite-eclogite from
northern Mali which is part of the Pan-African belt of West Africa. The most
recent discovery is of microdiamond in a kyanite-garnet-biotite-gneiss in Fjörloft island, Norway (Dobrzhinetskaya et al., 1993 a, b and discussed by Smith
in Chapter 9). All of these new localities have not yet been studied extensively.
and so the exact petrologic and tectonic settings are poorly known. So far,
most of the coesite and/or diamond localities are in the eastern hemisphere
and appear to be related to major continental orogenies that characteristically
involve continental collisions.

Geologic and Petrologic Nature of UHPM Occurrences

As a background for the following chapters in this book we have selected five
areas where UHPM rocks have been found and are well characterized. These
descriptions integrate published knowledge of the separate areas and will
include tectonic setting, lithology of the metamorphosed rocks and subjacent
lithic units. The present state concerning the age of metamorphism will be
summarized along with the important petrologic mineral assemblages in the
UHPM rocks. A geologic sketch map is presented for each of the five areas
discussed and these maps are all drawn at nearly the same scale so that it is
possible to make comparisons. A table provides various parameters that will
make it easy for the reader to evaluate each of these areas. The following

chapters will of course provide more detail and so this chapter is designed to give only a general summary.

Western Gneiss Region, Norway

The Western Gneiss Region (WGR) forms an elongate area along the southern coast of Norway from the Sognefjord in the south (61°N) to Trondheimsfjord on the north (63°N). The WGR is the basal tectonic unit of an assembly of thrust sheets that are emplaced onto the Baltoscandian continental margin during the Caledonian orogeny marking the continent–continent collision of the Greenland plate and the Baltic plate. The area affected by this continental collision of western Norway was at least 300 km long and 150 km wide (Fig. 1.2, Table 1.1). The WGR is considered to be autochthonous or parautochthonous crystalline basement of Middle Proterozoic or older age but contains infolded polymetamorphic nappes of Lower Paleozoic age that have been divided into two main units: (1) Fjordane complex, in a mixed association of supracrustal rock, anorthosite, and augen gneiss, on the west and northwest parts of the WGR; (2) Josuedl complex, a more coherent migmatitic orthogneiss unit to the east (Griffin et al., 1985; Cuthbert et al., 1983; Mørk and Krogh, 1987; Krogh and Carswell, Chapter 8, this volume).

The supracrustal rocks of the Fjordane gneissic complex consists of interlayered pelitic and igneous migmatite, marble, quartzite, and amphibolites. These rocks enclose gabbro, anorthosite, and peridotite which have been described as either tectonic inclusions or intrusions. Foliated intrusive rocks of intermediate to silicic composition invade the gneisses with the main intrusive type being augen gneiss (mangerite). These gneissic units predominantly exhibit amphibolite facies mineralogy, but relics of the high pressure assemblage such as clinopyroxene + garnet + quartz + K-spar + kyanite are present as coronas or are interstitial to the larger megacrysts. In restricted areas, relic high-pressure granulites showing variable amounts of retrogression to amphibolite facies are found in these gneisses (Cuthbert et al., 1983; Griffin et al., 1985).

Eclogites are widespread particularly within the Fjordane gneiss complex and occur as boudinaged pods or lenses from decimeters to tens of meters in length. Conformable contacts with the gneiss indicate recrystallization in the plastic zone of crustal deformation. Tabular bodies of eclogite suggest dike swarms of igneous origin. Gabbro and diabase intrusives deformed into lensoid shapes are concordant with gneissic foliation of the country rock which contains granulite or eclogite facies overprints (Griffin et al., 1985).

Table 1.1. *Comparisons of UHPM metamorphism from five selected areas*

	WESTERN GNEISS NORWAY	KOKCHETAV RUSSIA	DABIE MOUNTAINS CHINA	WESTERN ALPS	BOHEMIAN MASSIF
AGE OF THE PROTOLITH	1.6-1.8 Ga U/Pb zircon Torudbakken (1982)	1.8-2.6 Ga, > 2 Ga whole rock U/Pb Jagoutz (1990)	2.9 Ga, Dabie, 1.3-1.85 Ga (Zhang and Kang, 1989)	300-350 Ma Pb-Sr-Nd Tilton (1991)	550-600 Ma Cadomian ?
REGIONAL UHPM SUPRACRUSTAL ASSEMBLAGES	Gneiss ga+cpx+ksp+ky+plg+qtz	Zerendin gneiss plg+qtz+hlb+bi+ga+ph	Gneiss plg+qtz+hlb+bi+ph+ksp+ga	Gneiss and schist ab+ksp+ph+bi+qtz+ga	Gneiss granulite facies, variable
AGE OF UHPM	598 -107 Ma, Sm-Nd 433-437 ma, K/Ar cooling age (Krogh, 1992)	530 Ma, Sr-Nd-Pb Jagoutz (1990) 530 Ma U/Pb Claoue-Long et al (1991)	218 Ma, U-Pb Leslie Ames (1992)	100 to 38-40 Ma, Sr-Nd-Pb Tilton, Blueschist 90 - 60,100 - 80, 110-70 Ma	350 - 400 Ma
PT PEAK UHPM	32 ± 5 kbar, 800°±100°C 24 ± 5 kbar, 400°±100°C 14 -15 kbar, 720°±100°C 10 -18 kbar, 400-825°±100°C	40 kbar 1000°C	28 kbar 600 - 780°C	37 kbar 800°C	13 - 20 (37?) kbar 650 - 880°C
TECTONIC EVENT	Caledonian orogeny, collision of European plate with North American plate closing Iapetus	Caledonian orogeny, collision of Kokchetav massif with central Kazakhstan	Collision of Sino-Korean craton with Yangtze craton	Collision of Apulia with European craton; Western Alps 60 - 30 Ma Central Alps 40 - 0 Ma Insubrick Fault 30 - 0 Ma	Variscan orogeny collision of Laurasia with Gondwana 320 - 400 Ma
EXHUMATION DURATION	80 Ma	72 Ma	100 Ma	40Ma	100Ma
EXHUMATION RATE	1.4 mm / yr	1.8 mm / yr	1.3 mm / yr	2.5 mm / yr (Assuming 40 Ma for UHPM event)	1.4 mm/yr
SIZE OF THE UHPM MASSIF	350 x 150 km 2.1 x 10^6 km^3	300 x 150 km 1.8 x 10^6 km^3	450 x 100 km 4.5 x 10^6 km^3	200 x 75 km 0.6 x 10^6 km^3	50 x 100 km "Gfohl Terrane" 3.0 x 10^3 km^3

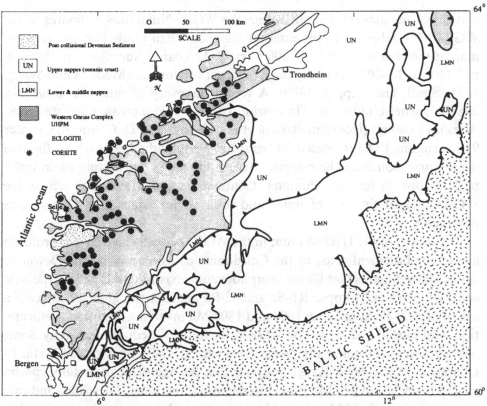

Figure 1.2. Map of Western Gneiss Region in Norway, showing distribution of eclogites. (Modified from Zwart, 1973; eclogite localities from Griffin and Mörk, 1981)

The eclogite mineral assemblages are mainly bimineralic omphacite and garnet with minor quartz, rutile, phengite, and kyanite. Coesite has been reported from eclogites collected at Grytting, Slje district in the northwestern part of the WGR in the Fjordane complex (Smith, 1984; Smith and Lappin, 1989); coesite coexists with garnet + pyroxene + quartz + dolomite + amphibole + rutile and apparently is very rare in the WGR. Orthopyroxene-bearing eclogites from garnet peridotites are referred to as internal eclogites and contain garnet + clinopyroxene + orthopyroxene + amphibole + rutile + quartz. These "internal" eclogites are thought to have undergone the same Caledonian high pressure history even though they may have had an ultimate mantle origin (Carswell and Harvey, 1985). Orthopyroxene eclogites have also been found within the gneissic rocks and are not exclusively present in the garnet peridotites.

The peak metamorphic conditions for the WGR is 700–850°C at 30 kbar where coesite is present. Structural consanguinity between the high pressure assemblages in the gneiss, peridotite, anorthosite, and augen gneiss indicates a regional metamorphic event within the WGR. Numerous estimates of the WGR *P-T* established in the external eclogites, when plotted on the geologic map of the area increase from 500°C at 10–12 kbar at Sunnfjord, in the east, to 800°C at 18–20 kbar along the west coast of Norway (Medaris and Wang, 1986; Smith and Lappin, 1989). A recent discovery of microdiamond in a kyanite-garnet-biotite-gneiss in Fjörloft Island would increase estimated pressures up to 40 kb (Dobrzhinetskaya et al., 1993a,b and D. C. Smith, Chapter 9, this volume.) The estimated retrograde *P-T* path indicates nearly isothermal depressurization from the eclogite to granulite facies, suggesting rapid uplift perhaps due to tectonic thinning. Continued nappe translation across the WGR changed the rate of uplift and some minor anatexis of the gneisses developed.

The timing of the UHPM event in the WGR is constrained by the protolith age of the basement prior to the Caledonian collision and by the Devonian age of the sediments that fill the morphotectonic basins developed on the west coast of Norway. Isotopic Rb-Sr and U-Pb ages on the gneisses indicate a major crustal event between 1600 and 1800 Ma involving intrusion, deformation, and partial melting of the WGR prior to the Caledonian orogeny. Some gabbro bodies intruding the gneiss have similar U-Pb ages of 1517 Ma. In Chapter 8 of this book Krogh suggests that the age of peak metamorphism ranges from 375 to 420 Ma based on numerous methods. These data indicate that approximately 80 Ma elapsed between the peak metamorphism and exhumation. It has been suggested that uplift is the result of erosion of the thickened bouyant crust. Deformation continued until Late Devonian when clastic deposits started to accumulate within extended basins of the WGR terrane (Cuthbert et al., 1983).

Kokchetav, NW Kazakhstan, Russia

The Kokchetav massif (see Chapter 12, this volume, for a more detailed discussion of this area, by Shatsky, Sobolev and Vailov) is located at the very northern boundary of the Kazakh Territory of Russia approximately 300 km southeast of Omsk. A low relief area is bounded on the west by the Turgi Basin and on the north by the West Siberian Basin. The Kokchetav massif lies within the Central Asian fold belt, and the Precambrian outcrops are part of a series of ancient blocks that form a single Kazakhstan–North Tienshan Precambrian massif modified by early Paleozoic deformation (Zohenshain et al., 1990). These basement rocks are complexly folded, faulted and intruded by Paleozoic diorite, tonalite and granodiorite plutons (Fig. 1.3, Table 1.1).

The Kokchetav massif may have been a fragment from a larger ancient massif consisting of gneiss domes, greenstone belts, and pure quartzite. Con-

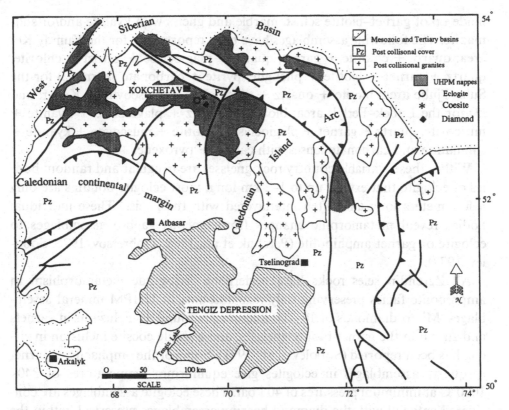

Figure 1.3. Map of Kokchetav massif and surrounding area showing distribution of diamond and coesite in UHPM rocks. (Modified from Zonenshain et al.,1990; and other details from Sobolev, 1977.)

solidation of the Kazakhstan followed the Late Precambrian breakup of Pangea. During this collision event between small continental blocks, diamond-bearing UHP rocks were formed within deeply subducted continental slices. Early Paleozoic arcs formed along the edge of the consolidating Kokchetav block, producing calc-alkaline intrusives and volcanics that invaded these ancient basement rocks. The sedimentary history in and around the Kokchetav massif indicates extensive erosion since the initial collision that produced the UHPM event. To the south of Kokchetav, the Tengiz depression consists of a thick section of flysch and molasse deposits occupying a foreland basin that developed as the Kokchetav massif thrust southward. Evidence for nappe structures is lacking within the Kokchetav massif, but the strong postmetamorphic deformation and retrogression indicate extensive tectonic transport. No record of "Post-Caledonian" sedimentation or igneous activity was found, indicating that the Kokchetav massif has been stable and near the Earth's surface since early Paleozoic.

The central part of the Kokchetav massif consists of the Zerendin series,

made up of garnet–biotite schist, marble and gneiss with kyanite and/or sillimanite. The following assemblages have been reported (1) for the Kumdy Kol
area: quartz + chlorite + garnet, quartz + plagioclase + garnet + chlorite,
quartz + garnet + biotite + phengite, pyrite + carbonate skarn; (2) for the
Sulu–Tjube area: biotite + quartz + plagioclase + muscovite + Kspar; and
(3) for the Enbek–Berlyk area: biotite + quartz + plagioclase + kyanite +
muscovite, quartz + garnet + plagioclase + biotite + muscovite + kyanite +
sillimanite and carbonate lenses with calcite + pyroxene.

Within these, variable country rock gneisses are irregular and random bodies of eclogite that extend up to 1–2 km long; most eclogite boudins are only
a few meters long and often interlayered with the gneiss. These individual
bodies reveal metamorphic changes from igneous gabbro assemblages to
eclogite or garnet amphibolite (Perchuk et al., 1969; Dobretsov, 1975; Sobolev, 1977).

All Zerendin seies rocks have undergone retrograde metamorphism to
amphibolite facies preserving only rare amounts of UHPM mineral assemblages. Micro-diamonds of 10–200 microns are found as inclusions in garnets
and zircon in the garnet-biotite gneisses, and a single coesite inclusion in zircon has been reported (Sobolev et al., 1991). Some of the omphacite + garnet
+ quartz assemblages in eclogites give equilibrium temperatures of 840–
1000°C at minimum pressures of 40 kbar. These eclogite assemblages are considered cofacial with the diamond-bearing assemblages preserved within the
garnets of the biotite-garnet gneiss which give values of 800–900°C at 40 kbar.
Other Kokchetav eclogites give lower P-T parameters of 580–780°C with P
of 12–14 kbar representing retrograde conditions (Sobolev and Shatsky, 1987,
1990; Claoue-Long et al., 1991).

Sm-Nd ages of the Zerendin diamond-bearing gneisses indicate an Archean
age of formation 2.5 Ga and an earlier metamorphic age of ~2 Ga. U-Pb and
Sm-Nd ages on the high-pressure eclogites indicate a Middle Cambrian age
of 530±7 Ma for the UHPM event (Jagoutz et al., 1989; Jagoutz, 1990; Claue-
Long et al., 1991).

Dabie Shan, Eastern China

The Dabie Shan is located in eastern China, 550 km west of Shanghai; it forms
an uplifted area around which the westward course of the Yangtze River is
diverted southward near Wuhan. (A more detailed description can be found
in Chapters 10 and 11 of this volume.) The intersection of borders among
Hubei, Henan, and Anhui provinces is centered within the Dabie Shan orogen
which forms the collision zone between the Sino–Korean and Yangtze cratons
(Fig. 1.4, Table 1.1).

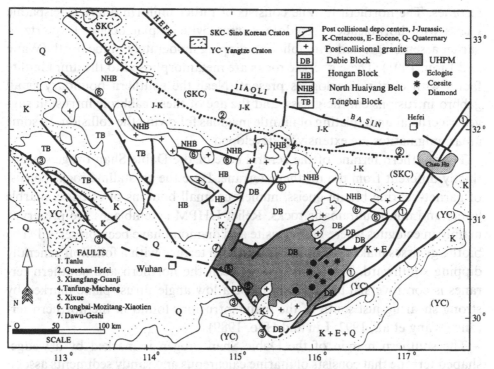

Figure 1.4. Map of Dabie Mountains orogenic belt showing distribution of UHPM rocks. (Modified from Geologists of Anhui Province, 1986; Geologists of Henan Province, 1989; Geologists of Hubei Province, 1990; and details from Wang et al., 1992.)

Arbitrary divisions of this collision zone are made according to the currently accepted ideas. The southern edge of the Sino-Korean craton is covered by Mesozoic–Tertiary sediments within the extensional Hefei–Jiaoli basin, and subsurface information indicates that the basement consists of Archean rocks similar to those exposed further north in the Sino-Korean craton. The eastern prolongation the Qinling Fold Belt forms the northern boundary of the Dabie Shan orogenic zone and is called the North Huaiyang Belt (NHB). Within this intensely deformed belt, Paleozoic active margin sediments and volcanics grade upward into Carboniferous coal-bearing sediments interlayered with marine sediments. These rocks have been recrystallized to the greenschist facies but lack the higher grades of metamorphism characteristic of the Qinling fold belt to the west. These deformed Paleozoic continental margin rocks are truncated by the Tongbo-Mozitan-Xiaotien fault which is interpreted as a high angle reverse fault having some strike-slip motion (Yang, 1986; Suo et al., 1987; Ma and Zhang, 1988; Liu and Hao, 1989; Shui and Xue, 1989; Wang et al., 1990; Ernst et al., 1991).

The central core of the Dabie Shan orogen consists of Late Archean to Early Proterozoic crystalline basement which has been divided into separate terranes. The northern terrane consists of rather uniform quartzo-felspathic gneiss with hornblende + epidote + muscovite + plagioclase + quartz ± garnet assemblages and with foliation dipping moderately to the north (Wang and Liou, 1991). These gneissic rocks are metamorphosed to the amphibolite facies with only rare eclogites present. Near the southern boundary small gabbro intrusions cut the gneiss, and the age of these gabbroic intrusion suggests a crustal underplating of mantle magmas following the collision. In some areas, migmatites are extensively developed.

The southern terrane of the central core of the Dabie Shan orogen is distinctly different from the northern terrane. Here the crystalline rocks consist of garnet-bearing biotite gneiss, numerous small boudins of eclogite, marble layers and minor ultramafic rocks. Relict UHPM assemblages are preserved mainly in eclogites where both coesite and diamond have been preserved. The biotite gneisses make up 80% of the southern terrane with a foliation generally dipping southward. The boundary between the northern and southern terranes is controversial but appears to be a low angle thrust plane marked by strong shearing just south of the mafic intrusions found in the northern terrane (Wang et al., 1992; Liu and Hao, 1989).

The southern margin of the Dabie Shan orogen is marked by a wedge-shaped terrane that consists of marine calcareous and sandy sediments associated with keratophyric and spilitic volcanic rocks of Sinian to Lower Paleozoic in age. This group of rocks are recrystallized mainly in the blueschist and greenschist facies, with a northward increase in metamorphic grade. This packet of rocks form nappe structures with south vergence and are thrust over the edge of the Yangtze continental margin resulting in a foreland fold belt within the Pre-Triassic continental margin sediments. Even though eclogite- and amphibolite-grade metamorphic rocks are present along the northern contact with the Dabie Shan core, a distinct gap in P-T conditions and lithologies exists between these two juxtaposed terranes. Low-angle thrusting of the Dabie Shan UHPM unit over the blueschist terrane indicates southward movement, with nearly 43% shortening of the foreland fold belt of the Yangtze craton (Dong et al., 1989; Wang and Liou, 1991).

The eastern boundary of the Dabie Shan orogen is truncated by the Tan-Lu strike-slip fault system comprised of a series of major faults that trend in a northeasterly direction and extend more than 5000 km northward into Korea and Russia. The Tan-Lu fault is considered to have developed after collision of the Yangtze and Sino-Korean cratons and has greatly modified the tectonic framework of the collision zone by left lateral displacement of more than 500–700 km. Extensional faulting and pull-apart structures in the early stages of the Tan Lu fault localized postcollisional Mesozoic granites.

These granites invaded the Dabie Shan orogen near intersections of the wrench faults and the earlier east-west trending thrust and reverse faults developed during the collision. Continued left lateral movement along the main Tan-Lu fault has offset parts of the Dabie Shan orogen nearly 500 km northward into the Jiangsu-Shandong areas (Wang and Liou, 1991; Wang et al., 1992) where coesite-bearing eclogites are reported (Hirajima et al., 1990; Zhang et al., 1990)

The country rock gneiss of the southern Dabie Shan UHPM terrane consists mainly of quartz + biotite + muscovite + garnet + epidote + Kspar + plagioclase + rutile. Quartz pseudomorphs after coesite have been identified as inclusions within the garnets of these gneisses. Eclogites are found as small isolated blocks (2 cm to 20 m) within the biotite gneiss and marble zones, as layers within banded biotite gneiss and garnet amphibolite, or within garnet pyroxenite. Coesite and diamond are present as inclusions in garnets from these eclogites. Wang et al. (1992) recognize three metamorphic stages in SDT starting with blueschist facies metamorphism preceding the coesite-diamond eclogite facies metamorphism and ending with an overprinting of retrograde amphibolite-greenschist facies metamorphism. The peak metamorphism for the Dabie Shan orogen diamond inclusions in eclogite is 40 kbar at 900°C, and for coesite inclusions in garnet is 28 kbar at 750°C. The retrograde assemblages of the eclogites formed at 6 kbar at 500°C produce an overall clockwise path (Okay et al., 1989; Wang et al., 1989, 1992; Xu et al., 1992). U-Pb ages from the country rock biotite gneiss range from 1.9 to 2.4 Ga indicating an Archean history and perhaps metamorphism up to amphibolite facies at this time. Age data on the diamond-coesite-bearing eclogites range from 212 to 249 Ma consistent with early Mesozoic collision established for the Sino-Korean and Yangtze cratons (Xie et al., 1983; Sang et al., 1987; Li et al., 1989; Li and Liu, 1990; Wang et al., 1992; Xu et al., 1992; Ames et al., 1993).

Western Alps (Dora-Maira)

The Dora-Maira massif is located in the Western Alps approximately 50 km west of Torino and forming the western boundary of the Po Basin. It consists of crystalline continental basement rocks that are exposed below the Upper Pennine nappes (ophiolitic and oceanic sediments). Continental basement rocks also occur at Gran-Paradiso and Monte Rosa as windows (Fig. 1.5, Table 1.1). These recrystallized and metamorphosed massifs represent the margin of the European plate which has been deeply subducted southward under the Apulian plate (Africa). The Apulian plate consists of the Austro-alpine nappes that form the highest structural unit of the eastern Alps. Just

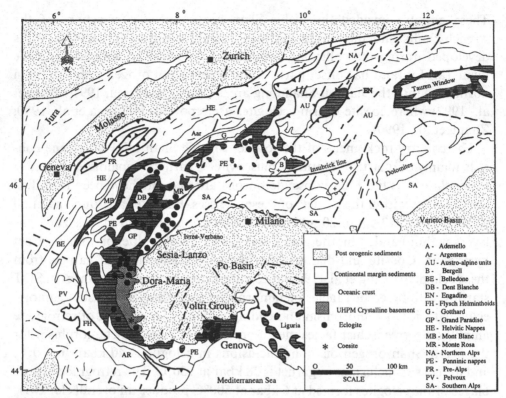

Figure 1.5. Map of Western Alps showing the distribution of eclogites and coesite bearing rocks. (Modified from Dietrich, 1979; distribution of eclogites after Bocquet, 1971; Dietrich et al., 1974.)

to the north of the Dora-Maira massif, an elongated Sesia Lanzo zone consists of deeply subducted, polymetamorphosed continental crust derived from the Apulian continental margin. The Sesia Lanzo zone overides the Pennine nappes to the west and is in transcurrent fault contact with the Hercynian continental basement of the Ivrea Zone (Compagnoni et al., 1977; Dal Piaz and Ernst, 1978; more detailed discussions are given in Chapters 4, 6 and 7 of this volume).

The Dora-Maira massif has been divided into several lens-shaped nappes separated by low-angle thrust faults. The basal nappe consists of Triassic and Carboniferous metasediments recrystallized in the blueschist facies and overprinted by greenschist assemblages. The Venesca unit overlies the basal nappe and consists of polymetamorphic schists and gneisses. The lower nappe of this unit contains coesite-bearing pyrope garnets within boudinaged quartzite layers. These layers are present within the fine-grained gneiss country rock near its contact with the Isasca formation consisting of metavolcanics and metasediments. Individual nappe units above the coesite-bearing units have

similar protoliths but were metamorphosed at lower pressure and temperature. Associated metasediments in these upper units are considered to be Permian to Middle Triassic in age. Pressure discontinuties between these units of
10–60 km of burial depth indicates considerable dislocation by faulting. Such
faulting has placed a thin slice of the upper Pennine oceanic nappes above
the Venesca unit, which in turn is overlain by another nappe of continental
basement consisting of polymetamorphic schist and orthogneiss associated
with Early Mesozoic metaclastics similar to those in the Venesca unit but also
containing Late Mesozoic calcschists. The metamorphic grade of this upper
nappe is coesite-free blueschist to eclogite facies rocks. The upper Pennine
nappes in this area are represented by the Mont Viso ophiolites which also
contain blueschist and eclogite facies and represent the highest structural unit
in the nappe pile of this area (Chopin, 1984; Chopin et al., 1991; Michard et
al., 1991; Schertl et al., 1991).

The pyrope-bearing quartzites containing coesite relics are rich in magnesia
(~7% MgO); the high pressure (HP) assemblage is pyrope + kyanite + talc +
phengite + coesite (quartz). The quartzite boudins are contained within fine-
grained gneiss with the assemblage albite + microcline + phengite + biotite
+ quartz + clinozoiste containing garnet and rutile relics. Eclogite mineral
asemblages are present within the Venesca unit and record the Alpine high
pressure event but do not contain coesite. The presence of grossular and rutile
as armored relics is evidence to suggest that the country rock gneiss attained
the same grade of metamorphism as the coesite-bearing pyrope quartzites
(Chopin, 1981, 1984, 1987; Goffe and Chopin, 1986; Pognante et al., 1987;
Vuichard and Ballevre, 1988; Chopin et al., 1991; Kienast et al., 1991; Reinecke, 1991).

The peak metamorphism of the Dora-Maira is considered to be 28–35 kbar
at 700–750°C for the coesite assemblage. Eclogite of the other Vanesca units
indicates 17–20 kbar at 550–580°C. Eclogite assemblages from the Monte Viso
meta-ophiolite nappe give 12 kbar at 500–550°C. The fine-grained gneiss containing the zones of pyrope quartzites has a retrogressive mineral assemblage
that indicates 6–3 kbar at 350–500°C, a record of the final adjustment during
exhumation. The clockwise *P-T* path for the Dora-Maira UHPM is interpreted to indicate that these rocks were at shallow levels at the beginning of the
pro-grade Alpine orogeny (Chopin, 1984; Vuichard and Ballévre, 1988; Pognante and Sandrone, 1989; Chopin et al., 1991; Pognante, 1991; Reinecke,
1991; Schertl et al., 1991). The age constraints on the UHPM event in the
Dora-Maira massif are not yet clarified; however, it is obvious that the protolith associated with the pyrope-bearing coesite underwent a Hercynian metamorphic event with formation of granites (303 Ma). The age of the UHPM is

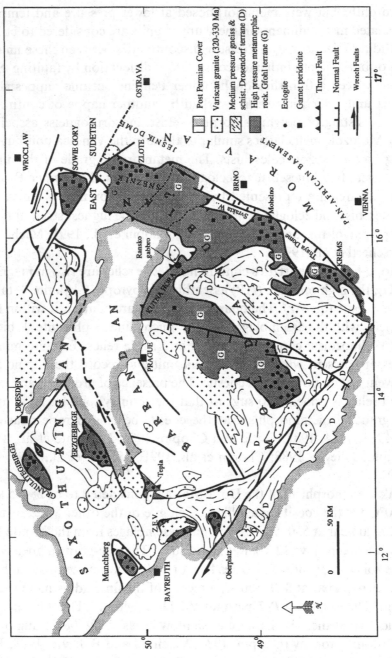

Figure 1.6. Map of Bohemian Massif showing the distribution of high pressure rocks
(Modified after Franke, 1989; Matte et al.,1990)

considered to have peaked between 90–125 Ma using $^{40}Ar/^{39}Ar$ on phengitic micas and a second age group using $^{40}Ar/^{39}Ar$ between 35–41Ma was interpreted as representing retrograde assemblages (Monié and Chopin, 1991; Monié and Philipot, 1989). In contrast to these ages, Tilton et al. (1991, 1989) reported age groupings of 40 Ma and 38 Ma using Pb, Nd, and Sr isotopes on zircon inclusions coexisting with coesite in the large pyrope crystals. Dates from other high pressure units such as the Sesia Lanzo and Zermatt-Zass zone (now known to contain coesite) indicate a high pressure event between 130–80 Ma; there appears to be good evidence of subduction and collision beginning in the Cretaceous. In contrast to these ages, the Glarus Alps has metamorphic ages of 29–36 Ma and the Lepontine event in the central Alps is considered to be 35–40 Ma (Becker, 1993). These results open the possibility that there may have been UHPM developing during the final collision between Africa and Europe rather than being confined to the Cretaceous (Bocquet et al., 1974; Hunziker, 1974; Chopin and Maluski, 1980; Hunziker and Martinotti, 1984; Stockhert et al., 1986; Biino and Pognante, 1989; Hunziker et al., 1989; Paquette et al., 1989; Tilton et al., 1990, 1991).

Bohemian Massif

The Bohemian massif is a large area (~90,000 km^2) of Pre-Permian rock that consists of mainly high-grade metamorphic rock, granite, and Paleozoic sedimentary rocks, all of which have been involved in the Variscan orogeny (Matte,1986; Franke,1989a,b; Matte et al., 1990). This orogeny is considered to have produced high-grade metamorphism within deeply subducted slabs of continental crust during the collision of Laurasia with Gondwana from the Devonian to Carboniferous.

The recent reports of quartz pseudomorphs after coesite at several localities where Variscan metamorphism has produced eclogites indicate that supracrustal rocks within the Bohemian massif may have undergone UHPM during collision. Matte et al. (1990) consider this major tectonothermal event within the Bohemian massif to be restricted between 300 and 290 Ma. They have divided the massif into various terranes whose boundaries are controlled by Variscan age thrusts and strike slip faults (Fig. 1.6). The eastern boundary of the Bohemian massif is marked by the Moravian terrane considered to have a Pan-African basement (containing 550 to 580 Ma granites) covered by Paleozoic platform sediments that have undergone a relatively low-grade metamorphism except where it is exposed along its present transcurrent fault boundary with the Moldanubian zone of the Bohemian massif (Matte, 1986; Franke, 1989a,b; Matte et al., 1990). Along this disturbed zone of dextral transpression a series of domes and windows expose narrow and highly deformed and metamorphosed parts of the Pan-African basement.

To the west of the Moravian terrane the familiar Moldanubian zone has been divided into two separate terranes that have undergone high-grade metamorphism as a result of the Variscan continental collision (Matte et al., 1990; Carswell, 1991). The Gföhl terrane is structurally the highest tectonic unit consisting of more than 2 km of high-grade ortho- and paragneisses showing some anatexis and containing lenses of garnet peridotite associated with eclogites and granulites. These mafic and ultramafic rocks are concentrated near the basal thrust of the Gföhl terrane. Eclogites are also found as boudins in amphibolites (Matte et al., 1990). The Gföhl terrane includes similar high-grade rocks to the north in the East Sudeten within the Sneznik Mountains along the Polish Czechoslovakian boundary (Matte et al., 1990; Bakun-Czubarow, 1991a,b).

The structurally lower Drosendorf terrane consists of nearly 6 km of pelitic sediments that are considered to represent passive margin deposits of Lower Proterozoic to Lower Paleozoic sediments that formed adjacent to the Moravian (Gondwana) continental margin. These meta-sediments are now biotite-garnet-staurolite $\pm$ kyanite $\pm$ sillimanite schist and gneiss all of which have undergone pervasive anatexis. The Drosendorf terrane is devoid of high-pressure garnet peridotites or eclogites that are so abundant in the overlying Gföhl terrane. Structural evidence shows that these two terranes have a polyphase metamorphic history with the climactic metamorphism taking place around 350 Ma (Matte et al., 1990). The present attitude of foliation in both terranes is generally at low angles that have been folded into large-amplitude synforms and antiforms.

The garnet peridotites of the Gföhl terrane retain a mantle assemblage that equilibrated at 26 kbar and 1000–1500°C overprinted by a lower temperature (1050°C), but a higher-pressure event at 31 kbars considered to represent the Variscan continental collision (Carswell, 1991). The presence of quartz pseudomorphs after coesite in the eclogite-granulite rocks of the Zlote Mountains indicates a deep subduction of the supracrustal rocks of the Gföhl terrane (Bakun-Czubarow, 1991a,b). Estimated P-T of 18 kbar and 880°C for the eclogite-granulite rocks of this terrane are somewhat higher than those reported from other coesite and diamond-bearing UHPM areas (Bakun-Czubarow, 1991a,b). The variation in P-T conditions over short distances between these terranes of the Moldanubian zone indicates tectonic juxtaposition after continental collision during the rapid exhumation of the Variscan orogen (Franke, 1989; Matte et al., 1990). Earlier reports on the HP metamorphism within the Bohemian massif suggested a Caledonian Age of metamorphism, but these ages probably represent protolith ages rather than the Variscan HP metamorphic event (Franke, 1989; Matte et al., 1990). Further age work is required to verify the age of the UHPM Variscan metamorphic event, but the contemporaniety between collision and UHPM is similar to other areas where an earlier medium pressure metamorphism and partial melting of the older supra crustal material can be related to magmatic underplating and continental extension.

Separating the SE-verging Moldanubian Zone from the NW-verging Saxo Thuringian terrane is the Barradian terrane which consists of a Proterozoic (Cadomian) basement of low-grade, medium pressure, metamorphosed Early Devonian volcanic pile and interlayered sediments (Franke, 1989; Matte et al., 1990)(Fig. 1.6). Overlying the Cadominian basement is a well-dated Cambrian to Middle Devonian sedimentary sequence that changes upward from shallow water marine clastics to shales and limestones. In contrast to the Moldanubian zone, the Barradian terrane shows much less intense deformation and metamorphism which is also considered to be Variscan (Matte et al., 1990). Separating the Barradian terrane on the NW side is the Tepla Zone, which is considered to be the root zone for the high-grade metamorphic Munchberg gneiss to the west, and Sowie Gory to the east; both are klippen that have been thrust over the Saxo Thuringian terrane. The Granulitegebirge and Erzgebirge are also thrust over the Saxo Thuringia terrane and in this synthesis are considered allochthonous rather than windows of Precambrian basement (Fig. 1.6). The Saxo Thuringia Terrane consists mainly of Cadomian basement overlain by Cambro-Ordovician to Devonian sediments, and no evidence has been found to show that the basement for this terrane is Pan-African (Gondwana).

The high-grade metamorphic eclogites and amphibolite eclogites form the highest nappe in the Munchberg klippe where they are conformable with the surrounding meta sediments (Fig.1.6). The equilibrium assemblage kyanite + omphacite indicates pressures in excess of 20 kbar at temperatures of 600–650°C. A radiometric date using Sm-Nd and Rb-Sr suggests an age of 395 Ma for the high pressure metamorphism with exhumation medium-presure events around 370–390 Ma (Okrusch et al., 1991). Just south of the Munchberg gneiss (Fig. 1.6) the Erbendorf-Vohenstraus zone (ZEV) contains eclogite relics overprinted by a 380-Ma medium-pressure event (Okrusch et al., 1991). Further south, in the Oberplatz klippen of the Gföhl terrane within the Moldanubian zone at Winklaren, eclogites associated with garnet peridotites show a peak high-pressure event at 710°C and ~15 kbar overprinted by granulite conditions from 700 to 800°C and at pressures from 7 to 12 kbar in thrust contact with a Drosendorf terrane cordierite-sillimanite gneiss dated at 330 Ma (Okrusch et al., 1991)

The Erzgebirge high-grade metamorphics, found in the Flöha syncline area, consist of eclogite, garnet peridotite, and kyanite-bearing gneiss (Schmadicke, 1991). Quartz pseudomorphs after coesite are found in eclogites associated with omphacite, phengite, quartz, and rutile indicating an equilibrium assemblage that formed at 700–800°C and 13–16 kbar (Schmadicke, 1991). These high-grade units form nappes above biotite gneisses and amphibolites that are barren of HP minerals.

The Granulitegebirge contains garnet peridotites and two-pyroxene granulites that are localized at the base of several imbricated nappes. These nappes have been folded into NE-trending antiforms. Matte et al. (1990) interpret the Granulitegebirge as a gneiss dome based on Precambrian ages from the metamorphics; however, the similarity of these rocks to the Moldanubian Gföhl terrane high-grade rocks suggests that the Granulitegebirge may also be a klippe of stacked nappes related to the Variscan continental collisional metamorphic event. Estimates of the early granulite metamorphism conditions are 900–1000°C and 8–11 kbar, and the associated eclogites in garnet peridotites reveal similar *P-T* conditions. (N.V. Dobretsov, personal communication)

The reassessment of ages for the high pressure event in the Bohemian massif indicates that this event is related directly to the final stages of continental collision in the Variscan Orogeny (~400–350 Ma). Older ages derived from the granulites and associated gneiss interlayered with the eclogites and garnet peridotites probably relate to the protolith age of the supracrustal rocks. Post-collisional granites (320–330 Ma) invade most of the high-grade terranes of the Bohemian massif. Clastic fragments of the Bohemian metamorphic core are present in sediments of Visean to Nammurain age found in the syntectonic foredeep basin within the Moravian Terrane, marking the exhumation of the deeply subducted, high-grade supra crustal rocks.

Comparisons of UHPM Areas

Size of Individual UHPM Blocks

The regional nature of UHPM is controversial because only relics of some UHP minerals are found in surface exposures or as inclusions in resistant minerals such as garnet and pyroxene within eclogite boudins. Prior to finding such relics in the country rocks enclosing the eclogites, it was considered that the eclogites had an exotic tectonic origin. The small size of these eclogites provided no great obstacle to their introduction by tectonic events. Careful attention to locating armored relics within the country rock gneiss enclosing the eclogites has revealed that these relics are widespread. Conformable structures between the eclogites and gneisses support the idea that the UHPM events were on a regional scale. The estimated size of these tracts (Table 1.1, Figs. 1.2–1.6) is large and suggests that the UHPM areas represent coherent packets of supracrustal continental rocks. The actual thickness of these packets is difficult to estimate because of complex folding and intrusive activity. It is not unreasonable to assume that these continental slices may be in excess of 30 km thick and could even represent former coherent complete sections of continental crust. The calculations used to arrive at the volume of these UHPM blocks assumed a thickness of 30 km except for the Bohemian massif (Table 1.1).

Protolith and Age of UHPM Rocks

All of the described UHPM areas consist predominantly of supracrustal rocks consisting of pelite, quartzite, marble, orthogneiss, granite, diorite, tonalite, and trondjhemite. UHP minerals have also been identified within ophiolite and mantle peridotites, thus it is evident that preservation of UHP assemblages is not restricted to old supracrustal rocks. The prevailing metamorphic grade of these rocks at their present level of exposure is amphibolite facies, but relics of granulite facies perhaps developed earlier in the lower continental crust prior to continental collision. Migmatite, restite, and mobilized intrusive granites attest to widespread anatexis and mobilization of these supracrustal rocks prior to their UHPM event. There is very little manifestation of widespread melting during the high-pressure metamorphism. Systematic geochronolgy of the supracrustal rocks reveals that they had undergone anatexis and granulite facies metamorphism long before the UHPM events; however, there may have been minor anatexis during exhumation. The presence of recognizable mafic and ultramafic intrusives within these supracrustal rocks suggests that during the early stages of cratonization mafic underplating was an important process. There is now evidence that ophiolite slabs (ancient oceanic crust) have undergone UHPM, but their relationships with the larger masses of supracrustal rock containing diamond and coesite is not clear. Reinecke (1991) reports coesite-bearing metasediments associated with ophiolites in the Zermatt-Sass Zone of the Pennine Nappes in the Alps. Relic igneous minerals and structures in some eclogites of UHPM areas support the idea that at least some of these eclogites were derived from earlier mafic intrusions. The high-

grade plastic deformation (boudinage) of these mafic intrusion has obscured
their igneous protolith. Garnet peridotites of some UHPM areas are thought
to represent tectonic slices of mantle introduced during collision; at least some
of these ultramafic rocks had their ultimate origin as cumulates crystallizing
during crustal underplating prior to the UHPM collisional event. In some
cases, eclogite layers within these peridotites indicate that they too may have
had an upper mantle history. More recently Pearson et al. (1993, and in Chapter 13) have described graphite pseudomorphs after diamond in the Beni
Bousera and Ronda orogenic peridotite bodies in the Betico-Rifean tectonic
belt of the western Mediterranean as well as microdiamonds in ophiolite peridotites in Tibet (Fang and Wenji, 1981). They suggest that these diamonds
were developed within large masses of peridotite (up to 300 km²) subducted
to mantle depths. The occurrence of crustal UHP minerals therefore does not
seem to be restricted only to older sialic crust.

The ages derived from UHPM mineral assemblages are consistently much
younger than the supracrustal protolith that contains them. Peak metamorphic ages record orogenies produced during continent–continent collisions
(Table 1.1). In some areas, particularly the Western Alps and Norway, subduction of oceanic crust may have preceded the collisions, as ophiolite nappes
are often structurally overthrust above the UHPM continental slices, and in
the Alps at least some of the subducted oceanic crust undergoes UHPM and
is returned to the surface. Ocean crust nappes are conspicuously missing
within the supracrustal UHPM slices, and the small mafic enclaves (eclogite)
seem to have their origins related to magmatic underplating or intrusions
prior to continental collision rather than representing fragments of oceanic
crust. Contemporaneous calc-alkaline volcanic or intrusive activity characteristic of active continental margins is very scarce within or on the periphery of
the UHPM collisional zones. These comparisons lead to a general scenario
for the formation of UHPM rocks.

1. Protoliths of the UHPM blocks consist of old, cold, and dry supracrustal rocks.
2. Widespread regional anatexis and mobilization during granulite grade metamorphism has dehydrated the lower crust, thus preventing fluid-enhanced crustal melting during the later UHPM event. Many of the UHPM occurrences have protoliths that have undergone such lower crust events prior to UHPM.
3. Terminal continent–continent collision accompanied by rapid crustal thickening
   and subduction is the prevalent orogenic event. Continued subduction of oceanic
   crust does not seem to provide the necessary tectonic setting for UHPM unless
   small microcontinents or sea mounts are introduced into the subduction zone.
4. Underthrusting of old, cold, and dry continental margin fragments at maximum
   strain rates enhances attainment of peak UHPM assemblages. Exhumation rates

need to be nearly equal to the subduction rates in order to preserve the UHPM mineral assemblages developed at depths greater than 100 km. Normal erosional exhumation is much too slow to allow preservation of the UHPM assemblages.

Geologic and Tectonic Problems

Prior to the plate tectonic paradigm geologists working in the Alps came to the conclusion that the continent–continent collision of the African and European margins had led to the consumption of continental crust (Ampferer, 1906; Laubscher, 1969; 1989). The missing parts of the lithosphere in the Alps required that at least some of the material be consumed during shortening and collision. Furthermore, the large amounts of consumed continental crust could explain major uplifts resulting from isostatic disequilibrium caused by consumption of the lighter continental crust. The UHPM areas described in this book are related to continent–continent collisions from Caledonian time to Late Tertiary (Table 1.1), and their peak metamorphic ages coincide very closely to the timing of the terminal continental collisions of each area. All of these collisions can be explained in terms of plate movements except for the Kokchetav massif, where poor exposures and low relief hamper efforts at a tectonic reconstruction.

Special conditions are required to develop the UHPM terranes, and these conditions constrain limits that define this style of metamorphism (Table 1.1, Fig. 1.7). Calculated P-T parameters for the prograde UHPM metamorphic assemblages indicate that the supracrustal rocks had to have low initial geothermal gradients combined with moderate rock conductivity prior to the collision event. Convergence and exhumation rates had to be high enough to subduct and unroof continental thicknesses before thermal relaxation overstepped the solidii of the supracrustal rocks and granitic melts began to form. The apparent lack of pervasive regional anatexis contemporaneous with the UHPM events is evidence that rapid and "cool" P-T-t paths were maintained during metamorphism. Evidence available from various UHPM areas indicates that these old gneiss areas remained "dry" and are only slightly heated above their earlier acquired steady-state lower crustal temperatures during the initial collision and ensuing subduction and exhumation (Fig. 1.7).

Exhumation by erosion combined with normal faulting related to extension or low-angle faulting at rates near 1.3–2.5 mm/yr are required to allow preservation of the UHPM assemblage following the crustal doubling and subduction of these blocks to depths in excess of 100 km. Such exhumation rates are greater than normal erosion and uplift rates of 0.5 to 1 mm/yr (Harrison, 1992). These exhumation rates can be maintained only by continued tectonic

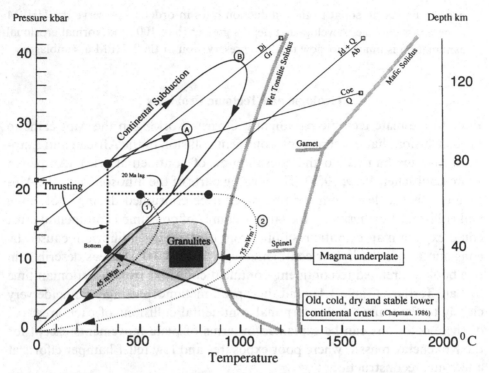

Figure 1.7. Pressure vs. temperature diagram showing *P-T* paths (solid lines with arrows) that could develop as a result of continent–continent collision if one assumes a crustal thickness of ~40 km and nearly instantaneous thickening of the crust by low-angle thrusting. The initial position is at the base of a ~40 km continental crust with a heat flow of <45 mW·m^{-2} and low thermal conductivity of 1.5 W·m^{-1}·K^{-1}. It is also assumed that this crust has undergone anatexis (or dehydrated) and is in thermal equilibrium for old crust (i.e., ~365°C; Chapman, 1986) at such depths. There is no time lag assumed after the subducted crust reaches its maximum depths where coesite (A) or diamond (B) could form. These curves are not calculated but are based on the thermal histories measured by coexisting mineral assemblages of the four UHPM areas discussed in the text. The time span estimated for these areas is 40–100 Ma giving rates of uplift for these five areas from 1.3 to 2.5 mm/yr (Table 1.1). The dotted curves represent Pressure–Temperature time paths calculated for two extreme cases: (1) Low heat flow of 45 m·W·m^{-1} and (2) High heat flow 75 m·W·m^{-1} both lagging 20 Ma after reaching maximum depth and arriving at the surface within 50 Ma. Data for these curves are from Thompson and England (1984); England and Thompson (1984). The *P-T* realm of old, cold, dry continental crust is highlighted showing boundaries only for granulite facies rocks. The wet solidus for tonalite is given as well as that for mafic magmas. The magma underplating line illustrates the possibility of introducing mafic enclaves into the lower crust prior to final consolidation and cooling. Stability fields for diamond → graphite, coesite → quartz, jadeite + quartz → albite are taken from Schreyer (1988). Pressure is assumed to be isotropic with continental crust 2.86 and upper mantle 3.33.

activity following the initial formation of the high pressure minerals. Slower exhumation rates would allow widespread anatexis-producing intrusive activity near the locus of collision, but such activity appears to be minimal (Table 1.1, Fig.1. 7). In some areas, granulite overprinting or anatexis of the UHPM rocks has been observed, indicating that it is possible that other exposed areas of exhumed granulite may have experienced UHPM, but the slow rate of exhumation developed higher temperatures that destroyed the earlier formed UHPM rock. Numerous tectonic schemes have been proposed for the exhumation of UHPM all of which require a dynamic tectonic setting. Intracontinental thrusting and wedging combined with erosion has been proposed by Avigad (1992), Cuthbert et al. (1983), Hsu (1991), Okay and Sengor (1992), and Nicolas et al. (1990) to be the main mechanism of exhumation. The presence of major detachment zones showing simple shear extensional faulting in the upper crust combined with pure shear and vertical shortening within the plastically deformed lower crust produced by collapsing orogenic welts of the continent–continent collision provides another mechanism of exhumation (Andersen and Jamveit, 1990; Ballevre et al., 1990; Blake and Jayko, 1990; Maruyama et al., 1993). Estimates of the time elapsed between the initial UHP event and first arrival of the metamorphosed crust at the Earth's surface for specific UHPM occurrences is about 40–100 Ma (Table 1.1). If these estimates are correct, then the average exhumation rates for UHPM assemblages containing coesite or diamond overlap the erosion rates 1 to ~5 mm/yr for modern eroding mountain belts (Copeland, 1993). The geochronology of retrograde overgrowths in UHPM assemblages indicates that the initial exhumation is often characterized by rapid isothermal decompression that could represent rapid vertical movements directly after the initial stage of collision and UHPM (Andersen and Jamtveit, 1990). In general, the initial rapid exhumation is diminished, and thus the final stages of uplift may be the result of normal denudation associated with orogenic belts.

Assumptions are made that the pressures developed during metamorphism are isotropic and are controlled by the lithostatic load. Recognition that the UHPM metamorphic blocks are present in gneissic rocks that are old-cold-dry compels the question, Can significant deviatoric stresses be developed during the short-term orogenies (40–100 Ma) observed for UHPM rocks? Strengths of quartz-rich, dry rocks such as these gneisses are not well known in the *P-T* realm of UHPM. Is it possible that the unusually fast rate of plate motion demonstrated for some continent–continent collisions could develop limited overpressures, thus reducing the amount of lithostatic burial required for such high pressures? No field or petrographic evidence for such overpressures has yet been found, but the dry and cold continental gneissic blocks in their initial stage of subduction and collision may have had enough strength to develop some overpressures prior to thermal weakening. Unreasonably high (~50 mm/yr) geologic strain rates might produce moderate overpressures, reducing the vertical distance required to exhume the UHPM assemblages described above (see D. C. Smith, Chapter 9, for further discussion of overpressures).

New Research Areas

These new finds of UHPM in Eurasia were brought to light only by the ability of mineralogists to identify mineral phases smaller than a few microns. The significant aspect of these discoveries is that they were not predicted in advance. Now there are new areas of future research that should lead to further unexpected results using modern mineralogical instrumentation. No doubt occurrences of UHPM are present in the Western Hemisphere of the Earth, and geologists should search for mineralogic evidence in known continental collison zones of this hemisphere.

Nearly all of the high pressure studies of deep mantle conditions has been carried out on minerals in the $FeO-MgO-SiO_2-Al_2O_3$ system. It is now critical to initiate high pressure studies on silica-rich assemblages of the granitic system in order to predict what might happen to such materials when they are transported down into the upper mantle.

Systematic studies of the stable and radiogenic isotopes combined with the rare earth element distribution in the country rock gneisses can lead to a better understanding of the metamorphic transformation of large lower crustal metamorphic blocks during UHPM.

Tomography is now able to outline smaller thermal anomalies in the crust and mantle, and a search for large blocks of continental crust in present-day deep zones of collision such as the Alps or Himalyas may reveal the actual geometry of subduction of these bouyant continental blocks into the upper mantle.

Careful field studies in known orogenic zones of continent–continent collisions will certainly bring to light other occurrences of UHPM. Field structural studies combined with careful determination of kinematic indicators are needed to establish the style of deformation within the UHPM as well as the associated metamorphic sequences. These measurements could shed light on the unusual strain rates developed during these UHPM events.

References

Ames, L., Tilton, G. R., and Zhou, G. 1993. Timing of collision of the Sino-Korean and Yangtse cratons: U-Pb zircon dating of coesite-bearing eclogites. *Geology* *21*, 339–342.

Ampferer, O. 1906. Ueber das Bewegungsbild von Faltengebirgen. *Jahrbuch der kaiserlichen und koniglishen Reichssanstalt (Wien) 56*, 539–622.

Andersen, T. B., and Jamtveit, B. 1990. Uplift of deep crust during orogenic extensional collapse: A model based on field studies in the Sogn–Sunnfjord region of Western Norway. *Tectonis 9*, 1097–1111.

Avigad, D. 1992. Exhumation of coesite-bearing rocks in the Dora-Maira massif (western Alps, Italy). *Geology 20*, 947–950.

Bakun-Czubarow, N. 1991. Geodynamic significance of the Variscan HP eclogite-granulite series of the Zlote Mountains in the Sudetes. *Publs. Inst. Geophysics Polish Acasd Sci. A–19 (236)*, 215–244.

Bakun-Czubarow, N. 1991. On the possibility of occurence of quartz psuedomorphs after coesite in the eclogite-granulite rock series of the Zlote Mountains in the Sudetes (SW Poland). *Archiwum Mineralogiczne 47*, 5–16.

Ballévere, M., Pinardon, J.-L., and Kienast, J. R. 1989. Reversal of Fe-Mg partitioning between garnet and staurolite in eclogite-facies metapelites from the Champtoceaux Nappe (Brittany, France). *Jour. Pet. 30*, 1321–1349.

Becker, H. 1993. Garnet peridotite and eclogite Sm-Nd mineral ages from the Lepontine dome (Swiss Alps): New evidence for Eocene high pressure metamorphism in the Central Alps. *Geology 21*, 599–602.

Biino, G., and Pognante, U. 1989. Paleozoic continental-type gabbros in the Gran Paradiso nappe (Western Alps): Early-Alpine eclogitisation and geochemistry. *Lithos 24*, 3–19.

Blake, M.C.J., and Jayko, A. S. 1990. Uplift of very high pressure rocks in the western Alps: evidence for structural attenuation along low-angle faults. *Mem. Soc. Geol. France 156*, 237–246.

Bocquet, J. 1971. Cartes de repartition de quelques mineraux du metamorphisme alpin dans les Alpes franco-italiennes. *Eclo. Geol. Helv. 64*, 71–103.

Bocquet, J., Delaloye, M., Hunziker, J. C., and Krummenacher, D. 1974. K-Ar and Rb-Sr dating of blue amphiboles, micas and associated minerals from the western Alps. *Contrib. Mineral. and Petrol.*, *47*, 7–26.

Bohlen, S. R., and Boettcher, A. L. 1982. The quartz-coesite transformation: a pressure determination and the effects of other components. *J. Geophys. Res. 87*, 7073–7078.

Caby, R. 1994. Precambrian coesite from northern Mali; first record and implications for plate tectonics in the trans-Saharan segment of the Pan African belt. *Euro. Jour. Mineral. 6*, 235–244

Carswell, D. A. 1991. Variscan high P-T metamorphism and uplift history in the Moldanubian Zone of the Bohermian Massif in Lower Austria. *Euro. Jour. Mineral. 3*, 323–342.

Carswell, D. A., and Harvey, M. A. 1985. The intrusive history and tectonometamorphic evolution of the Basal Gneiss Complex in the Moldefjord area, west Norway. *The Caledonide Orogen*. New York: J. Wiley, 843–857.

Chapman, D. S. 1986. Thermal gradients in the continental crust. *The Nature of the Lower Continental Crust*. London: Geological Society of London, Spec. Pub. No. 24, 63–70.

Chesnokov, B. V., and Popov, V. A. 1965. Increasing of volume of quartz grains in eclogites of the South Urals. *Dol. Akad. Nauk SSSR 162*, 176–178.

Chopin, C. 1984. Coesite and pure pyrope in high-grade blueschists of the western Alps: A first record and some consequences. *Contrib. Mineral. and Petrol. 86*, 107–118.

Chopin, C. 1981. Talc-phengite: A widespread assemblage in high-grade pelitic blueschists of the western Alps. *J. Petrol 22*, 628–650.

Chopin, C. 1987. Very-high-pressure metamorphism in the western Alps: implications for subduction of continental crust. *Philosophical Transaction of the Royal Society of London, Series A 321, no. 1557*, 183–197.

Chopin, C., Henry, C., and Michard, A. 1991. Geology and petrology of the coesite-bearing terrain, Dora-Maira massif, Western Alps. *Eur. Jour. Mineral. 3*, 263–291.

Chopin, C., Klaska, R., Medenbach, O., and Dron, D. 1986. Ellenbergite, a new high-pressure Mg-Al-Ti-Zr- silicate with a novel structure based on face-sharing octahedra. *Contrb. Miner. and Pet. 92*, 316–321.

Claoue-Long, J. C., Sobolev, N. V., Shatsky, V. S., and Sobolev, A. V. 1991. Zircon response to diamond-pressure metamorphism in the Kokchetav massif, USSR. *Geology 19*, 710–713.

Compagnoni, R., Dal Piaz, G. V., Hunziker, J. C., Gosso, G., Lombardo, B., and Williams, P. F. 1977. *The Sesia-Lanzo Zone, a slice of continental crust with Alpine high-pressure low-temperature assemblages in the western Italian Alps.* Torino, Italy, Consiglio Nazionale Ricerche.

Copeland, Peter. 1993. It happens in spurts; exhumation of mountain belts. *Abst. with Prog.- Geol. Soc. America, 25,* 176.

Cuthbert, S. J., Harvey, M. A., and Carswell, D. A. 1983. A tectonic model for the metamorphic evolution of the Basal Gneiss Complex, Western South Norway. *Jour. Metamorph. Geol. 1*, 63–90.

Dahlen, F. A., and Barr, T. D. 1989. Brittle frictional mountain building 1. Deformation and mechanical energy budget. *Jour. of Geophys. Res. 94*, 3906–3922.

Dal Piaz, G. V., and Ernst, W. G. 1978. Areal geology and petrology of eclogites and associated metabasites of the Piemonte ophiolite nappe, Breuil-St. Jacques area, Italian western Alps. *Tectonophysics, 51*, 99–126.

Dietrich, V., Vuagnat, M., and Bertrand, J. 1974. Alpine metamorphism of mafic rocks. *Schweiz. Mineral. Petrogr. Mitt. 54*, 291–333.

Dobretsov, N. L. 1975. Metamorphic belts of the northwestern Circum-Pacific region. *Geological Society of America Special Paper 151*, 133–144.

Dobrzhinetskaya, L. F., Posukhova, T., Tronnes, R., Korneliussen, A., and Sturt, B. 1993. A microdiamond from eclogite-gneiss area of Norway. *Terra Abstracts Suppl. 4 to Terra Nova 5*, 8.

Dobrzhinetskaya, L. F., Sheshkel, G. G., and Podkuiko, Y. A. 1993. The structural distribution of crustal microdiamonds within eclogite-gneiss formation (Northern Kazakhstan). *Terra Abstracts Suppl. 4 to Terra Nova 5*, 8.

Dong, S., Zhou, H., Zhang, W., and Cheng, G. 1989. A preliminary study of the movement of the Dabie Massif. *Bulletin of Institute of Geomechanics, Chinese Academy of Geological Science 12*, 99–112.

Eide, E. A., Wang, X., Maruyama, S., Liou, J. G., and Zhou, G. 1989. Mineral paragenesis of eclogite blocks from a serpentinite melange belt, northern Hubei, China. *EOS, Transaction American Geophysical Union 70*, 1379.

Enami, M., and Zang, Q. 1990. Quartz pseudomorph after coesite in eclogites from Shandong Province, east China. *Am. Mineralogist 75*, 381–386.

England, P. C., and Thompson, A. B. 1984. Pressure-temperature-time paths of regional metamorphism. Heat transfer during the evolutions of regions of thickened continental crust. *J. Petrol. 25*, 894–928.

Ernst, W. G., Zhou, G., Liou, J. G., Eide, E., and Wang, X. 1991. High-pressure and superhigh-pressure metamorphic terranes in the Qinling-Dabie Mountain Belt, central China; Early- to Mid-Phanerozoic accretion of the western paleo-Pacific rim. *Pacific Science Association Information Bulletin 43*, 6–14.

Fang, C., and Wenji, B. 1981. The discovery of Alpine-Type diamond bearing ultrabasic intrusion in Xizang (Tibet). *Int. Geol. Rev. 27*, 455–457.

Franke, W. 1989. Variscan plate tectonics in Central Europe – current ideas and open questions. *Tectonophysics 169*, 221–228.

Geologists of Anhui Province, 1986. Magmatic rock map of Anhui province of the People's Republic of China (1: 000,000). Beijing, Geological Map Printing House.

Geologists of Henan Province, 1989. Magmatic rocks map of Henan Province, the People's Republic of China (1: 000,000). Beijing, Geological Publishing House.

Geologists of Hubei Province, 1990. Tectonic map of Hubei Province of the People's Republic of China (1: 000,000). Beijing, Bureau of Geology and Mineral Resources, Hubei Province.

Gillet, P., Ingrin, J., and Chopin, C. 1984. Coesite in subducted continental crust: P-T history deduced from an elastic model. *Earth Planet. Sci. Lett. 70*, 426–436.

Goffe, B., and Chopin, C. 1986. High-pressure metamorphism in the western Alps: zoneography of metapelites, chronology and consequences. *Schweiz. Mineral. Petrog. Mitt. 66*, 41–52.

Griffin, W. L. 1986. 'On eclogites of Norway' – 65 years later. *Mineral. Mag. 66*, 41–52.

Griffin, W. L., Austrheim, H., Brastad, K., Bryhni, I., Krill, A. G., Krogh, E. J., Mörk, M.-B.E., Qvale, H., and Torudbakken, B. 1985. High-pressure metamorphism in the Scandinavian Caledonides. *The Caledonide Orogen*. New York, J. Wiley.

Griffin, W. L., and Mork, M.B.E. *Eclogites and basal gneisses in western Norway (Excursions in the Scandinavian Caledonides)*. Uppsala Caledonide Symposium (IGCP Project 27), Uppsala, 1981.

Harrison, C.G.A. 1992. Rates of continental erosion and mountain building. *Proceedings of the 21st IGC Kyoto, Japan 2*.

Hirajima, T., Ishiwatari, A., Cong, B., Zhang, R., Banno, S., and Nozaka, T. 1990. Coesite from Mengzhong eclogite at Donghai county, northeastern Jiangsu province, China. *Mineral. Mag. 54*, 579–583.

Hsu, K. J. 1991. Exhumation of high-pressure metamorphic rocks. *Geology 19*, 107–110.

Hunziker, J. C. 1974. Rb-Sr and K-Ar age determination and the Alpine tectonic history of the western Alps. *1st Geological Mineral. University Padova, Memoir. 31*, 54 pp.

Hunziker, J. C., Desmons, J., and Martinotti, G. 1989. Alpine thermal evolution in the central and the western Alps. *Geological Society of London Spec. Pub., 45*.

Hunziker, J. C., and Martinotti, G. 1984. Geochronology and evolution of the western Alps: a review. *Memorie della Societa geologica Italiana 29*, 43–56.

Jagoutz, E., Shatsky, V. S., Sobolev, N. V. 1990. Sr-Nd-Pb isotopic study of ultra high PT rocks from Kokchetav massif. *EOS Trans. Amer. Geophys. Union, 71*, 1707.

Kienast, J. R., Lombardo, B., Biino, G., and Pinardon, J. L. 1991. Petrology of very-high-pressure eclogitic rocks from the Brossasco-Isasca Complex, Dora-Maira Massif, Italian Western Alps. *Jour. Metamorph. Geol., 19–34*.

Laubscher, H. P. 1969. Mountain building. *Tectonophysics 7*, 551–563.

Laubscher, H. P. 1989. The tectonics of the southern Alps and the Austro-Alpine nappes: a comparison. *Geological Society of London, Special Publication, No. 45*.

Li, S., Ge, N., Liu, D., Zhang, Z., Yie, X., Zhen, S., and Peng, C. 1989. Sm-Nd age of C-group eclogites in northern Dabie Mountains and its tectonic significance. *Bulletin of Sciences 7*, 522–525 (in Chinese).

Li, S., Hart, S. R., Zhen, S., Guo, A., Liu, D., and Zhang, G. 1989. Sm-Nd age for the collision between the Sino-Korean and Yangtze cratons. *Academic Sinica 2B*, 312–319 (in Chinese).

Li, S., and Liu, D. 1990. Isotopic chronological evidence for Indosinian orogeny in Dabie Mountain. *Geotectonica et Metallogenia 14*, 159–163 (in Chinese with English abstract).

Liu, X., and Hao, J. 1989. Structure and tectonic evolution of the Tongbie-Dabie Range in the east Qinling collisional belt, China. *Tectonics 8*, 637–645.

Ma, B., and Zhang, Z. 1988. The features of the paired metamorphic belts and evolution of paleotectonics in the east part of Dabie Mountains. *Seismology and Geology 10*, 19–28 (in Chinese with English abstract).

Matte, P. 1986. Tectonics and plate tectonics model for the Variscan belt of Europe. *Tectonophysics, 126*, 329–374.

Medaris, L.G.J., and Wang, H. F. 1986. A thermal-tectonic model for high-pressure rocks in the Basal Gneiss Complex of Western Norway. *Lithos, Second International Eclogite Conference*, 323–332.

Michard, A., Chopin, C., and Henry, C. 1993. Compression versus extension in the exhumation of the Doria-Maira coesite-bearing unit, Western Alps, Italy. *Tectonophysics 221*, 173–193.

Mirwald, P. W., and Massonne, H. J. 1980. Quartz-coesite transition and the comparative friction measurements in piston-cylinder apparatus using talc-alsimagglass (TAG) and NaCl high pressure cell: a discussion. *Neues Jahrbuch fur Mineral Monatshefte hefte 10*, 469–477.

Monie, P., and Chopin, C. 1991. $^{40}Ar/^{39}Ar$ dating in coesite-bearing and associated units of the Dora-Maira massif, Western Alps. *Eur. Jour. Mineral. 3*, 239–262.

Monie, P., and Philipot, P. 1989. Mise en evidence de l'age eocene moyen du metamorphisme de haute-pression dans la nappe ophiolitique du Monviso (Alps Occidental) par la methode ^{39}Ar-^{40}Ar. *Paris, Academie des Sciences Comptes Rendus, ser II 309*, 245–251.

Mörk, M.B.E., and Krogh, E. J. 1987. *Excursion guide for the eclogite field symposium in western Norway.*

Okay, A. I., Xu, S., and Sengor, A.M.C. 1989. Coesite from the Dabie Shan eclogites, central China. *Eur. Jour. Mineral. 1*, 595–598.

Okrusch, M., Matthes, S., Klemd, R., O'Brien, P., and Schmidt, K. 1991. Eclogites at the north-western margin of the Bohemian Massif: A review. *Euro. Jour. Mineral. 3*, 707–730.

Paquette, J. R., Chopin, C., and Peucat, J. J. 1989. U-Pb zircon, Rb-Sr and Sm-Nd geochronology of high- to very-high-pressure meta-acidic rocks from the western Alps. *Contrib. Mineral. Petrol. 101*, 280–289.

Pearson, D. G., Davies, G. R., and Nixon, P. H. 1993. Geochemical constraints on the petrogenesis of diamond facies pyroxenites from the Beni Bousera peridotite massif, N. Morocco. *J. Petrol., 34*, 125–172.

Perchuk, L. L., Letnikov, F. A., Udovkina, N. G., Lennykh, V. I., and Mudrov, I. A. 1969. Origin of eclogite in the Kokchetav block. *Dokl. Akad. Nauk SSR (Translation), 186*, 441–444.

Pognante, U. 1991. Petrological constraints on the eclogite-and blueschist-facies metamorphism and P-T-t paths in the Western Alps. *Jour. Metamorph. Geol. 9*, 5–17.

Pognante, U., and Sandrone, R. 1989. Eclogites in the northern Dora-Maira nappe (western Alps, Italy). *Mineralogy and Petrology, 40*, 57–71.

Pognante, U., Talarico, F., Rastelli, N., and Ferrati, N. 1987. High-pressure metamorphism in the nappes of the Valle dell'Orco traverse (Western Alps collisional belt) *Jour. Metamorp. Geol. 5*, 397–414.

Reinecke, T. 1991. Very-high-pressure metamorphism and uplift of coesite-bearing metasediments from the Zermatt-Saas zone, Western Alps. *Eur. Jour. Mineral. 3,* 7–17.

Sang, B., Chen, Y., and Shao, G. 1987. The Rb-Sr ages of metamorphic series of the Susong group at the southeastern foot of the Dabie Mountains, Anhui province, and their tectonic significance. *Regional Geology of China 4,* 364–370.

Schertl, H.-P., Schreyer, W., and Chopin, C. 1991. The pyrope-coesite rocks and their country rocks at Parigi, Dora-Maira Massif, Western Alps: detailed petrography, mineral chemistry and PT-path. *Contrib. Mineral. Petrol. 108,* 1–21.

Schmadicke, E. 1991. Quartz psuedomorphs after coesite in eclogites from the Saxonian Erzgebirge. *Eur. Jour. Mineralogy 1,* 231–238.

Schreyer, W. 1985. Metamorphism of crustal rocks at mantle depths: High-pressure minerals and mineral assemblges in metapelites. *Fortschrrite der Mineralogie 63,* 227–261.

Schreyer, W. 1988. Subduction of continental crust to mantle depths: Petrological evidence. *Episodes 11,* 97–104.

Shatsky, V. S., Sobolev, N. V., Zayachkovsky, A. A., Zorin, Y. M., and Vavilov, M. A. 1991. New occurrence of microdiamonds in metamorphic rocks as a proof of regional character of ultra high pressure metamorphism in Kokchetav massif. *Dokl. Akad. Nauk, SSSR 321,* 189–193 (In Russian).

Shui, T., and Xue, H. 1989. The ductile shear zones at the southeastern foot of Dabie Mountain. *Bull. Nanjing Institute of Geology and Mineral Resources, Chinese Academy of Geological Sciences 10,* 52–63.

Smith, D. C. 1984. Coesite in clinopyroxene in the Caledonides and its implications for geodynamics. *Nature 310,* 641–644.

Smith, D. C., and Lappin, M. A. 1989. Coesite in the Straumen kyanite-eclogite pod, Norway. *Terra Research 1,* 47–56.

Sobolev, N. V., and Shatsky, V. S. 1987. Carbon mineral inclusions in garnets of metamorphic rocks. *Geologiya i Geofizika 7,* 77–80 (In Russian).

Sobolev, N. V., and Shatsky, V. S. 1990. Diamond inclusions in garnets from metamorphic rocks. *Nature 343,* 742–746.

Sobolev, N. V., Shatsky, V. S., Vavilov, M. A., and Goryainov, S. V. 1991. Coesite inclusion in zircon from diamondieferous gneiss of Kokchetav massif–first find of coesite in metamorphic rocks in the USSR territory. *Dokl. Akad. Nauk SSSR 321,* 184–188 (In Russian).

Sobolev, V. S. 1977. *Metamorphic Complexes of Asia (translated into English version by Brown, D.A, 1982).* Nauka Publication & Pergamon Press, Oxford.

Stockhert, B., Jager, E., and Voll, G. 1986. K-Ar age determinations on phengite from the internal part of the Sesia zone, western Alps, Italy. *Contrib. Mineral. Petrol., 92,* 456–470.

Suo, S., You, Z., Zhu, B., and Liu, W. 1987. Tectonic style and deformation sequence of Dabie metamorphic terrane. *Earth Science, Journal of China University of Geosciences 13,* 341–349 (in Chinese with English abstract).

Tagiri, M., and Bakirov, A. 1990. Quartz pseudomorph after coesite in garnet from a garnet-chloritoid-talc schist, Northern Tien-Shan, Kirghiz, USSR. *Proceedings of the Japan Academy 66, ser. B,* 135–139.

Thompson, A. B., and England, P. C. 1984. Pressure-temperature-time paths of regional metamorphism. II. Their inference and interpretation using mineral assemblages in metamorphic rocks. *J. Petrol. 24,* 929–955.

Tilton, G. R., Schreyer, W., and Schertl, H.-P. 1991. Pb-Sr-Nd isotopic behavior of deeply subducted crustal rocks from the Dora-Maira Massif, Western Alps-II:

what is the age of the ultrahigh-pressure metamorphism? *Contrib. Mineral. and Petrol. 108*, 22–33.

Tilton, G. R., Schreyer, W., and Schertl, H. P. 1989. Pb-Sr-Nd isotopic behaviour of deeply subducted crustal rocks from the Dora-Maira massif, Western Alps, Italy. *Geochim. et Cosmo, Acta 73*, 1391–1400.

Torudbakken, B. O. 1983. Ages of metamorphic and deformational events in the Beian nappe complex, Ordland, Norway. *Nor. Geol. Tidsskr. 399*, 27–39.

Van der Molen, I., and Van Roermund, H. L. M. 1986. The pressure path of solid inclusions in minerals: the retention of coesite inclusions during uplift. *Lithos 19*, 317–324.

Vuichard, J. P., and Ballévre, M. 1988. Garnet chloritoid equilibria in eclogitic pelitic rocks from the Sesia zone (Western Alps): their bearing on phase reklations in high-pressure metapelites. *Jour. Metamorph. Geol. 6*, 135–157.

Wang, X., Jing, Y., Liou, J. G., Pan, G., Liang, W., Xia, M., and Maruyama, S. 1990. Field occurrences and petrology of eclogites from the Dabie Mountains, Anhui, central China. *Lithos 25*, 119–131.

Wang, X. and Liou, J. G. 1991. Regional ultrahigh-pressure coesite-bearing eclogitic terrane in central China: evidence from country rocks, gneiss, marble, and metapelite. *Geology 19*, 933–936.

Wang, X., Liou, J. G., and Mao, H. K. 1989. Coesite-bearing eclogites from the Dabie Mountains in central China. *Geology 17*, 1085–1088.

Wang, X., Liou, J. G., and Maruyama, S. 1992. Coesite-bearing eclogites from the Dabie Mountains, central China: Petrogenesis, P-T paths and implications to tectonics. *J. Geol. 100*, 231–250.

Xie, D., Mao, C., Shi, H., Jiang, Y., and Fong, J. 1983. Study on the mineralogy of eclogites in the Dabieshan. *Acta Petrologica Mineralogica et Analytica 2*, 87–98.

Xu, S., Okay, A. I., Ji, S., Sengor, A.M.C., Su, W., Liu, Y., and Jiang, L. 1992. Diamond from the Dabie Shan metamorphic rocks and its implication for tectonic setting. *Science 256*, 80–82.

Yang, J., and Smith, D. C. 1989. Evidence for a former sanidine-coesite-eclogite at Lansantou, Eastern China, and the recognition of the Chinese 'Su-Lu coesite-eclogite province', East China [abs.]. *Terra Abstract, Third International Eclogite Conference 1*, 26.

Yang, S. 1986. Late Precambrian ancient tectonics of east Qinling and Dabie Mountain. *Precambrian Geology 3*, 273–282 (in Chinese with English abstract).

Zhang, R., Cong, B., Hirajima, T., and Banno, S. 1990. Coesite eclogite in Su-Lu region, eastern China. *EOS, Trans. Amer. Geophys. Union 71*, 1708.

Zhang, S., and Kang, W. 1989. The character of the blueschist belt and discussion of the formation age in central China. *Journal of Changchun College of Geology Special volume, Blueschist belt in Hubei and Anhui Province*, 1–9 (in Chinese with English abstract).

Zonenshain, L. P., Kuzmin, M. I., and Natapov, L. M. 1990. *Geology of the USSR: A plate-tectonic synthesis*. Washington, D. C., American Geophysical Union.

Zwart, H. J., Sobolev, V. S., and Lepezin, G. A. 1978. *Metamorphic map of Asia. 9 sheets (scale 1:5,000,000)*. Distributed Pergamon Press, Oxford, Institute of Geology, Novosibirsk, USSR.

2

Experimental and Petrogenetic Study of UHPM

H.-J. MASSONNE

Abstract

The first section of this chapter reviews previous experimental studies related to the stability fields of phases being indicative for UHPM. These phases are either natural minerals, such as coesite and diamond, or compositional end members of synthetic phases including K-cymrite, MgMgAl-pumpellyite, OH-topaz, high-pressure wollastonite, "piezotite," so-called phase A and 10 Å-phase, both hydrous Mg-silicates.

The second part describes numerous experimental studies in both simple model and complex natural systems. The former studies include primarily mineral assemblages typical for UHP and HP rocks such as Mg-chloritoid + talc, phengite + enstatite, and jadeite + talc. Reviewed experiments also include melting relations at high P, relatively low T conditions and solid solution phases in specified assemblages.

Furthermore, reference is made to recent progress leading to internally consistent thermodynamic data on sets of minerals. On the basis of thermodynamic data of Berman (1988) augmented by the data of Massonne (1992b) including activity models for the binary solid solutions chlorite, enstatite, sapphirine, Mg-staurolite and talc, a grid of the system $MgO-Al_2O_3-SiO_2-H_2O$ (MASH) is calculated for the P-T range 450°–900°C and 15–70 kbar. Moreover, a $Na_2O-FeO-MASH$ grid, calculated by Guiraud et al. (1990) on the basis of mineral data reported by Holland and Powell (1990), is presented. The application of these grids to define assemblages (not experimentally investigated) indicating UHPM rocks is discussed.

The above thermodynamic data sets can be used to calculate the P-T conditions of mineral equilibria relevant for natural rocks, as well. In the third section, these geothermobarometric applications are demonstrated for the common eclogitic assemblage garnet-omphacite-phengite using recent activity formulations of components in these minerals (Massonne, 1992c). Two examples, related to eclogites from the Franciscan complex, California, and the Greek Cyclades, which have reached *P-T* conditions of UHPM, are selected to show (1) the computational method, and (2) the potential of this method for detecting UHP metamorphic rocks and deciphering their *P-T* evolution.

Introduction

When Chopin (1983, 1984) first described coesite in regionally metamorphic rocks of the Western Alps, the *P-T* conditions of the coesite-quartz transition had already been well determined experimentally. *P-T* stabilities of other minerals and mineral assemblages relevant to UHPM were also known. Despite the fact that UHPM rocks are increasingly recognized throughout the world, few additional experiments in the *P* regime above 25 kbar and below that of the deeper mantle (roughly *P* < 70 kbar) have been completed relevant to these rocks.

The major goals of this chapter are to review existing experimental studies and to calculate phase relations at UHPM conditions using internally consistent thermodynamic data sets. Because rocks, witnessing their stay at mantle levels deeper than 200 km (e.g. Sautter et al., 1991), are extremely rare, the diagrammatic presentations of mineral stability fields and petrogenetic grids are restricted here to 70 kbar.

Finally, the real potential of internally consistent thermodynamic data sets, containing a considerable quantity of important silicate end members, is demonstrated by geothermobarometric applications. For that purpose, phengite-bearing eclogites from the northern part of the Franciscan Complex and from the island of Sifnos, Greek Cyclades, are selected as examples.

Experimental Studies on Index Minerals for UHPM

The most important minerals indicating UHPM are coesite and diamond; their *P-T* stabilities are well constrained. Other UHPM index minerals are very rare. However, experimental studies demonstrate that some phases, such as MgMgAl-pumpellyite and OH-topaz, are exclusively stable at UHPM conditions, but they have not been described in nature. Stability fields of minerals occurring in HP metamorphic rocks with compositions close to those of the model systems are also reported here (subsection on Stability Limits, below).

Coesite

The coesite-quartz transition (Fig. 2.1) has been extensively investigated; such studies include the detailed piston-cylinder apparatus studies with low-friction salt cells by Mirwald and Massonne (1980) and Bohlen and Boettcher (1982). The resulting transition curves, differing by about 1 kbar, lie at about 30 kbar and 1000°C with a gentle dP/dT slope of 10 bar/K. The lower P for the coesite stability boundary in the presence of Na, Al, and H_2O found by Mirwald and Massonne (1980) could be a possible reason for the lower P determined by Bohlen and Boettcher (1982) for the boundary of pure coesite because these authors used natural quartz from Brazil as starting material which may have contained some Na and Al. Unpublished reversal experiments with small amounts of Al, H_2O, and K instead of Na added to SiO_2 gave no significant difference to the coesite-quartz transition for pure SiO_2 given by Mirwald and Massonne (1980). The upper P stability of coesite is limited by the transition to stishovite at about 95.5 kbar and 1000°C (Jeanloz and Thompson, 1983) with a dP/dT slope of about 20 bar/K.

Diamond

After experimentalists were able to synthesize diamond in a high-P apparatus in the mid-fifties, P-T conditions of the diamond-graphite equilibrium were experimentally determined. Bundy et al. (1961) established the equilibrium curve at $P = 56$ kbar and 1500°C on the basis of diamond growth and graphitization in the presence of different catalysts. Kennedy and Kennedy (1976) extended the experimental T to 1100°C. Both transition curves show a relatively steep dP/dT slope of 28 bar/K, the curve of Kennedy and Kennedy lying at almost 2 kbar higher P.

On the basis of well-studied thermodynamic properties of graphite and diamond, the equilibrium curve was calculated by Berman (1979) at T ranging from 0° to above 2000°C (Fig. 2.1). At high T the calculated curve agrees with the experimentally determined curves. On a similar basis, the curve of Kennedy and Kennedy (1976) was extrapolated by Chatterjee (1991) down to 800°C; the result shown in Fig. 2.1 is a diamond-graphite transition curve lying at somewhat higher P than that of Berman (1979).

10 Å-Phase, Phase A, and Other Hydrous Mg-Silicates

A micaceous layer silicate with a basal spacing of 10 Å has been synthesized in the system MgO-SiO_2-H_2O by Sclar et al. (1965) at $P > 32$ kbar and $T < 525°C$. This 10 Å-phase is composed of a talc layer and an additional

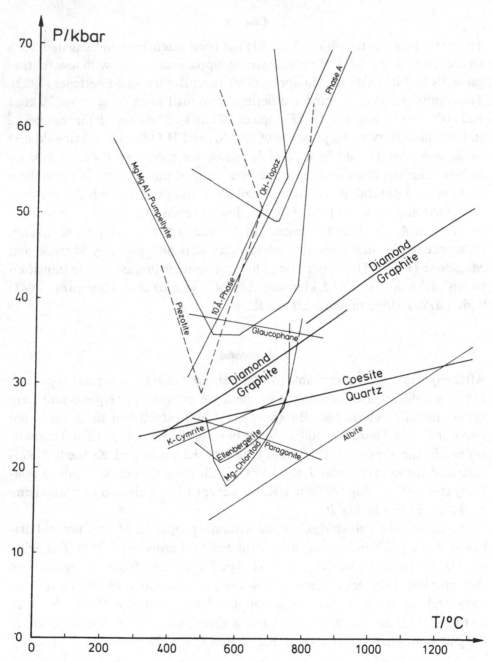

Figure 2.1. *P-T* diagram showing reaction curves limiting the stability of minerals and silicate phases that are indicative for UHPM or, at least, for high *P.* The reactions and the underlying experimental studies are given in detail in the text, but are repeated here concisely: coesite = quartz (Mirwald & Massonne, 1980), diamond = graphite (Berman, 1979 and at *T* >800°C: Chatterjee, 1991), 10 Å-phase and phase A (Yamamoto & Akimoto, 1977), OH-topaz and "piezotite" (Wunder & Schreyer, 1991),

layer of water molecules or oxonium ions yielding the bulk composition $Mg_3Si_4O_{10}(OH)_2 \cdot H_2O$ (Bauer and Sclar, 1981). Other hydrous Mg-silicates denoted as phases A, B and C were synthesized at high P by Ringwood and Major (1967) in excess of 100 kbar. Phase A, $Mg_7Si_2O_8(OH)_6$, an additional hydrous Mg-silicate [phase D = OH-chondrodite, $Mg_5Si_2O_8(OH)_2$] and OH-clinohumite, $Mg_9Si_4O_{16}(OH)_2$, have also been synthesized at clearly lower P (Yamamoto and Akimoto, 1974, 1977).

The upper thermal stability of the 10 Å-phase and the lower P stability of phase A (Fig. 2.1) were elaborated by Yamamoto and Akimoto (1977) on the basis of synthesis experiments. Some supporting evidence for that comes from the experimental studies of Khodyrev and Agoshkov (1986) and Khodyrev and Slutskiy (1987). The 10 Å-phase breaks down to talc + H_2O near 55 kbar and 730°C (Yamamoto and Akimoto, 1977). The reaction curve shows a steep dP/dT slope of 85 bar/K. At higher P the breakdown assemblage is enstatite + coesite + H_2O along a P-T curve with a negative dP/dT. This implies that the slopes of the curves of the talc breakdown to enstatite + coesite or 10 Å-phase are already clearly negative at P_{H_2O} = 55 kbar. According to Yamamoto and Akimoto (1977), enstatites of these breakdown assemblages are mono-clinic at P higher than 60 kbar. However, experimental reversals of the transition of orthoenstatite to high-P clinoenstatite with the space group C2/c undertaken by Pacalo and Gasparik (1990) point to P > 70 kbar for the appearance of clinoenstatite at T around 700°C. The minimum P for the formation of phase A is somewhat above 49 kbar (Yamamoto and Akimoto, 1977). With rising P and T phase A forms from brucite + forsterite along a reaction curve with a dP/dT of –23 bar/K. At 50 kbar and 720°C, phase A breaks down to form brucite + phase D. At P_{H_2O} > 64 kbar, the breakdown assemblage is periclase + phase D. Yamamoto and Akimoto (1977) assumed that the reaction curve of brucite decomposing to periclase + H_2O exhibits a clearly negative dP/dT slope. Both reaction curves limiting phase A to high T show a dP/dT close to 100 bar/K. OH-clinohumite is known to be unstable at T < 700°C (Yamamoto and Akimoto, 1977).

According to Akaogi and Akimoto (1980), phase B with the possible composition $Mg_{23}Si_8O_{36}(OH)_6$ (Akimoto and Akaogi, 1984) is stable only at P > 100 kbar. The same might be true for hydrous Mg-silicates, phases E and F, recently discovered by Kanzaki (1991). A phase with the composition

Caption for fig. 2.1 (cont.) MgMgAl-pumpellyite (Schreyer et al., 1991), K-cymrite (Massonne, 1991a), albite (Holland, 1980), glaucophane (Evans, 1990), and parago-nite (Holland, 1979). Dashed curves are preliminary as given by the authors cited above.

$Mg_{10}Si_3O_{14}(OH)_4$ (Pacalo and Parise, 1992), named superhydrous phase B (Gasparik, 1990), is stable at least above 150 kbar.

The most recent attempts by Wunder and Schreyer (1992) to determine the P-T stability of the 10 Å-phase using reversals point to a metastability at nearly the entire P-T conditions at which the 10 Å-phase was previously assumed to be stable. At the moment, this cannot be ruled out for other hydrous Mg-silicates mentioned here.

OH-Topaz and "Piezotite"

The hydroxyl end member of the topaz solid solution $Al_2SiO_4(F,OH)_2$ was first synthesized at 90–100 kbar and 900°–1000°C (Baller et al., 1990). At $T >$ 500°C pure OH-topaz forms from kyanite + H_2O with rising P. This reaction was given by Wunder and Schreyer (1991) to appear at 755°C and $P_{H_2O} = 60$ kbar with a dP/dT slope of 95 bar/K (Fig. 2.1).

An Al-silicate probably of the composition $Al_3Si_2O_7(OH)_3$ was first synthesized by Coes (1962) and named "piezotite." A preliminary stability field of this phase (Fig. 2.1) was recently determined (Wunder and Schreyer, 1991). The minimum P_{H_2O} for the stable appearance of "piezotite" in the system Al_2O_3-SiO_2-H_2O is 29 kbar at 490°C. At higher P and T the "piezotite" stability field is bounded by kyanite + coesite + H_2O, by OH-topaz + coesite at $P_{H_2O} > 50$ kbar and by diaspore + coesite at lower T.

MgMgAl-Pumpellyite

A synthetic pumpellyite phase with the composition $Mg_5Al_5Si_6O_{21}(OH)_7$ was reported by Schreyer et al. (1986). A preliminary stability field of this phase (Fig. 2.1). was determined by Schreyer et al. (1991). MgMgAl-pumpellyite forms from talc + chlorite + Mg-carpholite at 400° to 500°C. The lower P_{H_2O} stability at 37 kbar is defined by the reaction Mg-carpholite + Mg-chloritoid + talc = MgMgAl-pumpellyite. This phase successively breaks down to Mg-chloritoid + kyanite + talc + H_2O, kyanite + pyrope + talc + H_2O and coesite + kyanite + pyrope + H_2O with increasing P. The latter assemblage forms at $P_{H_2O} > 44$ kbar and T somewhat above 750°C. The upper thermal stability of MgMgAl-pumpellyite has also been investigated by Liu (1989) observing breakdown $T < 800$°C in the P range 60–220 kbar.

K-Cymrite

The potassium analogue to a rare mineral cymrite $BaAl_2Si_2O_8 \cdot H_2O$ is called here K-cymrite. Seki and Kennedy (1964) first synthesized K-feldspar hydrate at high P and determined its low-P stability, according to the reaction K-cymrite = K-feldspar + H_2O, between 300° and 800°C. This reaction has also been reported by Huang and Wyllie (1974). The most recent experimental study of the lower P stability of K-cymrite is by Massonne (1991a); the resulting curve shown in Fig. 2.1 has a dP/dT slope of about 30 bar/K.

Fe_2SiO_4 Spinel

A phase with fayalite composition but with spinel structure was first synthesized by Ringwood (1958) at 50 kbar and 400°C. The transition of olivine to spinel was located around 38 kbar at 600°C. According to further experimentation up to 1200°C by Akimoto and Fujisawa (1968), Ringwood and Major (1970), and Akimoto (1972), the slope of the transition curve is 33 bar/K. Increasing Mg content of olivine stabilizes this phase to even higher P. For instance, olivine with Fo_{50} breaks down at 69 kbar and 800°C to a spinel rich in Fe_2SiO_4 component and an olivine slightly richer in forsterite component (Akimoto, 1972).

Stability Limits of Minerals More or Less Diagnostic for UHPM

Some other silicates, which are, however, also stable within the P-T field of quartz, are indicative of HP metamorphic conditions. For instance, this holds for the magnesian end member of chloritoid, $MgAl_2SiO_4(OH)_2$. The minimum P_{H_2O} for the stability of this phase is as low as 18 kbar (Chopin and Schreyer, 1983; Chopin, 1985). Mg-chloritoid breaks down to chlorite + kyanite + corundum + H_2O and, below 570°C in the presence of H_2O, forms from chlorite + kyanite + diaspore with a P-T curve assumed to show a clearly negative dP/dT. At about 750°C and $P > 30$ kbar, Mg-chloritoid breaks down to form pyrope + corundum + H_2O (Chopin, 1985).

Rosenhahnite, $Ca_3Si_3O_8(OH)_2$, is a rare mineral stable up to P of at least 40 kbar (Chatterjee et al., 1984). It decomposes to wollastonite + H_2O at 27 kbar and 525°C along a curve with a dP/dT of 74 bar/K. At $P > 28.5$ kbar, rosenhahnite breaks down to a monoclinic polymorph of wollastonite + H_2O with a very steep dP/dT slope. The transition P for wollastonite to its triclinic HP polymorph was found to be 32.5 kbar at 1000°C by Huang and Wyllie (1975a) and about 28 kbar by Essene (1974) and Chatterjee et al. (1984). According to the latter authors, the triple point for the wollastonite polymorphs is located at 30 kbar and 850°C and the dP/dT slope of the transition of monoclinic wollastonite to the triclinic HP form is around 250 bar/K.

The lower P limit of an end member of ellenbergerite (Fig. 2.1), $Mg_{6.67}Ti_{0.66}$

$Al_6Si_8O_{28}(OH)_{10}$, was constructed by Chopin (1986) and is defined by ellen-bergerite = talc + chlorite + kyanite + rutile + H_2O and, at $T < 550°C$, talc + chlorite + Mg-carpholite + rutile = ellenbergerite + H_2O. Reconnaissance experiments in the system MgO-Al_2O_3-TiO_2-SiO_2-H_2O (Chopin, 1987; Chopin et al., 1992) broadly confirm the phase relations by Chopin (1986), but the P-T curve for ellenbergerite formation from talc, chlorite, kyanite, rutile and H_2O should have a negative dP/dT and, thus, the minimum P of the stability of ellenbergerite in MgO-Al_2O_3-TiO_2-SiO_2-H_2O could be as high as 26 kbar.

The P-T curve of the transition of the $CaCO_3$-polymorphs, aragonite = calcite, was determined at high P by Irvine and Wyllie (1975). It was located at 31.5 kbar and 1100°C with a dP/dT of 34 bar/K. Recent experimentation of Hacker et al. (1992) suggests a steeper slope and higher P, for example 24 kbar at 800°C.

Further indications for HP metamorphic conditions or even UHPM can also be due to the absence of a particular mineral. Instead, its HP decomposition assemblage appears. Important candidates for that are albite, paragonite, and glaucophane, the upper P stability limits of which have been determined experimentally and are shown in Fig. 2.1, as well. The mineral assemblages forming by the decomposition of these phases are discussed in subsequent sections.

Mineral Assemblages Indicating UHPM: Experiments and Thermodynamic Calculations

Particular mineral assemblages indicative for UHPM are rare, as well. Here again, experimental studies in model systems are cited that are not exclusively indicative for UHPM. Moreover, studies on the compositional change of solid solution phases in particular mineral assemblages with P and T are reviewed. Further experiments related to mineral assemblages + H_2O = melt at UHP conditions are considered. At least, occurrence of some features, such as feldspar-quartz inclusions in eclogitic garnets, in rocks could point to melting at high P (Massonne, 1991a).

In addition to the experimental studies, internally consistent thermodynamic data sets are used to calculate petrogenetic grids for UHPM. These grids supply important information on the stability of some mineral assemblages being indicative for UHPM.

Experimental Studies

System MgO-Al_2O_3-SiO_2-H_2O (MASH)

In addition to the results for MASH and its subsystems discussed in the previous sections, the experimental investigation of the reaction talc + Mg-chloritoid = chlorite + kyanite + H_2O by Chopin and Schreyer (1983) is of importance. The *P-T* curve was located at 25 kbar and 570°C with a *dP/dT* of 34 bar/K. Thus, there is only a moderate overlap of the stability field of the UHP assemblage talc + Mg-chloritoid with that of quartz.

Other experimental studies at $P \geq 25$ kbar yield results with limited significance for the diagnose of UHPM. Nevertheless, they can contribute to clarification of the *P-T* evolution of UHPM rocks. A good example for that is the recent study of the pyrope-bearing quartzite of the Dora-Maira massif by Schertl et al. (1991).

The upper thermal stability of the assemblage Mg-carpholite + quartz or coesite is due to the reaction to talc + kyanite + H_2O at $P > 19.5$ kbar. The *P-T* curve lies at 25 kbar and 558°C with a *dP/dT* close to 400 bar/K (Chopin and Schreyer, 1983). The breakdown of Mg-carpholite alone to talc + kyanite + Mg-chloritoid + H_2O takes place only some degrees above that of Mg-carpholite + SiO_2 (Chopin and Schreyer, 1983). At clearly higher *T* and $P > 30$ kbar, talc + kyanite breaks down to form pyrope + coesite + H_2O. Chopin (1985) determined a *P-T* curve lying at 30 kbar and 810°C with a *dP/dT* of almost –300 bar/K.

In the absence of SiO_2, pyrope + H_2O forms from talc + chlorite + kyanite at $P > 20$ kbar. According to Chopin (1985), the curve lies at 25 kbar and 755°C with a *dP/dT* of about –150 bar/K, thus, confirming the preliminary P_{H_2O}-*T* curve for pyrope formation given by Schreyer (1968). Further experiments of this author resulted in a reaction of chlorite + kyanite + corundum to form Mg-staurolite + H_2O at 25 kbar and 720°C. The reaction curve has a steep and negative *dP/dT* slope. The decomposition of Mg-staurolite to pyrope + kyanite + corundum + H_2O was determined by Schreyer (1968) to take place at 25 kbar and 980°C with a high positive value of *dP/dT* for the reaction curve. Later on, Schreyer (1988) suggested the formation of Mg-staurolite + H_2O from Mg-chloritoid + kyanite + corundum or diaspore at $P > 25$ kbar and from pyrope + kyanite + diaspore at $P > 48$ kbar. The curve of the latter reaction should have a positive *dP/dT* intersecting the Mg-staurolite decomposition curve at almost 90 kbar, thus being the maximum P_{H_2O} for Mg-staurolite.

According to Schreyer and Yoder (1968), a further MASH phase expected to be stable at UHP conditions is the magnesian end member of the rare mineral yoderite. However, recent experimentation by Fockenberg and Schreyer (1991) proves that yoderite is not stable in MASH. But in the presence of ferric iron, yoderite shows a considerable P-T stability. Nevertheless, the upper P stability of this mineral does not exceed 25 kbar (Fockenberg, 1991).

Among the MASH phases stable at UHP conditions, several phyllosilicates appear. Chlorite decomposes to pyrope + forsterite + spinel + H_2O at $P \geq$ 20 kbar and the equilibrium curve passes through 890°C at 20 kbar and 875°C at 40 kbar (Staudigel and Schreyer, 1977). In the presence of corundum, chlorite breaks down to pyrope + spinel + H_2O at about 30°C lower T and P = 28 kbar (Ackermand et al., 1975). According to Kitahara et al. (1966), breakdown of talc to enstatite + SiO_2 + H_2O in MSH occurs at the coesite-quartz transition and 830°C. Jenkins et al. (1991) reported decomposition T around 800°C at 28 kbar confirmed by unpublished reversals. In MASH, talc is stabilized through introduction of Al to somewhat higher T (Fig. 2.2).

Kitahara et al. (1966) also studied the reactions talc + forsterite = enstatite + H_2O, serpentine + brucite = forsterite and serpentine = forsterite + enstatite + H_2O up to 30 kbar. However, it is not clear whether serpentine appeared in the experiments as antigorite, the stable polymorph of serpentine at high P (O'Hanley et al., 1989). According to Kitahara et al. (1966), T close to 500° and 585°C were determined for the breakdown of serpentine + brucite and serpentine alone, respectively, at 30 kbar. The P-T curves are very steep at high P. The breakdown of forsterite + talc was located at 690°C and 30 kbar with a negative dP/dT for the reaction curve.

Melt originates in MASH at UHP conditions by reacting pyrope + kyanite + coesite + H_2O (Sekine and Wyllie, 1983). Massonne (1983) estimated T of the first appearance of melt at the coesite-quartz transition to be about 900°C. The melting curve has a dP_{H_2O}/dT of 68 bar/K.

Experimental studies on the reactions spinel + pyrope + corundum = sapphirine (Ackermand et al., 1975) and pyrope + forsterite = enstatite + spinel (MacGregor, 1964; Staudigel and Schreyer, 1977; Danckwerth and Newton, 1978; Perkins et al., 1981; Gasparik and Newton, 1984) at $T > 850$°C have demonstrated that sapphirine and enstatite + spinel are not stable at P-T of the coesite field. Sapphirine forms at 25 kbar and close to 1050°C with a dP/dT of 16 bar/K for the reaction curve. The assemblage pyrope + forsterite breaks down at P somewhat lower than that of the above reaction with sapphirine.

Some MASH phases discussed above are solid solutions, for instance, due to the Tschermako's substitution Mg + Si = 2Al. Pertinent information from experimental studies is rare. Therefore, we know little about the compositions

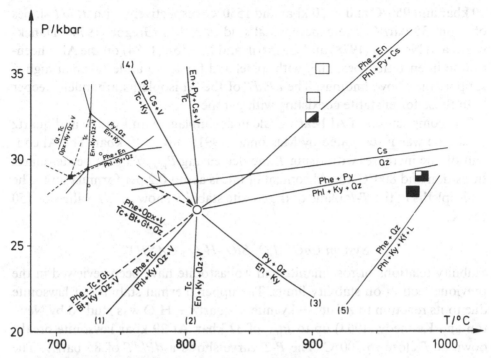

Figure 2.2 Phase relations in the systems K_2O-MgO-Al_2O_3-SiO_2-H_2O (solid curves) and K_2O-FeO-MgO-Al_2O_3-SiO_2-H_2O (dashed) with SiO_2 in excess. Abbreviations: Bt = biotite, Cs = coesite, En = enstatite, Gt = garnet, Ky = kyanite, Opx = orthopy-roxene, Phe = phengite, Phl = Phlogopite, Py = pyrope, Qz = quartz, Tc = talc, V = hydrous fluid. Open circles in the inlet are metastable invariant points. Boxes represent *P-T* conditions including 2σ reproducibility of critical experiments mentioned in the text (open = growth of Phe + En; closed = clear decomposition of Phe + Py; three-quarter closed = slight decomposition of Phe + Py; half closed = no reaction detected). The underlying experimental studies are those of Massonne & Schreyer, 1989 (1); Massonne, unpubl. (2); Massonne, 1983 (3); Chopin, 1985 (4); Massonne, 1985 (5). The dashed curve corresponding to (1), which was already given by Massonne & Schreyer (1989), was recalculated using slightly different $K_D = (Fe/Mg)^A/(Fe/Mg)^B$ values for a few mineral pairs A-B compared to their work. The *P-T* position of the other dashed curves were estimated in the same manner as the previous one, considering K_D values for orthopyroxene-biotite somewhat above unity according to Sengupta et al. (1990).

of the solid solution phases participating in most of the MASH reactions summarized above. Relatively sufficient, however, is our knowledge about the Al introduction in enstatite and talc even at $P > 25$ kbar. Boyd and England (1964), MacGregor (1974), Lane and Ganguly (1980) and Perkins et al. (1981) reported experiments on the variable composition of Al-bearing enstatite coexisting with pyrope in the *P-T* range 20–40 kbar and 900°–1600°C. The Al_2O_3 content of enstatite increases with rising T and decreasing P. The iso-pleths for 5 and 10 mol% Al_2O_3 in enstatite coexisting with pyrope lie at about

20 kbar and 950°C and at 30 kbar and 1550°C, respectively, with dP/dT slopes of about 35 bar/K. The experimental studies of MacGregor (1974), Danckwerth and Newton (1978) and Gasparik and Newton (1984) on the Al concentration in enstatite coexisting with spinel and forsterite undertaken at high P support the above findings. The dP/dT of the Al isopleths are much steeper than those for enstatite coexisting with pyrope.

The composition of Al-bearing talc in assemblage with kyanite and quartz or coesite was investigated by Massonne (1991a) up to 40 kbar. The Al content of talc increases with rising T and decreasing P_{H_2O}. At the coesite-quartz transition and 800°C, the Al content of talc is about 0.2 per formula unit. The Al-isopleths in the T-P plane of the coesite stability show dP/dT-values of 150 bar/K.

System CaO-Al_2O_3-SiO_2-H_2O ($CASH$)

Stability relations of rosenhanite and wollastonite have been reviewed in the previous section on stability limits. The upper thermal stability of lawsonite due to its reaction to zoisite + kyanite + quartz + H_2O was studied by Newton and Kennedy (1963) up to P_{H_2O} of 32 kbar. At 29 kbar lawsonite breaks down at T close to 700°C. The P-T curve shows a dP/dT of 85 bar/K. The investigation of the same reaction by Chatterjee et al. (1984) gave a lower T of 30°C at 20 kbar. The breakdown of grossular + kyanite + quartz to anorthite was most recently studied by Koziol and Newton (1988) confirming the earlier studies of Goldsmith (1980) and Gasparik (1984a, 1985a) undertaken up to 33 kbar. The reaction curve lies at 30 kbar and 1350°C with a dP/dT of 23 bar/K and, thus, entirely in the P-T field of quartz.

Extensive experimental work on melting in CASH at high P has been achieved by Boettcher (1970). The reaction of zoisite + kyanite + quartz or coesite leads to minimum melting T in CASH at P_{H_2O} between 13 and, at least, 35 kbar. At the coesite-quartz transition, the melting T is about 820°C.

System Na_2O-Al_2O_3-SiO_2-H_2O ($NASH$)

The stability of paragonite is limited towards high P by a reaction paragonite = jadeite + kyanite + H_2O reversed by Holland (1979). The P-T curve (Fig. 2.1) appears at 25 kbar and 585°C with a slightly negative dP/dT. At $T <$ 500°C the stability fields of coesite and paragonite overlap.

Albite and coesite coexist at $T >$ 1200°C based on the experimentally determined reaction jadeite + quartz = albite (Fig. 2.1; Holland, 1980). The results of Holland (1980) are in fairly good agreement with those obtained by Birch and LeComte (1960), Johannes et al. (1971), Hays and Bell (1973) and Mirwald and Massonne (1980). However, the latter authors reported, on the basis of one bracket at about 1100°C, an equilibrium P for albite breakdown that was almost 2 kbar higher than that given by Holland. The stability field of jadeite alone has a P about 8 kbar lower than that of jadeite + SiO_2 in a comparison of the results for the reaction jadeite = nepheline + albite, obtained by Robertson et al. (1957), with the albite breakdown reported by Holland (1980). Gasparik (1985a) observed a similar P difference at 1200°C. The difference reported by Newton and Kennedy (1968) is less than 5 kbar at $T > 800$°C.

Boettcher and Wyllie (1969) determined melting relations in the system $NaAlSiO_4$-SiO_2-H_2O up to 35 kbar. At high P_{H_2O} the breakdown of jadeite + SiO_2 leads, in their system, to minimum melting T. At the coesite-quartz transition it is close to 710°C. According to Boettcher and Wyllie (1969), both melting curves with quartz and coesite show a steep and significantly positive dP/dT.

System K_2O-MgO-Al_2O_3-SiO_2-H_2O (KMASH)

In contrast to the subsystem MASH, only a few experimental studies in KMASH have been completed at $P \geq 25$ kbar. These include the upper P stability of K-feldspar + H_2O reported in section on K-cymrite. The stability of phlogopite + SiO_2 + H_2O was examined by Massonne (1992a). Phengite, talc, and a siliceous fluid rich in K and Mg appear on the low-T side of the reaction located at the quartz-coesite transition at 550°C with a dP/dT slope of about 100 bar/K at UHP. The nature of the fluid might be that of a supercritical vapor phase, but it has, at least, the character of a peralkaline silicate melt (Massonne, 1992a). The breakdown of phlogopite + K-feldspar + H_2O leading to the formation of a similar fluid + phengite $\pm$ quartz occurs at 600°C and P as low as 19.5 kbar (Massonne, 1992a). The P-T curve is clearly flatter than the breakdown curves of phlogopite + SiO_2 + H_2O.

The equilibrium line of phengite + talc = phlogopite + kyanite + quartz + H_2O has been determined by Massonne and Schreyer (1989) at 25 kbar and 785°C with a dP/dT of 85 bar/K. However, this curve cannot be extended into the coesite stability field; it intersects with another curve in MASH at an invariant point at about 27.5 kbar. This reaction is either enstatite + kyanite = pyrope + quartz (studied by Massonne, 1983) or talc = enstatite + kyanite + quartz + H_2O (Massonne, unpubl. data, see section on MASH). Thus, two different arrangements of three invariant points are possible as examined by Massonne and Schreyer (1989). One of them is shown in Fig. 2.2. As a result of Schreinemaker's analysis, independent of the arrangement of invariant points selected, two stable reactions, leading to the assemblages phengite + pyrope and phengite + enstatite at high P, occur at $T > 810$°C. According to a reconnaissance study, the curve of the reaction phengite + pyrope = phlog-opite + kyanite + quartz lies subparallel to the coesite-quartz transition curve at about 1.5 kbar lower P (Fig. 2.2) and intersects with the phengite + quartz melting curve close to 1000°C. The reaction phengite + enstatite = phlogopite + pyrope + coesite probably occurs at 900°C and somewhat above 32 kbar. Thus, the slope of the P-T curve is clearly steeper than that of the phengite + pyrope reaction.

The anhydrous reaction of kalsilite + enstatite = sanidine + forsterite passes through 25 kbar and 1150°C with a dP/dT of 34 bar/K (Wendlandt and Eggler, 1980). The assemblage sanidine + forsterite reaches almost P-T conditions of stable coesite just before melting at about 1400°C (Wendlandt and Eggler, 1980).

The compositions of phengites participating in the reactions mentioned above are fairly well known. Massonne and Schreyer (1989) crystallized the assemblage phengite + talc + kyanite + quartz or coesite at 15–40 kbar and 550°–800°C. The Si contents of phengite in this assemblage increase clearly with rising P. The octahedral occupancy of phengite is somewhat above the ideal value of 2.00 per formula unit (pfu). According to Massonne and Schreyer (1989), a Si content of 3.6 pfu is reached at P_{H_2O} somewhat above 40 kbar. The Si isopleths in the T-P_{H_2O} plane show a somewhat higher dP/dT than the coesite-quartz curve. Further compositional results exist for phengite in paragenesis with phlogopite + K-feldspar + quartz on the basis of experiments up to P_{H_2O} = 25 kbar and 350 to 700°C (Massonne and Schreyer, 1987). At the maximum P of this study, the Si contents of phengite have already reached their upper limit of about 3.8 Si pfu. The Si isopleths are somewhat steeper in the T-P_{H_2O} plane than those of phengite coexisting with talc, kyanite and SiO_2.

In contrast to phengites, not much is known about the compositions of phlogopite in the above reactions. According to Massonne and Schreyer (1987), the Si contents of phlogopite, similar to that of phengites, vary with P. Octahedral occupancies might be as low as 2.8 pfu in assemblage with phengites. At P_{H_2O} > 20 kbar Si contents of phlogopite coexisting with phengite, K-feldspar, and quartz are clearly above 3.0 pfu.

Melting experiments in the subsystem KASH by Huang and Wyllie (1974) indicate the melting of K-feldspar + quartz + H_2O ± muscovite at T somewhat below 700°C for the P range 20–27 kbar. The melting curves are very steep in the T-P plane. The decomposition of K-feldspar + H_2O to K-cymrite and the transition of quartz to coesite clearly cause rising melting T with increasing P_{H_2O}. Similar melting experiments in the presence of MgO by Massonne (unpubl. data in Massonne and Schreyer, 1987) led to the conclusion that the melting T are lowered only by a few degrees through the introduction of this additional component to KASH.

Systems Na_2O-MgO-Al_2O_3-SiO_2-H_2O (NMASH) and Na_2O-K_2O-Al_2O_3-SiO_2-H_2O (NKASH)

On the basis of one experimental bracket of the reaction glaucophane = jadeite + talc (Carman and Gilbert, 1983), further experiments with glaucophane and various thermodynamic data, Evans (1990) recalculated the P-T position of this reaction limiting the stability field of glaucophane towards high P. The reaction curve appears in Fig. 2.1 at 800°C and 36 kbar with a slightly negative dP/dT. The curve of the same reaction was recalculated by Holland (1988) to occur at nearly the same P-T conditions but with a slightly positive dP/dT.

Huang and Wyllie (1975b) studied the minimum melting in the subsystem $NaAlSi_2O_6$-$KAlSi_3O_8$-SiO_2-H_2O up to P_{H_2O} of 35 kbar, at which melting occurs at about 700°C due to the reaction of jadeite + K-cymrite + coesite + H_2O. The reaction curve shows a significantly positive dP/dT. Minimum melting T are about 610°C at 17 kbar, at which albite reacts to jadeite + quartz, when one considers also the melting experiments at low P undertaken by Merrill et al. (1970).

Systems CaO-FeO-MgO-Al_2O_3-SiO_2 (CFMAS) and CaO-Na_2O-MgO-Al_2O_3-SiO_2 (CNMAS) and Exotic and More Complex Systems

Compositional relations of coexisting minerals. Experimental work in the subsystem CMAS at high P was aimed mainly at studying the compositional relationships of garnet, orthopyroxene, and clinopyroxene as a function of P and T. The solvus relations of enstatite and diopside in the Al-free system were investigated by Mori and Green (1976), Mori (1978), Perkins and Newton (1980), Brey and Huth (1984) and Nickel and Brey (1984) up to 50 kbar. At constant T the amounts of enstatite in diopside and diopside in enstatite decrease with rising P. The same is true in the presence of Al as was demonstrated by Perkins and Newton (1980) in experiments up to 40 kbar and by Yamada and Takahashi (1984) between 50 and 100 kbar. These authors also studied the introduction of Al in coexisting orthopyroxene and clinopyroxene buffered by pyrope-rich garnet. The same was achieved in experiments up to 50 kbar by Akella (1976), Nickel et al. (1985) and Brey et al. (1986). As shown for the system orthopyroxene-pyrope (see section on MASH), both orthopyroxene and clinopyroxene decrease in Al content with rising P at constant T. That rising T causes a definite increase in Al concentrations in the pyroxenes is also supported by the HP studies of Herzberg and Chapman (1976) and Gasparik (1984b), who investigated the Al contents in enstatite and diopside coexisting with spinel + forsterite. The content of the grossular component in garnet decrease, in particular, with rising P (Brey et al., 1986).

Coexisting orthopyroxene and garnet were investigated by Wood (1974), O'Neill and Wood (1979) and Harley (1984a, 1984b) in the systems FMAS and CFMAS up to 30 kbar. Wood (1974) and Harley (1984a) evaluated how the Al content in orthopyroxene varies with P and T as in the MAS system. O'Neill and Wood (1979) and Harley (1984b) concentrated on the T-dependent Fe^{2+}-Mg distribution among garnet and orthopyroxene, which is also a clear function of the Ca content in Al-garnet. Coexisting clinopyroxene and garnet were studied by Ellis and Green (1979) as well as Pattison and Newton (1989) in CFMAS up to 30 kbar. These authors evaluated their experimental data with respect to the Fe^{2+}-Mg distribution between garnet and clinopyroxene mainly as a function of T and grossular component in garnet. In addition, Ellis and Green (1979) presented data for the garnet-clinopyroxene pair buffered by kyanite + SiO_2.

The P dependence of the clinopyroxene-orthopyroxene solvus studied by Mori (1978) in the subsystem CFMS for compositions relatively rich in Fe^{2+} up to 30 kbar is not as sensitive as that in CMS (see above). The Ca-Mg exchange between olivine and clinopyroxene as well as olivine and orthopyroxene was investigated by Adams and Bishop (1982, 1986) in CFMS up to 41 kbar. The Ca concentration in olivines coexisting with clinopyroxene increase with rising T and falling P.

In NCAS, the introduction of Ca-Tschermaks pyroxene (CaTs) in jadeite coexisting with grossular and corundum was studied by Gasparik (1985a) up to 40 kbar. With increasing P and decreasing T, the amounts of CaTs decrease. Corresponding isopleths show a dP/dT of about 70 bar/K. The end member reaction grossular + corundum = CaTs was located at 25 kbar and 1435°C.

In NMCAS, the compositional P-T dependence of the diopside-jadeite solid solution series coexisting with albite and quartz/coesite was investigated by Kushiro (1969) and Gasparik (1985b) up to 34 kbar. Holland (1983) determined additionally the space groups C2/c and P2/n of pyroxene at 600°C. The Na contents of omphacite coexisting with albite and quartz or coesite decrease with increasing T and falling P. According to Gasparik (1985b), the isopleth for 20 mol% jadeite component lies at 800°C and 13.9 kbar with a dP/dT of 14.2 bar/K. In the system NFCAS, a similar study with respect to hedenbergite-jadeite solid solution was achieved by Perchuk (1990) leading to equivalent results.

In the system TiO_2-CFMS, the partitioning of Fe^{2+} and Mg between ilmenite and orthopyroxene as well as ilmenite and clinopyroxene was investigated by Bishop (1980) up to 36 kbar. The Fe^{2+}-Mg distributions of these mineral pairs are clear functions of T.

Nickel (1986) studied assemblages with two pyroxenes, garnet, spinel and forsterite in Cr_2O_3-CMAS up to 35 kbar focusing on the Al-Cr distribution among garnet and spinel, which is T-dependent. Cr is enriched in spinel.

Experiments similar to those reviewed above were undertaken by Brey et al. (1990) and Köhler and Brey (1990) up to 60 kbar with natural compositions to determine the distribution of Al, Ca, Cr, Mn, Na, Ti among coexisting orthopyroxene, clinopyroxene, garnet, olivine, and spinel. Qualitatively, their results are the same as those obtained in the simple model systems. The same is true for the Fe^{2+}-Mg partitioning between garnet and omphacite studied by Råheim and Green (1974) in a natural system up to 40 kbar. The distribution of Fe^{2+} and Mg between coexisting garnet and phengite was experimentally determined as a function of T by Green and Hellman (1982) between 20 and 40 kbar. Experiments with a natural lherzolite were undertaken by Akaogi and Akimoto (1979) above 45 kbar mainly to determine the Fe^{2+}-Mg partitioning between olivine, Mg,Fe^{2+},Si-spinel, pyroxenes, and garnet.

Mutual stability relations of minerals and assemblages. The upper P-T stabilities of several amphiboles with complex compositions, relevant for UHPM, were studied by several investigators. Typically, the reaction curves limiting the various amphibole stability fields show low and often negative dP/dT values at relatively low T (Hariya, 1984). The upper P limit of tremolite is due to a reaction to diopside + talc. Jenkins et al. (1991) located the curve in CMASH at 25 kbar and 600°C with a dP/dT of 9 bar/K. In contrast to that of glaucophane (see section NMASH and NKASH) and tremolite, the HP decomposition assemblages of other amphiboles are poorly defined and break down within a relatively wide P-T field. The decomposition P for various amphiboles are given subsequently at 800°C. Richterite and K-richterite $[KNaCaMg_5Si_8O_{22}(OH)_2]$ decompose between 28.5 and 36 kbar and between 22 and 33.5 kbar, respectively (Hariya et al., 1974). A 1:1 richterite-tremolite solid solution phase breaks down at about 38 kbar (Hariya and Terada, 1973). In the study of Pawley (1992) an amphibole rich in nyböite component is stable up to at least 32 kbar. A natural kaersutite disappears in the experiments of Merrill and Wyllie (1975) at about 30 kbar. Kaersutite investigated by Yagi et al. (1975) decomposes at about 20 kbar. Tschermakite breaks down between 16.5 and 23.5 kbar (Hariya et al., 1974) and a hornblende studied by Lambert and Wyllie (1968) decomposes already between 13 and 22.5 kbar. The maximum P stability of other hornblendes ranges between 22 and 28 kbar (Essene et al., 1970). However, some amphiboles exist at P much higher than given above. For example, Sudo and Tatsumi (1990) observed potassic amphiboles as one of the breakdown products of phlogopite + diopside at experiments between 50 and 130 kbar. At 1000°C the breakdown of this assemblage starts at about 75 kbar. At 120 kbar and 1000°C, phlogopite has totally disappeared but amphibole still exists. The corresponding P-T boundaries show relatively low and negative dP/dT values.

The breakdown of synthetic ferri-deerite, $Fe^{2+}_{12}Fe^{3+}_{6}Si_{12}O_{40}(OH)_{10}$, to ferrosilite + magnetite + quartz + H_2O was determined by Lattard and Schreyer (1981) up to 28 kbar and about 600°C. The reaction curve shows a high positive dP/dT. The breakdown of minerals similar to deerite, natural howieite and zussmanite, were also investigated by Lattard and Schreyer (1981) at high P. Related to the NiO/Ni buffer, maximum T are 530°C for zussmanite and 490°C for howieite at 25 kbar. The dP/dT of the breakdown curves is steep. In case of zussmanite, it is probably negative.

Experiments on the thermal stability of the $Fe^{2+}Mn^{3+}$-garnet calderite were undertaken by Lattard and Schreyer (1983) up to 30 kbar using the magnetite-hematite buffer. Calderite breaks down to pyroxmangite + magnetite at about 850°C and 30 kbar with a dP/dT of the reaction curve of 50 bar/K. Similarly, the thermal stability limit of almandine at oxidizing conditions occurs by breakdown to magnetite + kyanite + quartz at 30 kbar and about 950°C (Harlov and Newton, 1992). The reaction curve shows a dP/dT of 31 bar/K. The P-T stability of knorringite, $Mg_3Cr_2Si_3O_{12}$, was investigated by Irifune et al. (1982) up to 120 kbar. Its low P-limit characterized by decomposition to enstatite + eskolaite lies at 105 kbar and 1200°C with a dP/dT of 63 bar/K. The stability of the knorringite-pyrope solid solution series decomposing to enstatite and a corundum-eskolaite solid solution phase was investigated down to 30 kbar (Irifune et al., 1982). A garnet with 35 mol% knorringite component coexists with enstatite and an eskolaite-rich oxide nearly independent of T at about 67 kbar. Similar results were obtained by Woodland and O'Neill (1992) for skiagite-almandine solid solutions. The phases coexisting with garnet are, however, spinel and SiO_2.

The HP breakdown of pure chloritoid to almandine + diaspore in the presence of the iron-wüstite buffer was located at 29 kbar and 625°C with a flat negative dP/dT slope (Vidal and Theye, 1992).

The upper P stability of surinamite, $Mg_3Al_4BeSi_3O_{16}$, was studied by Hölscher et al. (1986) in the T range 700° to 1000°C. These investigators observed surinamite breakdown to pyrope + chrysoberyl at 45 kbar and 800°C. The curve shows a dP/dT of −12 bar/K.

The stability of boron-bearing kornerupine was recently investigated by

Krosse et al. (1992) in the B_2O_3-MASH system up to 40 kbar. Breakdown of this phase was only observed at 40 kbar and 800°C leading to the assemblage pyrope + Mg-staurolite + MgAl-borate. At 35 kbar and 800°C, kornerupine was found to be stable.

The reaction of magnesite + enstatite to form forsterite + CO_2 was studied by Newton and Sharp (1975) at high P and T. The reaction curve was located at 30 kbar and about 1240°C with a dP/dT of 44 bar/K.

Melting and solubility relations. In the presence of H_2O, melting relations were studied on numerous rock compositions up to P of 30 kbar and more. Subsequently, minimum melting T are given for 30 kbar. Granitic material melts in the range 685°–695°C (Stern and Wyllie, 1973,1981; Huang and Wyllie, 1981) similar to the melting T of the model system $NaAlSi_2O_6$-$KAlSi_3O_8$-SiO_2-H_2O (Huang and Wyllie, 1975b). The higher T of 730°C (Lambert and Wyllie, 1974) and 740°C (Stern and Wyllie, 1973) were determined for andesitic compositions. The determination of the solidus of basaltic material led to a T-range of 725°–760°C (Hill and Boettcher, 1970; Lambert and Wyllie, 1972; Stern and Wyllie, 1978). In the basaltic model system CaO-MgO-Al_2O_3-SiO_2-H_2O this T is 810°C (Sekine et al., 1981). Basanitic compositions gave solidus T of 830°C (Merrill and Wyllie, 1975) and 885°C (Millhollen and Wyllie, 1974). Increasing solidus T reviewed above can roughly be correlated with the decreasing SiO_2 contents of the compositions studied. Pelitic material with relatively low SiO_2 content, however, shows a minimum melting T of 680°C at 30 kbar as granites (Stern and Wyllie, 1973).

The solidus T of peridotites + H_2O at high P were studied by several authors who found apparently contradicting results. Kushiro et al. (1968), Green (1972) and Millhollen et al. (1974) reported about 1020°C at 30 kbar. Mysen and Boettcher (1975) observed distinctly lower solidus T between 770° and 890°C for four different peridotite nodules. However, various concentrations of alkalis in the peridotites studied have a major influence on the melting T. The higher the alkali contents, the lower is the peridotite solidus T (see Millhollen et al., 1974).

In the silicate systems, the dP_{H_2O}/dT of the solidus curves are generally positive at P of the coesite stability. In carbonate systems, the situation might be different at UHP conditions. For example, the slope of the melting reaction of aragonite, brucite, portlandite and vapor in the system CaO-MgO-CO_2-H_2O, determined by Boettcher et al. (1980) to lie at 500°C and 35 kbar, is steep and shows a negative dP/dT. Addition of SiO_2 to the system reduces the melting T by about 10°C (Boettcher et al., 1980).

The solubility of phlogopite, forsterite, pyroxenes, and various mineral assemblages in aqueous fluids were studied experimentally at or close to UHP. Ryabchikov and Boettcher (1980) observed K_2O concentrations in H_2O of about 25 wt% after dissolution of phlogopite in the presence of olivine $\pm$ orthopyroxene $\pm$ spinel at 1100°C and 30 kbar. K_2O/Al_2O_3 ratios of these solutions are close to unity. MgO and SiO_2 concentrations must be substantial. The solubility of diopside in an aqueous fluid at 30 kbar and 1200°C amounts to about 10 wt% (Eggler and Rosenhauer, 1978). About 25 wt% of jadeite is dissolved incongruently in H_2O at 30 kbar and 650°C (Ryabchikov and MacKenzie, 1985). Forsterite is nearly congruently dissolved at 30 kbar. At $T \leq 1100$°C, forsterite solubility is 2 wt% and less, whereas enstatite solubility is higher at these P-T conditions (Ryabchikov et al., 1982).

Petrogenetic Grids Calculated from Thermodynamic Data Sets

With the aid of well-known thermodynamic properties of minerals it is possible to calculate the P-T conditions of mineral equilibria at UHP conditions considering the $\Delta G = 0$ relation. A good example for that is the successful calculation of the diamond-graphite transition by Berman (1976). Thermodynamic data for many rock-forming minerals were compiled by Helgeson et al. (1978) and used by Delany and Helgeson (1978) to predict phase equilibria already for UHPM. Further progress with respect to the quality of such data sets was made in the 1980s by the simultaneous derivation of thermodynamic data from mineral equilibria leading to internally consistent data sets (e.g. Halbach and Chatterjee, 1984). Such data sets being most advanced are those of Berman (1988) and Holland and Powell (1990), which are considered here as well.

The thermodynamic data set of Berman (1988) contains ΔH_f°, S°, and V° values, including c_P, α_P, and β_T functions, of 67 minerals optimized by mathematical programming taking into account about 180 experimentally studied mineral equilibria, a few of which were presented in the previous sections, and numerous calorimetric data. Holland and Powell (1985) have also extracted an internally consistent data set from experimental studies, the number of which is similar to that evaluated by Berman (1988). However, these authors used a least-squares method (Powell and Holland, 1985) to refine ΔH_f° of the minerals only. Their most recent data set (Holland and Powell, 1990) includes additional experimental results and element distribution data, like those for Fe-Mg partitioning among mineral pairs, obtained from natural parageneses. The thermodynamic data of mineral end members in this new work is almost twice that found in the work of Berman (1988). Among them are end members to formulate mineral equilibria involving Tschermak's substitution and

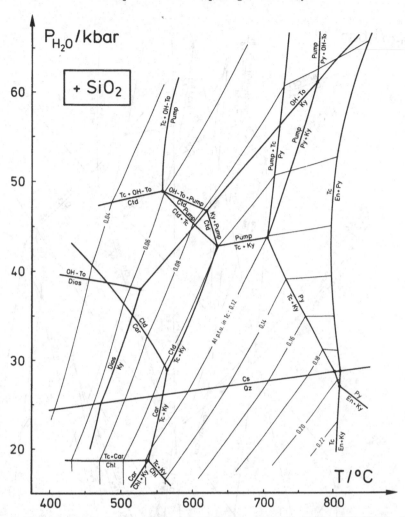

Figure 2.3. Calculated phase relations in the system MgO-Al$_2$O$_3$-SiO$_2$-H$_2$O with SiO$_2$ and H$_2$O in excess using the thermodynamic data set of Berman (1988) augmented by the data of Table 2.1. Isopleths for Al pfu in talc are given by thin lines. Abbreviations in addition to those of Fig. 2.2: Car = Mg-carpholite, Chl = chlorite composed of end members, Cli = clinochlore and Ame = amesite, Ctd = Mg-chloritoid, Dias = diaspore, OH-To = OH-topaz, Pump = MgMgAl-pumpellyite, end members of talc: Ta = ideally composed talc, Al-Tc = Al-talc. For chemical compositions see Table 2.1. Activity models used: a_{Cli} = (Si–2):(Mg/5)^6, a_{Ame} = (3–Si):(Al^{VI}/2)^6, a_{En} = X$_{En}$, a_{Al-En} = X$_{Al-En}$, a_{Ta} = (Si/4)^4, a_{Al-Tc} = (Al^{IV})^4.

Mg-Fe exchange in chlorite, talc, micas, and amphiboles assuming simple ideal ionic mixing models. However, the thermodynamic data of many of the corresponding end member components are preliminary (see Holland and Powell, 1990).

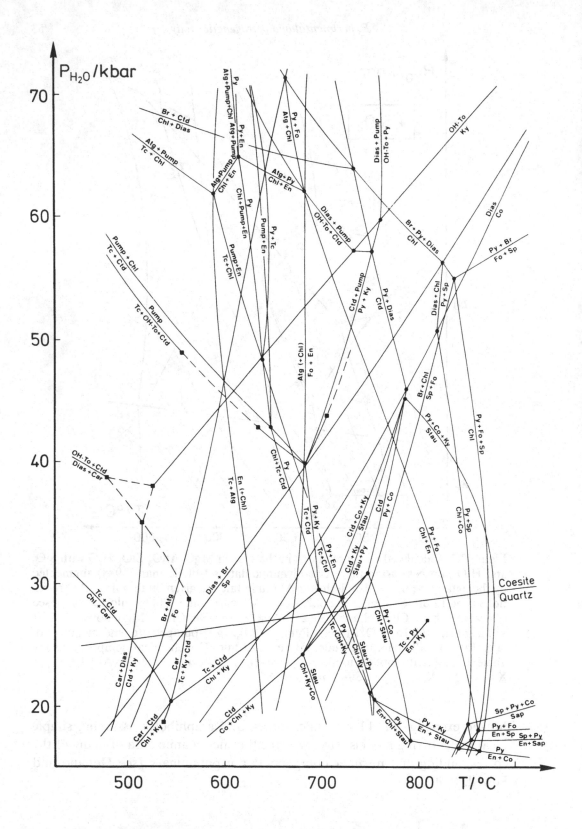

Subsystem *MASH*

A petrogenetic grid calculated for MASH in the presence of excess H_2O is shown in Figures 2.3 and 2.4. The grid is based on the thermodynamic data set of Berman (1988) and was calculated using the Ge0-Calc software package of Brown et al. (1988). However, the data set of Berman (1988) lacks a number of phases important for UHPM such as Mg-carpholite and the newly discovered OH-topaz. In addition, it is necessary to consider solid solution in some MASH-phases. The thermodynamic data set of Berman (1988) alone for this purpose is insufficient. The required new thermodynamic data, compatible to the data set of Berman (1988), were recently published by Massonne (1992b) and are given here in Table 2.1. For the derivation of this data set, which also comprises phases not relevant for UHPM (e.g, cordierite and boron-free kornerupine), the total of published experimental results obtained in MASH was broadly taken into account, including the experimental studies reported here in sections on Index Minerals and MASH. Thus, a good agreement between the calculated *P-T* curves and MASH curves, given above, results. The solid solution properties of enstatite and talc were also derived from experimental studies reported in the section on MASH, using simple but appropriate activity models (see legend of Fig. 2.3). Those of chlorite, sapphirine, and staurolite were partially estimated considering limited compositional information, for instance, those for chlorite are given by Massonne (1989). In the case of staurolite, a single substitution equation, $Al + H = Si$ (see Schreyer, 1968), was taken into account. The magnitude of the increase of hydrogen in MASH staurolite with rising P recalculated with the data of Massonne (1992b) is similar to that estimated by Holdaway et al. (1986) for Fe-staurolite in the presence of almandine + kyanite. The new data set of Massonne (1992b) has some phases included in Berman's set. However, reconsideration of the underlying experimental data as well as the regard for new

Figure 2.4. *P-T* grid for the system $MgO-Al_2O_3-SiO_2-H_2O$ with H_2O in excess calculated as relations in Fig. 2.3, which can be inserted here by considering dashed lines and solid squares (= invariant points in Fig. 2.3). Chl appearing in context with Atg breakdown reactions is given in parenthesis because the corresponding *P-T* curves are nearly the same as in the system MSH. Abbreviations in addition to those of Fig. 2.2 and 2.3: Atg = antigorite, Br = brucite, Co = corundum, Fo = forsterite, Sap = sapphirine composed of end members Sapp = ideally composed sapphirine and Al-Sap = Al-sapphirine, Sp = spinel, Stau = Mg-staurolite composed of end members Mg-Stau = ideally composed Mg-staurolite and Mg-H-Stau = ideally composed hydrogen-rich staurolite. For chemical compositions see Table 2.1. Activity models used: $a_{Sapp} = (Si-1)^2$, $a_{Al-Sap} = (2-Si)^2$, $a_{Mg-Stau} = (Si-7)^2$, $a_{Mg-H-Stau} = (8-Si)^2$ and those of Fig. 2.3.

Table 2.1 *Molar Properties At Standard Conditions For Some MASH Phases*

		H_f^0	S^0	V^0	$v_1 \cdot 10^6$	$v_2 \cdot 10^{12}$	$v_3 \cdot 10^6$	$v_4 \cdot 10^{10}$	cp
Coesite	SiO2	-907850	39.1	2.064a	-1.037a	3.000a	7.396a	43.605a	a
Al-enstatite	Al2O3	-1642600	58.6	c	-0.750	0.448	21.915	74.920	b
OH-topaz	Al2SiO4(OH)2	-2895530	107.8	5.359	-1.000	0.000	25.000	0.000	b
Pyrope	Mg3Al2Si3O12	-6281700	270.8	11.316a	-0.576a	0.442a	22.519a	37.044a	a
Sapphirine	Mg4Al8Si2O20	-11023000	430.0	19.733	-0.650	1.500	22.000	0.000	b
Al-sapphirine	Mg3Al10SiO20	-11123000	422.0	19.530	-0.650	1.500	22.000	0.000	b
Talc	Mg3Si4O10(OH)2	-5897387a	261.24a	13.655	-2.400	5.665a	29.447a	0.000a	a
Al-talc	Mg2Al2Si3O10(OH)2	-5955300	253.0	13.439	-2.400	5.700	29.450	0.000	b
Clinochlore	Mg5Al2Si3O10(OH)8	-8907500	434.0	21.147a	-1.819a	0.000a	26.452a	0.000a	a
Amesite	Mg4Al4Si2O10(OH)8	-8992500	426.0	20.920	-1.819	0.000	26.452	0.000	b
Mg-carpholite	MgAl2Si2O6(OH)4	-4778600	215.0	10.580	-1.440	3.000	30.000	0.000	b
Mg-chloritoid	MgAl2SiO5(OH)2	-3565090	124.0	6.874	-1.280	3.000	27.000	0.000	b
Mg-staurolite	Mg4Al18Si8O46(OH)2	-25299500	860.0	44.260	-0.650	1.500	24.000	0.000	b
Mg-H-staurolite	Mg4Al18Si7O42(OH)6	-24832000	1060.0	44.700	-0.650	1.500	24.000	0.000	b
MgMgAl-pumpellyite	Mg5Al5Si6O21(OH)7	-13893500	615.0	26.946	-0.600	0.000	25.000	0.000	b

Note: H_f^0 and S^0 of most components were refined, V^0 was taken from the literature and (v_1 - v_4) were estimated by Massonne (1992b). The data are compatible with the thermodynamic data set of Berman (1988, EOS of H2O after Haar et al.). Units are joule, bar, K. V p,t $=V^0$ $(1+v_1(P-1)+v_2(P-1)2+v_3(T-298)+v_4(T-298)2)$. a = Data according to Berman (1988), b = c p function estimated after Berman & Brown (1985), c = Variable, e.g. 2.7360 for 5 mol%, 2.7658 for 10 mol% Al-enstatite.

experiments led to a slight change of Berman's data for clinochlore, coesite, pyrope, and talc. With the new pyrope data, a better reproducibility of P-T curves at relatively low T is obtained in regard to, for instance, the stability curves of pyrope + H_2O and pyrope + coesite + H_2O determined by Chopin (1985, see section on MASH) and the chlorite stability at high P studied by Staudigel and Schreyer (1977). In order to calculate the compositional change of solid solution phases along univariant reaction curves, only simple activity models (see legends of Fig. 2.3 and 2.4) were used. In addition, the compositions of coexisting solid-solution phases were iteratively determined for the grid of Figures 2.3 and 2.4.

In the MASH grid of Fig. 2.3, the most prominent features are the following: Chlorite + SiO_2 is stable up to P_{H_2O} close to 20 kbar and, thus, absent at UHP conditions. Mg-chloritoid + SiO_2 is limited to P_{H_2O} between about 30 and 50 kbar and $T < 630°C$. The stability fields of talc + kyanite and pyrope + SiO_2 in the presence of H_2O are similar to those constructed by Schreyer (1988). Stability fields of MgMgAl-pumpellyite + SiO_2 and talc + OH-topaz being indicative for UHPM are newly defined. The upper P_{H_2O} stability limit of diaspore + coesite lies close to 40 kbar. However, it should be mentioned that a part of the grid is metastable with respect to the stable appearance of "piezotite" (see Fig. 2.1). Because volume data are lacking, this phase could not be introduced to the data set of Massonne (1992b).

In order to demonstrate the behavior of solid-solution phases that were considered throughout, an example is given in Fig. 2.3 involving Al introduction in talc. The dP/dT of the Al-isopleths for coexisting talc in the various assemblages are similar to those for the assemblage with kyanite + SiO_2 + H_2O for which experimental data exist (see section on MASH). An exception is the assemblage with pyrope + coesite + H_2O where the Al concentration in talc can be applied as geobarometer, because of the low dP/dT.

At SiO_2-deficient conditions (Fig. 2.4) the phase relations in MASH are more complex than those for SiO_2 in excess. The grid of Fig. 2.4 defines numerous new mineral reactions. At $P > 30$ kbar the lower thermal stability of pyrope + H_2O is due to such reactions that lead to the following assemblages given in the order of rising P_{H_2O}: chlorite + talc + Mg-chloritoid, chlorite + talc + MgMgAl-pumpellyite, chlorite + MgMgAl-pumpellyite + enstatite, chlorite + MgMgAl-pumpellyite + antigorite. The terminal reaction of chlorite to form pyrope + forsterite + spinel determined by Staudigel and Schreyer (1977) up to $P_{H_2O} = 40$ kbar can be extended further to about 55 kbar. At higher P, two new reactions forming brucite + pyrope + spinel + H_2O and brucite + pyrope + diaspore + H_2O occur. In contrast to that, the reactions limiting the P-T field of Mg-staurolite are those already expected by Schreyer (1988). However, the maximum P_{H_2O} is predicted to be only 45 kbar (see section MASH), owing to the fact that the upper thermal boundary of Mg-staurolite lies at T clearly lower than that determined by Schreyer (1968) and shows a dP/dT that becomes negative at $P > 30$ kbar.

Further new features of the SiO_2-deficient grid (Fig. 2.4) are the following: The MgMgAl-pumpellyite field differs from that given by Schreyer et al. (1991, see Fig. 2.1), although the original experimental data are reproduced. The difference is due to the assemblages from which MgMgAl-pumpellyite forms at low T and high P_{H_2O}. In the order of increasing T, talc + Mg-chloritoid + OH-topaz, talc + Mg-chloritoid + coesite (see Fig. 2.3) and talc + Mg-chloritoid + kyanite become successively stable. The formation of MgMgAl-pumpellyite from chlorite + Mg-carpholite ± talc experimentally determined by Schreyer et al. (1991) is metastable compared to the reaction talc + Mg-chloritoid + H_2O = MgMgAl-pumpellyite + chlorite, which also defines the upper P_{H_2O}-T limit of talc + Mg-chloritoid. The lower thermal boundary of this assemblage, which is important for UHPM, is due to the new reaction chlorite + Mg-carpholite = talc + Mg-chloritoid + H_2O. Pyrope + Mg-chloritoid is limited to P_{H_2O} as low as 29 kbar and, thus, at least unequivocally in MASH an UHP assemblage. Several new parageneses of UHPM appear in the grid of Fig. 2.4. These are diaspore + MgMgAl-pumpellyite, with a lower P_{H_2O} boundary at 57.5 kbar, antigorite + pyrope, above P_{H_2O} = 62 kbar, pyrope + brucite, above P_{H_2O} = 55 kbar; and Mg-chloritoid + brucite, at $P_{H_2O} > 64$ kbar. However, some reactions with brucite might be metastable with respect to stable hydrous Mg-silicates (see section on 10 Å-phases) + H_2O.

System Na_2O-FeO-MASH (NFMASH)

The second example for calculated petrogenetic grids concerns the system NFMASH under conditions of excess SiO_2 and H_2O. The grid was calculated by Guiraud et al. (1990) on the basis of the thermodynamic data set of Holland and Powell (1990) up to 50 kbar and 850°C. Part of the grid relevant for UHPM is reproduced in Fig. 2.5.

With respect to the subsystem MASH considerable differences exist between the grid of Guiraud et al. (1990) and that of Fig. 2.3. Apart from the P-T fields of OH-topaz and MgMgAl-pumpellyite due to data lacking in the set of Holland and Powell (1990), they are related to the stability fields of Mg-carpholite and chlorite in the presence of excess phases. Guiraud et al. (1990) predict a larger P-T field for Mg-carpholite and, thus, a smaller stability field for Mg-chloritoid + coesite than shown in Fig. 2.3. Future experiments might clarify this. However, with respect to the stability of chlorite +

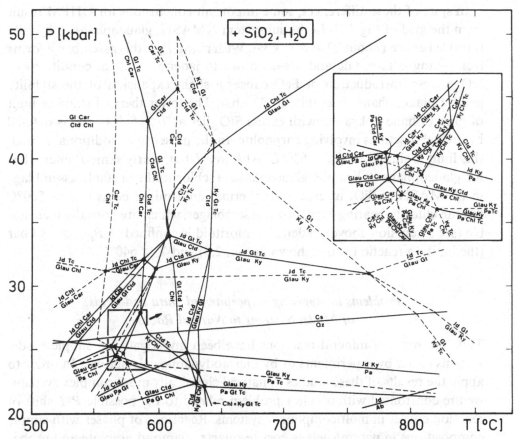

Figure 2.5. Phase relations in the system Na$_2$O-FeO-MgO-Al$_2$O$_3$-SiO$_2$-H$_2$O with H$_2$O and SiO$_2$ in excess, redrawn from the grid calculated by Guiraud et al. (1990) using the thermodynamic data set of Holland & Powell (1990). Dashed lines refer to the subsystem Na$_2$O-MgO-Al$_2$O$_3$-SiO$_2$-H$_2$O. Abbreviations as in Fig. 2.2 and 2.3 related here to Fe-Mg solid solution series, as well. Ab = albite, Glau = glaucophane, Jd = jadeite, Pa = paragonite.

SiO$_2$, experiments performed by Massonne (1989) prove that the thermodynamic data of Holland and Powell (1990) for binary MASH chlorites are insufficient, resulting in a decomposition T for chlorite + quartz higher than experimentally bracketed at high P. At P_{H_2O} = 20 kbar the difference amounts to about 100°C. In addition, chlorites along the terminal reaction of chlorite + quartz are clearly too high in Si at high P_{H_2O} but too low in Si at low P_{H_2O} with respect to compositional data given by Massonne (1989). The latter is also true for talc coexisting with kyanite + SiO$_2$ + H$_2$O. Talc compositions calculated by Guiraud et al. (1990) are, in general, too high in Al particularly at relatively low P and high T compared to the experimental study of Massonne (1991a, see section on MASH).

In spite of these differences, some important conclusions for UHPM result from the grid of Fig. 2.5. In the subsystem NMASH, glaucophane + kyanite is stable between about 22 and 31 kbar. With rising P_{H_2O} this assemblage forms from paragonite + talc and breaks down to jadeite + talc at conditions of SiO_2 excess. Introduction of FeO causes a slight expansion of the stability field of glaucophane + kyanite to 33 kbar. The lower thermal stability limit of glaucophane + kyanite with excess SiO_2 lies at $T > 560°C$ and is defined by several reactions involving carpholite. In the presence of additional garnet, this limit is shifted to about 630°C. At lower T but nearly same P interval as for glaucophane + kyanite, glaucophane + chloritoid is a stable assemblage in NFMASH at SiO_2 in excess. Its thermal limit might be as low as 500°C despite chlorite-bearing breakdown assemblages, which are partially metasta-ble (see discussion above). Jadeite + chloritoid is confined to $P_{H_2O} > 18$ kbar (the limiting reaction is not shown in Fig. 2.5) and $T < 640°C$.

Problems in Applying Experimental Data and Grids of Model Systems to Natural Rocks

Thus far, most of mineral reactions have been investigated in simple model systems either by experiments or by thermodynamic calculations. In order to apply the results of these studies to natural chemically more complex systems, we are confronted with a major problem related to the possible P-T shift of reaction curves in multicomponent systems. Reactions of phases with simple compositions in nature such as coesite-quartz, diamond-graphite and proba-bly wollastonite-HP wollastonite do not cause problems. However, most min-erals considered in the above sections are solid-solution phases. Subsequently, some examples are given to elucidate the problems.

Chloritoid is a solid solution phase composed mainly of the Mg and Fe^{2+} end members. In contrast to end member Mg-chloritoid (see sections on MASH and subsystem MASH), the Fe^{2+} end member is stable at low P (see Ganguly, 1969). Thus, both end members show quite contrasting P-T stability fields. This also concerns the stability fields of chloritoid coexisting with addi-tional phases (e.g., with SiO_2). Therefore, the conclusions made for assem-blages with chloritoid in the section on MASH and subsystem MASH hold true only for compositions close to the Mg-chloritoid end member. The stabil-ity field of ellenbergerite of the system TiO_2-MASH shows a minimum $P > 26$ kbar (see section on stability limits). Experimental introduction of phos-phorus into ellenbergerite, due to the simplified substitution P + Mg = Si + Al, expands the stability field of this phase down to 10 kbar (Beller, 1987).

The minerals of the above two examples show solid solution probably without miscibility gap at relevant metamorphic T. Other solid-solution phases behave differently. According to the compositions of natural pumpellyites, no significant Mg introduction on the Ca sites occurs. Thus, there is a considerable solvus in the system CaMgAl-pumpellyite and MgMgAl-pumpellyite. Under these circumstances, the stability field of MgMgAl-pumpellyite could be at least in CMASH nearly identical to that shown in Fig. 2.1. The same could be true for K-cymrite with regard to the system K-cymrite-cymrite, which is stable at relatively low P (Graham et al., 1992), although Ba and K substitute mutually in other tectosilicates. Here, we need further information either from nature or from experimental studies.

Similarly as described by Thompson (1976), Massonne (1989) determined the P-T shift of the stability field of MASH-chlorite + SiO_2 caused by introduction of FeO to MASH. The Fe-Mg distributions among the coexisting minerals were taken from observations on natural rocks reported in the literature. As an important result of Massonne's (1989) calculations, the upper P limit of chlorite + SiO_2 in FMASH lies about 1 kbar above that in MASH and, thus, still within the stability field of quartz. A further example concerns the P-T stability of phengite + orthopyroxene reported in the section on KMASH. The stability of this assemblage in KFMASH was estimated similarly as for chlorite + SiO_2. The result is a lower P limit for phengite + orthopyroxene as in KMASH (see Fig. 2.2). However, some univariant reactions in the NFMASH grid of Guiraud et al. (1990) show significant displacements after introduction of Ca or K to the model system. Most P-T shifts are difficult to predict quantitatively (see Guiraud et al., 1990), because, among others, such elements, in particular Ca, can be introduced in various NFMASH minerals, like garnet, glaucophane, jadeite and paragonite. A simple example is given here concerning the upper P limit of paragonite decomposing to kyanite + jadeite + H_2O. When K_2O is introduced into the system, paragonite contains significant amounts of muscovite component, thus reducing the paragonite activity. This leads to a P stabilization of paragonite. For example, the upper P limit of paragonite with an activity of 0.5 lies at P about 5 kbar higher than that for pure paragonite. If we would additionally introduce Ca in the system, leading to the formation of omphacite, a shift of the paragonite decomposition curve would occur in the opposite direction. Assuming a jadeite activity in omphacite of 0.5, the upper P limit of paragonite appears at the same P-T conditions as for the pure end members, when we ignore a further reduction of the paragonite activity by introduction of some Ca in paragonite.

The upper P stability of enstatite + spinel investigated in MAS to react to forsterite + pyrope (see section on MASH) is shifted to lower P in the presence of clinopyroxene in CMAS. Jenkins and Newton (1979) and Gasparik

(1984b) determined a P reduction of 4 kbar in the T range 900°–1400°C. In the system MAS-Cr_2O_3, however, a significant P stabilization of enstatite + spinel due to strong fractionation of Cr in spinel was calculated by Chatterjee and Terhart (1985) considering the nonideal behavior of the solid-solution phases. According to these authors, high Cr contents in the system can stabilize forsterite + spinel by more than 10 kbar so that this assemblage could theoretically appear at UHP conditions.

Thermodynamic Calculations of Multivariant Mineral Equilibria: A Key to Detect UHPM Rocks and Decipher Their *P-T* Evolution

The above discussion leads directly to the topic stressed in this section. In the past, the knowledge of the *P-T* shift of a univariant reaction or, more generally, of a particular mineral equilibrium was at least in some cases the good fortune for the petrologist, who used mineral equilibria as geobarometers or geothermometers to derive the *P-T* evolution of a metamorphic rock. However, such geothermobarometers were, in general, applied only when they had been calibrated either by experimental studies or empirically from observations on natural rocks.

A more general application of geothermobarometry is possible by thermodynamic calculations of equilibria with participating solid-solution phases. However, successful calculations require a reliable thermodynamic data base for mineral end members (see section on petrogenic grids) and a sufficient knowledge of the thermodynamic properties of solid solution phases (i.e., information about the activity of a mineral component in a solid solution phase). Ideal activity models frequently used for that purpose often do not describe properly the mixing properties of a solid-solution phase. Therefore, recent expressions of activity models related to real solid solutions, which are of relevance for UHPM, are summarized in the subsequent sections. The final sections describe applications of these models.

Thermodynamic Mixing Models

Diverse formulations for the deviation from the ideal behavior of solid solutions and, thus, for thermodynamic excess quantities are suggested in the literature (see e.g. Ganguly and Saxena, 1987). Among them, the Redlich-Kister equation is a prominent one. Here, we concentrate mainly on a special form, the Margules formulation, which is extensively used in the recent petrologic literature. A general form extending Wohl's formulation to a n-component system was given by Jackson (1989):

$$G^{Ex} = \sum_{i=1}^{n-1} \sum_{j=2}^{n} x_i x_j (x_i W_{ji} + x_j W_{ij}) + \sum_{i=1}^{n-2} \sum_{j=2}^{n-1} \sum_{k=3}^{n} x_i x_j x_k W_{ijk} \quad (i < j < k)$$

The *P-T* dependence of the Margules parameter *W* can be simply expressed by $W = W^H - TW^S + PW^V$.

Al-Garnet

Natural Al-garnet is composed mainly of four components given in Table 2.2; numerous experimental studies and calorimetric measurements have been undertaken to determine thermodynamic mixing properties of these components. Recent reviews of the data and proposals to express the excess quantities were published by Wood (1987), Ganguly and Saxena (1987), and Berman (1990); these data have been recently discussed on the basis of new experimental data by Koziol (1990), Berman and Koziol (1991), and Koziol and Bohlen (1992). With a few exceptions, the Margules parameters given by Berman (1990) were used (Table 2.3); for calculation examples see section on computation method. For pyrope-spessartine a symmetric model with a value around 13500 Joule/(mol garnet) for W^H was proposed by Cheng and Ganguly (1991); 1757 Joule/mol was given by Pownceby et al. (1987) for almandine-spessartine. The excess entropy for grossular-pyrope used by Berman (1990) is based on only one calorimetric measurement of Haselton and Westrum (1980). Therefore, Margules parameters for this binary solid solution (Table 2.3), together with excess data for clinopyroxene components (see next section) were redetermined by Massonne (1992c). According to these new data, the ternary interaction parameters W_{ijk} of Table 2.3 were revised with regard to those given by Berman (1990).

The Al-Fe^{3+} solid solution in the octahedral sites of the garnet structure is approximately ideal (Ganguly and Saxena, 1987). The same is true for Al-Cr according to data of Mattioli and Bishop (1984) and Wood and Kleppa (1984).

Clinopyroxene

Because of the intensive studies of clinopyroxene in various model systems (see e. g. section CFMAS), the nonideal behavior of mixing of clinopyroxene components is relatively well known. Excess quantities for solid-solutions involving end members diopside, hedenbergite, enstatite and ferrosilite were reported by Saxena et al. (1986) and Davidson and Lindsley (1989). The suggestions of the latter authors were used for the clinopyroxene model of Tables 2.2 and 2.3, which is related to omphacites occurring in metabasites and metapelites. This model applied in the section on computation methods is simpli-

Table 2.2. End Member Components And Kind Of Activity Models Used For Thermodynamic Calculations.

	Omphacite	Phengite (Ms)	Al-Garnet (Gr)	Amphibole
1	Diopside (Dp)	Muscovite (Ms)	Grossular (Gr)	$Ca/2 = X_{Ca}$
2	Hedenbergite	trioctahedral mica: $K(Mg,Fe^{2+})_3AlSi_3O_{10}(OH)_2$	Pyrope	$NaM4 \cdot [AlVI/(AlVI+Fe^{3+})]/2$
3	Jadeite	Mg-Al-celadonite	Almandine	$NaM4 \cdot [Fe^{3+}/(AlVI+Fe^{3+})]/2$
4	Orthopyroxene $(Mg,Fe^{2+})_2Si_2O_6$	Fe2+-Al-celadonite	Spessartine	
5	CaTs $CaAl_2SiO_6$	Paragonite		
6	Rest: $Na(Fe^{3+},Cr)Si_2O_6$ $NaTiAlSiO_6$ $CaMnSi_2O_6$	Rest: $(Ca,Ba)Al_4Si_2O_{10}(OH)_2$ $K(Fe^{3+},Cr)_2AlSi_3O_{10}(OH)_2$ $KTi(Mn,Mg,Fe^{2+})AlSi_3O_{10}(OH)_2$		

$$a_{Tr} = X_{Ca}^2 \cdot \gamma 1 \cdot OH/(OH+F) \cdot \blacksquare \cdot \gamma 2 \cdot Mg/(5-NaM4)$$

$$a_{Glau} = (NaM4 \cdot AlVI/(AlVI+Fe^{3+}))/2)^2 \cdot \gamma 1 \cdot OH/(OH+F) \cdot \blacksquare \cdot \gamma 2 \cdot Mg/(Mg+Fe^{2+}+Mn)$$

Tr = Tremolite; $Glau$ = Glaucophane; $\blacksquare$ = vacancy (A-site); and $a_{Dp} = X_{Dp} \cdot \gamma_{Dp}$, $a_{Ms} = X_{Ms} \cdot \gamma_{Ms} \cdot X_{OH}$

$a_{Gr} = X_{Ca}^3 \cdot X_{Al}^2 \cdot \gamma_{Gr}$

Table 2.3. *Mixing Properties (Margules Parameters) To Be Used With The General Form Of Wohl's Asymmetric Solution Model*

	P2/n / C2/c Omphacite			Phengite		
Parameter	WH	WS	WV	WH	WS	WV
12	-14550	-10	0	5000	-10	-.135
21	36500	15	0	5000	-10	-.135
13	-4340/17600	0/12.1	0	0	58.598	.735
31	13800/8600	0/7.6	0	0	15.92	.187
14	26125	0	-.0384	0	58.598	.735a
41	32301	0	-.0067	0	15.92	.187a
15	0	-12.26	-.17	38500	39	.32
51	41242	30.92	0	9500	-9	.152
16,61	10000	5	0 a	0	0	0 a
23	15090	8	0	15000	-30	-.20
32	3670	9	0	15000	-30	-.20
24	18682	0	-.0864	15000	-30	-.20 a
42	20001	0	.0271	15000	-30	-.20 a
25	0	-12.26	-.17 a	30000	0	0 a
52	41242	30.92	0 a	30000	0	0 a
26,62	10000	5	0 a	5000	-10	-.135a
34	22000	9	-.09 a	0	0	0 a
43	35000	8	.03 a	0	0	0 a
35	9740	7.15	0	30000	0	0 a
53	19800	7.15	0	30000	0	0 a
36,63	0	0	0 a		as 31,13	a
45,54	24000	0	-.026 a	30000	0	0 a
46,64	28000	8	0 a		as 41,14	a
56,65	15000	7	0 a		as 51,15	a
ijk ≈ 0.25.(ij+ji+ik+ki+jk+kj) a				0	0	0 a
123	11200	9.2	0 a			
124	24300	0	-.026 a			
125	16785	12.685	-.085			
126	10000	5	0 a			
134	0	0	0 a			
145	-36500	-58.5	-.067			

	Al-garnet			Amphibole
Parameter	WH	WS	WV	WH
12	69200	29.829	.10	27000 a
21	21560	-31.718	.10	27000 a
13	2620	5.08	.09	15000 a
31	20320	5.08	.17	15000 a
14	0	0	0 a	
41	0	0	0 a	
23	3720	0	.06	15000 a
32	230	0	.01	15000 a
24	13500	0	0	
42	13500	0	0	
34	1757	0	0	
43	1757	0	0	
ijk	0.5(ij+ji+ik+ki+jk+kj)a			as garnet a

Note: For numbers notating Margules parameters and for activity models see Table 2.2. Parameters for amphibole are related only to γ_1 of the amphibole activity model. Units are Joule/mol (WH), Joule/(K·mol) (WS), and Joule/(bar·mol) (WV). a means that WG=WH-T·WS+P·WV (with T in K, P in bar) is estimated. For sources of other Margules parameters, see text.

fied with respect to orthopyroxene components and other minor components (rest) containing Ti, Mn, Cr, and Fe^{3+}. The latter element is in acmite, the most important component of these minor components, because it can reach relatively high concentrations in omphacites of low-T eclogites. The nonideality of binary solid solutions with acmite is poorly understood. Liu and Bohlen (1992) suggest an ideal acmite-jadeite solid solution, whereas acmite-diopside behave nonideally; however, without a very significant deviation from ideality (Ganguly, 1973). Therefore, the Margules parameters in Table 2.3 involving the component rest (see Table 2.2) in omphacite were chosen to be zero for rest-jadeite and 10 kJ/mol and 5 J/(mol·K) for both rest-diopside and rest-hedenbergite. Rest-orthopyroxene and rest-CaTs were assumed to be symmetric solid solutions with Margules parameters being the average of the corresponding values for asymmetric jadeite-orthopyroxene and jadeite-CaTs solid solutions.

Relatively good information exists on thermodynamic properties of solid solutions with jadeite. Mixing properties of jadeite-CaTs, reported by Gasparik (1985a), are used here, after an approximate transformation from Redlich-Kister to Margules formalism. The same was undertaken for mixing data of jadeite-diopside given by Gasparik (1985b). This author also reports data for both, $C2/c$ and $P2/n$, omphacites. The study of Perchuk (1990) was considered for jadeite-hedenbergite. The mixing properties of jadeite-orthopyroxene were estimated by adding the values of Margules parameters for jadeite-hedenbergite and hedenbergite-ferrosilite as given by Davidson and Lindsley (1989).

Thermodynamic properties for diopside-CaTs (for Margules parameters see Table 2.3) were suggested by Ganguly and Saxena (1987) on the basis of Herzberg's (1978) experiments. The deviation from ideality for hedenbergite-CaTs was assumed to be that of diopside-CaTs. The orthopyroxene-CaTs solid solution is supposed to be symmetric with Margules parameters similar to clinopyroxene-orthopyroxene mixing. Therefore, 24 kJ/mol and −0.026 J/(bar·mol) were taken. The ternary interaction for diopside-orthopyroxene-CaTs was obtained by using the Margules parameters for diopside-CaTs and orthopyroxene-CaTs and fitting the Al-isopleths of clinopyroxene reported by Gasparik (1984b) after thermodynamic properties for CaTs had been recalculated (Table 2.4) and changed in Berman's (1988) data set. Other ternary interactions were estimated as given in Table 2.3 neglecting diopside-hedenbergite. The ternary interaction for diopside-jadeite-orthopyroxene was arbitrarily set to zero.

The mixing properties of diopside-hedenbergite have been suggested by Saxena et al. (1986), Davidson and Lindsley (1989), and Pattison and Newton

Table 2.4. *Thermodynamic standard state properties*

	H_f J/mol	S J/K·mol	V J/bar·mol	c_P	α_P, β_T
Muscovite	-5976740a	293.157a	14.080	a	a
Mg-Al-celadonite	-5832415	288.527	13.870	c	as muscovitea
Hedenbergite	-2845280	170.7	6.788	c	as diopsidea
Almandine	-5267212b	340.007b	11.511a	a	a
CaAl$_2$SiO$_6$	-3300400	140.0	6.356a	a	a

Sources:Determined by Massonne (1992c and unpubl.); and as given by *a* Berman (1988), and *b* Berman (1990), *c* according to Berman & Brown (1985).

(1989). Recently, Massonne (1992c) reevaluated the experimental data of Ellis and Green (1979) and Pattison and Newton (1989) and natural assemblages with garnet + clinopyroxene ± kyanite ± quartz/coesite, formed at relatively well known P and T, using the method of least-squares. The equilibria diopside + almandine = hedenbergite + pyrope, grossular = diopside + CaTs, grossular (or CaTs) + almandine + SiO_2 = hedenbergite + kyanite and CaTs + SiO_2 = kyanite + grossular were considered. The result includes thermodynamic properties of end member hedenbergite (Tab. 2.4), Margules parameters for diopside-hedenbergite, W^S for grossular-pyrope and ternary interaction parameters of diopside-hedenbergite-CaTs (Tab. 2.3). The new data are compatible with those reported in the previous sections on subsystem MASH and Al-garnet. With regard to the application to natural rocks, it is suggested that the clinopyroxene mixing model be used only for calculating the activity coefficients for diopside, hedenbergite, and jadeite. In the above model, differences among Margules parameters related to distinct space groups $C2/c$ and $P2/n$ concern solely the diopside-jadeite solid solution series. If the space group of omphacite were unknown, one could refer to the jadeite content and the estimated metamorphic T by applying the T-X-diagram of Gasparik (1985b, Fig. 1).

Phengite

Mixing properties of components in phengite (Table 2.2) are only partially known. Sufficient knowledge exists for muscovite-paragonite. Experiments of Flux and Chatterjee (1986) led to parameters describing the nonideality of this solid solution series (Chatterjee and Flux, 1986). They were used after approximate transformation from Redlich-Kister to Margules formalism. Experimental data in KMASH obtained by Massonne and Schreyer (1986, 1987, 1989) and unpublished experiments on the equilibrium Mg-Al-celadonite + kyanite = muscovite + pyrope + coesite were taken into account to estimate the mixing properties for the muscovite-trioctahedral mica component and the Mg-Al-celadonite-trioctahedral mica component and to determine the Margules parameters for muscovite-Mg-Al-celadonite and the thermodynamic properties of Mg-Al-celadonite (Table 2.4) using a least-squares method (Massonne, 1992c). The muscovite-Mg-Al-celadonite solid solution is characterized by significant excess volumes and excess entropies. The latter are, however, due to the activity model selected (see Table 2.2). Nonideality of Mg-Al-celadonite-paragonite is proven by a wide miscibility gap inferred from natural occurrences of coexisting phengite and paragonite (e.g. Katagas and Baltatzis, 1980; Mposkos and Perdikatzis, 1981; Shau et al., 1991). The Margules parameter of a symmetrical Mg-Al-celadonite-paragonite solid solution was assumed to be 30 kJ/mol (Massonne, 1992c). The behavior of muscovite-rest (comprising minor components in phengite, Table 2.3) is suggested to be ideal. The solid solution series with component rest and with Fe^{2+}-Al-celadonite are estimated to behave like the corresponding solid solutions with muscovite and Mg-Al-celadonite, respectively. The OH-F exchange is considered by X_{OH} in the activity model according to experiments of Massonne and Böving (1990).

Other Phases

Compared with solid solutions discussed above, experimental data on mixing of amphibole components are relatively poor. However, we have some additional information from solvus relations of natural amphiboles, particularly from coexisting sodic and calcic amphiboles. On the basis of that, Massonne (1992c) attempted to model the nonideal behavior of the components tremolite, glaucophane and Mg-riebeckite on the basis of symmetric solutions, although there is evidence from natural rocks that at least tremolite-glaucophane is slightly asymmetric (Reynard and Ballevre, 1988). The activity models extended to be applied to natural amphiboles are given in Table 2.2 and the derived Margules parameters are shown in Table 2.3. γ_2 in these models is related to a W^H (13.5 kJ/mol; see Massonne, 1992c) of a symmetric solution taking the occupation of the A-site through the edenite substitution into account. The OH-F exchange, the Tschermak's substitution and further cations substituting for Mg are simply considered by the introduction of X_{OH} and X_{Mg} in the activity formulation of Table 2.2 At least, the solid solution series of tremolite-tschermakite(Ts) $[Ca_2Mg_3Al_4Si_6O_{22}(OH)_2]$ seems to be nearly ideal with respect to the activity model $a_{Ts} = X_{Ts}$ according to experiments of Cho and Ernst (1991).

In the case of olivine and orthopyroxene, the nonideal mixing properties of Mg and Fe^{2+} end members play an important role. Many experimental investigations have contributed to the thermodynamic understanding of these binary solid solutions (see section CFMAS). The recent study of Koch-Müller et al. (1992) confirms a slight deviation from ideality for orthopyroxene (W^H = 3417 J/cat) as suggested previously by Davidson and Lindsley (1989). The deviation is stronger for olivine. Koch-Müller et al. (1992) estimated a W^H of 7124 J/cat. A slightly lower value arises by averaging the W^H values of an asymmetric model (16740 and 8370 J/mol) given by Wood (1987). A still lower value of 9000 J/mol is assumed by Hackler and Wood (1989), considering earlier experimental studies, but a clearly higher value of 20334 J/mol was given by Sack and Ghiorso (1989). The nonideality of orthopyroxene-clinopyroxene mixing is discussed above. Experimental studies (e.g. Gasparik and Newton, 1984) involving the solid solution of the orthopyroxene end members enstatite and Al_2O_3 suggest closely ideal mixing with respect to the activity model given in the legend of Fig. 2.3. However, a volume slightly in excess exists (Danckwerth and Newton, 1978). A linear extrapolation to the molar volume of the theoretical end member Al_2O_3 for example, through that of an aluminous enstatite of interest (see legend of Table 2.1), can be a reasonable approach (Gasparik & Newton, 1984). Thermodynamic properties of olivine solid solution series due to the introduction of Ca are known. Mukhopadhyay & Linsley (1983) studied the join kirschsteinite-fayalite, and Adams & Bishop (1985) determined the mixing properties of forsterite(fo)-monticellite(mo). Both series are clearly nonideal with Margules parameters W^H = 21866 J/mol for the iron system and W^G_{fomo} = 43100 J/mol - 6.01 J/(mol·K)·T and W^G_{mofo} = 61200 J/mol - 15.2 J/(mol·K)·T for the Mg system.

The thermodynamic properties of multicomponent solid solutions in oxides were a matter of numerous studies. Recent investigations are those by Nell & Wood (1989, 1991), Sack & Ghiorso (1991) and D. J. Andersen et al. (1991) for spinel and that by D. J. Andersen et al. (1991) for ilmenite.

The mixing properties of biotite end members annite and phlogopite were experimentally investigated by Schulien (1980), who concluded that this binary solid solution series behaves nearly ideally. However, on the basis of Fe^{2+}-Mg exchange experiments undertaken by Ferry & Spear (1978), Sack & Ghiorso (1989) proposed a clear nonideality of annite-phlogopite with a W^H of 13.43 kJ/mol. This value correlates with a W^H for almandine-pyrope of 16.11 kJ/mol, which is also clearly higher than given in Table 2.3. Mixing properties of solid solutions involving biotite components others than annite and phlogopite are hardly known. Indares & Martignole (1985) made the attempt to consider an aluminous component and a titaniferous component in a quaternary biotite solid solution. However, their data cannot be directly related to mixing properties of end members annite or phlogopite and one of these new components.

Table 2.5. *Eclogitic Rocks Described In The Literature And Used For Testing The Computation Method*

Locality	Authors	Sample No.	Minerals coexisting with Gt-Cpx-Phe	P-T given (kbar, °C)		Space group of Cpx	P-T calculated (kbar, °C)	
Dora Maira M., W. Alps	Kienast et al. (1991)	Br2	Am-Rt	31	715	P2/n	29.33	626
		To2	Rt	–	–	P2/n	30.57	714
		To6	Qz/Cs-Ky-Rt	–	–	P2/n	30.91	666
		Mf2	Am	–	–	P2/n	31.20	610
		Par6	Am-Rt	–	–	P2/n	20.13	571
Zermatt ophiol., W. Alps	Oberhänsli (1980)	PVB891	Glau-Para-Bt?	13	550	P2/n	24.63	533
		PB1296	Am-Para-Bt?	–	–	P2/n	23.36	556
		PB1544	Para	–	–	P2/n	25.03	592
Kokchetav M., Kazakhstan	Sobolev et al. (1986)	K21-79		>14	610	P2/n	23.60	567
Dhonghai county, China	Hirajima et al. (1990)	E88D-1c	Qz/Cs-Para-Kf-Rt	>28	810	P2/n	39.22	779
		r				P2/n	45.55	817
Bergen Arcs, Norway	Austrheim & Griffin (1985)	11A	Qz-Ky-Ep	>17.6	750	P2/n	20.54	683
		11B	Qz-Ky-Ep	–		P2/n	23.86	737
		11C	Qz-Ky-Ep	–		P2/n	18.75	776
		34/80	Qz-Ky-Ep	–		P2/n	16.56	669
Kristiansund, Norway	Krogh (1980)	MA2	Qz-Pl-Am-Bt-Ep-Rt	18.5	750	C2/c	15.64	752
Nordfjord, Norway	Green and Mysen (1972)	2520	Qz-Zs	>8	>600	P2/n	21.68	590
Seve Nappe, Sweden	Van Roermund (1985)	78c	Qz-Rt	>14	550	P2/n	21.41	580
		r				P2/n	14.72	564
Spitsbergen	Hirajima et al. (1988)	24-23	Qz-Glau-Ep-Para-Rt	19	610	P2/n	21.99	535
		43-3	Qz-Am-Ep-Para-Rt	–		P2/n	26.46	599
Armorican M., France	Godard (1988)	M1a	Qz-Ky	18	700	P2/n	19.51	781
Bitlis M., E Turkey	Okay et al. (1985)	MU12D	Ky-Am-Rt	16	625	P2/n	18.10	598

Key: c = core compositions or inclusion minerals, r = rim compositions. Mineral abbreviations: Am = amphibole, Bt = biotite, Cpx = clinopyroxene, Cs = coesite, Ep = epidote, Glau = glaucophane, Gt = garnet, Kf = K-feldspar, Ky = kyanite, Para = paragonite, Phe = phengite, Pl = plagioclase, Qz = quartz, Rt = rutile, Zs = zoisite.

The mixing properties of Fe^{2+} and Mg end members of chloritoid, chlorite and talc are unknown. However, the assumption can be made that these solid solutions are characterized by moderate deviations from ideality like other binary Fe^{2+}-Mg silicate solid solutions, because the thermodynamic behavior of solid solutions, related to mixing of a specified cation pair in a certain coordination polyhedron, is similar (e.g. Navrotsky, 1987). Solid solutions related to the Tschermak's substitution in chlorite and talc seem to be close to ideal with respect to activity models proposed in the legend of Fig. 2.3.

Examples for the Computation Method

Phengite-bearing eclogites from the Franciscan complex, California, and the Island of Sifnos, Greek Cyclades, were selected as examples to apply thermodynamic calculations of mineral equilibria for deciphering the *P-T* history of metamorphic rocks with UHP signature. Equilibria involving the almandine, pyrope, and grossular components of garnet, the diopside and hedenbergite component of omphacite, and the muscovite and Mg-Al-celadonite component of coexisting phengite were calculated. The corresponding equilibrium curves intersect and, thus, define a *P-T* point. In contrast to the thermodynamic calculations reported in the section on subsystem MASH, an advanced version of the Ge0-Calc software package distributed by Berman in 1989 was used. In particular, the programs PTAX and SOLN were applied. Because thermodynamic data and mixing models for the above components in garnet, omphacite, and phengite are sufficient, the computation method is expected to yield reliable results. In order to prove this, the calculation method was first applied to several eclogitic rocks reported in the literature (Table 2.5). The literature data contain analyses of coexisting garnet-omphacite-phengite, which were recalculated according to Table 2.6, and metamorphic *P-T* conditions.

Kienast et al. (1991) described coesite-bearing eclogites from the Dora-Maira massif. Four of five eclogites reevaluated yield metamorphic *P* around 30.5 kbar at an average *T* somewhat below 700°C. These conditions are compatible with the UHP mineral assemblage observed, although the nearby pyrope-quartzite was metamorphosed at somewhat higher *P* and *T* of 36 kbar and 730°C (Massonne, 1990). However, Kienast et al. (1991) did not report the maximum Si content of phengites but the average of high Si phengites. Sample Par6 (Tab. 2.5) yields a lower *P* and *T* of about 20 kbar and 570°C. This could reflect an equilibration stage after UHP metamorphism. A reevaluation of three eclogites from an ophiolite complex of the Zermatt-Saas area (Oberhänsli, 1980) leads to an average metamorphic *P* of 24.3 kbar. This value

Table 2.6. *Structural formulae of garnet, omphacite, phengite and amphibole somewhat simplified*

	Sifnos		Franciscan complex					
	225c	225r	TEC2	TIBBc	TIBBr	FF3c	FF3r1	FF3r2
Garnet (Al+Fe+Cr+Mn+Mg+Ca = 5 and Fe3+ = 2 - Al - Cr with maximum Al + Cr = 2):								
Al	1.956	1.938	2.000	2.000	2.000	1.934	1.934	1.951
Fe3+ +Cr	0.044	0.062	-	-	-	0.066	0.066	0.049
Ca	0.720	0.949	0.837	0.875	0.831	0.796	0.750	0.908
Fe2+	2.080	1.793	1.801	1.520	1.647	1.684	1.814	1.620
Mn	0.103	0.034	0.085	0.286	0.064	0.275	0.051	0.151
Mg	0.097	0.202	0.277	0.319	0.458	0.245	0.385	0.320
Omphacite (O=6 and 4 cations with maximum Si=2):								
Alt	0.002	-	0.017	-	0.020	0.006	0.010	0.007
Ti	-	-	0.002	0.001	0.003	0.002	0.002	0.001
Al°	0.392	0.430	0.430	0.123	0.431	0.396	0.462	0.402
Fe3++Cr	0.165	0.137	0.108	0.060	0.064	0.215	0.075	0.066
Fe2+	0.144	0.125	0.134	0.205	0.150	0.103	0.138	0.124
Mn	-	-	-	0.004	0.003	0.005	0.001	0.001
Mg	0.308	0.311	0.356	0.640	0.389	0.294	0.347	0.426
Ca	0.435	0.429	0.445	0.782	0.479	0.376	0.444	0.517
Na	0.556	0.568	0.525	0.185	0.481	0.609	0.531	0.463
Phengite (21-(Ca+Ba) 0.5 cation charges neglecting interlayer cations, octahedral occupancy is ≤ 2.05):								
Alt	0.441	0.531	0.660	0.783	0.611	0.421	0.587	0.668
Ti	0.010	0.010	0.016	0.043	0.014	0.001	0.002	0.009
Al°	1.400	1.485	1.555	1.577	1.519	1.344	1.448	1.484
Fe^{3+}+Cr	-	-	-	0.023	0.012	0.003	0.005	0.002
Fe2+	0.261	0.179	0.144	0.146	0.143	0.306	0.182	0.186
Mn	-	-	0.003	-	0.008	0.006	0.004	0.005
Mg	0.334	0.339	0.318	0.260	0.330	0.369	0.376	0.352
xK	0.986	0.919	0.924	0.840	0.936	0.982	0.954	0.915
xNa	0.014	0.081	0.076	0.160	0.064	0.008	0.015	0.057
Amphibole (13 cations without A-site cations, 46 cation charges):								
Alt	0.152	0.127		1.585	0.648			0.070
Ti	0.022	-		0.079	0.010			0.001
Alo	1.370	1.635		0.926	0.754			1.611
Fe3++Cr	0.317	0.382		0.649	0.486			0.327
Fe2+	2.279	1.168		1.099	0.718			1.168
Mn	-	-		0.022	0.017			0.013
Mg	1.103	1.815		2.225	3.011			1.880
Ca	0.109	0.089		1.329	1.249			0.109
Na	2.025	1.932		1.041	0.861			1.908
K+Ba	-	-		0.152	0.025			0.004

Note:Sources are representative electron microprobe analyses reported by Schliestedt (1980) for eclogite sample 225 from Sifnos, Greek Cyclades, given by Wakabayashi (1990) for samples TEC2 and TIBB and published by Massonne (1991b) for sample FF3. The latter samples are from the northern Franciscan complex. c = core composition or inclusion in garnet; r = rim composition (e.g., in the case of garnet: 1 = inner portion of rim zone, 2 = outermost rim).

is somewhat lower than those reported by Reinecke (1991; Reinecke et al., 1992) for coesite-bearing metasediments and eclogites from this area. An eclogite from the Kokchetav massif studied by Sobolev et al. (1986) yields almost 24 kbar. However, gneisses in this area contain microdiamonds (Sobolev & Shatsky, 1990) so that the peak metamorphic P could be clearly higher. Coexisting minerals in an eclogite from Dhonghai county reported by Hirajima et al. (1990) yield an average P of 42.5 kbar at T around 800°C, which is similar to that estimated by these authors. At the first glance, P seems to be too high; however, Hu & Liu (1992) observed microdiamonds in eclogites from this area pointing to minimum P of about 37 kbar. P determined for eclogitic rocks of the Bergen Arcs is hardly higher than the minimum P around 17.6 kbar estimated by Austrheim & Griffin (1985). T is only somewhat lower than given by these authors. The same T but 3 kbar lower P than given by Krogh (1980) were determined for a HP granulite with phengite enclosed in garnet from the Kristiansund area. P of almost 22 kbar was determined for an eclogite of the Nordfjord area. Although this value is clearly higher than the minimum P given by Green & Mysen (1972), coesite-bearing eclogites in the vicinity (Smith, 1988) point to still higher P. Rim compositions of minerals from an eclogite of the Seve nappe yield nearly the same P-T conditions as estimated by Van Roermund (1985). Core compositions point to higher P at an early metamorphic stage. The average P-T conditions for two samples from the Motalafjella HP metamorphic complex are about 24 kbar and 575°C, which are still compatible with the estimates of Hirajima et al. (1988), who expected somewhat lower P at higher T. The calculations of P and T for eclogites from the Armorican massif and the Bitlis massif are in good agreement with the estimates of Godard (1988) and Okay et al. (1985), respectively. With respect to the selected test objects, the new calculation method yields good results. There are no unexplainable discrepancies from the earlier P-T estimates.

For the above method, P-T uncertainties are estimated to be in the 2sigma range of ± 1 kbar and 30°C neglecting the problems in aligning mineral compositions with respect to equilibrium assemblages, particularly for early metamorphic stages. Moreover, a geohygrometric method was applied to the subsequent examples as reported by Massonne (1992c). This semiquantitative method requires the calculations of equilibria with H_2O; the above components of garnet, omphacite, and phengite; and, for instance, the tremolite component of amphibole and quartz/coesite or the clinochlore component of chlorite, the jadeite component of clinopyroxene and the glaucophane component of amphibole. The water activity is defined by the P-T datum determined from the vapor-absent equilibria.

Franciscan Complex, California

HP rocks in the Franciscan complex, such as eclogites, have been known for a long time and were relatively early related to the subduction processes of oceanic crust (Hamilton, 1969). The sample FF3 investigated is from an eclogite of the northern part of the Franciscan complex occurring a few kilometers SW of Healdsburg. It was described by Coleman et al. (1965) as Junction School eclogite. In this rock, inclusion-rich garnet appears in a matrix of finer

H.-J. Massonne

Figure 2.6. Photomicrograph (a) and element distribution maps for Mg (b–d) of corroded garnet and surrounding phengite, omphacite and chlorite in eclogite FF3. This sample is from the northern Franciscan complex (Junction School eclogite, SW Healdsburg). (a) Obtained with a polarizing microscope under crossed nicols. Scale bar represents 100 μm. (b–d) Resulted from work with the electron microprobe by scanning the object (56549 spots) in steps of 4 μm in *x*-direction and in *y*-direction. Counting time per spot was 2 seconds. Gray tones are related to counting rates per second as follows after smoothing: (b) < 240 and > 570 black, 240–319 dark, 320–399 medium light, 400–499 medium dark, 500–569 light; (c) < 570 black, 570–619 medium dark, 620–669 light, 670–729 medium dark, > 730 dark; (d) < 1000 black, 1000–1149 medium dark, 1150–1299 light, 1300–1449 medium light, > 1450 dark. Counting rates are related to the subsequent minerals as follows: garnet 240–570, phengite 560–730, omphacite 1000–1450, chlorite (which is zoned, as well) >1450.

grained, weakly foliated greenish omphacite and minor clinozoisite. Garnet is partially replaced by chlorite and phengite. However, phengite is an early eclogitic phase, as well, because it appears as inclusion phase in garnet. Further phases enclosed are omphacite, clinozoisite and quartz, which exclusively occurs as inclusion minerals. Na-amphibole and titanite are probably retro-

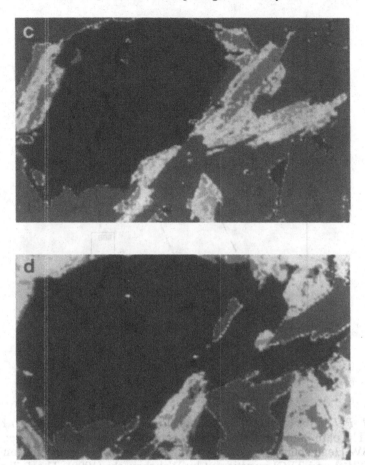

grade blueschist phases, because they replace omphacite and rutile, respectively. The minerals in FF3 are compositionally zoned (Table 2.6, Fig. 2.6). For instance, garnet shows an increase in pyrope component and a decrease in the Mn content from core to rim. However, the outermost rim of garnet also is enriched in Mn and depleted in pyrope component. This is the case, particularly at contacts to omphacite. The garnet zonation pattern was used to subdivide the metamorphic evolution in three stages related to core, inner rim and outermost rim of a completely developed garnet.

P-T conditions of an early eclogitic stage were calculated with consideration for the compositions of the garnet core and Si-rich phengite and acmite-rich omphacite enclosed in garnet. The result of 30.5 - kbar and 515°C, obtained with the geothermobarometric method explained in the previous section, reflects UHP conditions (Fig. 2.7). Subsequently, eclogite FF3 was somewhat heated, accompanied by an uplift process. The calculated result was obtained by using the compositions of the garnet inner rim, cores of

H.-J. Massonne

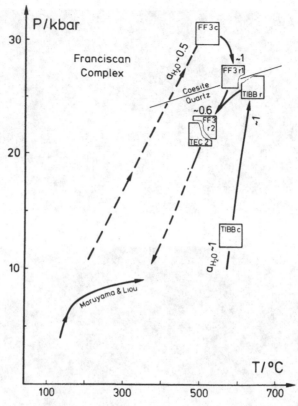

Figure 2.7. *P-T* paths (dashed = estimated) for eclogites of the northern Franciscan complex. The path in the upper left section is related to sample FF3 (Junction School eclogite, SW Healdsburg); that on the right side to eclogites from Tiburon peninsula (samples TIBB and TEC) investigated by Wakabayashi (1990). The *P-T* path in the lower left section was reported by Maruyama and Liou (1988). *P-T* conditions for various metamorphic stages, shown by 2σ error boxes, were obtained from thermodynamic calculations considering compositions of garnet, omphacite, and phengite reported in Table 2.6. Space groups for omphacites (*C2/c* for TIBBc, *P2/n* for others) were inferred from a graph given by Gasparik (1985b). Estimated H_2O activities are given by numbers (see text). The coesite-quartz transition is that of Fig. 2.1.

inclusion-free omphacites, which are generally smaller than those, being in the matrix as well, with inclusion-rich cores, and cores of relatively large phengites around garnet. The core compositions of such phengites are distinctly lower in Si than of those enclosed in garnet. In context with further uplift, eclogite FF3 was somewhat chilled (see Fig. 2.7). This result was based on compositions of phengite rims, the outermost rim of garnet and omphacite in contact with this particular garnet rim. In fact, at this metamorphic stage garnet and omphacite were partly corroded, but reequilibration of both minerals took place, too. Evidence for this reequilibration is the Mn enrichment in the outermost rim of garnet, because it is the result of the decomposition of a Mn-bearing mineral, which is garnet itself.

Further *P-T* data were calculated using the analytical data published by Wakabayashi (1990) for minerals from eclogites of Tiburon peninsula (samples TIBB and TEC) north of San Francisco. As already proposed by Wakabayashi, the mineral assemblage enclosed in garnet of TIBB reflects a lower metamorphic *P* than that of the eclogite mineral assemblage. The rim compositions of garnet, omphacite and phengite in TIBB yield 26 kbar, thus, clearly a higher *P* than assumed by Wakabayashi (1990). The *P* increase is accompanied by elevating *T* (see Fig. 2.7). Calculations with the mineral compositions given for TEC lead to *P-T* conditions similar to that of FF3 for the late eclogitic stage. It might be indeed that the investigated eclogites have a common metamorphic history concerning the late eclogitic stage and the subsequent exhumation as shown in Fig. 2.7. Moreover, it is conceivable that these eclogites were tectonically emplaced to depths of about 30 km and were in contact with rocks of very low-*T*, high-*P* origin, for instance, those investigated by Maruyama & Liou (1988). This is at least suggested by the *P-T* paths of Fig. 2.7.

Geohygrometry applied to FF3 and the two eclogite samples of Wakabayashi (1990) yields water activities displayed in Fig. 2.7. High values, suggesting the presence of a hydrous fluid phase, were obtained for the thermal peak conditions and for the early metamorphic stage of sample TIBB. Lower values result from calculations with mineral compositions related to the early UHP stage of FF3 and its late eclogitic stage. For the latter, a clinochlore activity of about 0.013 was considered.

Greek Cyclades

HP rocks occurring in the lower nappe unit of the Attic-cycladic belt of the Hellenides are related to a metamorphic event in the Eocene (Dürr, 1986). Such rocks, which crop out in a part of the Island of Sifnos, have been studied by Schliestedt (1980). Among the investigated rocks, nine phengite-bearing eclogites (see Massonne, 1991b) were reported with core and rim compositions of garnet, omphacite and phengite. Table 2.6 gives as an example the results obtained by Schliestedt (1980) from sample 225. This sample is the only one of the eclogites studied that also contained Na-amphibole and quartz.

The results obtained from the above geothermobarometric method are shown in Fig. 2.8. The prograde metamorphic evolution was inferred from the *P-T* data derived from core compositions. The *P-T* gradient amounts to 65 bar/K, which is a geotherm as low as 5 K/km, considering a mean rock density

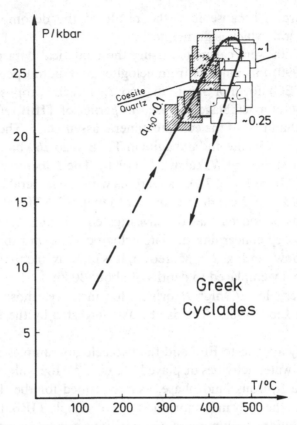

Figure 2.8. *P-T* path for HP rocks from the Island of Sifnos inferred from thermodynamic calculations taking into account the compositions of garnet, omphacite (*P2/n*) and phengite in eclogites (see Table 2.6) reported by Schliestedt (1980). Hatched boxes are related to calculation results for core compositions including 2σ error. Open boxes refer to rim compositions. Estimated water activities are given by numbers (see text). The coesite-quartz transition is that of Fig. 2.1.

of 3.25 g/cm³. The maximum *P* was 28 kbar, and, thus, UHP conditions were reached, although no coesite was detected so far in eclogites and related rocks from that area. After attaining maximum metamorphic *T* of 480°C, the eclogites were uplifted. The path inferred from the *P-T* data for rim compositions suggests slight cooling during exhumation similar to the eclogites of the Franciscan complex studied here. Probably, the UHPM rocks of Sifnos are genetically very similar to those of the Franciscan complex. This suggestion is also confirmed by consideration of the estimated a_{H_2O} changed during the *P-T* evolution as shown in Fig. 2.8. The H_2O activities during the prograde path, estimated from thermodynamic calculations with core compositions of sample 225, are low and point to the absence of a H_2O-rich fluid phase. Close to the metamorphic peak *P-T* conditions, such a fluid phase was probably pres-

Figure 2.9. Photomicrograph (a) and element distribution map for Al (b) of a cluster of phengites in sample 231 from the Island of Sifnos. (a) Obtained with a polarizing microscope under crossed nicols. Scale bar represents 100 μm. (b) Resulted from work with the electron microsonde by scanning the object (21010 spots) in steps of 3 μm in the x-direction and 2 μm in the y-direction.The counting time per spot was 2 seconds. Gray tones are related to counting rates per second as follows after smoothing: 3260–3799 dark, 3800–4249 light medium, 4250–4480 dark medium, 4480–4819 light, >4820 and <3260 black.

ent because high a_{H_2O} was determined from the rim compositions of sample 225. During the uplift of the HP rocks a_{H_2O} decreased. This result was obtained from calculations with the rim compositions of minerals in sample 263 of Schliestedt and Okrusch (1988) containing amphibole and quartz. The *P-T* data of 21.2 kbar and 415°C derived from this sample are typical of the late stage of HP metamorphism shown by the eclogites from Sifnos.

The *P-T* path of Fig. 2.8 was derived from nine eclogites yielding, in principle, distinct *P-T* dates, considering each sample separately. This problem was addressed by an investigation using Schliestedt's samples 231 and 455 that show fresh eclogitic mineral parageneses with phengite, but different rock fabrics. Sample 231 contains a few porphyroblasts of garnet and pyrite in a

Table 2.7. *Representative electron microprobe analyses of phengites in eclogites sampled by Schliestedt on Sifnos, Greek Cyclades*

	231c	231r	455c	455r
SiO_2	54.64	51.35	52.17	53.92
TiO_2	0.16	0.20	0.22	0.21
Al_2O_3	22.18	28.43	26.93	25.31
Cr_2O_3	0.09	0.07	0.08	0.10
FeO	3.80	2.45	2.25	2.96
MnO	0.02	0.03	0.02	0.09
MgO	4.70	3.39	4.03	4.53
BaO	0.39	0.30	0.28	0.14
Na_2O	0.04	0.51	0.41	0.22
K_2O	11.20	10.52	10.57	10.26
F	0.23	0.23	0.33	0.14
Si	3.608	3.360	3.417	3.488
Al_t	0.392	0.640	0.583	0.512
Ti	0.008	0.010	0.011	0.010
Al_o	1.333	1.553	1.496	1.418
Cr	0.005	0.003	0.004	0.005
Fe^{2+}	0.210	0.134	0.123	0.160
Mn	0.001	0.002	0.001	0.005
Mg	0.462	0.330	0.394	0.437
Ba	0.010	0.008	0.007	0.004
Na	0.005	0.064	0.052	0.027
K	0.943	0.878	0.883	0.847
F	0.048	0.047	0.068	0.029

Note: The analyzed grains are shown in Figs. 2. 9 and 2.10. c = core composition, r = rim composition.

fine-grained foliated matrix composed of omphacite, phengite, and clinozoisite. However, the matrix is bimodal in size. For instance, some coarser-grained phengites appear as clusters in the matrix. The major part of such a cluster is shown in Fig. 2.9. The element distribution map of Al shows that phengite cores are low in Al but relatively rich in Al along some grain margins. An increase in Al is strongly correlated with a decrease in Fe^{2+}, Mg and Si and an increase in Na (Table 2.7). This zonation pattern, related to the core-rim analyses of Schliestedt (1980) and the resulting *P-T* data, should be due to the retrograde path.

Typical for the clinozoisite-rich sample 455 are portions in the thin-section that show relatively large garnets as in sample 231, but medium-grained nearly equigranular amphibole, clinozoisite, omphacite, and phengite with numerous inclusions of fine-grained minerals. Small bands in the rock, with these minerals being fine-grained, are clearly foliated as in sample 231. Further phases, which were probably formed at a late stage of metamorphism, are carbonate and chlorite. Fig. 2.10 shows a cluster of phengites in the medium-grained part of the rock and their chemical zonation pattern (see Table 2.7). In contrast to the phengites of sample 231, these minerals are, along some rims, richer in Fe^{2+}, Mg, and Si and poorer in Al, Na, and F than in the relatively

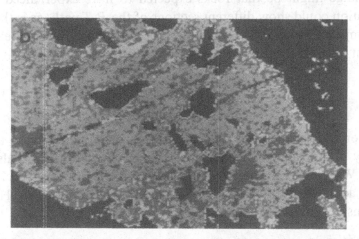

Figure 2.10. Photomicrograph (a) and element distribution map for Fe (b) of a relatively large phengite grain and some smaller ones in sample 455 from the Island of Sifnos. (a) Plane light. (b) See Fig. 2.9: 33540 spots, 4 μm in x-direction, 3 μm in y-direction, 1 second per spot, gray tones: 60–77 dark, 78–87 light medium, 88–97 dark medium, 98–119 light, >120 and <60 black.

homogeneous inner parts. This zonation pattern can be related to the prograde metamorphic path of the eclogites from Sifnos.

Concluding Remarks on Genetic Relationships

The deep burial of the investigated eclogitic rocks from Sifnos and the northern Franciscan complex to depths corresponding to UHP conditions is probably due to a geodynamic process related to subduction of oceanic crust. The prograde metamorphic paths for the Sifnos rocks and the Junction School eclogite as suggested in Fig. 2.7 and 2.8 could reflect *P-T* conditions near the surface of a subducted slab during ongoing subduction. In contrast to that, the *P-T* path for the eclogite TIBB from Tiburon must be explained in another way. Similar *P-T* histories of eclogites were related by Massonne (1991b) to stacking of continental crust. However, whether such a process would lead to UHP conditions as proposed by several authors, e. g. T. B. Andersen et al. (1991), is still questionable. A further problem is the uplift of the deeply buried rocks, because such eclogites of the subducted oceanic crust are relatively dense. In addition, the *T* decrease during exhumation has to be explained sufficiently. In spite of the lack of good tectonic models, it is important to decipher first the *P-T* evolution of the metamorphic rocks. The method applied here is suitable for that purpose. However, it turns out that a single rock sample must be investigated in great detail. Even the examples given here require still more work to refine the *P-T* paths shown in Figures 2.7 and 2.8. An advantage might be that rocks expected to have experienced the same *P-T* evolution could show different sections of the *P-T* path, as is the case for Sifnos samples 231 and 455. This difference is probably due to deformation events that affected these two eclogites in different ways.

Acknowledgments

The author thanks members of the Institute of Mineralogy at the Ruhr-University of Bochum for technical assistance. My thanks are also due to Professor W. Schreyer for the kind provision of the facilities of this institute. I gratefully acknowledge Professor J. G. Liou for his careful review. His helpful suggestions improved the manuscript.

References

Ackermand, D., Seifert, F., and Schreyer, W. 1975. Instability of sapphirine at high pressures. *Contrib. Mineral. Petrol., 50,* 79–92.

Adams, G. E., and Bishop, F. C. 1982. Experimental investigation of Ca-Mg exchange between olivine, orthopyroxene, and clinopyroxene: potential for geobarometry. *Earth Planet. Sci. Lett., 57,* 241–250.

Adams, G. E., and Bishop, F. C. 1985. An experimental investigation of thermodynamic mixing properties and unit-cell parameters of forsterite-monticellite solid solutions. *Am. Mineralogist, 70,* 714–722.

Adams, G. E. and Bishop, F. C. 1986. The olivine–clinopyroxene geobarometer: experimental results in the $CaO-FeO-MgO-SiO_2$ system. *Contrib. Mineral. Petrol., 94,* 230–237.

Akaogi, M., and Akimoto, S. 1979. High-pressure phase equilibria in a garnet lherzolite, with special reference to $Mg^{2+}-Fe^{2+}$ partitioning among constituent minerals. *Phys. Earth Planet. Int., 19,* 31–51.

Akaogi, M., and Akimoto, S. 1980. High-pressure stability of a dense hydrous magnesian silicate $Mg_{23}Si_8O_{42}H_6$ and some geophysical implications. *J. Geophys. Res., 85*, 6944–6948.

Akella, J. 1976. Garnet pyroxene equilibria in the system $CaSiO_3$-$MgSiO_3$-Al_2O_3 and in a natural mineral mixture. *Am. Mineralogist, 61*, 589–598.

Akimoto, S. 1972. The system MgO-FeO-SiO_2 at high pressures and temperatures-phase equilibria and elastic properties. *Tectonophysics, 13*, 161–187.

Akimoto, S., and Akaogi, M. 1984. Possible hydrous magnesian silicates in the mantle transition zone. *Materials Science of the Earth's Interior*, 477–480. Tokyo, Terrapub.

Akimoto, S., and Fujisawa, H. 1968. Olivine-spinel solid solution equilibria in the system Mg_2SiO_4-Fe_2SiO_4. *J. Geophys. Res., 73*, 1467–1479.

Andersen, D. J., Bishop, F. C., and Lindsley, D. H. 1991. Internally consistent solution models for Fe-Mg-Mn-Ti oxides: Fe-Mg-Ti oxides and olivine. *Am. Mineralogist, 76*, 427–444.

Andersen, T. B., Jamtveit, B., Dewey, J. F., and Swensson, E. 1991. Subduction and eduction of continental crust: Major mechanisms during continent-continent collision and orogenic extensional collapse, a model based on the South Norwegian Caledonides. *Terra Nova, 3*, 303–310.

Austrheim, H., and Griffin, W. L. 1985. Shear deformation and eclogite formation within granulite-facies anorthosites of the Bergen Arcs, Western Norway. *Chemical Geology (Chemistry and Petrology of Eclogites) 50*, 267–281.

Baller, T., Fockenberg, T., Ross, C., Rubie, D. C., Schreyer, W., Seifert, F., and Wunder, B. 1990. Hydroxyl-Topas, eine neue Hochdruckphase im System Al_2O_3-SiO_2-H_2O. *Ber. Dt. Mineral. Ges. Beih. Eur. J. Mineral., 2*, 10.

Bauer, J. F., and Sclar, C. B. 1981. The "10 Å phase" in the system MgO-SiO_2-H_2O. *Am. Mineralogist, 66*, 576–585.

Beller, U. 1987. Hochdruck-Synthese von phosphorhaltigem Ellenbergerit. Diploma thesis, Ruhr-Univ. Bochum, Inst. Mineralogie.

Berman, R. G. 1979. Thermal properties. *The properties of diamond*. London, Academic Press, 4–22.

Berman, R. G. 1988. Internally-consistent thermodynamic data for minerals in the system Na_2O-K_2O-CaO-MgO-FeO-Fe_2O_3-Al_2O_3-SiO_2-H_2O-CO_2. *J. Petrol., 29*, 445–522.

Berman, R. G. 1990. Mixing properties of Ca-Mg-Fe-Mn garnets. *Am. Mineralogist, 75*, 328–344.

Berman, R. G., and Brown, T. H. 1985. Heat capacity in the system Na_2O-K_2O-CaO-MgO-FeO-Fe_2O_3-Al_2O_3-SiO_2-TiO_2-H_2O-CO_2: representation, estimation, and high temperature extrapolation. *Contrib. Mineral. Petrol., 89*, 168–183.

Berman, R. G., and Koziol, A. M. 1991. Ternary excess properties of grossular-pyrope-almandine garnet and their influence in geothermobarometry. *Am. Mineralogist, 76*, 1223–1231.

Birch, F., and Lecomte, P. 1960. Temperature pressure plane for albite composition. *Am. J. Sci., 258*, 209–217.

Bishop, F. C. 1980. The distribution of Fe^{2+} and Mg between coexisting ilmenite and pyroxene with application to geothermometry. *Am. J. Sci., 280*, 46–77.

Boettcher, A. L. 1970. The system CaO-Al_2O_3-SiO_2-H_2O at high pressure and temperature. *J. Petrol., 11*, 337–379.

Boettcher, A. L., Robertson, J. K., and Wyllie, P. J. 1980. Studies in synthetic carbonatite systems: solidus relationships for CaO-MgO-CO_2-H_2O to 40 kbar and CaO-MgO-SiO_2-CO_2-H_2O to 10 kbar. *J. Geophys. Res., 85*, 6937–6943.

Boettcher, A. L., and Wyllie, P. J. 1969. Phase relationships in the system NaAlSiO₄-SiO₂-H₂O to 35 kilobars pressure. *Am. J. Sci., 267*, 875–909.

Bohlen, S. R., and Boettcher, A. L. 1982. The quartz-coesite transformation: a pressure determination and the effects of other components. *J. Geophys. Res., 87*, 7073–7078.

Boyd, F. R., and England, J. L. 1964. The system enstatite-pyrope. *Carnegie Inst. Washington Year Book 63*, 157–158.

Brey, G., and Huth, J. 1984. The enstatite-diopside solvus to 60 Kbar. *Kimberlites II: The mantle and crust-mantle relationships; Developments in Petrology 11B.* Amsterdam, Elsevier, 257–264.

Brey, G. P., Köhler, T., and Nickel, K. G. 1990. Geothermobarometry in four-phase lherzolites I. Experimental results from 10 to 60 kb. *J. Petrol., 31*, 1313–1352.

Brey, G. P., Nickel, K. G., and Kogarko, L. 1986. Garnet-pyroxene equilibria in the system CaO-MgO-Al₂O₃-SiO₂ (CMAS): prospects for simplified ("T-independent") lherzolite barometry and an eclogite-barometer. *Contrib. Mineral. Petrol., 92*, 448–455.

Brown, T. H., Berman, R. G., and Perkins, E. H. 1988. Ge0-Calc: Software package for calculation and display of pressure-temperature-composition phase diagrams using an IBM or compatible personal computer. *Computers and Geoscience, 14*, 279–289.

Bundy, F. P., Bovenkerk, H. P., Strong, H. M., and Wentorf, R. H. 1961. Diamond-graphite equilibrium line from growth and graphitization of diamond. *J. Chem. Phys., 35*, 383–391.

Carman, J. H., and Gilbert, M. C. 1983. Experimental studies on glaucophane stability *Amer. Jour. Sci., 283-A*, 414–437.

Chatterjee, N. D. 1991. *Applied mineralogical thermodynamics.* Berlin, Springer.

Chatterjee, N. D., and Flux, S. 1986. Thermodynamic mixing properties of muscovite-paragonite crystalline solutions at high temperatures and pressures, and their geological application. *J. Petrol., 27*, 677–693.

Chatterjee, N. D., Johannes, W., and Leistner, H. 1984. The system CaO-Al₂O₃-SiO₂-H₂O: new phase equilibria data, some calculated phase relations and their petrological application. *Contrib. Mineral. Petrol., 88*, 1–13.

Chatterjee, N. D., and Terhart, L. 1985. Thermodynamic calculation of peridotite phase relations in the system MgO-Al₂O₃-SiO₂-Cr₂O₃, with some geological applications. *Contrib. Mineral. Petrol., 89*, 273–284.

Cheng, W., and Ganguly, J. 1991. Some binary and ternary mixing relations in (Fe,Mg,Ca,Mn) garnets. *EOS Transactions of the American Geophysical Union, Fall Meeting, 72*, 565.

Cho, M. and Ernst, W. G. 1991. An experimental determination of calcic amphibole solid solution along the join tremolite-tschermakite. *Am. Mineralogist, 76*, 985–1001.

Chopin, C. 1983. High pressure facies series in pelitic rocks: a review. *Terra Cognita 3*, 183.

Chopin, C. 1984. Coesite and pure pyrope in high-grade blueschists of the western Alps: A first record and some consequences. *Contrib. Mineral. Petrol., 86*, 107–118.

Chopin, C. 1985. Les relations de phases dans les metapelites de haute pression (PhD thesis Univ. Paris et M. Curie Paris). *Mem. Sciences de la Terre, 85–11*, 84 pp.

Chopin, C. 1986. Phase relationships of ellenbergite, a new high-pressure Mg-Al-

Tisilicate in pyrope-coesite-quartzite from the Western Alps. *Geol. Soc. Amer. Mem., 164*, 31–42.

Chopin, C. 1987. Petrological, crystal-chemical and spectroscopic study of the ellenbergerite group. *Fortschr. Miner., 65*, 36.

Chopin, C., and Schreyer, W. 1983. Magnesiocarpholite and magnesiochloritoid; two index minerals of pelitic blueschists and their preliminary phase relations in the model system $MgO-Al_2O_3-SiO_2-H_2O$. *Am. J. Sci., 283-A*, 72–96.

Chopin, C., Schreyer, W., and Baller, T. 1992. Ellenbergite stability, a reappraisal. *Terra Nova Abst. Suppl., 4*, 10.

Coes, L. 1962. Synthesis of minerals at high pressures. *Modern very high pressure techniques*. London, Butterworths, 137–150.

Coleman, R. G., Lee, D. E., Beatty, L. B., and Brannock, W. W. 1965. Eclogites and eclogites; their differences and similarities. *Geol. Soc. Amer. Bull., 76*, 483–508.

Danckwerth, P. A., and Newton, R. C. 1978. Experimental determination of the spinel peridotite to garnet peridotite reaction in the system $MgO-Al_2O_3-SiO_2$ in the range 900–1100 °C and Al_2O_3 isopleths of enstatite in the spinel field. *Contrib. Mineral. Petrol., 66*, 189–201.

Davidson, P. M., and Lindsley, D. H. 1989. Thermodynamic analysis of pyroxene-olivine-quartz equilibria in the system $CaO-MgO-FeO-SiO_2$. *Am. Mineralogist, 74*, 18–30.

Delany, J. M., and Helgeson, H. C. 1978. Calculation of the thermodynamic consequences of dehydration in subducting ocean crust to 100kb and >800°C. *Am. J. Sci., 278*, 638–686.

Dürr, S. 1986. Das Altisch-Kykladische Kristallin. In: Jacobsen, V. (Ed.) *Geologie von Griechenland, Beitrage zur regionalen Geologie der Erde*, 19, 116–148.

Eggler, D. H., and Rosenhauer 1978. Carbon dioxide in silicate melts: II. Solubilities of CO_2 and H_2O in $CaMgSi_2O_6$ (diopside) liquids and vapors at pressures to 40 kb. *Am. J. Sci., 278*, 64–94.

Ellis, D. J., and Green, D. H. 1979. An experimental study of the effect of Ca upon garnet-clinopyroxene Fe-Mg exchange equilibria. *Contrib. Mineral. Petrol., 71*, 13–22.

Essene, E. 1974. High-pressure transition of $CaSiO_3$. *Contrib. Mineral. Petrol., 45*, 247–250.

Essene, E. J., Hensen, B. J., and Green, D. H. 1970. Experimental study of amphibolite and eclogite stability. *Phys. Earth Planet. Int., 3*, 378–384.

Evans, B. W. 1990. Phase relations of epidote-blueschists. *Lithos, 25*, 3–23.

Ferry, J. M., and Spear, F. S. 1978. Experimental calibration of the partitioning of Fe and Mg between biotite and garnet. *Cont. Mineral. and Petrol., 66*, 113–117.

Flux, S., and Chatterjee, N. D. 1986. Experimental reversal of the Na-K exchange reaction between muscovite-paragonite crystalline solutions and a 2 molal aqueous (NaK)Cl fluid. *J. Petrol., 27*, 665–676.

Fockenberg, T. 1991. Das Stabilitätsfeld von Yoderit. *Ber. Dt. Mineral. Ges., Beih. Eur. J. Mineral 3*, 80.

Fockenberg, T., and Schreyer, W. 1991. Yoderite, a mineral with essential ferric iron: its lack of occurrence in the system $MgO-Al_2O_3-SiO_2-H_2O$. *Am. Mineralogist, 76*, 1052–1060.

Ganguly, J. 1969. Chloritoid stability and related parageneses: Theory, experiments, and applications. *Am. J. Sci., 267*, 910–944.

Ganguly, J. 1973. Activity-composition relation of jadeite in omphacite pyroxene: Theoretical deductions. *Earth Planet. Sci. Lett., 19*, 145–153.

Ganguly, J., and Saxena, S. K. 1987. *Mixtures and mineral reactions.* Berlin, Springer.

Gasparik, T. 1984a. Experimental study of subsolidus phase relations and mixing properties of pyroxene in the system $CaO-Al_2O_3-SiO_2$. *Geochem. Cosmochim. Acta, 48,* 2537–2545.

Gasparik, T. 1984b. Two pyroxene thermobarometry with new experimental data in the system $CaO-MgO-Al_2O_3-SiO_2$. *Contrib. Mineral. Petrol., 87,* 87–97.

Gasparik, T. 1985a. Experimental study of subsolidus phase relations and mixing properties of pyroxene and plagioclase in the system $Na_2O-CaO-Al_2O_3-SiO_2$. *Contrib. Mineral. Petrol., 89,* 346–357.

Gasparik, T. 1985b. Experimentally determined compositions of diopside-jadeite pyroxene in equilibrium with albite and quartz at 1200–1350°C and 15–34 kbar. *Geochem. Cosmochim. Acta, 49,* 865–870.

Gasparik, T. 1990. Phase relations in the transition zone. *J. Geophys. Res., 95,* 15751–15769.

Gasparik, T., and Newton, R. C. 1984. The reversed alumina contents of orthopyroxene in equilibrium with spinel and forsterite in the system $MgO-Al_2O_3-SiO_2$. *Contrib. Mineral. Petrol., 85,* 186–196.

Godard, G. 1988. Petrology of some eclogites in the Hercynides: The eclogites from the southern Armorican massif, France. In D. C. Smith (ed.), *Eclogites and eclogite-facies rocks.* Amsterdam, Elsevier, 451–519.

Goldsmith, J. R. 1980. Melting and breakdown reactions of anorthite at high pressures and temperatures. *Am. Mineralogist, 65,* 272–284.

Graham, C. M., Tareen, A. K., and Lowe, B. M. 1992. An experimental and thermodynamic study of cymrite and celsian stability in the system $BaO-Al_2O_3-SiO_2-H_2O$. *Eur. Jour. Mineral., 4,* 251–270.

Green, D. H. 1972. Magmatic activity as the major process in the chemical evolution of the earth's crust and mantle. In: Ritsema, A. R. (ed.) The upper mantle. *Tectonophysics 13,* 47–71.

Green, D. H., and Mysen, B. O. 1972. Genetic relationship between eclogite and hornblende + plagioclase pegmatite in Western Norway. *Lithos 5,* 147–161.

Green, T. H., and Hellman, P. L. 1982. Fe-Mg partitioning between coexisting garnet and phengite at high pressure, and comments on a garent-phengite geothermometer. *Lithos 15,* 253–266.

Guiraud, M., Holland, T., and Powell, R. 1990. Calculated mineral equilibria in the greenschist-blueschist-eclogite facies in $Na_2O-FeO-MgO-Al_2O_3-SiO_2-H_2O$: methods, results and geologic applications. *Contrib. Mineral. Petrol., 104,* 85–98.

Hacker, B. R., Kirby, S. H., and Bohlen, S. R. 1992. Time and metamorphic petrology: Calcite to aragonite experiments. *Science, 258,* 110–112.

Hackler, R. T., and Wood, B. J. 1989. Experimental determination of Fe and Mg exchange between garnet and olivine and estimation of Fe-Mg mixing properties in garnet. *Am. Mineralogist, 74,* 994–999.

Halbach, H., and Chatterjee, N. D. 1984. An internally consistent set of thermodynamic data for twenty-one $CaO-Al_2O_3-SiO_2-H_2O$ phases by linear parametric programming. *Contrib. Mineral. Petrol., 88,* 14–23.

Hamilton, W. B. 1969. Mesozoic California and the underflow of the Pacific mantle. *Geol. Soc. Amer. Bull., 80,* 2409–2430.

Hariya, Y. 1984. H_2O in the Earth's interior. *Materials science of the Earth's Interior.* Tokyo, Terrapub, 463–475.

Hariya, Y., Oba, T., and Terada, S. 1974. *Stability relations of some hydro-silicate minerals at high pressure.* Proceedings 4th Int. Conf. High Pressure, 206–210; Kyoto, Japan.

Hariya, Y., and Terada, S. 1973. Stability of richterite50-tremolite50 solid solution at high pressures and possible presence of sodium calcic amphibole under upper mantle conditions. *Earth Planet. Sci. Lett., 18*, 72–76.

Harley, S. L. 1984a. The solubility of alumina in orthopyroxene coexisting with garnet in $FeO-MgO-Al_2O_3-SiO_2$ and $CaO-FeO-MgO-Al_2O_3-SiO_2$. *J. Petrol., 25*, 665–696.

Harley, S. L. 1984b. An experimental study of the partitioning of Fe and Mg between garnet and orthopyroxene. *Contrib. Mineral. Petrol., 86*, 359–373.

Harlov, D. E., and Newton, R. C. 1992. Experimental determination of the reaction 2 magnetite + 2 kyanite + 4 quartz = 2 almandine + O_2 at high pressure on the magnetite-hematite buffer. *Am. Mineralogist, 77*, 558–564.

Haselton, H. T., and Westrum, E. F. 1980. Low-temperature heat capacities of synthetic pyrope, grossular, and $pyrope_{60}grossular_{40}$. *Geochim. Cosmochim. Acta, 44*, 701–710.

Hays, J. F., and Bell, P. M. 1973. Albite-jadeite-quartz equilibrium: a hydrostatic determination. *Carnegie Inst. Washington Year Book 72*, 706–708.

Helgeson, H. C., Delany, J. M., Nesbitt, H. W., and Bird, D. K. 1978. Summary and critique of the thermodynamic properties of rock-forming minerals. *Am. J. Sci., 278-A*, 229.

Herzberg, C. T. 1978. Pyroxene geothermometry and geobarometry: experimental and thermodynamic evaluation of some subsolidus phase relations involving pyroxenes in the system $CaO-MgO-Al_2O_3-SiO_2$. *Geochim. Cosmochim. Acta, 42*, 945–957.

Herzberg, C. T., and Chapman, N. A. 1976. Clinopyroxene geothermometry of spinel-lherzolites. *Am. Mineralogist, 61*, 626–637.

Hill, R. E. T., and Boettcher, A. L. 1970. Water in the earth's mantle: melting curves of basalt-water and basalt-water-carbon dioxide. *Science, 167*, 980–981.

Hirajima, T., Banno, S., Hiroi, Y., and Ohta, Y. 1988. Phase petrology of eclogites and related rocks from the Motalafjella high-pressure metamorphic complex in Spitsbergen (Arctic Ocean) and its significance. *Lithos, 22*, 75–97.

Hirajima, T., Ishiwatari, A., Cong, B., Zhang, R., Banno, S., and Nozaka, T. 1990. Coesite from Mengzhong eclogite at Donghai county, northeastern Jiangsu province, China. *Mineral. Mag., 54*, 579–583.

Hölscher, A., Schreyer, W., and Lattard, D. 1986. High-pressure, high temperature stability of surinamite in the system $MgO-BeO-Al_2O_3-SiO_2-H_2O$. *Contrib Mineral. Petrol., 92*, 113–127.

Holdaway, M. J., Dutrow, B. L., and Shore, P. 1986. A model for the crystal chemistry of staurolite. *Am. Mineralogist, 71*, 1142–1159.

Holland, T. J. B. 1979. The experimental determination of the reaction paragonite = jadeite + kyanite + H_2O and internally consistent thermodynamic data for part of the system $Na_2O-Al_2O_3-SiO_2-H_2O$, with applications to eclogites and blueschists. *Contrib. Mineral. Petrol., 68*, 293–301.

Holland, T. J. B. 1980. The reaction albite = jadeite + quartz determined experimentally in the range 600–1200 °C. *American Mineralogist, 65*, 125–134.

Holland, T. J. B. 1983. The experimental determination of activities in disordered and short-range ordered jadeitic pyroxenes. *Contrib. Mineral. Petrol., 82*, 214–220.

Holland, T. J. B. 1988. Preliminary phase relations involving glaucophane and application to high pressure petrology: new heat capacity and thermodynamic data. *Contrib. Mineral. Petrol., 99*, 319–329.

Holland, T. J. B., and Powell, R. 1985. An internally consistent thermodynamic dataset with uncertainties and correlations: 2. Data and results. *Jour. Metamorph. Geol., 3*, 343–370.

Holland, T. J. B., and Powell, R. 1990. An enlarged and updated internally consistent thermodynamic dataset with uncertainties and correlations: The system K_2O-Na_2O-CaO-MgO-MnO-FeO-Fe_2O_3-Al_2O_3-TiO_2-SiO_2-C-H_2-O_2. *Jour. Metamorph. Geol., 8*, 89–124.

Hu, K., and Liu, X. 1992. Diamond-bearing eclogites in central China: An example of ultra-high-pressure metamorphism of crustal rocks. *29th Int. Geol. Congr. Kyoto Abstr. V.*, 599–600.

Huang, W. L., and Wyllie, P. J. 1974. Melting relations of muscovite with quartz and sanidine in the K_2O-Al_2O_3-SiO_2-H_2O system to 30 kilobars and an outline of paragonite melting relations. *Am. J. Sci., 274*, 378–395.

Huang, W.L., and Wyllie, P. J. 1975. Melting and subsolidus phase relationships for $CaSiO_3$ to 35 kilobars pressure. *Am. Mineralogist, 60*, 213–217.

Huang, W. L., and Wyllie, P. J. 1975. Melting reactions in the system $NaAlSi_3O_8$-$KAlSi_3O_8$-SiO_2 to 35 kilobars, dry and with excess water. *J. Geol., 83*, 737–748.

Huang, W. L., and Wyllie, P. J. 1981. Phase relationships of S-type granite with H_2O to 35 kbar: Muscovite granite from Harney Peak, South Dakota. *J. Geophys. Res., 86*, 10515–10529.

Indares, A., and Martignole, J. 1985. Biotite-garnet thermometry in the granulite facies: the influence of Ti and Al in biotite. *Am. Mineralogist, 70*, 272–278.

Irifune, T., Ohtani, E., and Kumazawa, M. 1982. Stability field of knorringite $Mg_3Cr_2Si_3O_{12}$ at high pressure and its implication to the occurrence of Cr-rich pyrope in the upper mantle. *Phys. Earth Planet. Int., 27*, 263–272.

Irving, A. J., and Wyllie, P. J. 1975. Subsolidus and melting relationships for calcite, magnesite and the join $CaCO_3$-$MgCO_3$ to 36 Kb. *Geochem. Cosmochim. Acta, 39*, 35–53.

Jackson, S. L. 1989. Extension of Wohl's ternary asymmetric solution model to four and *n* components. *Am. Mineralogist, 74*, 14–17.

Jeanloz, R., and Thompson, A. B. 1983. Phase transitions and mantle discontinuities. *Rev. Geophys. Space Phys., 21*, 51–74.

Jenkins, D. M., Holland, T.J.B., and Clarke, A. K. 1991. Experimental determination of the pressure-temperature stability field and thermochemical properties of synthetic tremolite. *Am. Mineralogist, 76*, 458–469.

Jenkins, D. M., and Newton, R. C. 1979. Experimental determination of the spinel peridotite to garnet peridotite inversion at 900°C and 1000°C in the system CaO-MgO-Al_2O_3-SiO_2 and at 900°C with natural garnet and olivine. *Contrib. Mineral. Petrol. 68*, 407–419.

Johannes, W., Chipman, D. W., Hays, J. F., Bell, P. M., Mao, H. K., Newton, R. C., Boettcher, A. L., and Seifert, F. 1971. An interlaboratory comparison of piston-cylinder pressure calibration using the albite-breakdown reaction. *Contrib. Mineral. Petrol., 32*, 24–38.

Kanzaki, M. 1991. Stability of hydrous magnesium silicates in the mantle transition zone. *Phys. Earth Planet. Int., 66*, 307–312.

Katagas, C., and Baltatzis, E. 1980. Coexisting celadonitic muscovite and paragonite in chlorite zone metapelite. *N. Jb. Mineral. Mh., 1980*, 206–214.

Kennedy, C. S., and Kennedy, G. C. 1976. The equilibrium boundary between graphite and diamond. *J. Geophys. Res., 81*, 2467–2470.

Khodyrev, O. Y., and Agoshkov, V. M. 1986. Phase transitions in serpentine in the MgO-SiO_2-H_2O system at 40–80 kbar. *Geokhimiya, 2*, 264–269.

Khodyrev, O. Y., and Slutskiy, A. B. 1987. Phase relationships in the MgO-SiO$_2$-H$_2$O-system and mantle ultrabasite petrology. *Geochem. Int. 24*, 122–127.

Kienast, J. R., Lombardo, B., Biino, G., and Pinardon, J. L. 1991. Petrology of very-high-pressure eclogitic rocks from the Brossasco-Isasca Complex, Dora-Maira Massif, Italian Western Alps. *Metamorph. Geol., 9*, 19–34.

Kitahara, S., Takenouchi, S., and Kennedy, G. C. 1966. Phase relations in the system MgO-SiO$_2$-H$_2$O at high temperatures and pressures. *Am. J. Sci., 264*, 223–233.

Koch-Müller, M., Cemic, L., and Langer, K. 1992. Experimental and thermodynamic study of Fe-Mg exchange between olivine and orthopyroxene in the system MgO-FeO-SiO$_2$. *Eur. Jour. Mineral., 4*, 115–135.

Köhler, T. P. and Brey, G. P. 1990. Calcium exchange between olivine and clinopyroxene calibrated as a geothermobarometer for natural peridotites from 2 to 60 kb with applications. *Geochim. Cosmochim. Acta 54*, 2375–2388.

Koziol, A. M. 1990. Activity-composition relationships of binary Ca-Fe and Ca-Mn garnets determined by reversed, displaced equilibrium experiments. *Am. Mineralogist, 75*, 319–327.

Koziol, A. M., and Bohlen, S. R. 1992. Solution properties of almandine-pyrope garnet as determined by phase equilibrium experiments. *Am. Mineralogist, 77*, 765–773.

Koziol, A. M., and Newton, R. C. 1988. Redetermination of the anorthite breakdown reaction and improvement of the plagioclase-garnet-Al$_2$SiO$_5$-quartz geobarometer. *Am. Mineralogist, 73*, 216–223.

Krogh, E. J. 1980. Compatible P-T conditions for eclogites and surrounding gneisses in the Kristiansund area, western Norway. *Contrib. Mineral. Petrol., 75*, 387–393.

Krosse, S., Daniels, P., Medenbach, O., and Schreyer, W. 1992. Hohe Druckstabilität von borhaltigem Kornerupin und ein neues Hochdruck-Mg-Al-Borat. *Ber. Dt. Mineral. Ges., Beih. Eur. J. Mineral., 4, No. 1, 166.*

Kushiro, I. 1969. Clinopyroxene solid solutions formed by reactions between diopside and plagioclase at high pressures. *Am. Min. Soc. Spec. Papers, 2*, 179–191.

Kushiro, I., Syono, Y., and Akimoto, S. 1968. Melting of a peridotite nodule at high pressures and high water pressures. *J. Geophys. Res., 73*, 6023–6029.

Lambert, I. B., and Wyllie, P. J. 1968. Stability of hornblende and a model for the low velocity zone. *Nature, 219*, 1240–1241.

Lambert, I. B., and Wyllie, P. J. 1972. Melting of gabbro (quartz eclogite) with excess water to 35 kilobars, with geological applications. *J. Geol., 80*, 693–708.

Lambert, I. B., and Wyllie, P. J. 1974. Melting of tonalite and crystallization of andesite liquid with excess water to 30 kilobars. *J. Geol., 82*, 88–97.

Lane, D. L., and Ganguly, J. 1980. Al$_2$O$_3$ solubility in orthopyroxene in the system MgO-Al$_2$O$_3$-SiO$_2$: a reevaluation and mantle geotherm. *J. Geophys. Res., 85*, 6963–6972.

Lattard, D., and Schreyer, W. 1981. Experimental results bearing on the stability of the blueschist-facies minerals deerite, howieite, and zussmanite, and their petrological significance. *Bull. Mineral., 104*, 431–440.

Lattard, D., and Schreyer, W. 1983. Synthesis and stability of the garnet calderite in the system Fe-Mn-Si-O. *Contrib. Mineral. Petrol., 84*, 199–214.

Liu, J., and Bohlen, S. R. 1992. Stability of jadeite-acmite in the presence of albite and quartz. *EOS Trans. Am. Geophys. Union, 73*, 601.

Liu, L.-G. 1989. Stability fields of Mg-pumpellyite composition at high pressures and temperatures. *Geophysical Research Papers, 16*, 847–849.

MacGregor, I. D. 1964. The reaction 4 enstatite + spinel → forsterite + pyrope. *Carnegie Inst. Washington Year Book, 63*, 157.

MacGregor, I. D. 1974. The system $MgO-Al_2O_3-SiO_2$: Solubility of Al_2O_3 in enstatite for spinel and garnet peridotite compositions. *Am. Mineralogist, 59*, 110–119.

Maruyama, S., and Liou, J. G. 1988. Petrology of Franciscan metabasites along the jadeite-glaucophane type facies series, Cazadero, California. *J. Petrol., 29*, 1–37.

Massonne, H.-J. 1983. Experiments on melting to 50 kbar in the system $MgO-Al_2O_3-SiO_2-H_2O$ (MASH) with excess SiO_2 and H_2O. *EOS Trans. Amer. Geophys. Union Abst., 64*, 875.

Massonne, H.-J. 1985. Continuous reactions involving micas and granitic melt in the system $K_2O-MgO-Al_2O_3-SiO_2-H_2O$ (KMASH). *Terra Cognita, 5*, 230.

Massonne, H.-J. 1989. The upper thermal stability of magnesian chlorite + quartz: an experimental study in the system $MgO-Al_2O_3-SiO_2-H_2O$. *J. Metamorph. Geol., 7*, 567–581.

Massonne, H.-J. 1990. Geothermobarometrie mit Al-Gehalten in Talk. *Ber. Dt. Mineral Ges., Beih. Eur. Jour. Mineral., 2*, 169.

Massonne, H.-J. 1991a. High-pressure, low-temperature metamorphism of pelitic and other lithologies based on experiments in the system $K_2O-MgO-Al_2O_3-SiO_2-H_2O$. Ph.D., Ruhr-Universität, Bochum.

Massonne, H.-J. 1991b. Druck-Temperatur-Entwicklung phengitführender Eklogite. *Mitt. Österr. Mineral. Ges., 136*, 55–77.

Massonne, H.-J. 1992a. Evidence for low-temperature ultrapotassic siliceous fluids in subduction zone environments from experiments in the system $K_2O-MgO-Al_2O_3-SiO_2-H_2O$ (KMASH). *Lithos, 28*, 421–434.

Massonne, H.-J. 1992b. Thermodynamische Eigenschaften von Phasen des Systems $MgO-Al_2O_3-SiO_2-H_2O$ (MASH) unter besonderer Berücksichtigung von Mischkristallreihen. *Ber. Dt. Mineral. Ges., Beih. Eur. J. Mineral. 4*, 186.

Massonne, H.-J. 1992c. *Thermochemical determination of water activities relevant to eclogitic rocks*. Water-Rock Interaction, Proc. 7th Int. Symp. Vol. 2, Moderate and high temperature environments, Park City, Utah, 1523–1526.

Massonne, H.-J., and Böving, R. 1990. Thermische Stabilisierung von Muscovit durch Einbau von Fluor. *Ber. Dt. Mineral Ges., Beih. Eur. J. Mineral., 2*, 170.

Massonne, H.-J., and Schreyer, W. 1986. High-pressure syntheses and X-ray properties of white micas in the system $K_2O-MgO-Al_2O_3-SiO_2-H_2O$. *N. Jb. Miner., 153*, 177–215.

Massonne, H.-J., and Schreyer, W. 1987. Phengite geobarometry based on the limiting assemblage with K-feldspar, phlogopite and quartz. *Contrib. Mineral. Petrol., 96*, 212–224.

Massonne, H.-J., and Schreyer, W. 1989. Stability field of the high-pressure assemblage talc + phengite and two new phengite barometers. *Eur. J. Mineral., 1*, 391–410.

Mattioli, G. S., and Bishop, F. C. 1984. Experimental determination of the chromium-aluminium mixing parameter in garnet. *Geochim. Cosmochim. Acta, 48*, 1367–1371.

Merrill, R. B., Robertson, J. K., and Wyllie, P. J. 1970. Melting reactions in the system $NaAlSi_3O_8-KAlSi_3O_8-SiO_2-H_2O$ to 20 kilobars compared with results for other feldspar-quartz-H_2O and rock-H_2O-systems. *J. Geol., 78*, 558–569.

Merrill, R. B., and Wyllie, P. J. 1975. Kaersutite and kaersutite eclogite from Kakanui, New Zealand – water-excess and water-deficient melting to 30 kilobars. *Geol. Soc. Amer. Bull., 86*, 555–570.

Millhollen, G. L., Irving, A. J., and Wyllie, P. J. 1974. Melting interval of peridotite with 5.7 per cent water to 30 kilobars. *J. Geol., 82*, 575–587.

Millhollen, G. L., and Wyllie, P. J. 1974. Melting relations of brown-hornblende mylonite from St. Paul's rocks under water-saturated and water-undersaturated conditions to 30 kilobars. *J. Geol., 82*, 589–606.

Mirwald, P. W., and Massonne, H.-J. 1980. The low-high quartz and quartz-coesite transition to 40 kbar between 600° and 1600°C and some reconnaissance data on the effect of $NaAlO_2$ component on the low quartz-coesite transition. *J. Geophys. Res., 85*, 6983–6990.

Mori, T. 1978. Experimental study of pyroxene equilibria in the system $CaO-MgO-FeO-SiO_2$. *J. Petrol., 19*, 45–65.

Mori, T., and Green, D. H. 1976. Subsolidus equilibria between pyroxenes in the $CaO-MgO-SiO_2$ system at high pressures and temperatures. *Am. Mineralogist, 61*, 616–625.

Mposkos, E., and Perdikatzis, V. 1981. Die Paragonit-Chloritoid führenden Schiefer des südwestlichen Bereiches des Kerkis auf Samos (Griechenland). *N. Jb. Mineral., 142*, 292–308.

Mukhopadhyay, D. K., and Lindsley, D. H. 1983. Phase relations in the join kirschsteinite ($CaFeSiO_4$)-fayalite (Fe_2SiO_4). *Am. Mineralogist, 68*, 1089–1094.

Mysen, B. O., and Boettcher, A. L. 1975. Melting of a hydrous mantle: I. Phase relations of natural peridotite at high pressures and temperatures with controlled activities of water, carbon dioxide, and hydrogen. *J. Petrol., 16*, 520–548

Navrotsky, A. 1987. Models of crystalline solutions. *Thermodynamic modeling of geological materials: Minerals, fluids and melts; Reviews in Mineralogy 17*, 35–69.

Nell, J., and Wood, B. J. 1989. Thermodynamic properties in a multicomponent solid solution involving cation disorder: $Fe_3O_4-MgFe_2O_4-FeAl_2O_4-MgAl_2O_4$ spinels. *Am. Mineralogist, 74*, 1000–1015.

Nell, J., and Wood, B. J. 1991. High-temperature electrical measurements and thermodynamic properties of $Fe_3O_4-FeCr_2O_4-MgCr_2O_4-FeAl_2O_4$ spinels. *Am. Mineralogist, 76*, 405–426.

Newton, M. S., and Kennedy, G. C. 1968. Jadeite, analcite, nepheline, and albite at high temperatures and pressure. *Am. J. Sci., 266*, 728–735.

Newton, R. C., and Kennedy, G. C. 1963. Some equilibrium reactions in the join $CaAl_2Si_2O_8-H_2O$. *J. Geophys. Res., 68*, 2967–2983.

Newton, R. C., and Sharp, W. E. 1975. Stability of forsterite + CO_2 and its bearing on the role of CO_2 in the mantle. *Earth Planet. Sci. Lett., 26*, 239–244.

Nickel, K. G. 1986. Phase equilibria in the system $SiO_2-MgO-Al_2O_3-CaO-Cr_2O_3$ (SMACCR) and their bearing on spinel/garnet lherzolite relationships. *N. Jb. Mineral. Abh., 155*, 259–287.

Nickel, K. G., and Brey, G. 1984. Subsolidus orthopyroxene-clinopyroxene systematics in the system $CaO-MgO-SiO_2$ to 60 Kb: a re-evaluation of the regular solution model. *Contrib. Mineral. Petrol., 87*, 35–42.

Nickel, K. G., Brey, G. P., and Kogarko, L. 1985. Orthopyroxene-clinopyroxene equilibria in the system $CaO-MgO-Al_2O_3-SiO_2$ (CMAS): new experimental results and implications for two-pyroxene thermometry. *Contrib. Mineral. Petrol., 91*, 44–53.

O'Hanley, D. S., Chernosky, J. V., and Wicks, F. J. 1989. The stability of lizardite and chrysotile. *Can. Mineral., 27*, 483–493.

O'Neill, S. C., and Wood, B. J. 1979. An experimental study of Fe-Mg partitioning between garnet and olivine and its calibration as a geothermometer. *Contrib. Mineral. Petrol., 70*, 59–70.

Oberhänsli, R. 1980. P-T Bestimmung anhand von Mineralanalysen in Eklogiten und Glaukophaniten der Ophiolithe von Zermatt. *Schweiz. Mineral. Petrogr. Mitt., 60*, 215–235.

Okay, A. I., Arman, M. B., and Göncüoglu, M. C. 1985. Petrology and phase relations of the kyanite-eclogites from eastern Turkey. *Contrib. Mineral. Petrol., 91*, 196–204.

Pacalo, R. E. G., and Gasparik, T. 1990. Reversals of the orthoenstatite-clinoenstatite transition at high pressures and high temperatures. *J. Geophys. Res., 95*, 15853–15858.

Pacalo, R. E. G., and Parise, J. B. 1992. Crystal structure of superhydrous B, a hydrous magnesium silicate synthesized at 1400°C and 20 GPa. *Am. Mineralogist, 77*, 681–684.

Pattison, D. R. M., and Newton, R. C. 1989. Reversed experimental calibration of the garnet-clinopyroxene Fe-Mg exchange thermometer. *Contrib. Mineral. Petrol., 101*, 87–103.

Pawley, A. R. 1992. Experimental study of the compositions and stabilities of synthetic nyböite and nyböite-glaucophane amphiboles. *Eur. Jour. Mineral., 4*, 171–192.

Perchuk, A. L. 1990. Equilibrium albite = jadeite + quartz: experiments, thermodynamics and application to petrology (in Russian). *Abstr. Dissertation, Inst. Ore Deposits, Petrography, Mineralogy & Geochemistry, Moscow, USSR Acad. Sci.*

Perkins, D., Holland, T.J.B., and Newton, R. C. 1981. The Al_2O_3 contents of enstatite in equilibrium with garnet in the system $MgO-Al_2O_3-SiO_2$ at 15–40 Kbar and 900°–1600°C. *Contrib. Mineral. Petrol., 78*, 99–109.

Perkins, D., and Newton, R. C. 1980. The compositions of coexisting pyroxenes and garnet in the system $CaO-MgO-Al_2O_3-SiO_2$ at 900°–1100°C and high-pressures. *Contrib. Mineral. Petrol., 75*, 291–300.

Powell, R., and Holland, T. J. B. 1985. An internally consistent thermodynamic dataset with uncertainties and correlations: 1. Methods and a worked example. *J. Metamorph. Geol., 3*, 327–342.

Pownceby, M. I., Wall, V. J., and O'Neill, H.S.C. 1987. Fe-Mn partitioning between garnet and ilmenite: experimental calibration and applications. *Contrib. Mineral. Petrol., 97*, 116–126.

Råheim, A., and Green, D. H. 1974. Experimental determination of the temperature and pressure dependence of the Fe-Mg partition coefficients for coexisting garnet and clinopyroxene. *Contrib. Mineral. Petrol., 48*, 179–203.

Reinecke, T. 1991. Very-high-pressure metamorphism and uplift of coesite-bearing metasediments from the Zermatt-Saas zone, Western Alps. *Eur. Jour. Mineral., 3*, 7–17.

Reinecke, T., Van Der Klauw, S., and Stöckhert, B. 1992. Ultra-high pressure metamorphism of oceanic crust in the Western Alps. *29th Int. Geol. Congr. Kyoto Abstr., 5*, 99.

Reynard, B. and Ballèvre, M. 1988. Coexisting amphiboles in an eclogite from the Western Alps: new constraints on the miscibility gap between sodic and calcic amphiboles. *Jour. Metamorph. Geol., 6*, 333–350.

Ringwood, A. E. 1958. The constitution of the mantle II. Further data on the olivine-spinel transition. *Geochim. Cosmochim. Acta, 15*, 18–29.

Ringwood, A. E. and Major, A. 1967. High-pressure reconnaissance investigations in the system $Mg_2SiO_4-MgO-H_2O$. *Earth Planet. Sci. Lett., 2*, 130–133.

Ringwood, A. E. and Major, A. 1970. The system $Mg_2SiO_4-Fe_2SiO_4$ at high pressures and temperatures. *Phys. Earth Planet. Int., 3*, 89–108.

Robertson, E. C., Birch, F., and Mac Donald, G. J. F. 1957. Experimental determination of jadeite stability relations to 25,000 bars. *Am. J. Sci., 255,* 115–137.

Ryabchikov, I. D., and Boettcher, A. L. 1980. Experimental evidence at high pressure of potassic metasomatism in the mantle of the Earth. *Am. Mineralogist, 65,* 915–919.

Ryabchikov, I. D., and Mackenzie, W. S. 1985. Interaction of jadeite with water at 20–30 kbar and 650°C. *Mineral. Mag., 49,* 601–603.

Ryabchikov, I. D., Schreyer, W., and Abraham, K. 1982. Compositions of aqueous fluids in equilibrium with pyroxenes and olivines at mantle pressures and temperatures. *Contrib. Mineral. Petrol., 79,* 80–84.

Sack, R. O., and Ghiorso, M. S. 1989. Importance of considerations of mixing properties in establishing an internally consistent thermodynamic database: thermochemistry of minerals in the system Mg_2SiO_4-Fe_2SiO_4-SiO_2. *Contrib. Mineral. Petrol. 102,* 41–68.

Sack, R. O., and Ghiorso, M. S. 1991. An internally consistent model for the thermodynamic properties of Fe-Mg-titanomagnetite-aluminate spinels. *Contrib. Mineral. Petrol., 106,* 474–505.

Sautter, V., Haggerty, S. E., and Field, S. 1991. Ultradeep (>300 kilometers) ultramafic xenoliths: petrological evidence from the transition zone. *Science 252,* 827–830.

Saxena, S. K., Sykes, J., and Eriksson, G. 1986. Phase equilibria in the pyroxene quadrilateral. *J. Petrol., 27,* 843–852.

Schertl, H.-P., Schreyer, W., and Chopin, C. 1991. The pyrope-coesite rocks and their country rocks at Parigi, Dora-Maira Massif, Western Alps: detailed petrography, mineral chemistry and PT-path. *Contrib. Mineral. Petrol., 108,* 1–21.

Schliestedt, M. 1980. Phasengleichgewichte in Hochdruckgesteinen von Sifnos, Griechenland. Ph.D. Techn. Univ. Braunschweig, 145 pp.

Schliestedt, M., and Okrusch, M. 1988. Meta-acidites and silicic meta-sediments related to eclogites and glaucophanites in northern Sifnos, Cycladic Archipelago, Greece. *Eclogites and eclogite-facies rocks; Developments in Petrology,* Amsterdam, Elsevier, 291–334.

Schreyer, W. 1968. A reconnaisance study of the system MgO-Al_2O_3-SiO_2-H_2O at pressures between 10 and 25 kbar. *Carnegie Inst. Washington Year Book, 66,* 380–392.

Schreyer, W. 1988. Experimental studies on metamorphism of crustal rocks under mantle pressures. *Mineral. Mag., 52,* 1–26.

Schreyer, W., Maresch, W. V., Medenbach, O., and Baller, T. 1986. Calcium-free pumpellyite, a new synthetic hydrous Mg-Al-silicate formed at high pressures. *Nature, 321,* 510–511.

Schreyer, W., Maresch, W. V., and Baller, T. 1991. A new hydrous, high-pressure phase with a pumpellyite structure in the system MgO-Al_2O_3-SiO_2-H_2O. In: Perchuk, L. L. (ed), *Progress in metamorphic and magmatic petrology.* Cambridge, Cambridge University Press, 47–64.

Schreyer, W., and Yoder, H.S. Jr. 1968. Yoderite: synthesis, stability, and interpretation of its natural occurrence. *Carnegie Inst . Washington Yearbook, 66,* 376–380.

Schulien, S. 1980. Mg-Fe partitioning between biotite and a supercritical chloride solution. *Contrib. Mineral. Petrol., 74,* 85–93.

Sclar, C. B., Carrison, L. C., and Schwartz, C. M. 1965. High-pressure synthesis and stability of a new hydronium-bearing layer silicate in the system $MgO-SiO_2-H_2O$. *EOS Trans. Am. Geophys. Union, 46,* 184.

Seki, Y., and Kennedy, C. 1964. The breakdown of potassium feldspar $KAlSi_3O_8$ at high temperatures and high pressures. *Am. Mineralogist, 49,* 1688–1706.

Sekine, T., Wyllie, P. J., and Baker, D. R. 1981. Phase relationships at 30 kbar for quartz eclogite composition in $CaO-MgO-Al_2O_3-SiO_2-H_2O$ with implications for subduction zone magmas. *Am. Mineralogist, 66,* 938–950.

Sekine, T., and Wyllie, P. J. 1983. Effect of H_2O on liquidus relationships in $MgO-Al_2O_3-SiO_2$ at 30 kilobars. *J. Geol., 91,* 195–210.

Sengupta, P., Dasgupta, S., Bhattacharya, P. K., and Mukherjee, M. 1990. An ortho-pyroxene-biotite geothermometer and its application in crustal granulites and mantle-derived rocks. *J. Metamorph. Geol., 8,* 191–197.

Shau, Y.-H., Feather, M. E., Essene, E. J., and Peacor, D. R. 1991. Genesis and sol-vus relations of submicroscopically intergrown paragonite and phengite in a blueschist from northern California. *Contrib. Mineral. Petrol., 106,* 367–378.

Smith, D. C. 1988. A review of the peculiar mineralogy of the "Norwegian coesite-eclogite province," with crystal-chemical, petrological, geochemical and geody-namical notes and an extensive bibliography. In: Smith, D. C. (ed.) *Eclogites and eclogite-facies rocks.* Amsterdam, Elsevier, 1–206.

Sobolev, N. V., Dobretsov, N. L., Bakirov, A. B., and Shatsky, V. S. 1986. Eclogites from various types of metamorphic complexes in the USSR and the problems of their origin. *Geol. Soc. Amer. Mem., 164,* 349–360.

Sobolev, N. V., and Shatsky, V. S. 1990. Diamond inclusions in garnets from meta-morphic rocks. *Nature, 343,* 742–746.

Staudigel, H., and Schreyer, W. 1977. The upper thermal stability of clinochlore, $Mg_5Al[AlSi_3O_{10}](OH)_8$, at 10–35 kb P_{H_2O}. *Contrib. Mineral. Petrol., 61,* 187–198.

Stern, C. R., and Wyllie, P. J. 1973. Melting relations of basalt-andesite-rhyolite-H_2O and a pelagic red clay at 30 kbars. *Contrib. Mineral. Petrol., 42,* 313–323.

Stern, C. R., and Wyllie, P. J. 1978. Phase compositions through crystallization inter-vals in basalt-andesite-H_2O at 30 kbar with implications for subduction zone magma. *Am. Mineralogist, 63,* 641–663.

Stern, C. R., and Wyllie, P. J. 1981. Phase relationships of I-type granite with H_2O to 35 kilobars: The Dinkey Lakes biotite-granite from the Sierra Nevada batho-lith. *J. Geophys. Res., 86,* 10412–10422.

Sudo, A., and Tatsumi, Y. 1990. Phlogopite and K-amphibole in the upper mantle: implication for magma genesis in subduction zones. *Geophys. Res. Letters, 17,* 29–32.

Thompson, A. B. 1976. Mineral reactions in pelitic rocks: II. calculation of some P-T-X(Fe-Mg) phase relations. *Am. J. Sci., 276,* 425–454.

Van Roermund, H. 1985. Eclogites of the Seve nappe, central Scandinavian Caledo-nides. In: Gee, J. G. and Sturt, B. A. (eds), *The Caledonide Orogen – Scandinavia and related areas.* New York, Wiley and Sons, 873–886.

Vidal, O., and Theye, T. 1992. Experimental study of the reactions between Fe-chloritoid and almandine at high pressure and low $f(O_2)$. *Terra Nova 4, Abst. Suppl., 1,* 45.

Wakabayashi, J. 1990. Counterclockwise P-T-t paths from amphibolites, Franciscan complex, California: relics from the early stages of subduction zone metamor-phism. *J. Geol., 98,* 657–680.

Wendlandt, R. F., and Eggler, D. H. 1980. Stability of sanidine + forsterite and its bearing on the genesis of potassic magmas and the distribution of potassium in the upper mantle. *Earth Planet. Sci. Lett., 51,* 215–220.

Wood, B. J. 1974. The solubility of alumina in orthopyroxene coexisting with garnet. *Contrib. Mineral. Petrol., 46*, 1–15.

Wood, B. J. 1987. Thermodynamics of multicomponent systems containing several solid solutions. *Thermodynamic modeling of geological materials: Minerals, fluids and melts; Reviews in Mineralogy, 17*, 71–95.

Wood, B. J., and Kleppa, O. J. 1984. Chromium-aluminum mixing in garnet: A thermochemical study. *Geochem. Cosmochim. Acta, 48*, 1373–1375.

Woodland, A. B., and O'Neill, H. 1992. Synthesis of skiagite ($Fe^{2+}_3Fe^{3+}_2Si_3O_{12}$) and phase relations with almandine ($Fe^{2+}_3Al_2Si_3O_{12}$)-skiagite solid solutions. *Terra Nova, Abstr. Suppl., 1*, 47.

Wunder, B., and Schreyer, W. 1991. "Piezotit," ein stabiles wasserhaltiges Hochdruck-Al-Silikat. *Ber. Dt. Mineral. Ges., Beih. Eur. J. Mineral., 3*, 302.

Wunder, B., and Schreyer, W. 1992. Metastability of the 10-Å phase in the system $MgO-SiO_2-H_2O$ (MSH). What about hydrous MSH phases in subduction zones? *J. Petrol., 33*, 877–889.

Yagi, K., Hariya, Y., Onuma, K., and Fukushima, N. 1975. Stability relation of kaersutite. *Jour. Fac. Sci. Hokkaido Univ., 16*, 331–342.

Yamada, H., and Takahashi, E. 1984. Subsolidus phase relations between coexisting garnet and two pyroxenes at 50 to 100 Kbar in the system $CaO-MgO-Al_2O_3-SiO_2$. In: Kronprobst, J. (ed) *Kimberlites II: The mantle and crust-mantle relationships; Developments in Petrology 11-b.* Amsterdam, Elsevier, 247–255.

Yamamoto, K., and Akimoto, S. 1974. High-pressure and high-temperature investigations in the system $MgO-SiO_2-H_2O$. *Solid State Chem., 9*, 187–195.

Yamamoto, K., and Akimoto, S. 1977. The system $MgO-H_2O-SiO_2$ at high pressures and temperatures: Stability field for hydroxyl-chondrodite; hydroxyl-clinohumite and 10 Å-phase. *Am. J. Sci., 277*, 288–312.

3

Principal Mineralogic Indicators of UHP in Crustal Rocks

CHRISTIAN CHOPIN AND N. V. SOBOLEV

Abstract

Reviewed are the characteristic mineralogic features that may be used to identify (both on exposure and in alluvial products) rocks bearing ultra-high-pressure (UHP) assemblages, or having undergone UHP. Among the uncommon features of diamond-bearing [crustal?] rocks is the presence of sodium in garnets coexisting with Na-rich clinopyroxene, and of potassium in clinopyroxene. Attention is drawn to zircon as a very safe container of UHP relics, and to microdiamonds and carbonados (and their isotopic properties) as tracers of "diamond-grade" metamorphism. Diamonds themselves, especially microdiamonds from metamorphic rocks, are different from kimberlitic and lamproitic diamonds in both their dominating cubic morphology and their anomalous isotopic composition of carbon depleted in ^{13}C. Caution is urged and some advice given for the identification in thin section of microdiamond and of quartz pseudomorphs after coesite. Uncommon or new minerals found as primary phases in coesite-bearing crustal rocks are bearthite, $Ca_2Al(PO_4)_2OH$, an accessory that is also stable at low pressures; ellenbergerite, $(Mg,Ti,Zr,\square)_2Mg_6Al_6Si_2Si_6O_{28}(OH)_{10}$, a rock-forming mineral showing complete solid solution with an isostructural new Mg-phosphate and so offering the first natural example of a complete silicate-phosphate series (the silicate is stable above 27 kbar and below 750°C, and the phosphate above 10 kbar, so providing geobarometric prospect for the Si-P substitution). Among phases that are or might be indicative of a decompression path starting under very high pressure conditions are, in decreasing order of appearance and pressure, magnesiostaurolite and magnesiodumortierite, as well as nyböitic amphiboles in Ca-poor eclogites, the tectosilicate lisetite, taramitic amphiboles, and phases allied to nepheline in Na-rich systems. Attention is finally drawn to a series of synthetic compounds the natural equivalent of which should be sought in UHP terranes or xenoliths.

Introduction

Most UHP metamorphic rocks were often first recognized as such through the presence of a few index minerals, usually high pressure (HP) polymorphs of simple compounds, like coesite or diamond (Chopin, 1984; Sobolev and Shatsky, 1990). Further investigation revealed that the whole mineralogy of these rocks may also be characteristic, in two ways: (1) by the presence of new minerals, with either entirely new structures like ellenbergerite or lisetite, or new compositions for already known structures like magnesiostaurolite or the amphibole nyböite; (2) by subtle chemical changes in otherwise common phases like pyroxene, garnet, or oxides.

The detailed study of diamond- and (or) coesite-bearing parageneses in Upper Mantle ultramafic (peridotitic) and eclogitic rocks which are known as xenoliths or as inclusions in diamonds revealed specific compositional features of the main rock-forming minerals such as garnets and pyroxenes formed at very high *P-T* conditions (Meyer, 1968; Sobolev, 1976). The graphite-diamond phase transition was proposed as the boundary between graphite-pyrope and diamond-pyrope facies of the upper mantle (Sobolev, 1974). The importance of such a mineralogic approach is its independence from most of the existing geobarometers and geothermometers which, without proper mineralogic support, may sometimes give unusually high estimates both for pressures and temperatures, as discussed by Pearson et al. in Chapter 13 of this book. We believe that both mineral data and *P-T* estimations have to support each other in order to have a realistic picture of the origin of some possible UHP peridotites and metamorphic rocks in the Earth's crust.

The significance of garnets from metamorphic rocks as containers of primary mineral inclusions is widely known and discussed in order to achieve a better understanding of the prograde metamorphic events, growth rate, and diffusion specific features (Brown, 1993). The coesite in crustal rocks was discovered as relics within polycrystalline quartz aggregate in pyrope grains (Chopin, 1984, 1986). However, in a majority of cases only quartz aggregates are left in the pyropes from the Dora-Maira type-locality for metamorphic coesite as well as in a number of other high pressure metamorphic rocks. Garnets as well as pyroxenes and other primary minerals of metamorphic rocks are very sensitive to retrograde alterations, being often completely replaced by aggregates of secondary minerals.

Special investigations have shown that zircon is the sole mineral of metamorphic rocks that may preserve coesite without any incipient transformation to quartz (Sobolev et al., 1991). The uniqueness of this most typical accessory mineral of metamorphic rocks as a possible and most promising container of primary mineral inclusions is also discussed in this chapter.

Diamonds are reported in a number of alluvial localities without visible relation to a traditional kimberlitic or lamproitic source (see Pearson et al., Chapter 13). Some of these localities may have as a source still undiscovered kimberlites or lamproites, with diamonds being transported a long distance from the primary source. But such manifestations in part might be related to peculiar deep-seated peridotites, or to UHPM rocks. Some specific features of microdiamonds, in particular their morphology and isotopic composition of carbon might be useful, in our opinion, as an independent signature of the possible presence of such rocks.

This chapter will consider specific features of coesite and microdiamond, as well as garnet, clinopyroxene, and zircon utilized to identify UHPM rocks of crustal origin. These unique aspects are based on our knowledge derived from kimberlites and UHPM rocks found in Kokchetav, Kazakshtan and Dora-Maira, Western Alps. New minerals, such as ellenbergerite, bearthite, and magnesiostaurolite are part of these highest pressure assemblages and their experimentally established stability fields indicate crystallization at high pressures. The diagnostic features of this special group of minerals will be helpful in discovery of new UHPM localities.

The present chapter is divided into two main sections, the first written by N. Sobolev and the second by Christian Chopin. Each section will unavoidably reflect each author's biases; therefore the reader is referred also to Smith (1988) or to chapter 9 written by Smith in this book for additional data (e.g., on mica chemistry), more recent developments being addressed here.

Minerals in Very-High-Pressure Rocks

Garnet

The unusual composition of magnesian garnets coexisting with diamonds as inclusions was first noted by Meyer (1968). These garnets appeared to contain much higher Cr and lower Ca compared to ordinary pyropes known from kimberlite pipes and from pyrope peridotites of several crustal occurrences in Europe (Sobolev, 1964). This observation was confirmed by the study of garnets included in diamonds from Yakutian kimberlites (Sobolev et al., 1969). Taking into account the discovery of diamondiferous pyrope serpentinite xenoliths in kimberlites of Yakutia containing coexisting low-Ca pyropes and Cr-spinels (Sobolev et al., 1969), as well as the initial discovery of pyrope and chromite grains of similar composition in kimberlites, Sobolev (1971) suggested that the presence of both minerals in kimberlite concentrates and in heavy mineral haloes within the areas of exploration for kimberlites was the best technique for distinguishing diamondiferous kimberlite pipes from barren pipes, that is, for distinguishing the depth of sampling in the Upper Mantle, within the diamond stability field or from shallower parts.

The low-Ca Cr-pyropes were interpreted (Sobolev et al., 1969; Sobolev, 1974) as originating in diopside-free parageneses. Based upon the detection of polymineralic inclusions in diamonds (low-Ca Cr-pyrope + Cr-rich spinel + olivine ± enstatite), the significance of a harzburgite-dunite paragenesis within the diamond stability field was proposed (Sobolev, 1971, 1974; Sobolev et al., 1971).

Experimental data on the solubility of Cr in garnets on the pyrope-knorringite join (Ringwood, 1977; Irifune et al., 1982; Brey et al., 1991) provided possibilities for the development of a new geothermobarometer for coexisting garnet and spinel, confirming that the most Cr-rich (knorringite rich) garnets coexist with chromites extremely rich in Cr only within the diamond stability field.

As mentioned in a number of publications the absolute majority of deep-seated xenoliths, even in diamond-bearing kimberlites, are represented by the pyrope-bearing rocks of the shallow, graphite-pyrope facies of the upper mantle. An impressive example confirming this is the result of a special testing of 20 tons of xenoliths sampled from peridotites and pyroxenites (griquaites) from the kimberlite pipes of the Kimberley field, South Africa, which did not yield a single diamond crystal (Davidson, 1967).

The variations in Cr contents of pyropes from the peridotites of Norway, Bohemian massif, and Alpe Arami are noted elsewere. Of special interest was to check for the possibility of an estimation of the whole range of variations in pyrope compositions from the same area or body, as was done in testing for the kimberlite concentrate. For this purpose a number of pyrope grains from T7 borehole of Lingorka massif were analyzed (Sobolev et al., 1986) and plotted on the same diagram as used for discrimination of diamondiferous and barren kimberlites (Fig. 3.1). As a result, the complete absence of low-Ca Cr-pyropes of harzburgite–dunite paragenesis was established. This observation confirms that all variations in Cr abundance of these pyropes were dependent only on the Cr/Al ratio in the rocks (Fiala, 1965; Fiala and Padera, 1977) and their lherzolitic (websteritic) paragenesis. However, the presence of pyropes unusually rich in Cr, previously not detected in crustal peridotites, was noted for the first time for the T7 borehole (Sobolev et al., 1986; Smith, 1988). Some grains in the bands, very poor in Al, contained up to 11 wt% Cr_2O_3 (i.e. about 20 mol% of the knorringite component, the Mg-Cr analogue of pyrope), which was very different from the Cr_2O_3 abundances (up to 8 wt%) reported previously. According to existing experimental data the formation pressures of such pyropes are rather close to 25–30 kbar (Brey et al., 1991).

This estimation of pressure extends the possibilities of applying this

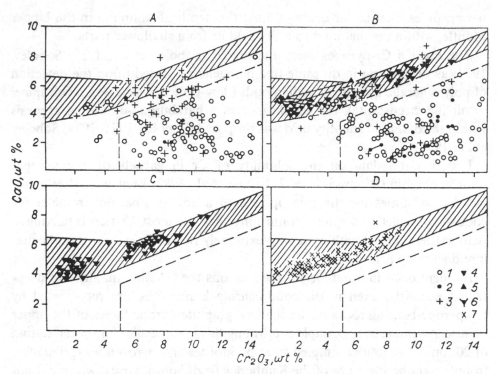

Figure 3.1. CaO versus Cr_2O_3 for magnesian garnets from two Yakutian high-grade kimberlite pipes: Aikhal (A) and Udachnaya (B) compared with the same plots for garnets from crustal pyrope peridotites of the Bohemian massif: pyrope peridotites from Lingorka massif, CSSR (C) and from heavy concentrate of sandstones of Stare region, CSSR (D). Garnets from diamonds (1); xenoliths of diamondiferous peridotites (2); xenoliths of megacrystalline peridotites (3); xenoliths of sheared peridotites (4); xenoliths of sheared ilmenite peridotites (5); xenoliths of sheared pyrope peridotites containing garnets of variable compositions in the same sample (6); heavy concentrate from sandstone of the Stare region, Lingorka massif, CSSR (7). (Data from Sobolev et al.,1986.)

approach to test crustal pyrope peridotites. Not only single peridotite bodies can be tested for Cr contents of pyropes, but also local alluvials containing eroded material may reveal the presence of random Cr-rich pyrope grains. Electron probe analysis of 170 pyrope grains extracted from carboniferous sandstones deposited on the Lingorka pyrope peridotite intrusion are plotted on Fig. 3.1. These results confirm the unique variability of Cr contents in pyropes of this intrusion and support the applicability of the proposed test for detrital grains from alluvials that overlie pyrope peridotites.

Sodium solubility in garnets coexisting with sodium-bearing minerals, mostly clinopyroxenes, is one of the most important mineralogic indicators of very high pressure (Sobolev et al., 1971). The low Na content in pyroxenes from diamondiferous metamorphic rocks of Kokchetav massif causes difficulties in finding some garnets containing sodium. However, a garnet-kyanite-muscovite-quartz rock with diamond inclusions in a kyanite, recently found at Kokchetav massif (Shatsky et al.,1993) contains garnet with 0.12 wt% Na_2O. A systematic examination of garnets for sodium content is required as an independent test for UHP mineral assemblages.

Clinopyroxene

The only useful indicator of high pressure formation of the clinopyroxenes from metamorphic rocks and deep-seated xenoliths from kimberlite pipes that was used until recent times was a limiting jadeite solubility (Dobretsov, 1964; Holland, 1980). However, a number of pyroxenes having very low sodium contents were documented as inclusions in diamonds, in xenoliths from kimberlites, and in diamondiferous metamorphic rocks of the Kokchetav massif (Sobolev, 1974; Sobolev et al., 1984; Meyer, 1987; Sobolev and Shatsky, 1990).

In discussing the problem of potassium abundance within the Upper Mantle minerals, the first experimental results testing the possibility of entering potassium into pyroxene structure are important (Erlank et al., 1970; Shimizu, 1971). The absence of detectable potassium in pyroxenes synthesized in the potassium saturated system up to pressures of 32 kbar (Erlank et al., 1970; Shimizu, 1971) made it possible to discriminate the pyroxenes from peridotitic and eclogitic parageneses in kimberlite pipes, with potassium contents in the range of 0.03–0.30 wt %, as measured by electron probe analyses (Sobolev et al., 1971). Most of these pyroxenes, completely fresh and unaltered grains included in diamonds, showed a homogeneous distribution of K within each grain. A second experimental result presented a positive correlation between K abundance in clinopyroxene and pressure in the range 40–100 kbar (Shimizu, 1971).

A detailed investigation of the first available sample of diamondiferous eclogite, containing a diamond crystal with inclusions, from the Mir pipe (Sobolev et al., 1972), revealed an important difference in K content between pyroxene included in diamond (0.30 wt % K_2O) and pyroxene of the host eclogite (0.08 wt % K_2O). This difference was explained as a loss of potassium during re-equilibration in subsolidus conditions at lower pressures.

The collection of abundant information on the presence of potassium in clinopyroxenes from diamond inclusions from all over the world (Sobolev, 1974; Meyer, 1987) showed a systematic character of contents of K in pyroxenes, and even negative correlation between Na and K contents in pyroxenes from diamonds of Argyle pipe (Jaques et al., 1989; Sobolev and Shatsky, 1989) (see Fig. 3.2). Such a negative correlation obtained by independent study of

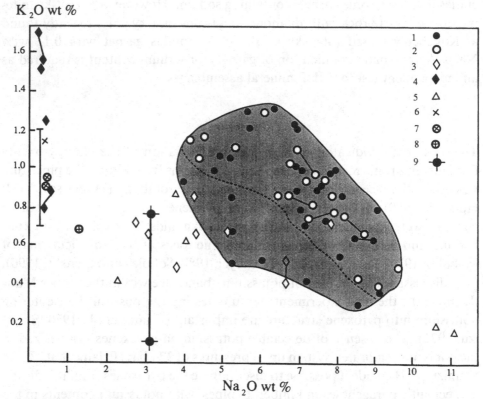

Figure 3.2. K_2O versus Na_2O for clinopyroxenes included in diamonds from kimberlites, lamproites, alluvial deposits, and diamondiferous metamorphic rocks. Pyroxene inclusions from Argyle diamonds (1,2); pyroxene inclusions from Udachnaya pipe diamonds (3); and from South African diamonds (4) and from alluvial diamonds of Siberia and the Urals (5); pyroxenes from diamondiferous metamorphic rocks, Kokchetav massif, Kazakhstan (6–9) including those from pyroxene-carbonate rocks (6,7) with inclusion in zircon (7); pyroxene inclusion in zircon from garnet-pyroxene rock (8); range of the composition of zoned pyroxene grain from diamondiferous gneiss (9). The plots of pyroxene compositions from Argyle diamonds are surrounded by a solid line. Data taken from: Jaques et al., 1989 (1); Sobolev et al., 1989 (2); Sobolev and Yefimova, unpublished (3); Harlow and Veblen, 1991 (4); Sobolev and Shatsky, 1990; Shatsky et al., Chapter 12, this volume and unpublished (6–9).

two groups provides significant support for a structural position of K in clinopyroxenes at very high pressure, with an increasing solubility of K toward diopsidic compositions (Sobolev and Shatsky, 1989). This was also confirmed by a transmission electron microscopic study of the diopside and omphacite inclusions (in diamonds) having the highest recorded abundance of K (Harlow and Veblen, 1991).

It is of special importance that potassium was also detected in clinopyroxenes from Kokchetav diamondiferous rocks (Sobolev and Shatsky, 1989). Before such a detection potassic-felspar exsolution lamellae were found in some clinopyroxene grains (Shatsky et al., 1985). The presence of potassium could be expected in those pyroxenes from Kokchetav diamondiferous rocks which are represented by inclusions both in garnets and in zircons. Such an occurrence protects the primary compositions of clinopyroxenes in the same way as for coesite grains (Sobolev and Shatsky, 1989). For some of these grains the zoned composition was expected. Examination of a number of clinopyroxene grains included in garnets and zircons shows considerable abundance of potassium, mostly close to 1 wt % K_2O and with a detectable zonal distribution of potassium in some grains, but without the differences in sodium contents (see Shatsky et al., Chapter 12, this volume, and Fig. 3.2). The pyroxenes from pyroxene-carbonate and garnet-pyroxene rocks are very poor in sodium and some grains contain even more potassium than sodium (see Shatsky et al., Chapter 12, Table 12.3). However, in diamondiferous gneisses a zoned pyroxene contains up to 3 wt % Na_2O (see Fig. 3.2). The pyroxene from eclogites of Kumdy Kol diamondiferous area contains up to 0.25 wt % K_2O (Sobolev and Shatsky, 1989). Such a potassium abundance makes it possible to ascribe these eclogites to the diamond stability field in spite of the fact that no diamond or graphite was detected in them (Sobolev and Shatsky, 1989).

Recent experimental studies related to K solubility in clinopyroxenes at very high pressures were performed in both dry and water-saturated conditions (Ryabchikov and Ganeev, 1990; Doroshev et al., 1992; Luth, 1992). Harlow (1992) and Ryabchikov and Ganeev (1990) suggested that diopside-rich pyroxenes should demonstrate a limited K solubility compared with jadeite, which contradicts the natural examples (see Fig. 3.2). As was demonstrated by Doroshev et al. (1992), the maximum solubility of K_2O along the join diopside – K-jadeite was found to be only 0.17 wt % at 1200°C and 65 kbar. However, even very low sodium content favors K solubility. In the water-saturated system at 1200°C and 80 kbar at supersolidus conditions, the K_2O contents of the clinopyroxenes are approximately 2 wt % (Luth, 1992). This result supports the conclusions on the participation of fluids in diamond formation (see Sobolev and Shatsky, 1990; Harlow and Veblen, 1991).

Recent experiments produced some controversial results related to Cr solubility in clinopyroxenes as a favorable factor for increasing K solubility (Harlow, 1992; Luth, 1992). However, a comparison with natural examples, with pyroxenes from diamonds and kimberlite heavy concentrates containing up to 45 mol % kosmochlor, showed no significant solubility of K in these pyroxene (Sobolev, 1974).

Further experimental studies on K solubility in clinopyroxenes at high pressure are required for establishing an independent geobarometer. At the same time a careful examination for K abundance of the clinopyroxenes from presumed very high pressure rocks, mostly the pyroxene inclusions in garnets and zircons, is of importance as a way to verify the *P-T* conditions even before verifying the findings of coesite and (or) diamond.

Zircon

Zircon is the most typical accessory mineral of metamorphic rocks and is widely used for geochronology. It is extremely stable and resistant over a wide temperature and pressure interval. The presence of zircons as inclusions in diamonds (Meyer and Svisero, 1975) testifies to their stability within the upper mantle conditions.

However, until recently very little attention has been paid to zircon as a possible container of primary mineral inclusions in UHPM rocks. A rare example was the brief note on the presence of omphacite, rutile, and quartz as inclusions in a zircon from Norwegian eclogites (Krogh et al., 1974).

The main difficulty in checking zircons for inclusions is, of course, the small grain size, not exceeding normally 150–200 micrometers (μm). This grain size of zircons is comparable with that of inclusions in diamonds, which is, on average, close to 100 μm, with some measureable inclusions having sizes as small as 10–20 μm (Sobolev, 1974; Meyer, 1987). The discovery of abundant microdiamonds in zircons from some metamorphic rocks of Kokchetav massif (Sobolev and Shatsky, 1990) led to the suggestion that zircon as a container of primary inclusions might have the same significance as garnet. This was tested in an attempt to check zircon grains from diamondiferous metamorphic rocks of Kokchetav massif for the presence of coesite, which had been unsuccessfully looked for within the garnet grains.

Zircon was selected in an attempt to find the coesite because it was considered to be the analogue closest to diamond in its hardness and resistivity. Diamond was previously recorded as the only mineral containing individual grains of coesite not even slightly retrograded to quartz (Harris, 1968; Sobolev and Shatsky, 1987, 1989; Jaques et al., 1989).

The primary identification of coesite in a zircon was made during the examination of a large number of zircon grains 100–200 μm in size and mounted in epoxy. Coesite was identified during the actual microprobe analysis by comparison with a quartz standard, because the fluorescence color excited by the electron beam and observed through the optical microscope made it possible to distinguish the bright-blue coesite from quartz whose fluorescence color is pink or grayish yellow (Sobolev et al., 1984). After applying Raman spectroscopy to selected grains, it was found that a very strong coesite peak along with a negligible quartz peak is present in the spectrum (Sobolev and Shatsky, 1990).

Investigation of a large number of extracted zircon grains from different types of diamondiferous rocks of Kokchetav massif led to identification of inclusions of coesite, garnets, pyroxenes, and micas. Of special significance is an inclusion of clinopyroxene grown with a polycrystalline coesite-diamond intergrowth, a large number of microdiamonds forming several clusters which filled most of the volume of the zircon, and a magnesite inclusion in a zircon from pyroxene-carbonate rock.

The list of primary mineral inclusions identified within the zircons from diamondiferous metamorphic rocks of Kokchetav massif is even more extensive than for the garnets of the same rocks. Of extreme importance is that all the minerals included in zircon are the same as in the host rock; this excludes the possibility that these zircons are of detrital character.

Zircons are considered the unique and best containers of primary UHP minerals, especially coesite and diamond, in metamorphic rocks. Careful examination of zircons for primary crystalline inclusions is of importance for identifying coesite, diamond, and coexisting minerals, especially for highly retrograded metamorphic rocks from collision zones of different ages, including those of Proterozoic and even Archean Age.

Coesite: a word of caution

Coesite in UHP rocks is essentially pure SiO_2. It is readily distinguished from quartz by its higher refringence, lower birefringence, occasional twinning, and blue luminescence under the electron beam. Coesite relics are most often found as monocrystalline inclusions in garnet, kyanite, or clinopyroxene, but also in zircon and diamond (this chapter), in wagnerite (Chopin, unpublished), dolomite (Okay, 1993) and tourmaline (Reinecke, 1991), a list that is certainly not exhaustive. A conspicuous feature is the presence of radial cracks in the host mineral around the coesite inclusions that have undergone a partial or a complete transformation to quartz. However tempting it may be, this feature observed around a quartz inclusion *cannot* be taken alone as evidence for the former presence of coesite. Indeed, the contrast in the elastic properties of quartz and garnet is such that a garnet having trapped a *quartz* inclusion at 15–20 kbar may well "explode" upon decompression, depending on the *P-T* path followed (Wendt et al., 1993). Therefore, in the absence of relics, *the only convincing evidence* for the previous presence of coesite may be the quartz texture in the inclusion: a radial rim of "palisade" quartz surrounding a core of either feathery or very finely grained polycrystalline quartz (e.g., see Smyth, 1977) is indisputable evidence for the transformation of coesite into quartz, as shown by Gillet et al. (1984). Subsequent annealing will inevitably tend to erase this characteristic feature, leading to a coarser and coarser-grained, less and less critical polycrystalline aggregate. The point is, then, whether the poly-crystalline nature of a well-annealed quartz inclusion may also be taken as evidence for a former coesite transformation (Smith, 1988); this would argue, for instance, that the inclusions studied by Wendt et al. (1993) that had rup-tured the host garnets in a coesite-free terrain are all monocrystalline. In the author's opinion this is going too far, because polycrystalline inclusions do exist in low pressure garnets and so could exist as well in garnets growing at high pressure (but outside the coesite field) and susceptible to "explosion" upon decompression. In summary, it is best to be very conservative in inter-preting of quartz inclusions when looking for coesite; this all the more because Nature seems to be able to preserve true coesite relics under conditions where one would not expect them, like those of a granulite-facies overprint (Wang et al., 1993).

Microdiamonds

Microdiamonds represent part of a diamond population from kimberlites, lamproites and alluvial sources. With a size less than 1 mm they were detected in kimberlites long ago (Varshavsky and Bulanova, 1974). Uniquely small microdiamonds with an average size less than 100 μm for almost 400 crystals have also been detected in some xenoliths of diamondiferous eclogites from Yakutian kimberlites (Sobolev et al.,1991). However, they were detected much earlier in alluvials, namely in Tertiary sands of different areas of the former Soviet Union, in Ukraine, and Northern Kazakhstan, prospected for titanium-zirconium minerals (Kashkarov and Polkanov, 1964; Kashkarov and Polkanov, 1972; Polkanov, 1967). The data available were summarized by Kvashitsa (1985). Wide variations in the morphologic types of microdiamonds were noted, representing practically all known types from kimberlites. A high proportion of cubic diamonds were noted in the sands that have been pro-cessed using a specially developed technique allowing the extraction of microdiamonds of 70–200 μm in size (Kashkarov and Polkanov, 1964, 1972).

Of special interest was the discovery of a number of microdiamond crystals by Kashkarov and Polkanov in 1967 in Tertiary sands from Kokchetav massif, Northern Kazakhstan. After 1.5 tons of sands were processed, 250 crystals of microdiamonds of both octahedral and cubic morphologies, ranging in size from 70 to 200 μm, were detected, as mentioned in a summary paper by

Essenov et al. (1968). Discussing the specific features of the discovered diamonds, Kashkarov and Polkanov (1972) noted that a number of very delicate polycrystalline aggregates were present among the microdiamonds, which indicated the closeness of their source. At that time a kimberlite source for these diamonds was the only possibility discussed (Essenov et al., 1968). The history of the discovery of microdiamonds in crustal rocks and introduction of the optical method for their diagnostics is discussed by Shatsky et al. in Chapter 13. The general morphologic features of microdiamonds are indistinguishable from macrodiamonds, but are very different from those of synthetic diamonds (Sunagawa, 1984; Shatsky et al., 1989).

Whereas microdiamonds and non-sintered delicate polycrystalline microdiamond aggregates are known at Kokchetav massif, sintered polycrystalline aggregates with randomly oriented microdiamonds having crystallites size in the range of 5–20 μm have been described and are called *carbonados* in Brazil (Trueb and Butterman, 1969; Trueb and De Wys, 1969) and *carbon* in Central Africa (Trueb and De Wys, 1971).

The published information on carbonado is very limited and is restricted to the above-mentioned papers and a few more discussions on their origin (Smith and Dawson, 1985) and the specific features of isotopic composition of carbon (Galimov et al., 1985). No attention is paid to carbonado in the most recent summary on diamond geology (Harris, 1968). However, it is possible to find in earlier descriptions that in some alluvial localities carbonado represents an important part of industrial-quality diamonds. These localities are known in Brazil, mainly in Bahia State, with some of less importance in Mato-Grosso State. Some of the discovered samples weighed up to 3167 carats, and carbonado samples weighing more than 100 carats were reported rather often. More limited alluvial localities of carbonado-type aggregates are known in Central Africa, documented as carbon (Trueb and De Wys, 1971). These localities are concentrated in two areas: West Ubangi and East Ubangi. Carbonados mined in the East are usually small and represent about 6–7% of alluvial diamond production, but in the West they are much larger and constitute about 30%. The largest known Central African carbon stone weighs 740.25 carats (Trueb and De Wys, 1971).

Polycrystalline diamonds known as framesite- or bort-type aggregates are described in a number of kimberlite pipes from all over the world. The only major difference from carbonados and carbons is in the relative size of the crystallites composing such aggregates, which are, at least, several orders of magnitude coarser on an average in kimberlites. Some pipes are especially rich in framesite-type aggregates, but some of them do not contain such aggregates at all. No aggregates comparable in crystallites size with Brazilian carbonados or Central African carbon were detected in either kimberlites or lamproites (Harris, 1968).

Because a very limited number of carbonado samples have been studied in detail, the presence of crystallites coarser than 10–20 μm in these aggregates was reported only occasionally, in particular, for a single sample from Central Africa (Trueb and De Wys, 1971). One sample from this locality contained an octahedral monocrystal diamond about 250 μm in size among an aggregate of much smaller crystallites.

Of special interest is the limited information on the abundance of inclusions within the porous aggregates of carbonado diamonds. Orthoclase constitutes up to 80% of all the inclusions observed (Trueb and De Wys, 1969). The presence of rutile along with orthoclase in a number of samples indicates the possibility of a metamorphic type of environment for carbonado growth.

One of the most important diagnostic features of diamonds in general is the isotopic composition of carbon, which correlates with the type of environment for diamond formation: ultramafic (peridotitic) or eclogitic (Sobolev, 1979). Among the eclogitic diamonds those depleted in ^{13}C are rather typical, indicating their crustal origin from organic carbon. Such diamonds are abundant among the eclogitic diamonds known in kimberlites and lamproites; these represent practically 100% of a diamond population in metamorphic rocks from Kokchetav massif (Galimov et al., 1994a; Sobolev and Shatsky, 1990).

All available information on the isotopic composition of carbon for diamonds from Kokchetav metamorphic rocks, adjacent alluvials, and microdiamonds of alluvial origin from the different areas of the world is summarized in Fig. 3.3. The total number of plots is more than 160. It was specially mentioned (see Shatsky et al., Chapter 12) that both Tertiary sands containing diamonds and diamondiferous metamorphics of the Kokchetav massif are located very close to each other. This situation might be considered as typical for such rocks containing microdiamonds. The common features of a δ^{13}C range for diamonds extracted both from rocks and sands are shown in Fig. 3.3 (1). The same wide range in δ^{13}C, but extended also into the typical "kimberlitic" area is typical of diamonds extracted from Ukrainian sands (Galimov et al., 1994a). This fact possibly indicates that some kimberlitic diamonds are also present in these sands [see Fig.3.3 (2)], which is supported by sporadic findings of pyrope grains.

Brazilian carbonados show a very compact restricted range of δ^{13}C (Galimov et al., 1994a) even expanded to the most δ^{13}C-depleted part of Fig. 3.3 (3), compared with previous data (Galimov et al., 1985). The available data on δ^{13}C of the microdiamonds from alluvials of Northern Territory, Australia

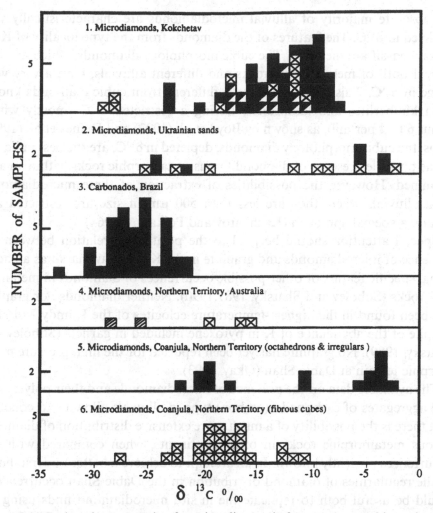

Figure 3.3. Carbon isotope ratios for microdiamonds from metamorphic rocks and alluvial deposits. (1) Microdiamonds from Kokchetav massif: from Tertiary sands, irregular and complex forms (black plots); cubic microdiamonds from sands (crossed plots); microdiamonds extracted from metamorphic rocks (half-black plots). (2) Microdiamonds from Ukrainian sands: octahedral, irregular and complex forms (black plots); cubic crystals (crossed plots). (3) Carbonados from Brazil. (4) Microdiamonds from Northern Territory, Australia: sample NT–1 (crossed plots), sample AS–1 (hatched plots). (5,6) Microdiamonds from Coanjula, Northern Territory, Australia: octahedrons and irregulars (5) and fibrous cubes (6). (Data from Galimov et al., 1994a,b; Sobolev et al., 1989; Lee et al., 1993).

[Fig. 3.3 (4–6)] also clearly show the predominance of $\delta^{13}C$-depleted samples. This especially relates to the cubic microdiamonds [Fig. 3.3 (4,6)].

To summarize the information on $\delta^{13}C$ variations of metamorphic and alluvial microdiamonds as well as carbonados, it is necessary to conclude that the absolute majority of alluvial microdiamonds are characteristically very depleted in $\delta^{13}C$. The features of the diamonds from the type locality of Kokchetav massif are the same. The cubic morphology diamonds, which are very typical both of metamorphic rocks and different alluvials, have a very wide range in $\delta^{13}C$. This specific feature is different from cubic diamonds known from kimberlites and lamproites, having a restricted $\delta^{13}C$ (mostly within minus 6 to 4 per mil), as shown by Boyd et al. (1992) and Deines et al. (1993). Thus, the cubic morphology diamonds, depleted in $\delta^{13}C$, are the best indicator of the possible presence of diamondiferous metamorphic rocks within an area examined. However, the possibilities of extraction of these microdiamonds from alluvials where they are less than 500 µm in size are restricted and require a special approach (Kashkarov and Polkanov, 1964).

Special attention should be paid to the positive correlation between the presence of microdiamonds and graphite in the Kokchetav massif as an additional specific feature of other possible occurrences of diamonds in metamorphic rocks (Sobolev and Shatsky, 1987,1990). Neither diamonds nor graphite has been found in the highest-temperature eclogites of the Kumdy Kol area, in spite of the abundance of K in pyroxene included in garnets (Sobolev and Shatsky, 1989). No graphite has yet been reported for the high pressure metamorphic terrain at Dabie Shan (Okay, 1993).

The available data on the presence of microdiamonds and their polycrystalline aggregates of carbonado (carbon) type in alluvials allow us to conclude that there is the possibility of a much more extensive distribution of diamondiferous metamorphic rocks in other continents, when compared with two occurrences presently known. In an attempt to achieve a better understanding of the regularities of diamond distribution in the Dabie Shan occurrence, it would be useful both to replicate the in situ microdiamond finds using the optical technique utilized for the Kokchetav rocks (Sobolev and Shatsky, 1990) and to check the local alluvials in Dabie Shan area for the presence of microdiamond. Besides Kazakhstan and China, diamondiferous metamorphic rocks might be expected within the southern boundary of the Ukrainian Shield, within the metamorphic terrain of the eastern part of Brasilian Shield, in Central Africa and Northern Australia (Sobolev and Shatsky, 1989). In spite of microdiamond identification in the metamorphic rocks, some peridotitic massifs and alluvial occurrences are very important for understanding the very high pressure character of specific rocks, or predicting their occurrences; it is therefore especially important to examine critically any new possible occurrence of microdiamonds in any crustal rock. This relates to descriptions of diamonds from crushed rocks without verification by optical methods in thin sections and doubly polished plates 200–300 μm thick as used for Kokchetav metamorphic rocks (Sobolev and Shatsky, 1987, 1990). Such a verification has to be accompanied by Raman spectroscopy data. In some cases only visual estimations are available, even from the most recent publications where a group of faceted crystals having a high refractive index were identified as diamonds (Okay, 1993). The general appearance of such crystals could easily allow them to be classified as another accessory phase but not as a diamond. Xu et al. (1992) have reported diamond from the same area using Raman spectra, but this has not been verified by other mineralogists. The requirement of additional verifications also relates to reports of macrodiamonds and microdiamonds present in a number of ultramafic rocks as discussed by Pearson et al. (see Chapter 13, this volume). The problem of identification was also carefully discussed for some of the listed occurrences, especially in Canada (Sobolev, 1951), and here special attention was also paid to possible contamination of the crushed or dissolved rock samples during processing (Sobolev, 1979).

Besides the two above-mentioned localities of microdiamonds which have considerably different levels of proof for the presence of diamond (i.e., optical method with an identification of many hundreds of microdiamonds in a number of thin sections for Kokchetav massif and diamond verified in only one specimen from Dabie Shan locality; Xu et al., 1992), several more occurrences of microdiamonds in different metamorphic rocks have been reported in Russian scientific journals (Dergachev, 1986; Biryukov and Kosygin, 1989). Unfortunately, in all these cases, a limited number of microdiamonds in several samples have been detected after processing and dissolving whole rock samples in the same laboratories that deal with processing the diamondiferous rocks from Kokchetav massif. All microdiamonds extracted by this approach, even those from some xenoliths of deep-seated rocks, are completely similar to the Kokchetav diamonds. Under such conditions it is totally impossible to avoid a contamination with a limited number of extremely tiny microdiamonds. Thus duplicate examinations are required for such samples in independent laboratories with reproducible results. The same problem appeared recently in regard to the identification of several microdiamonds of Kokchetav type in some samples of Norwegian metamorphic rocks in a Moscow laboratory that deals with examination of Kokchetav diamondiferous rocks (Dobrzhinetskaya et al., 1993). Another problem is the contamination of thin sections by diamond paste during polishing procedures, where a diamond peak on Raman spectrum was reported (Garanin et al., 1988). Most of these listed artifacts were specially discussed (Sobolev and Shatsky, 1989). Proper attention to the probability of contamination should be considered in any attempts to find and verify any new, in situ microdiamond occurrences.

New Minerals in Very-High-Pressure Rocks

In actuality all of these new minerals were reported from the two coesite-bearing terranes discovered first, namely in the southern Dora-Maira massif, Italian Western Alps (geologic overview in Chopin et al., 1991; Compagnoni et al., Chapter 7, this volume) and the Western Gneiss Region, Norway (overview in Smith, 1988). Several of these minerals belong to the highest pressure assemblage in the rock and so certainly formed at high pressure; the relevant question is then whether or not their stability field is confined to these high pressures; this was the incentive for a number of experimental studies, with very contrasting results depending on the mineral considered. Some other new phases are texturally later than the highest pressure assemblage and so formed during decompression; they often have attracted less experimental work. These different aspects are reviewed in the following sections, starting with the structurally and petrologically most promising ellenbergerite group.

High-pressure phases

The ellenbergerite group

Ellenbergerite is a silicate, ideally $(Mg,Ti,Zr,\square)_2Mg_6Al_6Si_2Si_6O_{28}(OH)_{10}$, found as purple hexagonal prisms included in pyrope megablasts from the coesite-bearing unit in the southern Dora-Maira massif (Chopin et al., 1986). It was reported to incorporate variable but unusually high amounts of phosphorus. At that time, up to 16 wt% P_2O_5 were known, corresponding to about 2.4 P atoms per formula unit (pfu); they could be essentially accounted for by the $Al^{VI}\,Si^{IV} = Mg^{VI}\,P^{IV}$ substitution, with a change in space group from $P6_3$ to $P6_3mc$.

The structure of the silicate end-member ($P6_3$; Chopin et al., 1986; Comodi and Zanazzi, 1993a) consists of two types of octahedral chains parallel to the c axis: single chains of face-sharing octahedra partly occupied by Mg and Ti + Zr, leaving about every third octahedron vacant, and double chains composed of edge-sharing pairs of face-sharing octahedra in which one Al and one Mg are completely ordered. These chains are interconnected by isolated SiO_4 tetrahedra, two of which (per cell), on the three-fold axes, have a free OH apex. Six additional protons could be located along the double chain, the two remaining ones being delocalized along the single chain. With increasing P content and so decreasing Al content, the Mg-Al ordering in the double chain disappears, the two face-sharing octahedra become indistinguishable, resulting in a new symmetry plane, hence the space group $P6_3mc$. The connection of octahedra by face-sharing instead of edge- or vertex-sharing is unusual for silicates and results in a dense structure ($D = 3.15$ g/cm^3).

A silicate-phosphate series. During the following years, the further study of mineral inclusions in the pyrope megablasts revealed the presence of a blue hexagonal magnesium phosphate, isostructural with $P6_3mc$ ellenbergerite, with zoning that demonstrates the existence of a complete solid-solution series between the silicate and phosphate end-members, with correlative complex chemical and optical changes along the series. An optical absorption spectrometric study showed the purple color of the silicate and the blue-green of the phosphate to be due, respectively, to Fe^{2+}-Ti^{4+} and Fe^{2+}-Fe^{3+} charge transfer along the c-axis (Chopin and Langer, 1988).

Occurrence. In the Dora-Maira coesite-bearing terrane, the ellenbergerite-type minerals occur exclusively as inclusions in pyrope megacrysts of the phengite-quartzite ("whiteschist") from which metamorphic coesite was first reported (Chopin, 1984; Chopin, 1986; Schertl et al., 1991). This characteristic rock type occurs as discontinuous layers and boudin trails within orthogneiss, on either side of (i.e., structurally above and below) a varied series of metapelite, metabasite, and marble (Chopin et al., 1991), the Brossasco-Isasca Unit of Biino and Compagnoni (1992; see also Compagnoni et al., Chapter 7, this volume).

Depending on the structural position of the enclosing quartzite body, two types of zonation and chemical range are found in the "ellenbergerite" crystals. Macroscopically purple crystals, sometimes with a pale pink or grayish core, occur exclusively in the structurally lower trail, with the key outcrop at Case Ramello, near Parigi, Martiniana Po. These crystals may be zoned, from *at most* 2.7 P pfu in the core to a P-free rim (Fig. 3.4). Macroscopically blue "ellenbergerite" crystals occur exclusively in the structurally higher trail(s), in Val Gilba and Val Varaita (map and profile in Chopin et al., 1991). They may be zoned with a core containing *at least* 2.9 P pfu to an almost Si free blue-green rim (Fig. 3.4). Most significantly, this contrast in chemistry and zoning pattern is also reflected in the parageneses of the other inclusions: kyanite, clinochlore, talc, coesite, rutile, zircon, monazite, and rare dumortierite occur in both groups, but glaucophane occurs exclusively in the lower-trail pyropes, whereas magnesiostaurolite, enstatite, gedrite, corundum and sapphirine

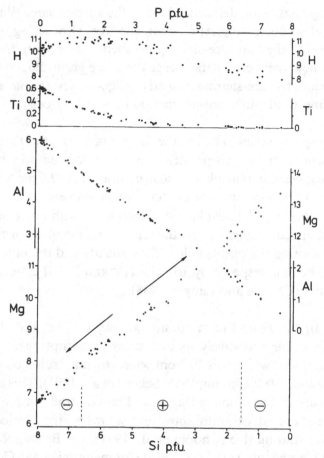

Figure 3.4. Chemistry of the isostructural ellenbergerite-Mg-phosphate series. Atomic contents of Al, Mg + Fe (Mg), Ti + Zr (Ti), and H$^+$ as a function of Si or P contents, all calculated on the basis of Si + P = 8 tetrahedral cations and 38 oxygens pfu. Arrows represent the zonation patterns observed depending on occurrence (see text); circled (+) and (−) indicate the optical sign, changing twice along the series.

exclusively occur in the upper one(s). Wagnerite, $Mg_2(PO_4)(F,OH)$, is locally abundant in the upper trail(s); likewise, apatite is more common in the upper-trail pyropes.

Chemical variations. To illustrate these changes, the results of microprobe analyses obtained on about 10 zoned crystals are reported in Fig. 3.4, in which the number of the octahedral cations and of protons are plotted against the Si and P contents (on a Si + P = 8 tetrahedral site basis). The first important feature in Fig. 3.4 is the apparently *continuous* change from a pure Mg-Al-(Ti,Zr)-silicate to an almost pure Mg-phosphate. The number of Al atoms decreases from 6 to about 0, that of Ti + Zr decreases from 0.65 to 0, whereas

Mg (plus minor Fe and Ca) increases from 6.7 to about 13.5, suggesting little change in the total number of vacancies. By extrapolation to 0 P and 0 Si, the end-members would be close to $(Mg_{1/3},(Ti,Zr)_{1/3}, \square_{1/3})_2 Al_6Mg_6Si_2Si_6O_{28}(OH)_{10}$ (Chopin et al., 1986) and $(Mg_{0.75,} \square_{0.25})_2Mg_{12}P_2P_6O_{29} (OH)_9$. The obvious and indeed major mechanism $Al^{VI} Si^{IV} = Mg^{VI} P^{IV}$ can account for the substitution of 6 Si by P; the replacement of the two remaining tetrahedral cations is charge-compensated along the octahedral single chain by replacement of Ti by Mg [or of TiO_2 by $Mg(OH)_2$] for P < 3 and by subtle changes in the number of protons and/or of vacancies.

This continuous series between isostructural silicate and phosphate is a unique feature, even if numerous examples exist of isostructural silicates and phosphates (e.g., quartz-berlinite, zircon-xenotime, cerite-whitlockite, britholite-apatite).

Stability. The restricted occurrence of the silicate ellenbergerite (in spite of its common chemical constituents) suggested that it could be due to a restricted stability field. Indeed, an experimental study on the synthetic silicate end-member (Chopin et al., 1992) has shown that it has a very high pressure, wedge-shaped stability field, limited at pressures in excess of 28 kbar and temperatures lower than 725 ± 25°C by the two reactions:

$$talc + chlorite + kyanite + rutile + H_2O = ellenbergerite \qquad (3.1)$$
$$ellenbergerite = pyrope + talc + kyanite + rutile + H_2O \qquad (3.2)$$

Uncommon in reaction 3.1 is that ellenbergerite actually dehydrates toward ` the *low* temperatures. The stability field so outlined (Fig. 3.5) makes ellenbergerite both an index for very high pressures and a fine indicator of negative thermal anomaly at mantle depths. In addition, the phosphate end-member is stable down to much lower pressures (near 8 kbar; Beller, 1987) in the same temperature range and so gives to the Si/P ratio of zoned crystals great interest and potential for geobarometry. As a matter of fact, the Mg-phosphate "ellenbergerite" also occurs in Norway, at the Modum deposit (G. Raade, personal communication. 1990), whereas the Si-rich members of the series are so far known only from the Dora-Maira coesite-bearing rocks. Looking for magnesian pelitic-type rock compositions in other very-high-pressure terranes would certainly be rewarding in this respect.

Bearthite

Occurrence. This new mineral (Chopin et al.,1993) occurs as an accessory phase in several rock-types of the Dora-Maira coesite-bearing unit: in normal

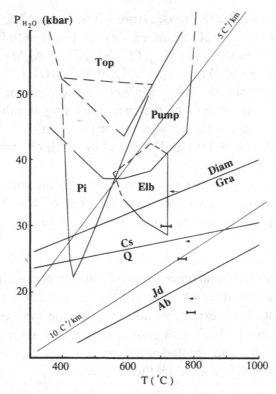

Figure 3.5. Stability fields of some high pressure phases or assemblages. Ab = albite; Cs = coesite (Mirwald and Massonne, 1980); Diam = diamond (Bundy, 1980); Elb = ellenbergerite (Chopin et al., 1992); Gra = graphite; Jd = jadeite (Holland, 1980); Pi = phase Pi (Wunder et al., 1993a); Pump = Mg-Mg-Al-pumpellyite (Schreyer et al., 1991); Q = quartz; Top = topaz-OH (Wunder et al., 1993b). The three segments and arrows represent respectively experimental reversals of the reaction chlorite + kyanite = talc + Mg-staurolite + H_2O and upper bounds to the reaction talc + Mg-staurolite = pyrope + kyanite + H_2O (Chopin, unpub.). The 5° and 10°C/km isotherms are indicated, calculated for a density of 3.3 g/cm^3.

metapelites (with almandine, kyanite, $Si_{3.5}$ phengite-3T, quartz, apatite, rutile and pseudomorphs after jadeite and coesite) and in the pyrope-phengite-quartzite, both in the matrix (with quartz/coesite, phengite, kyanite and pyrope) and as inclusion in pyrope megablasts. It is clearly part of the highest-pressure assemblage (formed at about 30 kbar, 700–750°C). Bearthite actually also occurs in quartz-lazulite veins of the Monte Rosa massif, Western Alps, which formed at about 15 kbar, 500°C, and at the Vestan deposit, Waermland, Sweden, in quartz-kyanite-lazulite veins.

Physical and chemical properties. In thin section the mineral is colorless and, though not abundant, may be rather easily spotted through its high refringence and moderate birefringence (n_x 1.662, n_y 1.671, n_z 1.696, 2V 65°, biaxial positive). When present, a narrow selvage of apatite around lozenge-shaped sections or rounded grains of bearthite (with no or only one poor cleavage) may be an additional hint for recognition.

Chemically, it may be pure $Ca_2Al(PO_4)_2OH$, but Sr often substitutes part of Ca toward the Sr-analogue goedkenite (Moore et al., 1975). The Dora-Maira material may also incorporate up to 11 wt% rare earth elements (REE) through the charge-balanced substitution Ca Al = REE Mg, as is well known in allanite.

Bearthite is monoclinic, space group $P2_1/m$. The structure (Chopin et al., 1993) is that of the brackebushite group, which comprises several arsenates, vanadates, and sulfates of Pb, Ba, and Cu, Mn, etc., all of low-pressure and low-temperature origin. The main feature is infinite chains of edge-sharing AlO_5OH octahedra extending along b. Four PO_4 tetrahedra per unit-cell are connected via common corners to each octahedral chain to form the fundamental building block $[Al_2(PO_4)_4(OH)_2]$. The same building block $[Al_2(SiO_4)_4(OH)_2]$ is present in clinozoisite, in tornebohmite, and, by condensation of tetrahedra, in the dense phases lawsonite and vuagnatite.

Stability. Since bearthite has, among the brackebushite-group minerals, the smallest cation in each structural position, it could be expected to be stable to higher pressures than are the other members of the group (as is clear from its occurrences) and to have its stability restricted to high pressures. In fact, an experimental study has just shown that its stability field encompasses the whole pressure range from at least 25 kbar down to 500 bars or less, and does not exceed about 500°C at 1 kbar and 850°C at 10 kbar (Brunet, 1992). So it appears to be fortuitous that bearthite has been discovered first in high-pressure rocks!

Magnesiostaurolite

Occurrence. The Mg-dominant member of the staurolite group (IMA decision 92–035, Chopin et al., in prep.) occurs as inclusions in the Dora-Maira pyrope megablasts (together with talc, kyanite and clinochlore, sometimes enstatite) with extremely high values of the Mg/(Mg + Fe) ratio (= Xmg), between 0.85 and 0.95. This staurolite is colorless and devoid of pleochroism, which makes its identification just through the high refringence (ca. 1.71) and low birefringence a little difficult. Although associated with kyanite in the Dora-Maira pyropes, magnesiostaurolite does not occur in parallel growth with it, as is common for Fe-rich or Zn-rich staurolites.

Actually, other Mg-rich, rarely Mg-dominant staurolites were already known from a number of high-pressure rocks, usually eclogitic metagabbros or metabasites, in which staurolite sometimes appears as a secondary phase during decompression [Xmg 0.43–0.56 (Ward, 1984); Xmg 0.68–0.74 (Enami and Zang, 1988); Xmg 0.5 (Smith, 1984); Xmg 0.74 and $Cr_2O_3 > 6$ wt % (Gil Ibarguchi et al., 1991)] and also from high-pressure granulites [Xmg 0.49 (Schreyer et al., 1984); Xmg 0.52 (Nicollet, 1986)]. Mg-rich staurolite (Xmg 0.53-0.57) was also produced at 740–760°C, 24–26 kbar, from olivine tholeiitic material by Hellman and Green (1979). Unfortunately, few of these staurolites have been fully characterized, in particular for their H_2O and Li contents.

Properties. There are three interesting features in the Dora-Maira highest-pressure magnesiostaurolite. One is the presence of high Li content, up to 1.2 wt% Li_2O in a zoned crystal, coupled with numbers of vacancies and of protons among the highest recorded (Hawthorne et al., 1993). However, high Li contents are not specific to this HP staurolite and can also be found in low-pressure, iron-rich staurolite (Dutrow et al., 1986; Holdaway and Goodge, 1990).

The second feature is the absence of tetrahedral Al (Hawthorne et al., 1993), as can be expected for a HP phase. This also holds for the Fe-rich staurolite inclusions in almandine porphyroblasts of the Dora-Maira coesite-bearing metapelites (Chopin et al., 1991); this is one of the reasons why the present author considers these inclusions as relics of the prograde Alpine high-pressure metamorphism, not as pre-Alpine relics (cf. Compagnoni et al., Chapter 7, this volume).

The key feature of the Dora-Maira magnesiostaurolite resides not so much in the high number of vacancies (and of protons) but rather in their location, which offers a solution to the "staurolite paradox." Staurolite is indeed one of the few silicates to contain Fe (or Mg) in tetrahedral coordination. Since high pressures are known to favor increasing coordination, it was indeed paradoxical that Mg-rich staurolites or synthetic Mg-staurolite could be HP phases, stable at higher pressures than alternative assemblages with Mg exclusively octahedrally coordinated. Actually, the crystal structure refinement of Dora Maira magnesiostaurolite (Hawthorne et al., 1993) reveals that the relevant tetrahedral site T(2) is occupied to less than 70% (by Mg and Li) against 90–95% in most staurolites; conversely, Mg occupies about 25% of the octahedral M(4) sites, which have only a few percent occupancy in most staurolites. To emphasize this, the formula may be written

$$(Mg_{2.1}Li_{0.9}\square_1)^{T2} (Mg_1\square_3)^{M4} (Al_{17.7}Mg_{0.1}\square_{0.2})_{\Sigma=18} Si_8O_{44.4} (OH)_{3.6}$$

This shift of the divalent cation from a tetrahedral to an octahedral position is a new and significant feature, which has very important bearing on the thermodynamics and mixing properties of the staurolite group. It may well be the reason why this group has for so long resisted thermodynamic analysis. In the present state of knowledge it is admittedly necessary to refine the structure of an Fe-rich UHP staurolite to make sure that this structural feature is indeed related to high pressures of formation and not to the presence of Mg instead of Fe or Zn. Yet it remains a very appealing possibility.

Petrologically, there are two points of interest, namely Fe/Mg partitioning between staurolite and garnet, and the coexistence of staurolite with talc. Ballèvre et al. (1989) have emphasized that Fe/Mg partitioning between (Fe-rich) staurolite and garnet is reversed in high-pressure rocks, with staurolite being the more Fe-rich phase. The underlying mechanism is not understood but could be related to the changing structural position of the divalent cation just mentioned. Although the actual picture may be not so simple (there are a few instances of high-pressure pairs without reversals and vice versa), there is indeed a clear reversal in partitioning between the almandine porphyroblasts and their staurolite inclusions in Dora-Maira metapelite (Chopin, 1985; Pinardon, 1990; Chopin et al., 1991). As for the magnesiostaurolite included in pyrope, the partitioning is so close to one that repeated analyses have not allowed the author to decide in favor of a reversal or not. The magnesiostaurolite described by Enami and Zang (1988) is secondary after garnet and so of little help in this respect.

The talc-staurolite coexistence may be considered parallel with that of talc and chloritoid; in both cases a mineral reputed to be very Al and Fe rich should coexist with an Al-free Mg-silicate. Whereas the latter coexistence is now well established and a fine HP indicator (e.g., Chopin and Schreyer, 1983; Korikovskiy et al., 1983; Chopin, 1991), the former has, to the author's knowledge, been reported only twice, by Grew and Sandiford (1984) in Antarctica (Xmg st = 0.40–0.42), and for the Dora-Maira magnesiostaurolite, which coexists with talc, clinochlore, and kyanite (not with corundum, unlike most Mg-rich staurolites).

The possibility of high-Li content renders the application of the experimental results obtained on Mg-staurolite, all in the system $MgO-Al_2O_3-SiO_2-H_2O$ (MASH), less straightforward than could be hoped. In this system, Schreyer and Seifert (1969) showed that synthetic Mg-staurolite is a HP phase that breaks down to pyrope + kyanite + corundum + H_2O on the high-temperature side, and hydrates most likely to chlorite + kyanite + corundum on the low-temperature side. The resulting wedge-shaped stability field starts near 14–15 kbar, 750–800°C, and opens towards higher pressures, well in excess of 25 kbar (see also Schreyer, 1988). One of the key problems is whether the tie lines chlorite-kyanite or pyrope-kyanite remain stable in this stability field, or whether talc-staurolite can become stable instead. Experiments in the MASH system by the author on the two reactions

$$\text{chlorite} + \text{kyanite} = \text{talc} + \text{staurolite} + H_2O$$
$$\text{talc} + \text{staurolite} = \text{kyanite} + \text{pyrope} + H_2O$$

(both written with the low-temperature assemblage on the left-hand side), show them to be hardly distinguishable within experimental uncertainty (from about 17 kbar, 780 $\pm$ 15°C, to 30 kbar, 710 $\pm$ 20°C), that is, leaving virtually no stability field to talc + Mg-staurolite in the system MASH (Fig. 3.5). Admittedly, adding Li to the system will evidently favor the staurolite-bearing pair with respect to the other assemblages, and also extend the stability field of Mg-staurolite, because Li is preferentially incorporated in this phase. An ongoing study carried out in Bochum on the MASH system on the effect of P and T on the number of OH groups and vacancies in the staurolite structure suggests that both, and especially the former, are favored by increasing P (Fockenberg, 1992), in keeping with the features of the Dora-Maira near end-member magnesiostaurolite.

Decompression-Related Phases

Magnesiodumortierite. The borosilicate dumortierite, ideally $(Al_{1-x}\square_x)_2$ $Al_4Al_8 B_2 Si_6O_{36}H_{6x}$ is the only silicate to bear some structural similarities with ellenbergerite, as emphasized by the above formula (actually twice the formula unit). The SiO_3OH tetrahedra on the threefold axes of ellenbergerite are replaced by planar BO_3 groups; the single chain of face-sharing octahedra remains unchanged but is dominantly occupied by Al (plus some Fe, Ti or Mg); one-third of the octahedral double chains also remain unchanged (with the Mg-Al pairs replaced by Al-Al), and two-thirds of them are replaced by double chains in which the connection of the octahedra is achieved purely by edge-sharing, that is, without face-sharing (Moore and Araki, 1987).

Pink dumortierite is rather common in the Dora-Maira pyrope megablasts, less so in their quartzite matrix. Its status is texturally ambiguous; in some instances it is fine-grained and clearly secondary, filling cracks in pyrope or associated with phases formed by the breakdown of highest-pressure assemblages. However, some larger crystals included in pyrope seem to have been trapped during garnet growth and stably coexist with ellenbergerite, so that dumortierite formation may have spanned a large range of metamorphic conditions. In any event both types are chemically similar and characterized by unusual Mg (and Ti) contents (up to 8.5 wt % MgO). The structure refinement (Chopin et al., 1995) shows that the orthorhombic pseudohexagonal symmetry of dumortierite is retained (space group *Pmcn)* but that the octahedral single chain is free of Al and partly occupied by Ti and dominant Mg (just as in ellenbergerite), hence the introduction of the new species *magnesiodumortierite*. The remaining Mg substitutes Al in the face-sharing octahedra of the ellenbergerite-type double chain. The simplified formula-unit may then be written: $(Mg,Ti,\square)$ $(Al,Mg)_2 Al_4$ BSi_3 O_{18-x} $(OH)_x$ with x between 2 and 3.

In summary, all the octahedra that share faces and are normally occupied by Al in dumortierite tend to be occupied by Mg in magnesiodumortierite as they are in ellenbergerite. The question is to what extent this feature may be related to high pressures of formation. Admittedly, replacing high-charge cations by lower-charge ones reduces the repulsion across the shared faces and so may stabilize the structure at high pressures, which are expected to reduce the cation-cation distances (see Comodi and Zanazzi, 1993b, for a high-pressure refinement of the ellenbergerite structure). On the other hand, dumortierite with up to 3.4 wt % MgO has been reported as a relatively low-pressure secondary phase in high-grade cordierite-orthoamphibole rocks of the Bamble area, south Norway (Visser and Senior, 1991).

Lisetite and Related Phases. Lisetite, $CaNa_2Al_4Si_4O_{16}$, is a tectosilicate chemically midway between the composition of anorthite and the pure Na-equivalent of nepheline; structurally it is neither a feldspar nor a feldspathoid, but its framework is based upon the connection of four-membered rings of tetrahedra whose apical oxygens point *alternatively* up and down with respect to the (100) plane (cf. feldspars), with an ordered distribution of Si and Al (Rossi et al., 1986). It occurs in strongly retrograded jadeite-clinopyroxenite layers of the Liset eclogite pod, in the coesite-eclogite province of the Western Gneiss Region; it is part of a complex corona pattern formed in Al-rich Si-poor domains (originally represented by eclogite-facies kyanite) and essentially related to the breakdown of jadeite-rich pyroxene (Smith et al., 1986). Interestingly, the density of lisetite (2.73) is higher than that of the chemically equivalent 1:1 mixture of anorthite plus pure Na-nepheline (2.66); likewise the combination of lisetite + 4 quartz is denser than the isochemical counterpart andesine, $An_{33.33}$ (Smith et al., 1986). For this reason and on the basis of textural evidence, Smith et al. (1986) proposed that lisetite was one of the first phases formed upon incipient decompression and that its stability field would straddle the limit between the high P/T eclogite facies and the adjacent greenschist, amphibolite, and granulite facies. Yet quantification still awaits experimental work.

Phases forming the outer part of the lisetite-bearing corona are an amphibole (magnesio-alumino-taramite) and a unique K-poor Ca-rich tectosilicate that has the tetrahedral framework of nepheline, ca. $Ca_{11}Na_6Al_8Si_8O_{32}$, but with an unusual distribution of the large-size cations (Rossi et al., 1989). Although the change in chemistry from lisetite to this phase is small, there is a complete change in the structure, from one more compact than that of the feldspars (lisetite) to the relatively loose structure of the feldspathoids, which should be favored by lower pressure and/or higher temperature (Rossi et al., 1989).

Both tectosilicates have so far only been recorded from (retrograded) UHP rocks of south Norway, but their occurrence should be much more common on both chemical and physical grounds. The problem is that the two minerals are optically so similar to plagioclase – lisetite having the refringence of calcic plagioclase (1.56–1.57) and the birefringence of quartz (0.009), the nepheline-type phase having somewhat lower refringence and birefringence–that they can be very easily overlooked. As a matter of fact, a new related phase is just being investigated, that is chemically close to the pure Na-equivalent of nepheline and occurs as a breakdown product of jadeite in jadeitite veins of Burma (Kiénast, personal communication, 1993).

Nyböite and Other Funny Amphiboles. Amphiboles of unusual composition have been consistently reported in eclogites from the UHP terranes, be it southern Norway (Ungaretti et al., 1981; Smith, 1988), eastern China (Smith et al., 1990; Hirajima et al., 1992) or the Western Alps (Chopin et al., 1991; Hirajima and Compagnoni, 1993). They approach in composition the ideal nyböite/ferro-nyböite end-members, $NaNa_2(Mg,Fe)_3Al[Si_7Al]O_{22}(OH)_2$ as defined in Norway (Ungaretti et al., 1981), but span a whole range toward alumino-katophorite, $NaCaNa(Mg,Fe)_4Al[Si_7Al]O_{22}(OH)_2$, and alumino-taramite, $NaCaNa(Mg,Fe)_3Al_2[Si_6Al_2]O_{22}(OH)_2$, the latter also first recorded from Norway (Ungaretti et al., 1981).

On textural grounds, the nyböitic to katophoritic porphyroblasts or poikiloblasts probably grew slightly later than the eclogitic assemblage (which they often include); they also may coexist with albite + omphacitic pyroxene (Hirajima et al., 1992). Taramitic amphibole is clearly formed later than the others, either rimming them or as part of symplectite, and it may coexist with albite + aegirine-rich pyroxene.

There is a clear chemical control on the occurrence of these amphiboles, which are restricted to Na-rich Ca-poor rocks or microsystems, like metagranite (Hirajima and Compagnoni, 1993) or former jadeite-rich pyroxene (Smith et al., 1986). Indeed, in common, more Ca-rich eclogites of the same terranes, a pargasitic amphibole occupies the same textural position as nyböite or katophorite do in Ca-poor eclogites (e.g. Henry, 1990; Kienast et al., 1991), while near-end-member glaucophane occurs in Ca-free Mg-rich rocks, where it is close to its upper stability limit as shown by its coexistence with (or secondary formation from) talc + jadeite (cf. Chopin et al., 1991; Kienast et al., 1991). Yet, the restriction of nyböite occurrence to UHP terranes suggests that there are also physical controls on its stability. Experimental work in the system Na_2O-MgO-Al_2O_3-SiO_2-H_2O at high pressures (Pawley, 1992) shows the following. 1. Neither the glaucophane nor the nyböite end-member can be synthesized, but amphiboles with less than about 75 mol% nyböite can be. 2. "A likely continuous reaction relating glaucophane to nyböite is glaucophane + jadeite = nyböite + 3 quartz, involving a change in Al coordination from 6 to 4, with a corresponding entropy increase. The volume change is also positive; therefore, the slope of the reaction will be positive, such that the amphiboles become progressively more nyböitic with decreasing pressure and increasing temperature" (Pawley, 1992). 3. In this system, amphibole disappears below about 15-kbar pressure through continuous reactions to form albite, sodium phlogopite, and either nepheline or talc, depending on the bulk silica content. These results are consistent with the picture arising from the natural assemblages, with nyböite appearing as a lower-pressure and possibly higher-temperature substitute for glaucophane. Therefore, the occurrence of nyböite attests to temperature and pressure conditions that typically prevail during the early decompression of UHP rocks and so may be a precious indication of the former presence of such rocks (in the event that relics of the highest-pressure assemblages have disappeared).

On the other hand, taramitic amphiboles seem to be relatively high-temperature but definitely lower-pressure products, which are not restricted to UHP terranes; some occur, for instance, as reaction rim between glaucophane and almandine in the Gran Paradiso massif (Western Alps; Chopin, 1979) in which pressure is unlikely to have exceeded 15 kbar and temperature 500–550°C.

Conclusion

It is the authors' hope that these sections dealing with both common and uncommon minerals will provide the reader with a variety of tools for discovering new occurrences of UHP rocks or identifying relics or for retrogression products that can be safely attributed to them. Conversely, in such rocks once identified, attention should be paid to the potential occurrence of a few high-pressure phases so far known only from experiment.

Possible candidates are the fully hydrated analogue of topaz, $Al_2SiO_4(OH)_2$, a low-temperature and high-pressure equivalent of kyanite + H_2O, or at least OH-rich members of the series. These should occur at conditions near 50–60 kbar and 600–750°C (Fig. 3.5), that is, along cold geotherms of less than 5°C/ km, which can be expected at depth in long-lasting subduction zones (Wunder et al., 1993a). Likewise, hydroxyl-clinohumite is a high-pressure phase in contrast with the fluorine end-member (Yamamoto and Akimoto, 1977); OH-rich clinohumite indeed occurs in xenoliths and in UHP metamorphic rocks (Okay, 1994), yet it represents a sink for fluorine rather than H_2O at mantle depth (Engi and Lindsley, 1980). Another candidate is phase Pi, $Al_3Si_2O_7(OH)_3$, which is stable between about 400 and 700°C, 20 and 55 kbar (Wunder et al., 1993b), and so overlaps the low-pressure end of the stability field of OH-topaz (Fig. 3.5). It is therefore a low-temperature alternative to kyanite + coesite + H_2O. One more candidate, also produced by the high-pressure group in Bochum, is a phase with the pumpellyite structure in which sevenfold coordinated Ca is entirely replaced by Mg, that is, $Mg_4MgAl_5Si_6O_{21}(OH)_7$. Its stability field (Fig. 3.5) starts at about 37 kbar, 550°C, and expands toward higher pressures to both lower and higher temperatures (ca. 775°C), at which it breaks down to pyrope, kyanite, coesite, and H_2O (Schreyer et al., 1991). Again, if one ignores possible stabilization to lower pressures by Ca incorporation, "cold" geotherms of about 5°C/km are required for this phase to occur at depth. Since such conditions are essentially transient and will inevitably be followed by a period of thermal relaxation, all these low-temperature phases have virtually no chance to survive the temperature increase on a way back to the surface through erosional and tectonic processes. One possibility is yet offered by inclusion and armoring in high-mechanical-strength containers like garnet, zircon, or diamond, as discussed in this chapter. An alternative is immediate transport to the surface as xenoliths in magmas generated at still greater depth.

Acknowledgments

We thank Robert G. Coleman for his help and considerable patience in editing this manuscript.

References

Ballèvre, M., Pinardon, J.-L., and Kienast, J. R. 1989. Reversal of Fe-Mg partitioning between garnet and staurolite in eclogite-facies metapelites from the Champtoceaux Nappe (Brittany, France). *Journal of Petrology, 30,* 1321–1349.

Becker, H. 1993. Evidence from ultra-high-pressure marbles for recycling of sediments into the mantle. *Nature, 358,* 745–748.

Beller, U. 1987. Hochdruck-Synthese von phosphorhaltigem Ellenbergerit. Diploma thesis, Ruhr-Univ. Bochum, Inst. Mineralogie.

Biino, G., and Compagnoni, R. 1992. Very high pressure metamorphism of the Brossasco coronite metagranite, southern Dora-Maira massif, western Alps. *Schweiz. min. und petr. Mitt., 72,* 347–362.

Biryukov, V. M., and Kosygin, Y. A. 1989. On find of accessory diamonds in drusite-eclogites of some banded complexes of the Pre-Baikalian region. *Dokl. Akad. Nauk. SSSR 306,* 1204–1208.

Boyd, S. R., Pilinger, C. T., Milledge, H. J., and Seal, M. J. 1992. C and N isotopic composition and the infrared absorption spectra of coated diamonds: Evidence for the regional uniformity of CO_2-H_2O rich fluids in lithospheric mantle. *Earth Planet. Sci. Lett., 108,* 139–150.

Brey, G., Dorshev, A. M., and Kogarko, L. N. 1991. *The join pyrope-knorringite: Experimental constraints for a new geothermo-barometer for coexisting garnet and spinel.* International Kimberlite conference, Araxa, Brazil.

Brown, M. 1993. P-T-t evolution of orogenic belts and the causes of regional metamorphism. *Journal of the Geological Society of London, 150b,* 227–241.

Brunet, F. 1992. Synthèse et stabilité de la bearthite $CaAl(PO_4)_2OH$. Diploma thesis, Université Paris 6.

Bundy, F. R. 1980. The P, T phase reaction diagrams for elemental carbon. *J. Geophys. Res., 85,* 6930–6936.

Chopin, C. 1979. De la Vanoise au massif du Grand Paradis, une étude pétrologique et radiochronologique, Ph.D., Université Paris 6.

Chopin, C. 1984. Coesite and pure pyrope in high-grade blueschists of the western Alps: A first record and some consequences. *Contrib. Mineral. Petrol., 86,* 107–118.

Chopin, C. 1985. Les relations de phases dans les métapélites de haute pression (Thèse d'Etat Univ. Paris et M. Curie Paris). *Memoires Sciences de la Terre 85–11,* 84 p.

Chopin, C. 1986. Phase relationships of ellenbergerite, a new high-pressure Mg-Al-Ti-silicate in pyrope-coesite-quartzite from the Western Alps. *Geol. Soc. Amer. Mem., 164,* 31–42.

Chopin, C. 1991. Eclogite-facies mineral parageneses in pelitic rock compositions. *Eclogite facies rocks.* Glasgow, Blackie, 35–42.

Chopin, C., Brunet, F., Gebert, W., Medenbach, O., and Tillmanns, E. 1993. Bearthite, $Ca_2Al(PO_4)_2OH$, a new mineral from high pressure terranes of the western Alps. *Schweiz. min. und petr. Mitt., 73,* 1–9.

Chopin, C., Ferraris, G., Ivaldi, G., Compagnoni, C., Davidson, C., and Davis, A. 1994. Magnesiodumortierite: Crystal chemistry and structure of a new mineral from very-high pressure rocks, Western Alps. *Eur. J. Mineral.* (in press).

Chopin, C., Henry, C., and Michard, A. 1991. Geology and petrology of the coesite-bearing terrain, Dora-Maira massif, Western Alps. *Eur. J. Mineral., 3,* 263–291.

Chopin, C., Klaska, R., Medenbach, O., and Dron, D. 1986. Ellenbergerite, a new high-pressure Mg-Al-(Ti, Zr)- silicate with a novel structure based on face-sharing octahedra. *Contrib. Mineral. Petrol., 92,* 316–321.

Chopin, C., and Langer, K. 1988. Fe^{2+}-Ti^{4+} charge transfer between face-sharing octahedra: polarized absorption spectra and crystal chemistry of ellenbergerite. *Bull. Mineralogie, 111,* 17–27.

Chopin, C., and Schreyer, W. 1983. Magnesiocarpholite and magnesiochloritoid; two index minerals of pelitic blueschists and their preliminary phase relations in the model system $MgO-Al_2O_3-SiO_2-H_2O$. *Am. J. Sci., 283-A*, 72–96.

Chopin, C., Schreyer, W., and Baller, T. 1992. Ellenbergerite stability, a reappraisal. *Terra Nova Abst. Suppl., 4*, 10.

Claoue-Long, J. C., Sobolev, N. V., Shatsky, V. S., and Sobolev, A. V. 1991. Zircon response to diamond-pressure metamorphism in the Kokchetav massit, USSR. *Geology, 19*, 710–713.

Comodi, P., and Zanazzi, P. F. 1993a. Structural study of ellenbergerite. Part I: effects of high temperatures. *Eur. J. Mineral., 5*, 819–830.

Comodi, P., and Zanazzi, P. F. 1933b. Structural study of ellenbergerite. Part II: effects of high pressure. *Eur. J. Mineral., 5*, 831–838.

Davidson, C. F. 1967. The so-called "cognate xenoliths" of kimberlite. *Ultramafic and related rocks*. New York, John Wiley & Sons.

Deines, P., Harris, J. W., and Gurney, J. J. 1993. Depth related carbon isotope and nitrogen concentration variability in the mantle below the Orapa kimberlite, Botswana, Africa. *Geochim. Cosmochim. Acta, 57*, 2781–2796.

Dergachev, D. V. 1986. Diamonds in metamorphic rocks. *Dokl. Akad. Nauk. SSSR 291*, 189–191.

Dobretsov, N. L. 1964. The jadeite rocks as indicators of high pressure in the Earth's crust. *22nd International geological Congress (India), 16*, 177–196.

Dobrzhinetskaya, L. F., Posukhova, T., Tronnes, R., Korneliussen, A., and Sturt, B. 1993. A microdiamond from eclogite-gneiss area of Norway. *Terra Abst., 5*, 8.

Doroshev, A. M., Sobolev, N. V., and Brey, G. 1992. Experimental evidence of high-pressure origin of the potassium-bearing pyroxenes. *29th Inter. Geol. Congress, 2*, 602.

Dutrow, B. L., Holdaway, M. J., and Hinton, R. W. 1986. Lithium in staurolite: its petrologic significance. *Contrib. Mineral. Petrol., 94*, 496–506.

Enami, M., and Zang, Q. 1988. Magnesium staurolite in garnet-corundum rocks and eclogite from the Donghai district, Jiangsu Province, east China. *Am. Mineralogist, 73*, 48–56.

Engi, M., and Lindsley, D. H. 1980. Stability of titanian clinohumite: Experiments and thermodynamic analysis. *Contrib. Mineral. Petrol., 72*, 415–424.

Erlank, A. J., and Kushiro, I. 1970. Potassium contents of synthetic pyroxene at high temperatures and pressures. *Carnegie Inst. Washington Yearbook, B68*, 439–442.

Essenov, S. E., Yefimov, I. A., Shlygin, E. D., Abdulkabirova, M. A., Vedernikov, N. N., and Nurlybaev, A. N. 1968. On the problem of diamond prospecting in Northern Kazakhstan. *Vestnik. Akad. Nauk. Kazakh. SSR 24*, 37–45 (In Russian).

Fiala, J. 1965. Pyrope of some garnet localities of the Czech massif. *Kristallinikum, 3*.

Fiala, J., and Padera, K. 1977. The chemistry of the minerals of the pyrope dunite from borehole T7 near Stare (Bohemia). *Tscherm. Min. Petr. Mitt., B24*.

Fockenberg, T. 1992. Neue experimentelle untersuchungen zur Stabiliät von Pyrop und Chemismus von Mg-Staurolith im System $MgO-Al_2O_3-SiO_2$ (MASH). *Mitt. der Deutschen Miner. Gesellschaft 4, 1*, 82 (abs).

Galimov, E. M., Kaminosky, F. V., and Kodina, L. A. 1985. New data on isotopic composition of carbon of carbonado. *Geokhimija, 5*, 723–726.

Galimov, E. M., Sobolev, N. V., Polkanov, Y. A., Maltsev, K., and Yefimova, E. S.

1994a. Isotopic composition of carbon of microdiamonds from Ukranian sands. *Geokhimiya* (in press).

Galimov, E. M., Sobolev, N. V., Shatsky, V. S., Zorin, Y. M., Zayachkovsky, A., Maltserv, K. A., and Yefimova, E. S. 1994b. Isotopic composition of microdiamonds from Kokchetav metamorphic rocks and Tertiary sands. *Geokhimiya* (in Russian) (in press).

Garanin, V. K., Guseva, E. V., Dergachev, D. V., Kudrjavtseva, G. P., and Orlov, R. Y. 1988. Diamond crystals in garnets from slightly gneissic granites. *Dokl. Akad. Nauk. SSSR 298*, 190–194.

Gil Ibarguchi, J. I., Mendia, M., and Girardeau, J. 1991. Mg- and Cr-rich staurolite and Cr-rich kyanite in high pressure ultrabasic rocks (Cabo Ortegal, northwestern Spain). *Am. Mineralogist, 76*, 501–511.

Gillet, P., Ingrin, J., and Chopin, C. 1984. Coesite in subducted continental crust: *P-T* history deduced from an elastic model. *Earth Planet. Sci. Lett., 70*, 426–436.

Grew, E. S., and Sandiford, M. 1984. A staurolite-talc assemblage in tourmaline-phlogophite-chlorite schist from northern Victoria Land, Antarctica, and its petrogenetic significance. *Contrib. Mineral. Petrol., 87*, 337–350.

Harlow, G. E. 1992. Potassium in clinopyroxene at high pressure. *GSA Abstract with Program, 24*, 129.

Harlow, G. E., and Veblen, D. R. 1991. Potassium in clinopyroxene inclusion from diamonds. *Science, 251*, 652–655.

Harris, J. W. 1968. The recognition of diamond inclusions P.1 Syngenetic mineral inclusions. *Industr. Diam. Rev., 334*, 402–410.

Hawthorne, F. C., Ungaretti, L., Oberti, R., Cauca, F., and Callegari, A. 1993. The crystal chemistry of staurolite. I: Crystal structure and site occupancies. *Can. Mineral., 31*, 551–582.

Hellman, P. L., and Green, T. H. 1993. The high pressure experimental crystallisation of staurolite in hydrous mafic compositions. *Contrib. Mineral. Petrol., 68*, 369–372.

Henry, C. L'unité à coesite du massif Dora-Maira dans son cadre pétrologique et structural (Alpes occidentales, Italie) Ph.D., Université de Paris VI.

Hirajima, T., and Compagnoni, R. 1993. Petrology of a ferro-nyböite-bearing jadeite-quartz/coesite-almandine-phengite fels from the Brossasco-Issaca unit, southern Dora-Maira massif, western Alps, Italy. *Eur. J. Mineral., 5*, 943–955.

Hirajima, T., Zhang, R., Li, J., and Cong, B. 1992. Petrology of the nyböite-bearing eclogite in the Donghai area, Jiansu Province, eastern China. *Mineral. Mag., 56*, 37–46.

Holdaway, M. J., and Goodge, J. W. 1990. Rock pressure vs. fluid pressure as a controlling influence on mineral stability: An example from New Mexico. *Am. Mineralogist, 75*, 1043–1058.

Holland, T. J. B. 1980. The reaction albite = jadeite + quartz determined experimentally in the range 600–1200 °C. *Am. Mineralogist, 65*, 125–134.

Irifune, T., Ohtani, E., and Kumazawa, M. 1982. Stability field of Knorringite $Mg_3Cr_2Si_3O_{12}$ at high pressure and its implication to the occurrence of Cr-rich pyrope in the upper mantle. *Phys. Earth Planet. Int., 27*, 263–272.

Jaques, A. L., Hall, A. E., Sheraton, J., Smith, C. B., Sun, S-S., Drew, R. M., Foudoulis, C., and Ellingsen, K. 1989. Composition of crystalline inclusions and C-isotope composition of Argyle, and Ellendale diamonds. *Geol. Soc. America Special Pub., 14*, 966–989.

Kashkarov, I. F., and Polkanov, Y. A. 1964. On diamond find in titanite-zircon sands. *Dokl. Akad. Nauk. SSSR 157*, 1129–1130 (In Russian).

Kashkarov, I. F., and Polkanov, Y. A. 1972. Some specific features of diamonds from titaniferous placers of Northern Kazakhstan. *Novye dannye o monerlakh, 21*, 183–185 (In Russian).

Kienast, J. R., Lombardo, B., Biino, G., and Pinardon, J. L. 1991. Petrology of very-high-pressure eclogitic rocks from the Brossasco-Isasca Complex, Dora-Maira Massif, Italian Western Alps. *J. Metamorp. Geol., 9*, 19–34.

Kinny, P. D. and Meyer, H. O. A. 1993. Late Proterozoic diamonds; evidence from zircon in diamond. *Absts. with Prog. Geol. Soc. America, 25, 6*, 322.

Korikovskiy, S. P., Talitskiy, V. G., Boronikhin, V. A., and Ivanov, V. P. 1983. Paraenezis talk-khloritoid v metapelitakh i yego petrologicheskoye znacheniye (na primere Makbalskogo antiklinoriya Tyan-Shania). *Doklady Akad. Nauk. SSSR, 268*, 1453–1457.

Krogh, T. E., Mysen, B. O., and Davis, G. L. 1974. A Paleozoic age for the primary minerals of a Norwegian eclogite. *Carnegie Institute Washington Yearbook, 73*, 575–576.

Kvashitsa, V. N. 1985. *Small diamonds.* Kiev, Naukova Dumka.

Lee, D. C., Boyd, S. R., Griffin, B. J., Griffin, B. W., and Reddicliffe, T. 1991. Coanjula diamonds, Northern Territory, Australia. In Proceedings of the International Conference, 5, 231–233.

Lee, D. C., Boyd, S. R., Griffin, B. J., Griffin, W. I., and Reddicliffe, T. 1993. Coanjula diamonds, Northern Territory, Australia. *Proc. 5th Kimberliote Conf., Aruxu, Brazil.*

Luth, R. W. 1992. Potassium in clinopyroxene at high pressure: Experimental constraints. *EOS Trans. Am. Geophys. Union*, 608.

Meyer, H. O. A. 1968. Chrome pyrope: An inclusion in natural diamond. *Science, 160*, 1446–1447.

Meyer, H. O. A. 1987. Inclusions in diamonds. *Mantle xenoliths.* New York, John Wiley and Sons, 501–522.

Meyer, H. O. A., and Svisero, D. C. 1975. Mineral inclusions in Brazilian diamonds. *Phys. Chem. of the Earth 9*, 785–795.

Mirwald, P. W., and Massonne, H.-J. 1980. The low-high quartz and quartz–coesite transition to 40 kbar between 600° and 1600°C and some reconnaissance data on the effect of $NaAlO_2$ component on the low quartz-coesite transition. *J. Geophys. Res., 85*, 6983–6990.

Moore, P. B., and Araki, T. 1987. Dumortierite, $Si_3B[Al_{6.75}, \square_{0.25} O_{17.25}(OH)_{0.75}]$: a detailed structure analysis. *Neues Jahrbuch für Miner. Abhand., 132*, 231–241.

Moore, P. B., Irving, A. J., and Kampf, A. R. 1975. Foggite, $CaAl(OH)_2(H_2O)[PO_4]$; goedkenite,$(Sr,Ca)_2Al(OH) [PO_4]_2$; and samuelsonite, $(Ca,Ba)Fe_2^{2+}Mn_2^{2+}Ca_8$-$Al_2(OH)_2 [PO_4]_{10}$; three new species from the Palermo No. 1 Pegmatite, North Groton, New Hampshire. *Am. Mineralogist, 60*, 957–964.

Nicollet, C. 1986. Saphirine et staurotide riche en magnésium et chrome dansles amphibolites et anorthosites á corindon du Vohbory Sud, Madagascar. *Bull. de Minéralogie 109*, 599–612.

Okay, A. I. 1993. Petrology of a diamond and coesite-bearing terrain: Dabie Shan, China. *Eur. Jour. Mineral., 5*, 659–675.

Okay, A. I. 1994. Sapphirine and Ti-clinohumite in ultra-high-pressure garnet-pyroxenite and eclogite from Dabie Shan, China. *Contrib. Mineral. Petrol. 116*, 145–155.

Pawley, A. R. 1992. Experimental study of the compositions and stabilities of synthetic nyböite and nyböite-glaucophane amphiboles. *Eur. J. Mineral. 4*, 171–192.

Pinardon, J. L. 1990. La staurotide dans les metapélites de haute pression. Relations de phases et approche thermodynamique. Ph.D., Université Paris 6 (153 p.).

Polkanov, Y. A. 1967. Diamonds of cubic habit from Tertiary placers of Pre-Dneper region. *Dokl. Akad. Nauk. SSSR, 173*, 901–902.

Reinecke, T. 1991. Very-high-pressure metamorphism and uplift of coesite-bearing metasediments from the Zermatt-Saas zone, Western Alps. *Eur. J. Mineral., 3*, 7–17.

Ringwood, A. E. 1977. Synthesis of the pyrope-knorringite solid solution series. *Earth Planet. Sci. Lett., 36*, 443–448.

Rossi, G., Oberti, R., and Smith, D. C. 1989. The crystal structure of a K-poor Ca-rich silicate with the nepheline framework and crystal-chemical relationships in the compositional space $(K,Na,Ca)_8 (Al,Si)_{16}O_{32}$. *Eur. J. Mineral., 1*, 59–70.

Rossi, G., Oberti, R., and Smith, D. C. 1986. Crystal structure of lisetite, $CaNa_2Al_4$-Si_4O_{16}. *Am. Mineralogist, 71*, 1378–1383.

Ryabchikov, I. D., and Ganeev, I. I. 1990. Isomorphic substitution of potassium in clinopyroxene at high pressures. *Geokhimija, 1*, 3–12.

Schertl, H.-P., Schreyer, W., and Chopin, C. 1991. The pyrope-coesite rocks and their country rocks at Parigi, Dora-Maira Massif, Western Alps: detailed petrography, mineral chemistry and *PT*-path. *Contrib. Mineral. Petrol., 108*, 1–21.

Schreyer, W. 1988. Experimental studies on metamorphism of crustal rocks under mantle pressures. *Mineral. Mag., 52*, 1–26.

Schreyer, W., Horrocks, P. C., and Abraham, K. 1984. High-magnesium staurolite in a sapphirine-garnet rock from the Limpopo belt, southern Africa. *Contrib. Mineral. Petrol., 86*, 200–207.

Schreyer, W., Maresh, W. W., and Baller, T. 1991. A new hydrous, high-pressure phase with a pumpellyite structure in the system $MgO-Al_2O_3-SiO_2-H_2O$. *Progress in metamorphic and magmatic petrology*. Cambridge, Cambridge University Press, 47–64.

Schreyer, W., and Seifert, F. 1969. High-pressure phases in the system $MgO-Al_2O_3$-SiO_2-H_2O. *American Jour. Sci., 267-A*, 407–443.

Shatsky, V. S., Jagoutz, E., Kozmenko, O. A., and Sobolev, N. V. 1993. The geochemistry of ultrahigh pressure rocks from Kokchetav massif. *Geochem. Cosmochim. Acta*, in review.

Shatsky, V. S., Sobolev, N. V., and Efimova, E. S. 1989. *Morphological features of accessory microdiamonds from metamorphic rocks of the Earth's crust*. International Geologic Congress, Washington, D. C., USA.

Shatsky, V. S., Sobolev, N. V., and Stenina, N. G. 1985. Structural pecularities of pyroxenes from eclogites. *Terra Cognita, 5*, 436–437.

Shimizu, N. 1971. Potassium contents of synthetic cliopyroxenes at high pressures and temperatures. *Earth Planet. Sci. Lett., 11*, 374–380.

Smith, D. C. 1984. Remarques cristallochimiques et pétrogénétiques sur des minéraux inhabituels dans les eclogites de Liset et de Rekvika, Norvége: Dixiéme réunion annuelle des sciences de la terre, Bordeaux 1984. *Société géologique de France ed (Abs.)*

Smith, D. C. 1988. A review of the peculiar mineralogy of the "Norwegian-eclogite province," with crystal-chemical, petrological, geochemical and geodynamical notes and an extensive bibliography. *Eclogites and eclogite-facies rocks*. Amsterdam, Elsevier, 1–206.

Smith, D. C., Kechid, S.-A., and Rossi, G. 1986. Occurrence and properties of lisetite, $CaNa_2Al_4Si_4O_{16}$, a new tectosilicate in the system Ca-Na-Al-Si-O. *Am. Mineralogist, 71*, 1372–1377.

Smith, D. C., Yang, J., Oberti, R., and Previda-Massara, E. 1990. A new locality of nyböite and taramite, the Jianchang eclogite pod in the "Chinese Su-LU coesite eclogite province" compared with nyböite- and taramite-bearing Liset eclogite pod in the "Norwegian coesite-eclogite province." *International Mineralogical Association meeting in Beijing Abst.*, 889–890.

Smith, J. V., and Dawson, J. B. 1985. Carbonados; diamond aggregates from early impacts of crustal rocks. *Geology, 13*, 342–343.

Smyth, J. R. 1977. Quartz pseudomorphs after coesite. *Am. Mineralogist, 62*, 828–830.

Sobolev, N. V. 1964. *Paragenetic types of garnets.* Moscow, Nauka.

Sobolev, N. V. 1971. On mineralogical indicators of diamond potential of kimberlites. *Geologija i Geofizika 3* (in Russian).

Sobolev, N. V. 1974. *Deep seated inclusions in kimberlites and a problem of composition of the upper mantle.* Novosibirsk, Nauka.

Sobolev, N. V. 1976. Coesite, garnet and omphacite inclusions in Yakut diamonds – first finding of coesite paragenesis. *Dokl. Akad. Nauk. SSSR, 230*, 1442–1444 (in Russian).

Sobolev, N. V., Yefimova, V., Koptil, V. I., Lavrentiev, Y. G., and Savolev, V. S., Bakumento, I. T., Efimova, E. S., and Pokhilenko, N. P. 1991. Morphological features of microdiamond sodium in garnets and potassium in clinopyroxenes contents, of two eclogite xenoliths of the Udachnaja kimberlite pipe (Yakutia). *Dokl. Akad. Nauk. SSSR , 31*, 585–591 (In Russian).

Sobolev, N. V., Efimova, E. S., Lavrent'ev, Y. G., and Sobolev, V. 1984. Dominant calcsilicate association of crystalline inclusions in diamonds from alluvial deposits of South Eastern Australia. *Dokl. Akad. Nauk SSSR, 274*, 13–135.

Sobolev, N. V., Galimov, E. M., Ivanovskaya, I. N., and Yefimova, E. S. 1979. Isotopic composition of carbon of diamonds containing cyrstalline inclusions. *Dakl Akad. Nauk. SSR, 249*, 1217–1220 (in Russian).

Sobolev, N. V., Galimov, E. M., Ivanovskaya, I. N., Yefimova, E. S., Maltsev, K. A., Hall, A. E., and Usova, L. V. 1989. A comparative study of the morphology, inclusions, and carbon isotopic composition of diamonds from alluvials of the King George River and Argyle lamproite mine (Western Australia) and of cube microdiamonds from Northern Australia. *Soviet Geology and Geophysics, 30*, 1–19.

Sobolev, N. V., Kharkiv, A. D., and Pokhilenko, N. P. 1986. Kimberlites, lamproites and the problem of composition of the upper mantle. *Geologija i Geofizika, 7*, 18–27.

Sobolev, N. V., Lavrent'ev, Y. G., and Pospelova, L. N. 1971. *The peculiarities of minor element content in minerals of xenoliths from kimberlitic pipes as a criteria of depth of their origin.* International Geochem. Congress, Moscow.

Sobolev, N. V., Lavrent'ev, Y. G., Pospelova, L. N., and Sobolev, E. V. 1969. Chrome pyropes from Yakutian diamonds. *Dakl. Akad. SSSR, 189*, 162–165 (in Russian).

Sobolev, N. V., and Shatsky, V. S. 1987. Carbon mineral inclusions in garnets of metamorphic rocks. *Geologiya i Geofizika, 7*, 77–80 (In Russian).

Sobolev, N. V., and Shatsky, V. S. 1990. Diamond inclusions in garnets from metamorphic rocks. *Nature 343*, 742–746.

Sobolev, N. V., and Shatsky, V. S. 1989. Mineralogical indicators of the ultrahigh-pressure and high-temperatures of metamorphic records. *Eclogites and glaucophane schists in folded zones.* Novosibirsk, Nauka, 184–193.

Sobolev, V. S. 1951. *Geology of diamond deposits of Africa, Australia, North America and Borneo Island.* Moscow, Gosgeoltekhizdat.

Sobolev, V. S. 1979. New danger of disinformation in results of contamination of the samples with foreign materials and industrial products. *Zapiski Vses. Mineral Obsch, 108,* 691–695.

Sobolev, V. S., Nai, B. S., Sobolev, N. V., Lavrent'ev, Y. G., and Pospelova, L. N. 1969. Xenoliths of diamond-bearing pyrope serpentinites from the Aikhal pipe, Yakutia. *Dokl. Akad. Nauk. SSSR, 188,* 1141–1143.

Sobolev, V. S., Sobolev, N. V., and Lavrent'ev, Y. G. 1972. Inclusions in diamond from diamondiferous eclogite. *Dokl. Akad. Nauk. SSSR, 207,* 164–167. (in Russian).

Sunagawa, I. 1984. Morphology of natural and synthetic diamond crystals. *Material Science of the Earth's Interior,* Tokyo, 303–331.

Trueb, L. F., and Butterman, W. C. 1968. Carbonado: a microstructural study. *Am. Mineralogist, 54,* 412–425.

Trueb, L. F., and De Wys, E. C. 1971. Carbon from Ubangi – a microstructural study. *Amer. Mineralogist, 56,* 1252–1268.

Trueb, L. F., and De Wys, E. C. 1969. Carbonado: natural poly-crystalline diamond. *Science, 165,* 799–802.

Ungaretti, L., Smith, D. C., and Rossi, G. 1981. Crystal-chemistry by x-ray structure refinement and electron microprobe analysis of a series of sodic-calcic to alkali-amphiboles from the Nybö eclogite pod, Norway. *Bull. de Minéralogie, 104,* 400–412.

Varshavsky, A. V., and Bulanova, G. P. 1974. Microcrystals of natural diamond. *Dokl. Akad. Nauk. SSSR, 217,* 1069–1072.

Visser, D., and Senior, A. 1991. Mg-rich dumortierite in orthoamphibole-bearing rocks from the high grade Bamble Sector, south Norway. *Mineral. Mag., 55,* 563–577.

Wang, Q., Ishiwatari, A., Zhao, Z., T., H., Enami, M., Zhai, M., Li, J., and Cong, B. 1993. Coesite-bearing granulite retrograded from eclogite in Weihai, Eastern China. *Eur. J. Mineral., 5,* 141–151.

Ward, C. M. 1984. Magnesium staurolite and green chromian staurolite from Fiorland, New Zealand. *Am. Mineralogist, 69,* 531–540.

Wendt, A. S., D'Arco, P., Goffé, B., and Oberhansli, R. 1993. Radial cracks around α-quartz inclusions in almandine: constraints on the metamorphic history of the Oman mountains. *Earth Planet. Sci. Lett., 114,* 449–461.

Werding, G., and Schreyer, W. 1990. Synthetic dumortierite: Its PTX-dependent compositional variations in the system Al_2O_3-B_2O_3-SiO_2-H_2O. *Contrib. Mineral. Petrol., 105,* 11–24.

Wunder, B., Medenbach, O., Krause, W., and Schreyer, W. 1993. Synthesis, properties and stability of $Al_3Si_2O_7(OH)_3$ (phase Pi), a hydrous high-pressure phase in the system Al_2O_3-SiO_2-H_2O (ASH). *Eur. J. Mineral., 5,* 637–649.

Wunder, B., Rubie, D. C., Ross, C.R.I., Medenbach, O., Seifert, F., and Schreyer, W. 1993. Synthesis, stability, and properties of $Al_2SiO_4(OH)_2$: A fully hydrated analogue of topaz. *Am. Mineralogist, 78,* 285–297.

Xu, S., Okay, A. I., Ji, S., Sengor, A.M.C., Su, W., Liu, Y., and Jiang, L. 1992. Diamond from the Dabie Shan metamorphic rocks and its implication for tectonic setting. *Science, 256,* 80–82.

Yamamoto, K., and Akimoto, S. 1977. The system MgO-H_2O-SiO_2 at high pressures and temperatures: Stability field for hydroxyl-chondrodite; hydroxyl-clinohumite and 10 Å phase. *Am. J. Sci., 277,* 288–312.

4

Structures in UHPM Rocks: A Case Study from the Alps

ANDRÉ MICHARD, CAROLINE HENRY, AND
CHRISTIAN CHOPIN

Abstract

Structures in UHP rocks may provide some insight into the rheology and deformation regime of continental rocks subducted at mantle depth, as well as on their exhumation tectonics. We focus on the Dora-Maira case study (Western Alps), with references to a few other areas of UHP or eventually HP-eclogite facies metamorphism. In southern Dora-Maira, UHP Eoalpine rocks are found as a coherent, 1000-m-thick lensoid unit within a pile of HP eclogite and blueschist-facies slices. The whole pile was affected by a common, Mesoalpine greenschist-facies overprint. From the local preservation of Pre-Alpine textures, and from the various UHP and HP relic structures (either irrotational or rotational), we infer a strong deformation partitioning of ductile flow in the subducting continental crust down to a 100-km depth. The blueschist, then greenschist-facies overprint developed during westward thrusting of the UHP-HP eclogitic slices onto more external zones, then during the collapse of the thickened orogenic wedge. The early exhumation of the UHP rocks up to the 30-km depth is controversial. If a continuous convergence between Apulia and Europa is accepted, then a forced-flow, or extrusion tectonics of imbricate slices in the subduction wedge would be hypothesized. If not, an extensional, metamorphic core complex stage could be invoked during the Eoalpine-Mesoalpine interval.

Introduction

While a number of occurrences of ultrahigh pressure metamorphic (UHPM) rocks have been described from various Phanerozoic belts in the last decade, detailed structural studies dedicated to these rocks are still lacking. Two exceptions are the south Norwegian Caledonides (Western Gneiss Region), and the Dora-Maira massif in the Western Alps. During their long trip back to the surface, UHPM rocks are dramatically overprinted by deformation at lower *P-T* conditions. However, one can reasonably benefit from structural analysis of UHP metamorphic rocks by (1) unraveling relic structures formed under UHP conditions (UHP structures in the following), to gain insight into the rheology and deformation regime in crustal rocks buried at asthenospheric depths in a subduction zone; and (2) deciphering the superimposed structures formed under lower-grade conditions in the UHP rocks, and their structural relationships with the juxtaposed, lower-grade rock units, to constrain models for exhumation (in the sense of England and Molnar, 1990) of UHPM rocks.

In this chapter we report mainly on our structural studies in the southern Dora-Maira massif. After coesite-bearing rocks were discovered there (Chopin, 1984), a number of papers were dedicated to the area, mostly with emphasis on mineralogy-petrology or isotopic geochemistry, but also including detailed mapping (Henry, 1990; Compagnoni et al., Chapter 7, this volume) and structural analysis (Henry, 1990; Philippot, 1990; Henry et al., 1993). The area, which is part of the most studied mountain belt in the world, the Alps, might be regarded as a case study. References will be also made in the following to other UHP-HP areas, either from the Alps or from other countries (Norway, Oman).

Geologic Setting of an UHP Terrain: The Dora-Maira Case Study

Figure 4.1A indicates the location of the Penninic Dora-Maira massif, in the western part of the Europe–Africa (Apulia) collisional zone. The massif consists of crystalline nappes with rare continental metasediments, exposed beneath ophiolites and associated metasediments (Upper Penninic "Schistes lustrés" nappes), which are in turn overlain by the Austroalpine crystalline nappes. Geologic observations and seismic reflection experiments (Nicolas et al., 1990) demonstrate that these nappes are extensively thrust westward onto the subducting European lithosphere (Fig. 4.1B). In most palinspastic restorations, the Tethyan suture results from a SE-dipping subduction zone located between the Penninic and Austroalpine domains, and the Dora-Maira massif is ascribed to the distal part of the European Mesozoic margin, east of the Middle Penninic Briançonnais domain (Lemoine et al., 1986; Coward and Dietrich, 1989; Stampfli and Marthaler, 1990). Alternative restorations were proposed, in particular models with two subduction zones, either on each side of a Dora-Maira microblock (Hunziker et al., 1989; Philippot, 1990), or both to the SE of Dora-Maira (Avigad et al., 1993).

In relation with the Europa–Apulia convergence, HP-LT metamorphism affected most of the Penninic zone and part of the juxtaposed Austroalpine

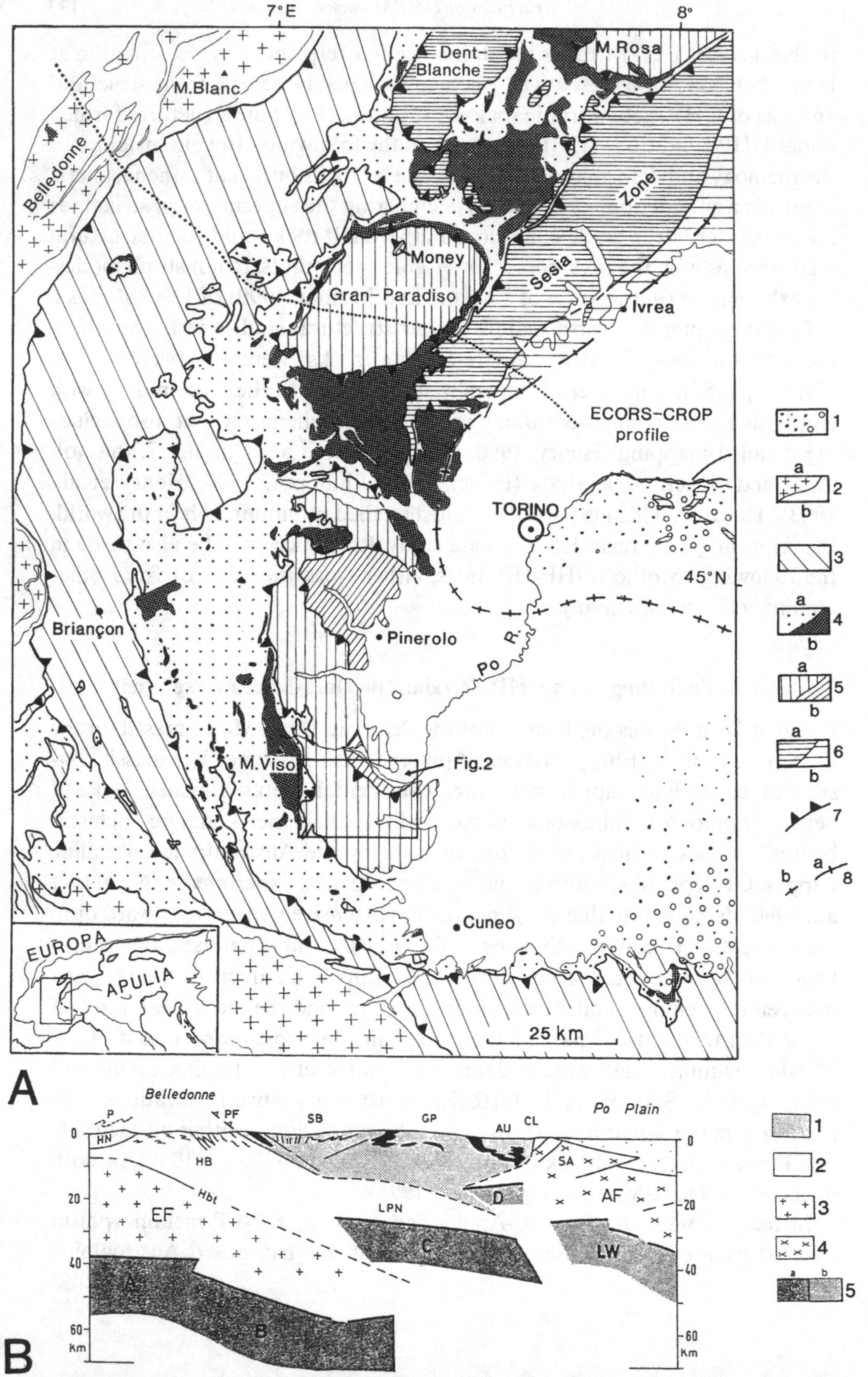

7°E
8°
M.Blanc
Dent-Blanche
M.Rosa
Belledonne
Zone
Money
Sesia
Gran-Paradiso
Ivrea
ECORS-CROP profile
TORINO
45°N
Briançon
Pinerolo
Po R.
M.Viso
Fig.2
Cuneo
EUROPA
APULIA
25 km
A
1
a
2
b
3
a
4
b
a
5
b
a
6
b
7
b
a
8
Belledonne
P
PF
SB
GP
AU
CL
Po Plain
HN
HB
SA
Hbt
EF
D
AF
LPN
C
LW
A
B
0
20
40
60
Km
B
1
2
3
4
a
b
5

terrains. Eclogitic assemblages are widespread in the internal Penninic and some of the adjacent Austroalpine nappes (Dal Piaz and Lombardo, 1986; Goffé and Chopin, 1986). UHP assemblages have been identified up to now only from three areas of the suture zone, (1) the southern Dora-Maira massif (here considered); (2) the Zermatt ophiolitic nappe (Reinecke, 1991); and (3) the Hohe Tauern Penninic window in Eastern Alps (Holland, 1979). At most places in the internal Alps, the early, high P/T metamorphism is strongly over-printed by a low-pressure, low- to medium-grade greenschist-facies metamor-phism. The HP-UHP mineral assemblages and structures are therefore mostly preserved as relics within mainly low-P rocks. The greenschist-facies event is fairly well dated at 40–35 Ma, wheras the ages of the UHP-HP assemblages are scattered between 130 Ma and 35 Ma (Hunziker et al., 1989; Monié and Philippot, 1989; Monié and Chopin, 1991; Tilton et al., 1991). The latest ^{39}Ar/^{40}Ar and Sm-Nd studies on the Monviso and Zermatt metagabbros yielded ages in the range 60–70 Ma (Barnicoat et al., 1993). The underlying continen-tal massifs such as the Dora-Maira would have undergone their peak meta-morphism at the same time (Bouffette, 1993). The Cretaceous ages are usually referred to as "Eoalpine," and would correspond at least partly to a precolli-sional phase (Hunziker et al., 1989). The Eocene ages correspond to the main Alpine collisional phase, or "Mesoalpine" phase.

The Dora-Maira massif forms an elliptic structural dome, outlined by the dip of the regional (mainly greenschist-facies) foliation, and by the distribu-tion of rock units (Figs. 4.1 and 4.2). The eastern flank of the dome is partly

Figure 4.1. A: Geologic setting of the Dora-Maira coesite-bearing area (framed) in the Western Alps, mainly after Bigi et al. (1990). (1) Molasses of the Po basin (coarse signature = Oligocene; dots = Oligo-Miocene; white = Plio-Quaternary); (2) Dauphiné-Helvetic external zones (2a = basement; 2b = cover); (3) Lower and Mid-dle Penninic (Sub-Briançonnais and Briançonnais-Grand Saint Bernard nappes); (4a) Helminthoid Flysch (SW of 3) and Schistes lustrés metasediments (NE of 3); (4b) main ophiolitic massifs; (5) Internal Penninic massifs (5a = upper units, mainly poly-metamorphic basement; 5b = graphitic schists and associated meta-intrusives, proba-bly Briançonnais windows); (6a) Austroalpine nappes; (6b) South-Alpine domain; (7) main thrust contacts; (8a) Ivrea gravimetric and magnetic anomaly and (8b) prolon-gation of the magnetic anomaly south of Torino, after Bozzo et al. (1992). (B) Inter-pretation of the ECORS-CROP profile (located on A), after Polino et al. (1990). (1) Uplifted and eroded remnant of the Cretaceous-Eocene wedge; (2) deformed Euro-pean passive margin; (3) European foreland and its prolongation at depth (EF); (4) uplifted Adria crust and South-Alpine foreland (AF); (5a) European upper mantle (A-B-C); (5b) Adriatic upper mantle (D-LW). (6) Ivrea gravimetric and magnetic anomaly. AU = western Austroalpine high-P–low-T nappes; CL = Canavese (Insu-bric) line; GP = Gran Paradiso; SB = Grand St Bernard nappe; PF = Penninic thrust front; P = PreAlpine nappes; HB = Helvetic and Ultrahelvetic basement; HBt: = Helvetic-Jura basal thrust; HN = Helvetic cover nappes; LPN = lower Penninic nap-pes; SA = South-Alpine basement; black = ophiolitic units.

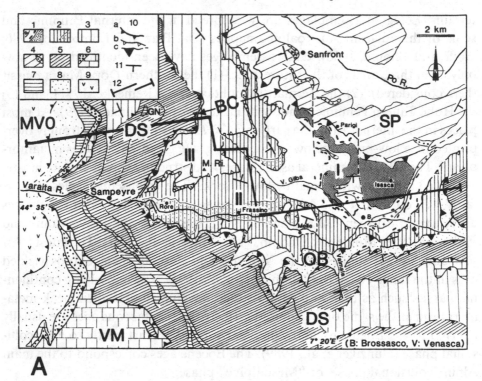

A

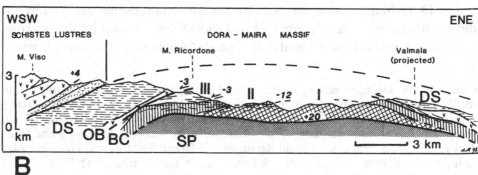

B

Figure 4.2. Structure of the southern Dora-Maira massif, after Henry (1990) and Michard et al. (1993). (A) Geologic sketch map. (1) UHP unit I (black = pyrope-coesite bearing quartzite; white = orthogneiss; crosses = Brossasco metagranite; dotted = Isasca varied schists, gneisses, marbles, and eclogites). (2) HP eclogitic unit II (ruled = gneiss; ruled and dotted = varied polymetamorphic series. (3) polymetamorphic schists from the HP eclogitic unit III (M. Ri.: M. Ricordone) and the blueschist Dronero-Sampeyre (DS) unit. (4) Permian (?) silvery schists from unit III, Permo-Carboniferous (?) graphitic schists from the Sanfront-Pinerolo unit (SP), Permian-Triassic conglomeratic quartzite and carbonate (dotted). (5) Permian–Triassic (?) silvery schists, quartzite and orthogneiss from the Dronero-Sampeyre unit (DS). (6) Triassic carbonates (briks), mainly from the Val Maira unit (VM); Permian-Triassic slices (dotted). (7) Quartzic limestones (Garitta Nuova, GN), massive calcschists with

collapsed through steeply dipping faults beneath the Oligocene-Miocene molasses of the Po basin. Three major continental nappes have been recognized there by early workers (Argand, 1934; Michard, 1965, 1967; Vialon, 1966; Borghi et al., 1984, 1985). The lowermost, Sanfront-Pinerolo (SP) unit involves granodioritic gneiss, Carboniferous (?) graphitic-conglomeratic schists, Permian-Triassic metaclastics, and Middle Triassic carbonates. This unit is usually considered as a Middle Penninic window (Briançonnais-Grand Saint Bernard nappe). The SP unit is overlain by the Venasca or Basement Complex (BC) unit, which consists of orthogneiss and polymetamorphic schists derived from a Pre-Alpine basement, and associated with rare Permian-Triassic metasediments. The BC complex is overlain, either directly or with intercalation of a band of ophiolitic slices and calcschists (OB), by the Dronero-Sampeyre unit (DS), including polymetamorphic schists, Permian–Triassic metaclastic and metaigneous rocks. Post-Triassic metasediments are suspected on top of the DS unit and of its northern equivalents, including Jurassic carbonates and Upper Cretaceous olistolithic calcschists (Marthaler et al., 1986; Philippot, 1988; Henry, 1990).

Defining a UHP Unit and Its Metamorphic Environment

As the first UHP rock occurrences were discovered in the southern Dora-Maira BC unit in the form of a few boudins of pyrope-bearing quartzite in the Parigi gneiss (Fig. 4.2, A), the problem arose with respect to their tectonic setting: were they allochthonous blocks within country rocks having never reached UHP conditions, or preserved parts of a coherent, UHPM unit, elsewhere thoroughly retrogressed? Detailed mapping of the whole southern Dora-Maira (Henry, 1990; Chopin et al., 1991) revealed numerous other occurrences of UHP assemblages in various rock types, scattered in a restricted area in the lowest part of the BC, while more "normal" (HP) eclogi-

Caption for fig. 4.2 (cont.) dolomite boulders (Jurassic–Upper Cretaceous ?). (8) Serpentinites, metabasalts, calcschists, and dark metashales from the Lower Varaita ophiolitic band (OB) and the lowest Monviso (MVO) unit. (9) Monviso ophiolites. (10) Low-angle tectonic contacts (teeth down-dip) [10a = in the part of the pile including ophiolites; 10b = within the Basement Complex (BC); 10c = footwall and hanging-wall limits of the BC units]. (11) Dip of the greenschist-facies foliation. (12) Trace of the cross-section, B. (B) Schematic cross-section. Abbreviations as in A. Boldface numbers: difference in peak metamorphic pressure between the juxtaposed units (kbar, positive if increasing upward). Half arrows: Sense of motion on the low-angle ductile faults, inferred from the fault–foliation obliquity (Sm greenschist-facies foliation, schematically shown close to the corresponding faults).

tic assemblages occur throughout the rest of the BC. It proved possible to define a limit between the areas of UHP and HP eclogite–facies occurrences, within a few hundred meters across strike (about 100 m at some places). In the field, this limit runs parallel to the regional, greenschist-facies foliation within a monotonous mylonitic gneiss series in which the only UHP and HP pinpoints are represented by discrete lenses of schist and metabasite. For mapping purposes, we drew this limit at the base of the lowest trail of HP schist lenses. This is the contact between units I and II on Fig. 4.2. Although not conspicuous in the field, such a boundary between two units or complexes of contrasted metamorphic grade must be seen as a major tectonic contact, the nature of which will be discussed below.

The coesite-bearing unit I (mostly equivalent to the Brossasco-Isasca unit of Compagnoni et al., Chapter 7) involves two main lithological series (Fig. 4.2 A): (1) a series of mylonitic, augen, medium or fine-grained gneiss, within which the phyllitic, pyrope-bearing quartzite (whiteschist in Compagnoni et al., Chapter 7) is included in the form of several trails of lenses; the gneiss series is considered to be derived from igneous protoliths; and (2) a varied series of schists, fine-grained gneiss, marbles, basic or quartz-bearing eclogites, which is regarded as polymetamorphic (Henry, 1990; Compagnoni et al., Chapter 7). In the gneiss series, UHP mineral indicators are scarce (outside of the pyrope-bearing lenses and some "jadeite-garnet fels"; Compagnoni et al., Chapter 7, this volume), and locally may have been represented by grossular-rich garnet-rutile, or quartz (coesite)-jadeite-phengite-garnet associations (Chopin et al., 1991; Compagnoni et al., Chapter 7). In contrast, numerous occurrences of UHP rocks were found throughout the varied polymetamorphic series, such as quartz (coesite)-kyanite-jadeite-garnet metapelites or quartz (coesite)-kyanite bearing eclogites. The peak *PT* conditions determined from the pyrope-bearing quartzite and the various UHP rocks of the polymetamorphic series are similar, close to 28–35 kbar, 700–750°C (Chopin, 1987; Massonne and Schreyer, 1989; Chopin et al., 1991; Kienast et al., 1991; Schertl et al., 1991; Compagnoni et al., Chapter 7, this volume). Then we may conclude that the whole unit I, about $1 \times 10 \times 15$ km in size, was affected by UHP metamorphism as a coherent terrain.

In the overlying, HP eclogitic part of the Basement Complex, two units can be separated on the base of lithologic, rather than metamorphic evidence. Unit II includes protoliths similar to those of unit I, but its "cold," HP eclogites and talc-chloritoid-kyanite magnesian schists equilibrated at only 16–20 kbar, 550–580°C (Henry, 1990; Chopin et al., 1991). Unit III includes polymetamorphic schists associated with Permian–Lower Triassic metaclastic rocks and Middle Triassic dolostone; mineral assemblages equilibrated at 12–

20 kbar, 500–550°C were found in the northern part of unit III (B. Patrick, in Henry, 1990; Chopin et al., 1991). The three eclogitic units (I-III) possess similar lenticular shape with kilometric thickness. On top of the eclogitic BC pile, the ophiolitic band (OB), the DS continental unit, and the overlying Schistes lustrés equilibrated essentially under blueschist-facies conditions. However, the Monviso ophiolites include eclogitic slices, equilibrated at 500–550°C, and at least 12–15 kbar (Lombardo et al., 1978; Nisio et al., 1987; Philippot, 1988). In contrast, the SP unit, which underlies the whole tectonic pile, shows peak metamorphic conditions not higher than 10–12 kbar, 500°C (Borghi et al., 1984; Henry, 1990; Cadoppi, 1990).

The Dora-Maira UHP unit, and all of the juxtaposed units show a pervasive ductile fabric, developed under blueschist (?), and then mainly under greenschist-facies conditions. This common regional fabric likely developed during and after the juxtaposition of the considered units at a similar crustal level. It consists of a shallow-dipping, mylonitic foliation S_m, associated with a widespread stretching lineation L_m. In most gneisses (units I, II, DS), large relic phengite crystals deformed along the S_m planes, forming mica fishes (Lister and Snoke, 1984). In these porphyroclasts, the phengitic substitution decreases from high values in the core (Si 3.5) to low values in the rim, similar to those from the small intra-S_m flakes (Si 3.1). This indicates that the mylonitic fabric developed essentially during a *P-T* decrease from ca. 12 kbar, 550°C to ca. 3 kbar, 350–400°C (Chopin et al., 1991). In all rocks, most of the L_m lineations also correspond to the orientation of greenschist mineral assemblages, but blue amphibole preferred orientation was also observed with the same direction in the SP, DS, and Monviso units (Philippot, 1988) and in the northern part of the SP unit (Mahwin et al., 1983; Cadoppi, 1990). In central Dora-Maira, much of the dominant foliation and stretching lineation also formed at pressure of the order of 11 kbar before their greenschist-facies latest evolution (Wheeler, 1991).

Relict UHP Structures

Structures that can be considered as formed under UHP are scarce in the Dora-Maira unit I, and preserved only within boudins of moderately overprinted rocks, or at a lower scale as internal structures within some UHP minerals. In the pyrope-bearing quartzite and juxtaposed layers, a deformed eclogitic foliation S_e can be observed at the outcrop scale (Figs. 4.3A,B). S_e is outlined mainly by the bedding-parallel orientation of quartz aggregates, large phengite flakes, and kyanite crystals. The foliation existed already during the peak *P-T* conditions since (1) the pressure shadows on the pyrope megablasts

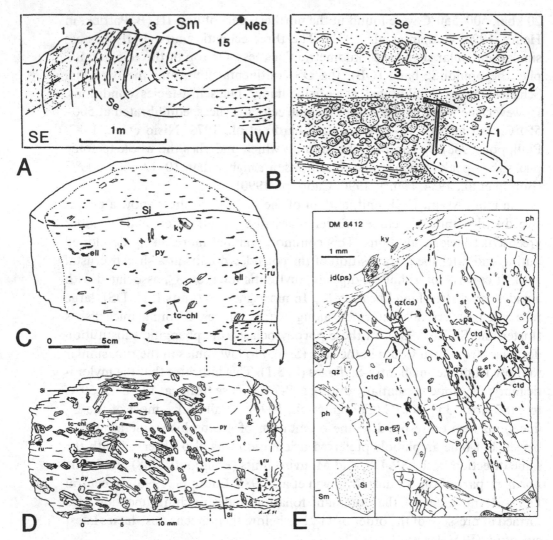

Figure 4.3. Relict UHP structures from the Dora-Maira coesite-bearing unit. (A) UHP layering and foliation affected by a late, syn-greenschist fold, Gilba area; (1) kyanite-bearing phengitic quartzite; (2, 3) fine-grained gneiss layers with various quartz amount; (4) phengitic quartzite; Se = eclogitic foliation; Sm = mylonitic, greenschist-facies foliation. (B) Bedding-parallel eclogitic foliation (Se) in a pyrope-bearing phengitic quartzite, Gilba area; (1) quartz-pyrope-rich layer with subequant structure; (2) phengite-rich foliae; (3) boudinaged pyrope megablast. (C) Polished section of a pyrope megablast from Parigi, val Po; py: = pyrope; tc-chl = talc-chlorite; ell = ellenbergerite; ky = kyanite; ru = rutile; Si = internal UHP foliation. (D) Close-up of (C), thin section; qz = quartz pseudomorph after coesite. (E) Foliation of a prograde stage preserved in the core of an almandine crystal, Isasca metapelites; abbreviations as above, with ctd = chloritoid; jd (ps) = albite pseudomorph after jadeite; pa = paragonite; ph = phengite; st = staurolite.

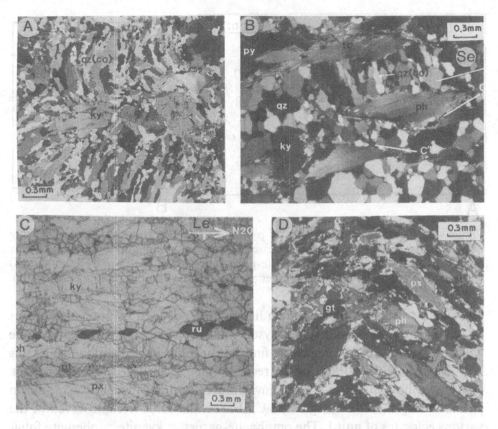

Figure 4.4. Micrograph of structures from the UHP and HP early retrograde stages in the southern Dora-Maira unit I. Abbreviations as in Fig.4.3. (A) Palisade texture of quartz after coesite (co) in the matrix of pyrope-bearing quartzite, Parigi, val Po; courtesy W. Schreyer; crossed nicols. Quartz fibers are disposed normally to any interface of other minerals, here kyanite and phengite grains, underlining the UHP foliation. Incipient recrystallization into HT isodiametric quartz grains (right). (B) Palisade texture overprinted by high-temperature recrystallization of quartz, and conjugate shear planes (C', C''); pyrope-bearing quartzite, val Po; crossed nicols. (C) Eclogitic foliation and lineation (Le) in a quartz-rich eclogite, lower val Gilba; plane polarized light; gt = garnet; px = omphacitic pyroxene; ru = rutile. (D) Hinge zone of an UHP-HP isoclinal fold in an eclogite from the Isasca formation; crossed nicols.

include the talc-phengite-kyanite association (see Fig. 4.5A, left corner); (2) in the quartz-phengite matrix, a palisade fabric of fibrous quartz is observed by place (Fig. 4.4 A, B), which must have formed by retrogression of a coesite-phengite-kyanite foliation (the growth of the quartz fibers was controlled by the geometry of the coesite-phengite or coesite-kyanite grain boundaries; Chopin et al., 1991).

A UHP foliation can also be found in the eclogites, and particularly in the quartz-mica-kyanite-bearing ones. S_e was observed in some outcrops to be

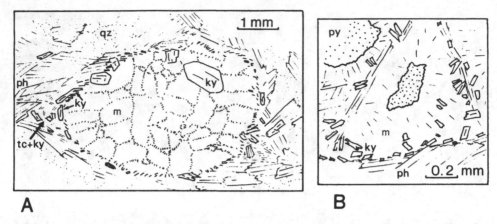

Figure 4.5. Cellular pseudomorph of pyrope megablasts from the pyrope-bearing quartzite, east of Parigi, val Po. (A): General aspect; m = phlogopite-microcrystalline quartz mixture; tc-ky-ph = talc-kyanite-phengite pressure shadow (UHP stage). (B) Close-up of the cellular fabric in another pseudomorph with relic pyrope patches (py).

associated with a strong stretching lineation L_e, defined by elongated grains of fresh kyanite and omphacite crystals, and by elongated garnet and rutile aggregates (Fig. 4.4C). The prolate finite strain ellipsoid corresponding to this L > S fabric (Flinn, 1965) might record either a coaxial deformation, or a convergent flow in a non-coaxial environment (among varied possible strain paths) during the UHP stage. Syn-UHP isoclinal folds were observed in the various eclogites of unit I. The omphacite-garnet ± kyanite ± phengite foliation and lineation are seen rotated, but the eclogitic minerals remain stable in the hinge zone of these early folds (Fig. 4.4D). Folding probably occurred by shearing on the foliation plane (dispersion of the poles of the lineation on a great circle of the Wulff stereonet; Hansen, 1971).

Additional information on the local strain regime can be obtained from the observation of inclusion patterns in garnets. Within the pyrope megablasts, the mineral inclusions (coesite, talc, kyanite, Mg-chlorite, ellenbergerite, rutile) are often randomly oriented, thus recording a static UHP recrystallization. However, an internal schistosity S_i was observed to be formed by the preferred orientation of these inclusions in some of the megablasts (Fig. 4.3C,D), and then interpreted as relict UHP foliation. Its sigmoidal curvature in the rim of the pyrope crystal points to the initiation of a non-coaxial flow at the end of the pyrope growth. Moreover, within the core of the almandine crystals of the garnet-kyanite-metapelites, trails of staurolite, Mg-rich chloritoid and paragonite inclusions also define sometimes an internal foliation, whereas coesite inclusions are found in the crystal rims (Fig. 4.3E). The mineral assemblage of the inner part of these garnet crystals corresponds to an early prograde, HP stage of the metamorphism (Chopin et al., 1991). This suggests that the UHP foliation itself transposed an earlier, HP eclogitic foliation developed during the prograde evolution.

HP Eclogite Structures

Besides the rare UHP structures reported above, most of the pre-greenschist-facies structures in the UHP unit I developed under HP ("normal") eclogitic conditions, during the retrogression of the UHP asemblages. We already mentioned the palisade quartz which probably formed in the UHP foliation by coesite inversion at the very beginning of the decompression history. In the same samples (Fig. 4.4 A,B), or in the juxtaposed jadeite-kyanite gneiss, the polygonal quartz texture often indicates a recrystallization of the palisades by grain boundary migration at high temperature and low strain rate (Schmid and Handy, 1991). Some pyrope megablasts, although preserving their idioblastic habit, have been pseudomorphically replaced by a "cellular" aggregate made up of a kyanite-phlogopite-quartz assemblage (Fig. 4.5). This is also an early, and almost static stage of retrogression (Schertl et al., 1991). A high-Si phengite stretching lineation is commonly found in these rocks, normal to the intersection of conjugate, although unequally developed shear planes (Fig. 4.4 B). This might denote a regime of almost coaxial flow during the early decompression of these UHP rocks. In some of the fine-grained, UHP basic eclogites, poikilitic amphibole prisms with light blue, Na-rich core, and greenish, Ca-rich rim are oriented roughly parallel to the eclogitic lineation. This could result from syntectonic crystallization and/or mimetic growth during the HP decompression history.

We may reasonably hypothesize that the UHP unit I underwent its early decompression under conditions close to the HP units II and III, should it be while the latter units were still under HP conditions, or partly unloaded (see last section). Then additional data on the Eoalpine, regional strain regime might be gained from the HP structures in units II and III. In their HP eclogites, L-S tectonite fabrics and isoclinal folds are found, as in the UHP metabasites. By contrast, in the Permian metasediments, the quartz-rich layers often present a subequant, random orientation of chloritoid and phengite crystals (Fig. 4.6). In other Permian samples, however, the fabric is again that of L-S tectonites, with conjugate, symmetrical shear planes. This mostly coaxial deformation likely occurred close to the peak *P-T* conditions of the rock, as suggested by the preservation of fresh chloritoid and kyanite, and by the occurrence of quartz with serrated grain boundaries.

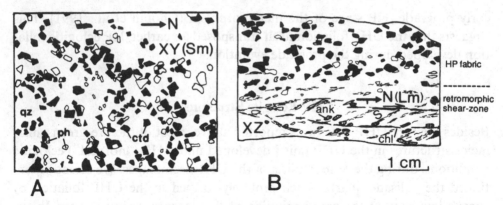

Figure 4.6. HP subequant fabric in a Permian quartz-rich micaschist, southern Dora-Maira unit III, val Varaita. Sketch of two polished surfaces: (A) section parallel to Sm; (B) section normal to Sm (note N-verging S/C structures in the retromorphic shear zone at the bottom of the sample). Abbreviations as Fig. 4.5, with chl = chlorite; ank = ankerite.

Preservation of Pre-UHP Structures

A surprising observation can be made in the Dora-Maira UHP unit: magmatic textures are locally preserved amidst the orthogneiss series. This phenomenon can be observed in the so-called Brossasco metagranite, the initial porphyritic texture of which is found almost undeformed by place, although the magmatic minerals underwent pseudomorphic to coronitic reequilibration reactions (Fig. 4.7 A) (Henry, 1990; Biino and Compagnoni, 1992; Compagnoni et al., Chapter 7, this volume). In regards to the lack of noticeable ductile deformation within domains of this metagranite, the question arises as to whether or not it was involved in the dive down to asthenospheric depths as was the rest of unit I. And if it actually did, then the fact that Pre-Alpine textures might be preserved in so deeply a subducted rock body would be significant information on the rheologic behavior of crustal rocks under UHP conditions.

The mapping pattern of the Brossasco metagranite (Fig. 4.2A) suggests an emplacement in the form of sills or dykes intrusive into the orthogneiss and the varied polymetamorphic series, and at least prior to the Mesoalpine, greenschist-facies deformation (transposition of the intrusion boundaries in the regional foliation plane). The age of intrusive emplacement must be regarded even as Pre-Alpine, since (1) relict, pre-HP minerals of contact metamorphism were found in relation with some Brossasco metagranite outcrops (Compagnoni and Hirajima, 1991; Compagnoni et al., Chapter 7, this volume); and (2) the metagranite yielded a 303 $\pm$ 1 Ma by the U/Pb method on zircon (J. Paquette, personal communication, 1989). Relict jadeite and high-

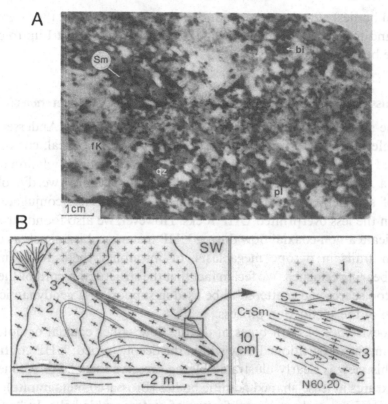

Figure 4.7. Heterogeneous deformation of the Brossaco metagranite, Dora-Maira UHP unit (location: Fig. 4.2). (A) Close-up of a polished sample with preserved magmatic texture. Dark = green biotite-phengite-garnet-rutile-epidote pseudomorphs after magmatic biotite (bi); gray = polygonal granoblastic quartz aggregates (qz); light gray = microcline phenocrysts (fK) with inclusions of any other mineral phases; white = albite-epidote-phengite-(jadeite)-garnet pseudomorphs after zoned plagioclase crystals (pl); Sm = incipient greenschist-facies foliation. (B) Mylonitic boundary of a virtually undeformed lense, vertical cliffs along the Varaita river south of Brossasco; (1) preserved porphyritic granite; (2) mylonitic orthogneiss with passively folded quartz veins (4) and minor isoclinal fold (detail); (3) ultra-mylonite (top-to-the-SW S/C structures in the transitional zone 1–3).

Si phengite from a HP stage were actually found in the metagranite (Biino and Compagnoni, 1992; Compagnoni et al., Chapter 7, this volume). Moreover, polygonal granoblastic quartz aggregates are interpreted by the latter authors as derived from inversion of coesite, formed itself on the site of coarse-grained igneous quartz. Therefore we may accept that the Brossasco metagranite together with its country rocks was driven down to asthenospheric depths.The structurally preserved, porphyritic granitoid outcrops correspond to meter-sized lenses limited by mylonitic shear-zones (Fig. 4.7B). It must be hypothesized that these shear-zones (or similar shear-zones bounding undeformed granitoid bodies) already operated during the UHP event, but they were reactivated and totally overprinted during the exhumation of unit I up to crustal level (see below).

Discussion 1: Strain in Continental Rocks During UHP Metamorphism

From the structural study of the Norwegian UHP eclogites, Andersen et al. (1991) inferred that continental crust experienced near vertical, coaxial constrictional strain when subducting at asthenospheric depth, in relation to slab-pull by a cold lithospheric root. In southern Dora-Maira, we did observe traces of symmetric, constrictional strain (L > S fabrics, conjugate shear bands) in the less overprinted UHP rocks. However, we also found structures that indicate a non-coaxial flow during UHP metamorphism, such as curved inclusion trails in pyrope megablasts, or intrafolial shear folds in fresh kyanite-bearing eclogites. We feel in fact that the structural record of the UHP stage is too scarce and scattered to be used in support of a geodynamic interpretation of the subduction process.

The coexistence of both types of ductile flow, either coaxial or not, additionally implies a partitioned flow within the deeply buried, UHP continental crust. This view is clearly illustrated by the preservation of Pre-Alpine magmatic textures in lense-shaped granitic bodies (Brossasco metagranite). Strain partitioning is a nearly universal feature of metamorphic belts. In "normal," HP eclogite-facies continental rocks, strain partitioning was demonstrated, for instance, in the Austroalpine Sesia zone by Lardeaux et al. (1982) and Vuichard (1989). We also found records of both coaxial and non-coaxial ductile flow (which proves partitioning of strain) in the HP structures from units II and III. Then the main rheologic characteristics and boundary conditions that control partitioning of strain in the subducting continental crust down to lithospheric mantle depths remain operative down to asthenospheric depths. Such partitioning could be favored by the lenticular structure of the extended, marginal crust (Hamilton, 1987) prior to its subduction.

Late Deformation of the Dora-Maira UHP and HP Units

In every HP-LT belt, late, mostly greenschist-facies structures form the main tectonic grain at the regional scale, as well as the main fabric at the sample scale. This rule applies to the Dora-Maira UHP-HP pile, in which the most conspicuous penetrative structures, i. e., the S_m foliation and L_m lineation developed under blueschist (?), then mainly under greenschist-facies condi-

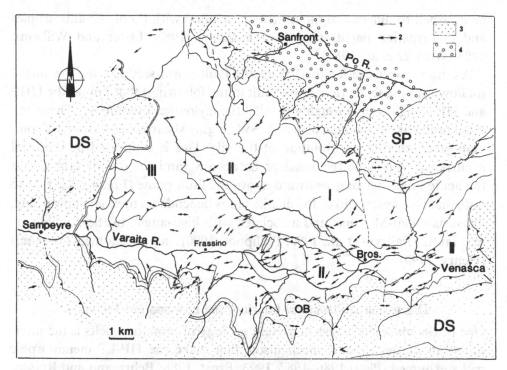

Figure 4.8. Direction of the greenschist-facies stretching lineation, and polarity of the associated kinematic indicators in the southern Dora-Maira massif (Henry et al., 1993). Abbreviations as in Fig. 4.2. (1) Unambiguous shear sense; (2) ambiguous results; (3) blueschist-facies units; (4) quaternary alluvium.

tions during the latest, Eocene stages of retrogression of the UHP-HP rocks. These structures are worthy of consideration here as they might provide some insight into the exhumation tectonics of the UHP terrain.

In contrast with the highly dispersed eclogitic lineations, L_m shows a nearly constant direction, close to N60–70°E (Fig. 4.8). Asymmetric microstructures on XZ sections (parallel to L_m and normal to S_m) contain various kinematic indicators (Berthé et al., 1979; Simpson and Schmid, 1983; Lister and Snoke, 1984). A large majority of data indicate a top-to-the-WSW shear direction (Henry, 1990; Philippot, 1990). Ambiguous criteria are also found, that indicate a domainal pure shear in a dominantly non-coaxial environment. A mainly non-coaxial flow regime is also shown by the occurrence of syn-S_m isoclinal or sheath folds, the axes of which parallel L_m, and by late-S_m asymmetric west-vergent folds (Henry et al., 1993). Therefore the top-to-the-WSW sense of shear on the greenschist foliation can be interpreted as the regional direction of transport during the common decompression of the SP-BC-DS units. The local inversion of shear direction (e.g. Fig. 4.8, Frassino area) might be ascribed to differential movements in relation with rheologic anisotropies and deformation partitioning (Garcia-Celma, 1982; Lister and Williams, 1983; Wheeler et al., 1987).

As discussed above for the I/II ductile fault (third section), the low-angle, shallow-dipping faults (low-angle fault in the following) that bound the UHP and HP eclogitic units roughly parallel the mylonitic foliation S_m. Therefore, these faults likely operated also as WSW-verging shear zones during the common greenschist-facies evolution of the pile. This is supported by the local occurrence of oblique relationships between S_m and the ductile faults, with the acute angle opening westward above the fault plane (Fig. 4.2 B). It is to note that the present dip of the ductile faults depends on their late upwarping, dated as Oligo-Miocene, and associated with high-angle, brittle-ductile normal faults (Ballèvre et al., 1990; Philippot, 1990) and pseudotachylite seams (Henry, 1990).

Discussion 2: Structural Imprints and Exhumation Tectonics

One of the main challenges in present geology of mountain belts is the interpretation of the tectonic mechanisms that make the HP-LT metamorphic rocks exhumed (Platt, 1986, 1987, 1993; Ernst, 1988; Behrmann and Ratschbacher, 1989; Avigad and Garfunkel, 1991). The challenge is clearly paroxysmal when diamond (!!) or coesite-bearing rocks are involved (Polino et al., 1990; Blake and Jayko, 1990; Andersen et al., 1991; Wheeler, 1991; Avigad, 1992; Merle and Ballèvre, 1992; Okay and Sengör, 1992; Michard et al., 1993). Most of the arguments presented in support of the variety of selected models rely, on one hand, on petrologic and geochronologic evidence (P-T-t paths, cooling curves), and, on the other hand, on large-scale structural evidence (architecture of the belt, especially of its metamorphic parts). May we draw constraints from the smaller-scale structural description of the Dora-Maira UHP-HP pile, which would help us in modeling the exhumation of these rocks?

Special attention must be paid to the low-angle fault that bound the various UHP-HP units, since normal low-angle fault plays a critical role in most current models, based on extensional tectonics. The I/II ductile fault, in particular, is a good candidate as a normal fault responsible for the "omission" or "excision" of a 40-km-thick rock section in a hypothetical, normally ordered metamorphic sequence (Fig. 4.2 B). This important, mylonitic ductile fault operated with a top-to-the-SW sense of shear during the Eocene greenschist event (as indicated by the syn-greenschist-facies kinematic indicators). We cannot use the fact that motion is down-dip with respect to the shear planes on the southwestern side of unit I to support a conclusion that the fault operated as a low-angle *normal* fault, since these shear planes and the fault itself have been upwarped afterward. However, a convincing evidence of an Eocene tectonic thinning of the Dora-Maira overburden (Avigad, 1992) is that the juxtaposed units suffered their common ductile deformation under conditions of decreasing *P*, and low *P/T* ratio, consistent with thermal modelling of crustal extension. At a regional scale, this can be interpreted as the result of the collapse (Dewey, 1988) of the Penninic wedge thickened by the Mesoalpine collision (syn-blueschist thrusting of the upper Dora-Maira nappes over the SP unit). The collapse of the inner part of the wedge might have been coeval with the outward progression of thrusts in its more external parts (External Briançonnais, Sub-Briançonnais, Ultra-Dauphinois zones; e.g., Tricart 1984; Fry, 1989). Collapse and foreward spreading of the orogenic wedge (Platt, 1986, 1987, 1993; Behrmann and Ratschbacher, 1989) were possibly accompanied by hinterland-directed blind extensional excision, as defined by Hodges and Walker (1992) in the Sevier orogen. This hypothesis, similar to that of Wheeler (1991) in central Dora-Maira, could explain the top-to-the-east sense of syn-greenschist shearing reported by Philippot (1990) in eastern Dora-Maira. In this interpretation, the role of the low-angle normal faults in the exhumation of the UHP-HP rocks during Eocene is negligible with respect to the distributed thinning of the orogenic wedge, and to the additional effect of erosion (Michard et al., 1993).

However, this Mesoalpine period corresponds only to the minor part of the decompression and cooling path of the UHP and HP units (Fig. 4.9). The major part of this path was performed during the Eoalpine-Mesoalpine interval, and was achieved when the UHP and HP eclogitic units of the BC were juxtaposed at about a 30-km-depth on top of the SP unit. Unfortunately, the structures related to this early and major unloading evolution are rare, deeply reoriented, and then kinematically almost meaningless. The I/II and II/III shear zones had most likely operated prior to the greenschist-facies event. We reported above on the occurrence of overprinted UHP-HP shear zones in the orthogneiss series (Fig. 4.7). The I/II and II/III faults probably operated through the subducted crust at least as early as the initiation of the decompression history of the UHP and HP eclogitic units that they bound, but we have no structural record left of their early sense of shear, and no more of their early dip. Therefore, these omission-creating faults can be integrated in contrasting hypothetic models (Fig. 4.10). As suggested by Avigad (1992), these faults could belong to a sytem of large, low-angle, normal faults (Fig. 4.10B,1), also invoked in the models proposed by Butler (1986), Deville (1991), and Merle and Ballèvre (1992) for the Gran Paradiso transect. Alternatively

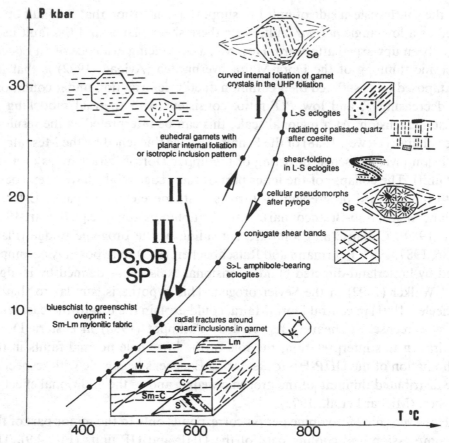

Figure 4.9. *P-T* path and structural record in the southern Dora-Maira coesite-bearing unit I, and *P-T* paths of the juxtaposed units (see Fig. 4.2), after Henry (1990) and Henry et al. (1993).

(Fig. 4.10B,2), the early juxtaposition of the I/II and II/III units could result from more steeply dipping shear zones operating in a subduction wedge between imbricated slices (Michard et al., 1993). The latter model of forced-flow, or extrusion tectonics is inspired by those of Ernst (1977, 1988), Cowan and Silling (1978), England and Holland (1979), and Cloos (1982, 1986). In the Zermatt area, Fry and Barnicoat (1987) and Reinecke (1991) also favor a buoyancy-controlled, tectonic uplift of imbricate slices in a subduction setting for the exhumation of the UHP-HP meta-ophiolites. The low-angle normal faults postulated by Deville (1991) and Merle and Ballèvre (1992) would have operated transversally to the Alps during an extensional phase intercalated between the Eoalpine and Mesoalpine contractional phases. The main argument against this view (and in support of a purely contractional, forced-flow model) comes from the plate tectonic restorations, which suggest an appar-

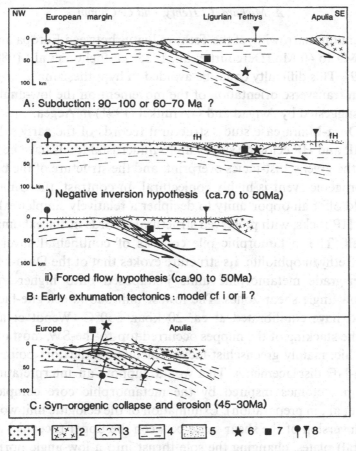

Figure 4.10. Current tectonic scenarii for the origin and exhumation of the Dora-Maira UHP unit. (1) Thinned continental crust of the European margin; (2) Apulian continental crust; (3) Tethyan oceanic crust; (4) lithospheric mantle. (5) Upper Cretaceous–Paleocene flysch deposits (Helminthoid flyschs, fH); (6) Dora-Maira UHP rocks; (7) associated HP eclogite-facies rocks; (8) pinpoints linked to the European and Tethyan crusts, respectively. No vertical exaggeration, except for the thickness of the individual UHP and HP slices. (A) Hypothetic position of the Dora-Maira protoliths in an obduction–subduction zone bounding the European margin to the south, during their peak metamorphism, either at about 90–100 Ma (Hunziker et al., 1989; Monié and Chopin, 1989; Scaillet et al., 1990) or 60–70 Ma (Tilton et al., 1991; Barnicoat et al., 1993). The simultaneous occurrence of another subduction zone driving the Tethyan lithosphere beneath the Apulian margin is featured after Avigad et al. (1993). (B) Two alternative models for the exhumation of the UHP rocks up to deep crustal levels: (1) extensional phase with negative inversion of the obduction–subduction shear zone and isostatic rebound of the lower plate (cf Butler, 1986; Deville, 1991; Merle and Ballèvre, 1992); lithospheric-scale low-angle shear zones could also operate longitudinally rather than transversally, in an ongoing convergence setting (cf. Avigad, 1992); (2) forced flow of imbricate slices in an active, deep accretionary wedge, or extrusion tectonics (cf. Fry and Barnicoat, 1987; Wheeler, 1991; Michard et al., 1993). (C) Exhumation of the UHP rocks and associated units up to shallow crustal levels by distributed extensional deformation of the active orogenic wedge, normal faults, and erosion (cf Platt, 1986, 1987; Philippot, 1990; Avigad, 1992; Michard et al., 1993); the Dora-Maira complex reaches the surface at ca. 20 Ma in a similar tectonic setting (cf. present state, Fig. 4.1B).

ently continuous narrowing of the Tethyan hiatus between Europa and Apulia from 110 Ma to 40 Ma (Dercourt et al., 1985; Le Pichon et al., 1988; Dewey et al., 1989). This difficulty could be avoided in hypothesizing a longitudinal, rather than transverse orientation of the movement on the low-angle normal faults, as suggested by Avigad and Garfunkel (1991) in Aegea.

In the Dora-Maira case study, structural records of the early exhumation of the UHP and HP eclogitic slivers up to midcrustal levels have been almost erased by the greenschist-facies overprint, and the structure of the belt before this late-orogenic event is highly conjectural. In contrast, the Oman Mountains would offer an opportunity to decipher a relatively simpler exhumation history of HP rocks, with preserved structural records of the exhumation tectonics itself. The metamorphic pile consists of continental slivers stacked below the Tethyan ophiolite. Its structure evokes that of the Dora-Maira pile, with lower-grade metamorphic units juxtaposed onto higher-grade ones through low-angle shear zones. The lowest unit shows eclogite-facies rocks that would have equilibrated at ca. 20 kbar, 550°C (Wendt et al., 1993). Whereas the stacking of the nappes occurred top-to-the-SW, most of the late-metamorphic, mainly greenschist-facies kinematic indicators point to major top-to-the-NE displacements. This would support an interpretation of the exhumation tectonics inspired by the metamorphic core complex model (Michard et al., in preparation). Exhumation of the eclogitic unit would result from the inversion of the shear sense between the upper (oceanic) and lower (continental) plates, changing the sole thrust into a low-angle normal fault. The whole orogenic cycle (from contraction to extension and exhumation of the HP-LT rocks) is bracketed there between 80 Ma and 65 Ma (cf. Eoalpine phase). The preservation of the structural record of the inversion tectonics was permitted owing to the lack of later (Mesoalpine) continental collision, in contrast with the Dora-Maira case study. This is the reason why Oman is worth being considered here. Although the metamorphic grade of the Oman eclogites does not actually reach the UHP grade, the exhumation tectonics restored there might be reasonably extrapolated to UHP cases. It must be noted that in the Oman case, plate tectonics restorations for the Late Cretaceous–Paleocene interval do not mention any remission of the Africa–Eurasia convergence (Dercourt et al., 1985; Le Pichon et al., 1988). This would encourage those geologists who resist the forced-flow concept, to follow Butler (1986), Deville (1991), Avigad (1992), and Merle and Ballèvre (1992) when they suggest (despite the rather contradictory plate-tectonic data) that an extensional event occurred in the west Alpine realm after the Eoalpine precollisional phase.

Conclusions

Structures in UHP rocks are still poorly known. In this chapter we focused on the Dora-Maira case study. We feel the following conclusions can be emphasized:

1. UHP rocks may form coherent terrane of kilometer-scale size, bounded by polyphase shear zones, and tectonically included amidst units of similar size and various metamorphic grade.

2. A few UHP structures, and even pre-UHP textures can be found preserved despite the later, lower-pressure overprint; the UHP structures (and the preservation of earlier textures) document a partitioned strain regime, with no significant difference with respect to the strain regime under "normal" eclogitic conditions; the rheology of the subducted continental crust does not seem to change deeply from HP to UHP conditions.

3. The only significant kinematic indicators are related to the late metamorphic, mainly greenschist overprint; in the Dora-Maira case, they document no more than the latest and minor part of the exhumation of the UHP terrane, likely by gravity collapse; the earlier, and major part of the unloading process is open to discussion, and cannot be constrained by local structural analysis. In contrast, the Oman case, tectonically simpler, could illustrate an exhumation tectonics by inversion of the relative plate motion, with preservation of the related structural imprints in the HP (not UHP) rocks. This model could possibly be accepted in the Western Alps too, for the earliest part of the exhumation of the coesite-bearing terrane.

Acknowledgments

This study was made possible by the Institut National des Sciences de l'Univers, Programme DBT, and the Centre National de la Recherche Scientifique, URA 1316. We are indebted to Robert Compagnoni and B. Patrick for comments and advice on an early draft.

References

Andersen, T. B., Jamtveit, B., Dewey, J. F., and Swensson, E. 1991. Subduction and eduction of continental crust: Major mechanisms during continent-continent collision and orogenic extensional collapse, a model based on the South Norwegian Caledonides. *Terra Nova, 3,* 303–310.

Argand, E. 1934. La zone pennique. *Guide Géol. Suisse. Wepf. Bâle, 3,* 149–189.

Avigad, D. 1992. Exhumation of coesite-bearing rocks in the Dora-Maira massif (western Alps, Italy). *Geology, 20,* 947–950.

Avigad, D., Chopin, C., Goffé, B., and Michard, A. 1993. Tectonic model for the evolution of the western Alps. *Geology, 21,* 659–662.

Avigad, D., and Garfunkel, Z. 1991. Uplift and exhumation of high-pressure metamorphic terrains: the example of the Cycladic blueschist belt (Aegean Sea). *Tectonophysics, 187*, 1–15.

Ballèvre, M., Lagabrielle, Y., and Merle, O. 1990. Tertiary ductile normal faulting as a consequence of lithospheric stacking in the western Alps. *Mém. Soc. géol. France, 156*, 227–236.

Barnicoat, A. C., Cliff, R. A., Guise, P. G., Inger, S., and Rex, D. C. 1993. Assessing the age of high-pressure metamorphism in the western Alps using Sm-Nd and $^{40}Ar/^{39}Ar$ studies. *Terra Abst., 5*, 380.

Behrmann, J. H., and Ratschbacher, L. 1989. Archimedes revisited: a structural test of eclogite emplacement models in the Austrian Alps. *Terra Nova, 1*, 242–252.

Berthé, D., Choukroune, P., and Gapais, D. 1979. Orientations préférentielles du quartz et orthogneissification progressive en régime cisaillant: l'exemple du cisaillement sud-arméricain. *Bull. Minéral., 102*, 265–272.

Bigi, G., Castellarin, A., Coli, M., Dal Piaz, G. V., Sartori, R., Scandone, P., and Vai, G. B. 1990. Structural model of Italy, sheet n°1, 1/500.000 Cons. Naz. Ric. Italy.

Biino, G., and Compagnoni, R. 1992. Very high pressure metamorphism of the Brossasco coronite metagranite, southern Dora-Maira massif, western Alps. *Schweiz. min. und petr. Mitt., 72*, 347–362.

Blake, M.C.J., and Jayko, A. S. 1990. Uplift of very high pressure rocks in the western Alps: evidence for structural attenuation along low-angle faults. *Mem. Soc. geol. France, 156*, 237–246.

Borghi, A., Cadoppi, P., Porro, A., Sacchi, R., and Sandrone, R. 1984. Osservazioni geologiche nella Val Germanasca e nella media Val Chisone (Alpi Cozie). *Boll. Mus. Reg. Sci. Nat. Torino, 2*, 503–530.

Borghi, A., Cadoppi, P., Porro, A., and Sacchi, R. 1985. Metamorphism in the northern part of the Dora-Maira massif (Cottian Alps). *Museo Regionale di Scienze Naturali di Torino, Bullettino, 3*, 369–380.

Bouffette, J. 1993. Evolution tectonométamorphique des unités océaniques et continentales au Nord du massif Dora-Maira (Alpes occidentales) Univers. Claude Bernard, Lyon I, 163 pp.

Bozzo, E., Campi, S., Capponi, G., and Giglia, G. 1992. The suture between the Alps and Apennines in the Ligurian sector based on geological and geomagnetic data. *Tectonophysics, 206*, 159–169.

Butler, R. W. H. 1986. Thrust tectonics, deep structure and crustal subduction in the Alps and Himalayas. *J. Geol. Soc. London, 143*, 857–873.

Cadoppi, P. 1990. Geologia del basamento cristallino nel settore settentrionale del massiccio Dora-Maira (Alpi occidentali). PhD., 208 pp. University of Torino.

Chopin, C. 1984. Coesite and pure pyrope in high-grade blueschists of the western Alps: A first record and some consequences. *Contrib. Miner. Petrol., 86*, 107–118.

Chopin, C. 1987. Very-high-pressure metamorphism in the western Alps: implications for subduction of continental crust. *Philosophical Transactions of the Royal Society of London, Series A, 321, no. 1557*, 183–197.

Chopin, C., Henry, C., and Michard, A. 1991. Geology and petrology of the coesite-bearing terrain, Dora-Maira massif, Western Alps. *Eur. Jour. Mineral., 3*, 263–291.

Cloos, M. 1982. Flow melanges: Numerical modelling and geologic constraints on their origin in the Franciscan subduction complex, California. *Geol. Soc. Amer. Bull., 93*, 330–345.

Cloos, M. 1986. Blueschists in the Franciscan Complex of California: Petrotectonic constraints on uplift mechanisms. *Geol. Soc. Amer. Mem., 164*, 77–93.

Compagnoni, R., and Hirajima, T. 1991. Geology and petrology of the Brossasco-Isasca Complex, southern Dara-Maira massif, Western Alps. *Terra Abstract, 3*, 84.

Cowan, D. S., and Silling, R. M. 1978. A dynamic model of accretion at trenches and its implications for the tectonic evolution of subduction complexes. *J. Geophys. Res., 83*, 5389–5396.

Coward, M., and Dietrich, D. 1989. Alpine tectonics–an overview. In: Alpine Tectonics, *Geol. Soc. London, Spec. Pub.*, No. 45, 1–32.

Dal Piaz, G. V., and Lombardo, B. 1985. Review of radiometric dating in the western Italian Alps. *Rendiconti, 40*, 125–138.

Dal Piaz, G. V., and Lombardo, B. 1986. Early Alpine eclogite metamorphism in the Penninic Monte Rosa-Gran Paradiso basement nappes of the northwestern Alps. *Geol. Soc. Amer. Mem., 164*, 249–265.

Dercourt, J., and others. 1985. Présentation de neuf cartes paléogéographiques au 1/20 000 000 s'étendant de l'Afrique au Pamir pour la période du Lias à l'Actuel. *Bull. Soc. Géol. Fr. 1*, 637–653.

Deville, E. 1991. Within-plate type meta-volcaniclastic deposits of Maastrichtian-Paleocene age in the Grande-Motte unit (French Alps, Vanoise). *Geodinamica Acta, 4*, 199–210.

Dewey, J. F. 1988. Extensional collapse of orogens. *Tectonics, 7*, 1123–1140.

Dewey, J. F., Helman, M. L., Turco, E., Hutton, D.W.H., and Knott, S. D. 1989. Kinematics of the western Mediterranean. *Alpine Tectonics, Geological Society Special Publication, 45*, 265–283.

England, P. C., and Holland, T.J.B. 1979. Archimedes and the Tauern eclogites: the role of buoyancy in the preservation of exotic eclogite blocks. *Earth Planet. Sci. Lett., 44*, 287–294.

England, P., and Molnar, P. 1990. Surface uplift, uplift of rocks, and exhumation of rocks. *Geology, 18*, 1173–1177.

Ernst, W. G. 1977. Mineral parageneses and plate tectonic settings of relatively high-pressure metamorphic belts. *Fortschr. Miner. 54*, 192–222.

Ernst, W. G. 1988. Tectonic history of subduction zones inferred from retrograde blueschist P-T paths. *Geology. 16*, 1081–1084.

Flinn, D. 1965. On the symmetry principle and the deformation ellipsoid. *Geol. Mag. 102*, 36–45.

Fry, N. 1989. Southwestward thrusting and tectonics of the western Alps. *Alpine tectonics. Geol. Soc. London, Spec. Publ., 45*: 83–109.

Fry, N., and Barnicoat, A. C. 1987. The tectonic implications of high-pressure metamorphism in the western Alps. *J. Geol. Soc. London, 144*, 653–659.

Garcia-Celma, A. 1982. Domainal and fabric heterogeneities in the Cap de Creus quartz mylonites. *Journal of Structural Geology, 4*, 443–445.

Goffé, B., and Chopin, C. 1986. High-pressure metamorphism in the western Alps: zoneography of metapelites, chronology and consequences. *Schweiz. Mineral. Petrog. Mitt. 66*, 41–52.

Hamilton, W. 1987. Crustal extension in the Basin and Range. *Geol. Soc. London Spec. Pub., 28*, 155–176.

Hansen, E. 1971. *Strain facies.* Berlin: Springer Verlag.

Henry, C. 1990. L'Unité à coesite du massif Dora-Maira dans son cadre pétrologique et structural (Alpes occidentales, Italie). TheSe Universite de Paris VI.

Henry, C., Michard, A., and Chopin, C. 1993. Geometry and structural evolution of ultra-high-pressure and high-pressure rocks from the Dora-Maira massif, Western Alps, Italy. *Journal of Structural Geology, 15,* 965–982.

Hodges, K. V., and Walker, J. D. 1992. Extension in the Cretaceous Sevier orogen, North American Cordillera. *Geol. Soc. Amer. Bull., 104,* 560–569.

Holland, T.J.B. 1979. High water activities in the generation of high-pressure kyanite eclogites of the Tauern window. *Austria Journal of Geology, 87,* 1–28.

Hunziker, J. C., Desmons, J., and Martinotti, G. 1989. Alpine thermal evolution in the central and the western Alps. *Geol. Soc. London Spec. Paper, 45,* 353–367.

Kienast, J. R., Lombardo, B., Biino, G., and Pinardon, J. L. 1991. Petrology of very-high-pressure eclogitic rocks from the Brossasco-Isasca Complex, Dora-Maira Massif, Italian Western Alps. *Jour. Metamorph. Geol., 9,* 19–34.

Lardeaux, J. M., Gosso, G., Kienast, J. R., and Lombardo, B. 1982. Relations entre le métamorphisme et la déformation dans la zone de Sesia-Lanzo (Alpes occidentales) et le problème de l'éclogitisation de la croûte continentale. *Bull. Soc. géol. France, 4,* 793–800.

Le Pichon, X., Bergerat, F. and Roulet, M. J., 1988. Plate kinematics and tectonics leading to the Alpine belt formation; a new analysis. *Geol. Soc. Amer. Spec. Paper., 218,* 111–131.

Lemoine, M., Bas, T., Arnaud-Vanneau, A., Arnaud, H., Dumont, T., Gidon, M., Bourbon, M., De Graciansky, P. C., Rudiewicz, J. L., Megard-Galli, J., and Tricart, P. 1986. The continental margin of the Mesozoic Tethys in the western Alps. *Marine Petrol. Geol., 3,* 179–199.

Lister, G. S., and Williams, P. F. 1983. The partitioning of deformation in flowing rock masses. *Tectonophysics, 92,* 1–33.

Lister, G. S., and Snoke, A. W. 1984. S-C mylonites. *J. Struct. Geol., 6,* 617–638.

Lombardo, B., Nervo, R., Compagnoni, R., Messiga, B., Kienast, J. R., Mevel, C., Fiora, L., Piccardo, G. B., and Lanza, R. 1978. Osservazioni preliminari sulle ofiolite metamorfiche del Monviso (Alpi occidentali). *Rend. Soc. It. Miner. Petrol., 34,* 253–305.

Marthaler, M., Fudral, S., Deville, E., and Rampnoux, J. P. 1986. Mise en évidence du Crétacé supérieur dans la couverture septentrionale de Dora-Maira, région de Suse, Italie (Alpes occidentales). Conséquences paléogéographiques et structurales. *C. R. Acad. Sci. Paris, 302–2,* 92–96.

Massonne, H. J., and Schreyer, W. 1989. Stability field of the high-pressure assemblage talc + phengite and two new phengite barometers. *Eur. Jour. of Mineral., 1,* 391–410.

Mawhin, B., Jeannette, D., and Tricart, P. 1983. Relations entre structures longitudinales et transverses au coeur de l'arc alpin occidental: l'exemple de Val Germanasca (Massif cristallin de Dora-Maira). *C.R. Acad. Sci. Paris, 260,* 4012–4015.

Merle, O., and Ballèvre, M. 1992. Late Cretaceous-early Tertiary detachement fault in the Western Alps. *C. R. Acad. Sci. Paris, 315,* 1769–1776.

Michard, A. 1965. Une nappe de socle dans les Alpes cottiennes internes? Implications paléogéographiques et rôle éventuel des mouvements crétacés. *C.R. Acad. Sci. Paris, D–260,* 4012–4015.

Michard, A. 1967. *Etudes géologiques dans les zones internes des Alpes cottiennes.* Paris: CNRS, 447 pp.

Michard, A., Chopin, C., and Henry, C. 1993. Compression versus extension in the exhumation of the Dora-Maira coesite-bearing unit, Western Alps, Italy. *Tectonophysics, 221*, 173–193.

Michard, A., Goffé, B., Saddigi, O., Oberhänsli, R., and Wendt, A. S. 1994. Late Cretaceous exhumation of the Oman blueschists and eclogites: a two-stage extensional mechanism. *Terra Nova, 6,* (in press).

Monié, P., and Chopin, C. 1991. ^{40}Ar/^{39}Ar dating in coesite-bearing and associated units of the Dora-Maira massif, Western Alps. *Eur. Jour. Mineral., 3*, 239–262.

Monié, P., and Philipot, P. 1989. Mise en évidence de l'âge éocène moyen du metamorphisme de haute-pression dans la nappe ophiolitique du Monviso (Alps Occidental) par la méthode 39Ar–40Ar. *Paris, Académie des Sciences Comptes Rendus, ser. II, 309*, 245–251.

Nicolas, A., Polino, R., Hirn, A., Nicolich, R., and Group, E.-C.W. 1990. ECORS-CROP traverse and deep structure of the western Alps, a synthesis. *Soc. Géol. de France, Mémoires, 156*, 15–27.

Nisio, P., Lardeaux, J. M., and Boudeulle, M. 1987. Evolutions tectonométamorphiques contrastées des éclogites dans le massif du Viso; conséquences de la fragmentation de la croûte océanique lors de l'orogenèse alpine. *C. R. Acad. Sci. Paris, 304*, 355–360.

Okay, A. I., and Sengör, A.M.C. 1992. Evidence for intracontinental thrust-related exhumation of the ultra-high-pressure rocks in China. *Geology, 20*, 411–414.

Philippot, P. 1988. Déformation et éclogitisation progressives d'une croûte océanique subductée: le Monviso, Alpes occidentales. Contraintes cinématiques durant la collision alpine. Montpellier, Travaux Centre Géol. Géophys. Montpellier, 19: 269 pp.

Philippot, P. 1990. Opposite vergence of nappes and crustal extension in the French-Italian Western Alps. *Tectonics, 9*, 1143–1164.

Platt, J. P. 1986. Dynamics of orogenic wedges and the uplift of high-pressure metamorphic rocks. *Geol. Soc. Amer. Bull., 97*, 1037–1053.

Platt, J. P. 1987. The uplift of high-pressure, low-temperature metamorphic rocks. *Philosophical transaction of Royal Society of London, 321*, 87–103.

Platt, J. P. 1993. Exhumation of high-pressure rocks: a review of concepts and processes. *Terra Nova, 5*, 119–133.

Polino, R., Dal Piaz, G. V., and Gosso, G. 1990. Tectonic erosion of the Adria margin and accretionary processes for the Cretaceous orogeny of the Alps. *Soc. Géol. de France, Mémoires, 156*, 15–27.

Reinecke, T. 1991. Very-high-pressure metamorphism and uplift of coesite-bearing metasediments from the Zermatt-Saas zone, Western Alps. *Eur. Jour. Mineral., 3*, 7–17.

Scaillet, S., Feraud, G., Lagabrielle, Y., Ballevre, M., Ruffet, G. 1990. ^{40}AR/^{39}AR laser probe dating by step heating and spot fusion of phengites from the Dora Maire Nappe of the western Alps, Italy. *Geology 18,* 719–744.

Schertl, H.-P., Schreyer, W., and Chopin, C. 1991. The pyrope-coesite rocks and their country rocks at Parigi, Dora-Maira Massif, Western Alps: detailed petrography, mineral chemistry and PT-path. *Contrib. Mineral. Petrol., 108*, 1–21.

Schmid, S. M., and Handy, M. R. 1991. Towards a genetic classification of fault rocks: geological usage and tectonophysical implications. *Controversies in modern geology.* London: Academic Press, 339–361.

Simpson, C., and Schmid, S. M. 1983. An evaluation of criteria to deduce the sense of movement in sheared rocks. *Geol. Soc. Amer. Bull., 94*, 1281–1288.

Stampfli, G. M., and Marthaler, M. 1990. Divergent and convergent margins in the North-Western Alps confrontation to actualistic models. *Geodinamica, 4,* 159–184

Tilton, G. R., Schreyer, W., and Schertl, H.-P. 1991. Pb-Sr-Nd isotopic behavior of deeply subducted crustal rocks from the Dora-Maira Massif, Western Alps-II: what is the age of the ultrahigh-pressure metamorphism? *Contributions to Mineralogy and Petrology, 108,* 22–33.

Tricart, P. 1984. From passive margin to continental collision: a tectonic scenario for the western Alps. *Am. J. Sci., 284,* 97–120.

Vialon, P. 1966. Etude géologique du massif cristallin Dora-Maira, Alpes cottiennes internes, Italie. Thèses d'Etat, Univ. de Grenoble, 293 pp.

Vuichard, J. P. 1989. La marge austroalpine durant la collision alpine: évolution tectono-métamorphique de la zone Sesia-Lanzo. *Mém. Doc. C.A.E.S., Rennes, 24,* 1–307.

Wendt, A. S., D'Arco, P., Goffé, B., and Oberhänsli, R. 1993. Radial cracks around α-quartz inclusions in almandine: constraints on the metamorphic history of the Oman mountains. *Earth Planet. Sci. Lett.,* 449–461.

Wheeler, J. 1991. Structural evolution of a subduction continental sliver: the northern Dora-Maira massif, Italian Alps. *J. Geol. Soc. London, 148,* 1101–1113.

Wheeler, J., Windley, B. F., and Davies, F. B. 1987. Internal evolution of the major Precambrian shear belt at Torridon, NW Scotland. *Geol. Soc. Lond. Spec. Pub., 27,* 153–163.

5

Creation, Preservation, and Exhumation of UHPM Rocks

B. R. HACKER AND S. M. PEACOCK

Abstract

Coesite-bearing eclogites exposed in the Western Alps and China record peak metamorphic temperatures of 550–900°C at pressures $\geq$ 2.5 GPa ($\geq$25 kbar). Their presence as regionally metamorphosed rocks requires subduction of >400 km² of upper crustal material to depths $\geq$90 km and subsequent exhumation. Existing experimental kinetic data suggest that pure coesite rock should not survive exhumation in the presence of a fluid, and this is corroborated by field observations. Partial survival of coesite is linked to its occurrence as inclusions in porphyroblasts that can maintain high internal pressures and prevent ingress of fluids. In the absence of coesite, reliable indicators of its former presence include subparallel, sometimes curving quartz subgrains and sets of quartz subgrains that truncate other sets. Polycrystalline quartz aggregates and cracks radiating outward from inclusions are not *a priori* evidence of ultrahigh pressure (UHP) metamorphism.

Petrologic constraints demand that coesite-bearing regional metamorphic rocks cooled during exhumation. Cooling during exhumation requires either (1) continued subduction beneath the eclogites that effectively chills the overlying UHP rocks during exhumation, or (2) transport toward the surface in the lower plate of an extensional structure such as a low-angle normal fault or shear zone. Both processes may occur; neither process demands rapid exhumation rates, although radiometric dating of different portions of the Dora-Maira pressure-temperature path indicate average long-term exhumation rates of 3.00 km Ma^{-1}.

Introduction

Coesite is a high-pressure polymorph of SiO_2 stable at pressures $\geq$ 2.5 GPa for metamorphic temperatures $\geq$500°C (Fig 5.1). In 1984, Chopin (1984) and

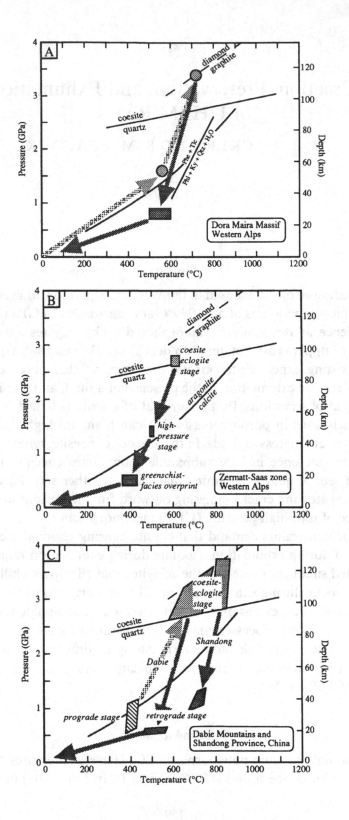

Smith (1984) reported the discovery of coesite in regional high-pressure metamorphic rocks in the western Alps and Norway, respectively. The discovery in China of coesite in metasedimentary rocks (Wang et al., 1989) of regional extent (>400 km^2) demonstrates that rocks deposited at the surface are subducted to depths ≥90 km and subsequently exhumed by erosional and tectonic processes.

In this chapter, we first review the geologic and petrologic settings of coesite occurrences in the Western Alps and China. We then examine natural coesite → quartz reaction textures and the petrologically determined pressure–temperature (*P-T*) paths in light of existing kinetic data and thermal models. In this review, we document the following conclusions:

1. The coesite-bearing eclogites record relatively low temperatures at high pressures, consistent with formation in a subduction zone environment.
2. During exhumation, complete reversion of coesite to quartz can be prevented only if the coesite occurs as an inclusion within a porphyroblast capable of maintaining high internal pressures and excluding fluids that would otherwise catalyze retrogression.
3. Coesite-bearing eclogite terrains record cooling during exhumation which requires that exhumation occurred while subduction/underflow continued to chill the overlying lithosphere or that the terrains were transported to the surface in the lower plate of extensional structures.

Geologic and Petrologic Setting

There are at least six worldwide occurrences of UHP rocks, but we focus on the better described samples from the Western Alps and China. For more detailed descriptions of these rocks and the methods used to evaluate their metamorphic history, we refer the reader to other chapters in this book.

Western Alps

Coesite occurs in the Dora-Maira massif (Chopin, 1984) and in the Zermatt-Saas zone (Reinecke, 1991) of the Western Alps. The Dora-Maira massif is a

Figure 5.1. Pressure-temperature (*P-T*) paths for coesite-bearing eclogites from (A) Dora-Maira massif, Western Alps (Chopin, 1984; Schertl et al., 1991; Sharp et al., 1992); (B) Zermatt-Saas zone, Western Alps (Reinecke, 1991); and (C) Sino Korean–Yangtze collision zone, China (Enami and Zang, 1990; Wang and Liou, 1991, 1992). Quartz-coesite reaction from Bohlen and Boettcher (1982), calcite-aragonite reaction from Hacker et al. (1995), and graphite-diamond reaction from Bundy (1980). Phengite + talc = phlogopite + kyanite + quartz + H$_2$O reaction from Schertl et al. (1991). Depth-pressure relations based on a density of 3000 kg/m^{-3}.

domal internal crystalline massif of the Penninic domain and may represent the southeastern edge of the European plate or a microcontinent (Platt, 1986). The coesite-bearing rocks occur as boudins in a "polymetamorphic" series that is bounded above and below by clastic Paleozoic–Mesozoic rocks also metamorphosed at high pressure (Chopin et al., 1991). The polymetamorphic series consists of the coesite-bearing unit and overlying eclogite-facies "micaschiste amygdalaire" and "cold-eclogite" units (Chopin et al., 1991). An orthogneiss derived from Hercynian granites makes up the bulk of the coesite-bearing unit (Chopin et al., 1991; Tilton et al., 1991) and forms country rock to the blocks of coesite rock and a unit of paragneiss, mafic eclogite, and marble (Chopin et al., 1991). The coesite occurs in magnesian boudins perhaps derived from evaporites or by metasomatism (Schreyer, 1977).

Coesite crystals occur in kyanite and garnet, and quartz pseudomorphs after coesite occur in omphacite, kyanite, and garnet (Chopin et al., 1991). The coesite-bearing blocks reached pressures of 3.2–3.6 GPa and temperatures of 700–750°C (Massone and Schreyer, 1989; Chopin et al., 1991; Schertl et al., 1991; Sharp et al., 1993) (Fig 5.1A). Relict minerals in garnet cores indicate that the prograde P-T path passed near $P \approx 1.5$ GPa and $T \approx 600°$C, precluding the possibility of granulite-facies metamorphism immediately prior to the UHP metamorphism (Chopin et al., 1991). Because coexisting talc and phengite remained stable during decompression and did not break down to the low-pressure assemblage phlogopite + kyanite + quartz + H_2O, temperatures during decompression must have remained below 750°C at $P \approx$ 2.0 GPa and below 600°C at $P \approx 1.0$ GPa (Schertl et al., 1991).

Clastic rocks beneath the polymetamorphic series reached metamorphic conditions of $P < 1.0$ GPa and $T \approx 500°$C (Chopin et al., 1991). The "cold eclogite" and "micaschiste amygdalaire" that make up the rest of the polymetamorphic series reached $P \approx 1.5$ GPa and $T = 500$–$550°$C (Chopin et al., 1991). Clastic rocks overlying the polymetamorphic series reached $P \approx 1.0$–1.2 GPa and $T \approx 500°$C (Chopin et al., 1991). Late deformation under greenschist-facies conditions, perhaps at $P \approx 0.7$–0.9 GPa and $T \approx 500$–600°C, caused intermixing of units and variable development of a foliation and lineation (Chopin et al., 1991; Schertl et al., 1991).

There is conflicting information about the activity of H_2O during the Alpine UHP metamorphism. Massone (1990) used the Al_2O_3 content of talc to infer that a_{H_2O} was ≥ 0.8. Probable partial melting textures in the country rock gneiss have also been used to suggest that the activity of H_2O was near unity (Schertl et al., 1991). Note, however, that H_2O partitions into silicate melt at high pressures, and thus partial melting will reduce the H_2O activity in solid phases coexisting with melt. Rossman et al. (1989) used infrared spectrophotometry to infer that the pyropes contain 0.002–0.003 wt% hydroxyl, a measurement that reportedly implies garnet growth at low H_2O activity. The preservation of UHP assemblages outside their stability field also argues for low aqueous fluid activity. On the basis of thermodynamic calculations of the position of the reaction pyrope + coesite + H_2O = kyanite + talc and temperatures estimated from oxygen isotope measurements, Sharp et al. (1993) inferred that a_{H_2O} was 0.4–0.75. Moreover, they proposed that the fluid diluent was silicate melt (and specifically not CO_2).

It is uncertain whether the UHP metamorphism in Dora-Maira occurred at ≈100 Ma or ≈40 Ma. High-pressure metamorphism in the Alps is generally recognized to have occurred ≈100 Ma (e.g., Hunziker et al., 1989). Dora-Maira coesite quartzite has yielded ≈100 Ma phengite $^{40}Ar/^{39}Ar$ ages (Monié and Chopin, 1991) and ≈121 Ma U-Pb zircon ages (Paquette et al., 1989). UHP rocks overprinted by greenschist-facies assemblages, as well as structurally adjacent greenschist-facies units, yield $^{40}Ar/^{39}Ar$ phengite ages of 30–50 Ma (Scaillet et al., 1990; Monié and Chopin, 1991). These data suggest that UHP metamorphism occurred at ≈100 Ma and was followed by retrogression at ≈40 Ma. In contrast, comprehensive Sm-Nd and U-Th-Pb dating by Tilton et al. (1991) produced markedly younger, 38–30 Ma, ages for the UHP metamorphism. Zircon inclusions in four pyrope crystals (stable only at pressures greater than ≈1.2 GPa; Schreyer, 1988) and one pyrope quartzite yielded a chord with a lower intercept age of ≈38 Ma (Tilton et al., 1991). Ellenbergerite (a phase stable only at pressures >2.0 GPa; Chopin, 1986) and monazite from the same rock types yielded U-Pb and Th-Pb ages of 30-34 Ma (Tilton et al., 1991). Sm-Nd dating of 13 pyrope, zircon and phengite crystals produced an errorchron of 38 Ma (Tilton et al., 1991). Other ultrahigh-pressure rocks in the Alps from the Cima Lunga nappe have yielded five Sm-Nd garnet-pyroxene-whole rock ages of 38–42 Ma (Becker, 1993) and two U-Pb zircon ages of 32 and 36 Ma (Gebauer et al., 1991). Thus, there is an apparent 60 Ma discrepancy among different radiometric attempts to date the UHP metamorphism. The discrepancy may be attributable to an effect of deformation, pressure, phase chemistry, or some other factor on isotopic system(s) in one or more high-pressure phase(s). Given the increasing recognition that phengite may incorporate excess ^{40}Ar (e.g., Hacker and Wang, in review), we favor the Sm-Nd and U-Th-Pb ages of ≈38 Ma.

Coesite also occurs in the Zermatt-Saas zone, the uppermost part of the Pennine nappes, which consists of HP meta-sedimentary and meta-ophiolitic rocks representing portions of the Tethyan ocean basin subducted southward beneath the Apulian (African) plate in late Mesozoic time. Coesite inclusions occur in garnet and tourmaline in piemontite-bearing quartzite of the oceanic "schistes lustrés" (Barnicoat and Fry, 1986). Conditions for coesite crystallization are estimated as $T \approx 600°C$ and $P \approx 2.7$ GPa. The retrograde P-T path is judged to have passed first through $T \approx 500°C$ and $P \approx 1.4$ GPa, and then through $T \approx 400°C$ and $P \approx 0.5$ GPa (Reinecke, 1991) (Fig 5.1B).

China

Coesite-, aragonite-, and diamond-bearing eclogite occur in the collision zone between the Yangtze and Sino-Korean cratons in China (Ernst et al., 1991; Wang and Liou, 1991; Xu et al., 1992). Mafic eclogite occurs as blocks and rare layers in marble, schist, and quartzofeldspathic gneiss (Wang and Liou, 1991). Coesite pseudomorphs occur in garnet in the latter three rock types, and in omphacite and epidote in eclogite (Wang et al., 1990; Wang and Liou, 1991; R. Zhang, personal communication, 1992). Peak pressures and temperatures for coesite-bearing rocks are estimated as $T \approx 740$–$840°C$, $P > 2.8$ GPa on the Shandong Peninsula at the eastern end of the collision zone (Enami and Zang, 1990; Hirajima et al., 1990) (Fig 5.1C), and $T \approx 550$–$700°C$, $P > 2.6$–2.8 GPa in the Dabie Mountains farther west in the collision zone (Wang et al., 1990). Partial melting of gneiss in the northern Dabie Mountains may indicate that the H_2O activity was near unity at ultrahigh pressures or the partial melting may be related to decompression following the UHP metamorphism. Retrograde parageneses indicate amphibolite-facies and greenschist-facies overprints (Wang and Liou, 1991) at $T \approx 500°C$ and $P \approx 0.6$ GPa in the Dabie Mountains (Wang et al., 1990), and $T \approx 670$–$750°C$ and $P \approx 0.9$–1.4 GPa in Shandong Province (Enami and Zang, 1990). These rocks also contain partially transformed aragonite inclusions in garnet and omphacite (Wang and Liou, 1991; 1993) and diamond inclusions in garnet (Xu et al., 1992). Zircons from UHP gneisses yield U-Pb ages of ≈ 209 Ma (Ames et al., 1993), and hornblende, muscovite and biotite record cooling through 500–300°C by 180 Ma (Hacker and Wang, in review). Hornblende barometry on metamorphism associated with emplacement of Cretaceous plutons (≈ 105–135 Ma; Li and Wang, 1991) suggests crystallization at pressures of ≈ 500 MPa (Liou et al., 1992). Thus, the UHP rocks reached depths as shallow as ≈ 15 km by ≈ 125 Ma.

Regional Extent of UHP Metamorphism

The distribution of key mineral parageneses and phase compositions indicates that the area metamorphosed at ultrahigh pressure in China exceeds 400 km². UHP metamorphism may have occurred on a similar scale in the Alps, even though the outcrop area is limited by overlapping postorogenic sediments. In the Dora-Maira massif, coesite-bearing boudins are distributed over an area of 10×15 km. Jadeite pseudomorphs and garnet compositions suggest that the gneisses containing the coesite-bearing blocks were also metamorphosed at UHPs (Chopin et al., 1991). Tectonic models for these areas must account for exhumation of "blocks" exceeding 20 km in two dimensions. In the Zermatt-Saas zone, where the coesite quartzite unit is only several hundred meters long (Reinecke, 1991), further work is needed to determine the regional extent of UHP metamorphism.

Coesite → Quartz and Aragonite → Calcite Retrogression Textures

Silica inclusions in garnet from UHP regional metamorphic rocks are as large as 300 μm (Smith, 1984; Enami and Zang, 1990; Reinecke, 1991). Partially preserved coesite inclusions are invariably single crystals. The amount of retrogression from coesite to quartz varies from ≈25 to 100 vol%. Partially retrogressed coesite inclusions are surrounded by as many as several thousand elongate quartz grains (ranging down to less than 1 μm wide) arranged radially in an ellipsoidal shell around the coesite (Chopin, 1984; Smith, 1988). The quartz grains occur in groups of similarly shaped crystals with long axes subperpendicular to the host coesite crystal grain boundary (Fig 5.2). Individual thin sections show not only relict coesite crystals surrounded by thousands of quartz grains, but also coesite pseudomorphs composed of tens to only a few coarse grains (Smith, 1984). These textures are taken to indicate that after a host coesite crystal was consumed, the quartz grains coalesced and coarsened (Smith, 1988).

Ingrin and Gillet (1986) reported a lack of topotaxy between coesite and quartz from the Dora-Maira massif, based on transmission electron microscopy (TEM). We measured, with a universal-stage microscope, the *c*-axis orientations of quartz crystals grown on the largest coesite crystal reported by Wang and Liou (1991; their Figs 5.2A and 5.2B). The fine grain size and pervasive undulatory extinction make measurement difficult, but the *c*-axes of measurable grains are neither orthogonal to the host grain boundary nor parallel to the quartz crystal long axes. Thus, the crystallographic orientation of the quartz crystals appears not to have been controlled by the host garnet and coesite crystals, and the shape of the quartz crystals is related chiefly to the retreat of the unstable coesite crystal rather than to crystallographic controls on growth rate. The high density of atopotactic quartz crystals likely reflects a high nucleation to growth rate ratio. Such rate ratios are favored during transformations by rapid changes in reaction driving potential (Rubie,

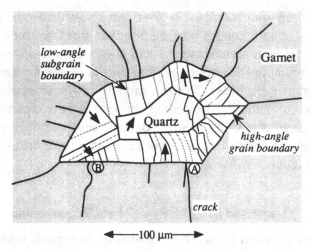

Figure 5.2. Line drawing of garnet (gray) including coesite pseudomorph retrogressed to quartz (unlabeled). Solid lines within garnet are cracks formed during expansion of the inclusion. Solid lines within quartz are high-angle grain boundaries and pale lines are low-angle subboundaries. Most grain boundaries are straight, but a minority are gently or sharply curved (A). Note the areas of grain impingement near the corners of the garnet grain boundary, where grains of one orientation truncate grains of another orientation (B). Not shown is a pervasive undulatory extinction within the quartz grains. Small arrows indicate traces of c-axes of grains measured with a u-stage microscope; the c-axes show no consistent relationship with the host coesite-garnet grain boundary. This is sample MW37 of Wang and Liou (1991; see their fig. 2A, 2B).

1983). The coesite → quartz reaction has relatively low $\Delta P/\Delta T$, thus, changes in potential are more easily effected by decompression, rather than by heating. If decompression caused coesite → quartz reaction, it might have been the result of erosional or tectonic removal of overlying rocks or cracking of the enclosing garnet crystal.

Aragonite single-crystal inclusions in garnet and omphacite are surrounded by aggregates of calcite crystals, much like the textures associated with the breakdown of coesite to form quartz (Wang and Liou, 1993). The calcite occurs as fine-grained aggregates surrounding single aragonite crystals.

Minerals that host inclusions of coesite and aragonite invariably contain radiating cracks (Fig 5.2; Wang and Liou, 1993). These radiating cracks are typically inferred to have formed during the polymorphic phase transformation to form the low-pressure polymorph quartz or calcite (van der Molen and Paterson, 1986). However, the expansion may instead be a result of differential volume change of the host and inclusion phases during decompression or cooling (Wendt et al., 1993). In the absence of coesite or other supporting evidence, radial cracks emanating from quartz inclusions do not mandate UHP metamorphism. The most reliable indicators of the earlier presence of coesite are the textures illustrated in Fig 5.2, particularly the subparallel, sometimes curving quartz subgrains (Fig 5.2 label A) and the truncation of one set of subgrains by another (Fig 5.2 label B).

Experimental Kinetic Studies

There have been a modest number of experimental studies on the equilibrium boundary and rate of the quartz-coesite transformation (see Bohlen and Boettcher, 1982, for a review); however, these data are not generally applicable to natural metamorphism. The physical state under which quartz grew from coesite in these experiments differs substantially from nature. In nature, the transformation was of coesite single crystals enclosed in single-crystal garnet, whereas the laboratory experiments were conducted on powdered coesite often in the presence of powdered quartz and H_2O. Powdering results in an experimental transformation rate that is faster than the natural transformation, and can provide only an upper bound on the natural case (e.g., Rubie and Thompson, 1985; Hacker et al., 1992).

However, experimental data do suggest that the coesite → quartz reaction is sluggish in the absence of water; Dachille et al. (1963) found that dried, powdered coesite persisted for at least 550 hours when heated at 1070°C and room pressure. Transformation experiments on more "natural" material must be conducted to provide data that can be applied quantitatively to the retrogression and exhumation of coesite-bearing rocks. One possibility is to conduct experiments on coesite single crystals contained within garnet crystals, but synthesizing such material is fraught with difficulty.

Significantly better experimental data are available for the transformation of aragonite to calcite. Carlson and Rosenfeld (1981) measured the growth rate of calcite from aragonite single crystals at 1 atm, and Carlson (1983) calculated a nucleation rate from Kunzler and Goodell's (1970) experiments on powdered aragonite. Extrapolation of Carlson's data to UHP metamorphic rocks meets with difficulty, however. Even for small reaction free energies (<1 kJ mole^{-1}), Carlson's data predict that 1000 μm aragonite grains will revert completely to calcite in less than 1 Ma at temperatures as low as 200°C. In obvious conflict with this is the local preservation of aragonite in UHP rocks in China that evidently entered the calcite stability field at temperatures in excess of 500°C (Wang and Liou, 1993). Fundamental differences must therefore exist between the experimental configuration and the natural situation. Possible differences include:

1. The retrograde *P-T* path following UHP metamorphism may have been such that the free energy potential to drive nucleation and growth was very

small or zero. In other words, the depressurization and cooling path may have followed the conditions of the calcite–aragonite equilibrium boundary to low temperatures. Calculations based on Carlson's data suggest that reaction free energies of 100 J are sufficient to drive complete transformation in less than 1 Ma at temperatures of 200°C. Such a small reaction free energy means that the pressure during decompression cannot have deviated from the equilibrium pressure by more than ≈50 MPa. However, field study (Wang and Liou, 1991, 1993) indicates that retrogression occurred in the calcite stability field at pressures ≈0.5 GPa below the equilibrium boundary and at temperatures of ≈500°C.

2. The natural nucleation or growth rates may be very different from the experimental situation, because the natural aragonite crystals are enclosed as single crystals within host garnet and omphacite single crystals. A large rate difference is plausible because nucleation in powders and single crystals is controlled by the free energy of the aragonite crystal surface in contact with vapor or fluid, whereas nucleation in the natural case must occur at the crystalline interface between the aragonite and host crystal. Surface and grain-boundary free energies differ by an order of magnitude in metals (Murr, 1975; Porter and Easterling, 1981), and similar differences between aragonite-fluid and aragonite-garnet interfaces would yield notable differences in transformation kinetics. Evaluating the importance of such effects is experimentally possible but difficult.

3. The transformation of aragonite to calcite involves dilatation, and the host mineral may act as a pressure vessel that confines the expanding inclusion. The aragonite → calcite transformation can be inhibited if the pressure buildup causes the inclusion to remain in its HP stability field, even while the outside of the host crystal is at substantially lower pressure (Gillet et al., 1984; van der Molen and van Roermund, 1986).

Gillet et al. (1984) and van der Molen and van Roermund (1986) postulated that the transformation of coesite to quartz was inhibited during cooling and decompression because the garnet surrounding the coesite acted as a pressure vessel that maintained high pressure on the silica inclusion even as the rock pressure and temperature moved well into the quartz stability field. Modeling isotropic, linear elastic behavior of garnet and coesite, they calculated the rate of quartz formation based on the assumption that any increase in inclusion pressure required to stay out of the quartz stability field was instantly achieved by the volume increase produced when coesite transformed to quartz. In spite of the assumptions required to make the problem tractable, the models lead to the interesting prediction that the pressure vessel effect restricts the coesite → quartz retrogression to only moderate amounts (≈25–30 vol%) before the garnet fractures. Van der Molen and van Roermund (1986) calculated that when the temperature has decreased to ≈400°C, the inclusion pressure will have built up to more than three times the external pressure, and the garnet may fracture, allowing ingress of fluid and more complete retrogression. Their model requires that the host crystal be able to sustain tensile stresses of ≈500 MPa at temperatures of ≈500°C. Of the phases that include coesite and aragonite, only the mechanical behavior of clinopyroxene has been investigated at high temperatures. Kollé and Blacic (1982) found that clinopyroxene single crystals can deform by twinning at significantly lower differential stresses, about 100–140 MPa, at strain rates of 10^{-4}–10^{-8} s^{-1} and temperatures of 500–1000°C. This implies that clinopyroxene crystals might not be able to preserve coesite inclusions, but would instead deform by mechanical twinning.

P-T Paths and Thermal Evolution

The occurrence of coesite-bearing eclogite derived from sedimentary protoliths requires (1) subduction of crustal material to depths of ≈100 km, (2) detachment from the downgoing slab (transfer to the overriding plate), and (3) exhumation by a combination of erosional and tectonic processes.

Prograde P-T *Path*

Coesite-bearing eclogites record relatively cool temperatures of 550–900°C for depths of 90–125 km (Fig 5.3A). Typical geotherms from stable continental and oceanic settings suggest that temperatures exceed 1000°C at 100 km depth (Pollack and Chapman, 1977). In convergent plate margins, temperatures at depth can be substantially cooler than in intraplate settings because the subduction of oceanic lithosphere carries cold near-surface rocks downward, depressing isotherms at depth. The peak metamorphic pressures and temperatures recorded by coesite-bearing eclogites are consistent with *P-T* paths calculated for subduction zones (e.g., Peacock, 1990, 1993).

In thermal steady state, the *P-T* path followed by the top of the subducting oceanic crust coincides with the *P-T* conditions along the subduction shear zone. In the absence of significant radiogenic heating, steady-state temperatures along a subduction shear zone are approximated by (Molnar and England, 1990, eqns. 16 and 23):

$$T = \frac{(Q_0 + \tau V)\, z_f / k}{S} \qquad (5.1)$$

where T = temperature (K), Q_0 = heat flux into the base of the subducting lithosphere (W m^{-2}), τ = constant shear stress (Pa), V = convergence rate (m s^{-1}), z_f = depth to the fault (m), k = thermal conductivity (W m^{-1} K^{-1}), and

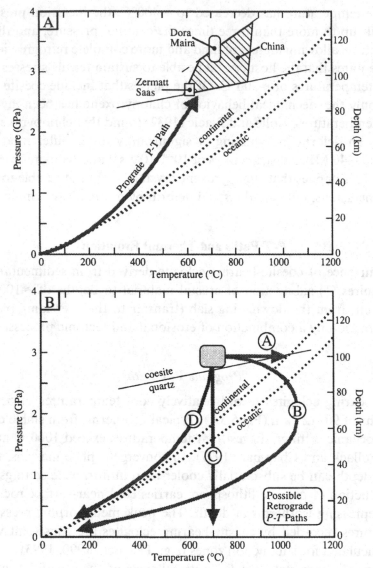

Figure 5.3. (A) Representative prograde subduction zone *P-T* path consistent with metamorphic conditions recorded by coesite-bearing eclogites; this path is one of a family of transient and steady-state subduction-zone *P-T* paths consistent with coesite-eclogite-facies metamorphism. This particular *P-T* path was calculated for a steady-state subduction zone with a convergence rate = 3 cm·yr⁻¹, a constant shear stress = 80 MPa, a dip of 25°, and average thermal parameters. See text for discussion. (B) Four possible *P-T* paths subsequent to coesite-eclogite facies metamorphism at $P \approx 2.9$ GPa and $T \approx 700°$C. Dotted lines represent average oceanic and continental geotherms [surface heat flow = 50 mW m⁻² (for both cases), Pollack and Chapman, 1977]. Depth-pressure relation based on a density of 3000 kg·m⁻³. Path (A): No exhumation after peak metamorphism results in isobaric heating toward the steady state geotherm. Path (B): Exhumation at a moderate rate (≈1 km·Ma⁻¹) results in

S = a divisor that accounts for advection (heat transported by the subducting oceanic lithosphere):

$$S = 1 + b \sqrt{\frac{V z_f \sin \delta}{\kappa}} \qquad (2)$$

where b = a constant (≈ 1 based on numerical experiments), δ = angle of subduction, and κ = thermal diffusivity ($m^2 s^{-1}$).

The prograde P-T path in Fig 5.3A represents the calculated steady state P-T path for the case of $Q_0 = 0.05$ W m^{-2}, $\tau = 80$ MPa, $V = 3$ cm yr^{-1}, $k = 2.5$ W m^{-1} K^{-1}, $\delta = 25°$, and $\kappa = 10^{-6}$ m^2 s^{-1}. This representative P-T path is but one of a family of subduction zone P-T paths that pass through the P-T conditions recorded by coesite-bearing eclogites; similar steady-state P-T paths may be calculated for different combinations of convergence rate, shear stress, and other parameters.

Alternatively, the coesite-bearing eclogites may have formed during the early stages of subduction prior to the establishment of thermal steady state. At 100 km depth and $V = 3$ cm yr^{-1}, achievement of thermal steady state requires several tens of Ma (see equations in Molnar and England, 1990). If coesite-bearing eclogites form during the first few tens of Ma of plate convergence while the subduction zone is still warmer than the steady-state thermal structure, then lower rates of shear heating (and lower shear stresses) are required to achieve the conditions of coesite–eclogite metamorphism. Shear heating may have been negligible (shear stresses ≈ 0), if the coesite-bearing eclogites formed during the first few million years of subduction while the hanging wall was still hot (Peacock, 1992, 1993).

Exhumation Mechanism

Exhumation of metamorphic rocks, defined as the displacement of rocks toward the Earth's surface (England and Molnar, 1990), requires removal of the rock overburden that existed at the time of metamorphism. Exhumation may involve sedimentary erosion, tectonic thinning, or some combination of the two processes.

Caption for Fig. 5.3 (cont.) moderate heating during decompression followed by cooling as the rocks approach the surface. Path (C): "Infinitely fast" exhumation results in nearly isothermal (adiabatic) decompression. Path (D): Exhumation occurring at moderate rates while subduction continues beneath the exhuming rocks results in cooling during decompression. Continued subduction chills the system, allowing rocks to be returned to the surface without heating toward the steady-state geotherm at high pressures. See text for discussion.

Large syn- to postorogenic Alpine sedimentary deposits document the importance of sedimentary erosion in exhuming Alpine metamorphic rocks. However, it is unlikely that sedimentary erosion alone could have been responsible for the removal of ≈100 km of overburden that existed above the Dora-Maira massif. The volume of Alpine syn- to postorogenic sedimentary deposits is dramatically smaller than that predicted by an erosion model (Platt, 1986), although some of the sedimentary record may have been subducted. Moreover, the pressure difference (≈1.5 GPa) between the coesite-bearing unit and adjacent units cannot be explained by local differential erosion but must record displacement between units.

Metamorphic rocks can also be exhumed by tectonic thinning of overburden. Extensional pure shear of the lithosphere displaces rocks toward the Earth's surface. In extensional simple shear, rocks in the footwall are exhumed as they are drawn out from beneath the hangingwall. Several workers have proposed that thrust faulting alone can exhume metamorphic rocks, but in the absence of erosion or tectonic thinning of the hangingwall, rocks in the upper plate do not approach the surface (i.e., they are not exhumed). In contrast to the views of some workers, exhumation of UHP rocks by extensional tectonics can occur within an active convergent plate margin. Platt (1986, 1992) has argued for extensional exhumation of high-pressure metamorphic rocks, pointing out that extension is occurring today in modern continental collision zones such as Tibet (Molnar and Lyon-Caen, 1989), and in modern accretionary wedges such as the Hikurangi prism of New Zealand (Walcott, 1987). Paleo-extensional structures have been reported from virtually every orogenic belt. Recently, Ballevre et al. (1990) and Blake and Jayko (1990) documented a west-dipping ductile normal fault overlying the Dora-Maira massif that appears to have accommodated some of the Dora-Maira exhumation in mid-Tertiary time, and Avigad (1992) similarly interpreted data collected by Henry (1990). Wheeler (1991) reported that thrusting may have occurred during exhumation as well, as HP rocks in the Dora-Maira massif are in thrust contact with underlying lower pressure rocks. Fossen and Rykkelid (1992) proposed that exhumation of Norwegian eclogites occurred partly by subhorizontal extension. Thus, tectonic thinning by subhorizontal extension, perhaps with the aid of thrusting at deeper levels and erosion above, seems the most likely mechanism for exhumation of coesite-bearing eclogite.

Cooling During Exhumation

The great depth and high P/T ratio of UHP metamorphism indicates that the deep burial of coesite-bearing rocks must have occurred by subduction. Away from plate boundaries, inferred oceanic and continental geotherms pass through temperatures in the range 1000–1120°C at depths of $\approx$100 km (Fig. 5.3; Pollack and Chapman, 1977; surface heat flux of 50 mW·m^{-2}). The coesite-bearing eclogites record markedly cooler temperatures (550–900°C) at similar depths. As discussed above, these relatively cool metamorphic conditions are consistent with transient and steady-state P-T paths calculated for subducting lithosphere.

A greater problem is posed by the preservation of low temperatures during exhumation of coesite-bearing eclogites, as illustrated in Fig 5.3B. Most exhumation scenarios result in heating during decompression, in contrast with the petrologic evidence discussed above, which requires that the coesite-bearing eclogites cooled during decompression.

P-T path A (Fig 5.3B) represents the trajectory followed by coesite-bearing eclogites if subduction ceases and no exhumation occurs. In this scenario, the eclogites heat isobarically as the thermal structure of the former subduction zone relaxes toward a new, warmer steady state. Once subduction ceases, the isotherms rise toward the surface because the isotherms are no longer deflected downward by the subducting lithosphere. If the coesite-bearing eclogites followed P-T path A, new, higher temperature metamorphic assemblages would be expected to form.

If subduction ceases and the coesite-bearing eclogites are exhumed by erosion at a moderate rate of $\approx$0.5–1.0 km Ma^{-1}, they will follow a P-T trajectory like path B (Fig 5.3B). This "clockwise" P-T path is similar to the calculated paths followed by metamorphic rocks exhumed by erosion subsequent to crustal thickening (e.g., England and Thompson, 1984). During the early stages of decompression, the rocks heat up as the thermal structure reequilibrates and approaches the post-subduction steady-state geotherm. As exhumation continues, the rocks cool as they approach the surface. Because the rocks become hotter during decompression, new mineral assemblages may form to reflect the higher temperatures achieved at lower pressures. For P-T path B, a new mineral assemblage in the eclogites is likely to form near the maximum temperatures reached during decompression: temperatures of $\approx$1050°C at pressures of $\approx$2 GPa.

P-T path C (Fig 5.3B) represents the hypothetical trajectory followed by coesite-bearing eclogites if subduction ceases and the rocks are exhumed instantaneously (i.e., an infinitely rapid erosion rate). In this end-member situation, the eclogites will undergo adiabatic (nearly isothermal) decompression. During decompression, the rocks would cross many metamorphic reactions that have a positive P-T slope, such as the reaction phengite + talc $\rightarrow$ phlogopite + kyanite + quartz + H_2O (Fig 5.1A). There is no petrologic evidence for such reaction having occurred during decompression of Dora-Maira rocks (e.g., Chopin, 1984).

The retrograde *P-T* paths determined for coesite-bearing eclogites suggest cooling during decompression (Fig 5.1) in contrast to the heating (paths A and B) and isothermal (path C) decompression paths depicted in Fig 5.3B. Maximum temperatures appear to have been reached during peak coesite-eclogite facies metamorphism and not during decompression. Cooling during decompression requires that the coesite-bearing eclogites lost heat to the surrounding rocks throughout the exhumation process. Heat may have been conducted downward into cooler underlying rocks if subduction continued beneath the coesite-bearing eclogite terrains during exhumation of the eclogites. Alternatively, if the coesite-bearing eclogites were exhumed by tectonic extension of overlying rocks, then eclogites lying beneath the normal fault (i.e., in the upper part of the lower plate) might cool as heat is conducted into the cooler upper plate of the detachment system. Cooling during decompression would be most pronounced if subduction continued to occur during exhumation and if eclogites were transported toward the surface in the lower plate of extensional structures.

P-T path D (Fig 5.3B) represents the trajectory followed by coesite-bearing eclogites if exhumation occurred while subduction continued beneath the eclogites, rather than after subduction ceased. Continued subduction or underplating of additional units would keep isotherms depressed in the region of the subducting slab and rocks exhumed during this period would cool during decompression (e.g., Rubie, 1984; Davy and Gillet, 1986; Gillet et al., 1986; Ernst, 1988). If coesite-bearing eclogites formed and were exhumed in a steady-state subduction zone thermal structure, then the retrograde, decompression *P-T* path could closely track the prograde, subduction *P-T* path. In thermal steady state, temperatures within the lithosphere of the upper plate increase monotonically from the surface down to the subduction shear zone; no inverted thermal gradients exist in the upper plate. If exhumation took place under approximately steady-state thermal conditions, we cannot constrain the location of the eclogites with respect to the subduction shear zone during exhumation. Alternatively, if the eclogite protoliths were subducted prior to the development of thermal steady state, then the retrograde *P-T* path could be cooler than the prograde *P-T* path.

In the second heat loss scenario, coesite-bearing eclogite terrains in the lower plate of an extensional structure, such as a normal fault or shear zone, cool during exhumation as a result of conduction of heat into the upper plate. Prior to exhumation, rocks lying above the eclogites will be generally cooler than a steady-state geotherm because of the depression of isotherms by the subduction process. If the eclogites are exhumed by movement along a low-angle normal fault that cuts through these cooler rocks, then the upper portions of the footwall will lose heat to the cooler hangingwall during movement along the fault. Rocks located at the top of the footwall will cool more than rocks located deeper in the footwall. If exhumation occurs after subduction ceases, then there will be competition between cooling due to contact with the cooler hangingwall and heating from below as isotherms move toward the surface during postsubduction thermal relaxation.

In the case of the Western Alps, the hypothesis of exhumation during continued subduction is supported by plate reconstructions which suggest that the Mesozoic Tethyan ocean basin was slowly subducted beneath the Apulian (African) plate between 120 and 40 Ma (e.g., Dewey et al., 1989; Hsü, 1989). Eoalpine HP metamorphism in the Dora-Maira region occurred during the early stages of plate convergence, followed by thrusting of the Pennine nappes beneath the Dora-Maira. Consumption of the Tethyan ocean basin led to collision of the European and Apulian plates at ≈40 Ma, emplacement of the Helvetic nappes, and widespread regional metamorphism. At ≈40 Ma, according to Monié and Chopin (1991), the Dora-Maira massif was undergoing greenschist-facies retrogression at depths of 15–20 km. Subduction of the Tethyan ocean basin was apparently occurring beneath the Dora-Maira massif from 120 to 40 Ma while the eclogites were being exhumed from depths of 100–120 km to 15–20 km.

While this scenario appears to be accepted by most workers, the recent radiometric data of Tilton et al. (1991), Gebauer et al. (1991), and Becker (1993) raise the possibility that UHP Dora-Maira metamorphism may be as young as 40 Ma, and would thus correspond to the onset of continental collision. In this case, the exhumation of UHP rocks must have occurred without the refrigerative effects of continuing, deeper-level subduction, and cooling the eclogites during decompression is even more problematic.

Exhumation Rates

In certain situations, the shape of the decompression portion of a metamorphic *P-T* path can be used to constrain the exhumation rate of the metamorphic rock. For example, Draper and Bone (1981) showed that if subduction ceases, blueschist-facies rocks in a subduction zone will heat up and be overprinted by greenschist- or amphibolite-facies assemblages unless very rapid exhumation rates of 10–100 km Ma^{-1} prevail. For the case of the coesite-bearing eclogites, as well as the many noncollisional blueschist terrains that cooled during exhumation (e.g., Ernst, 1988), the thermal structure during exhumation was probably controlled by deeper-level subduction coincident with exhumation. Therefore, it is not possible to constrain the exhumation rate solely on the basis of the retrograde P-T path (Rubie, 1984, 1990). Exhumation may have been relatively rapid or relatively slow as long as subduction continued to chill the overlying rocks.

Average rates of exhumation may be calculated by dating different portions of a metamorphic P-T path. For example, if the Dora-Maira coesite-bearing eclogites underwent UHP metamorphism at $\approx$38 Ma (Tilton et al., 1991) and metamorphic pressures of $\approx$3.2 GPa (Schertl et al., 1991), the average long-term exhumation rate was a moderate 3.0 km Ma^{-1} (3.0 mm·yr^{-1}). Exhumation rates for Dabie Mountains coesite-bearing eclogites constrained by data discussed earlier are $\approx$0.75 km m·$y.^{-1}$. Fault-zone parallel displacement rates of 10–20 mm·yr^{-1} have been inferred for Basin and Range extensional faults by Davis (1988), Hacker et al. (1991), and Spencer and Reynolds (1991). With a displacement rate of 10–20 mm·yr^{-1}, an extensional fault dipping 30° could bring rocks from a depth of 100 km to the surface in 10–20 Ma.

Conclusions

Coesite-bearing eclogites record relatively low-temperature, HP metamorphic conditions consistent with formation in a subduction zone environment. Existing kinetic data suggest that in the presence of fluid the retrograde coesite $\rightarrow$ quartz and aragonite $\rightarrow$ calcite reactions should progress to completion during exhumation of UHP rocks. The preservation of coesite and aragonite only as inclusions in other minerals suggests that retrogressive phase changes can be inhibited within a porphyroblast capable of maintaining high internal pressures and excluding catalytic volatiles. More experimental kinetic data on the coesite $\rightarrow$ quartz and aragonite $\rightarrow$ calcite reactions are needed; the effects of fluids, of powdering, and of inclusions within other phases must be assessed quantitatively before experimental data can be extrapolated to the Earth.

Careful examination of natural samples that reached different peak temperatures at UHPs may reveal additional information about the rate and mechanisms of the coesite $\rightarrow$ quartz and aragonite $\rightarrow$ calcite transformations. UHP occurrences in central China are a good candidate for such study, because peak temperatures vary from $\approx$550°C to 850°C along the length of the collisional orogen. Cathodoluminesence and microscale isotopic and chemical determinations might also be used to identify different stages in the formation of retrogressive quartz or calcite.

Polycrystalline quartz aggregates and cracks radiating outward from inclusions are not, by themselves, reliable indicators of UHP metamorphism. In the absence of relict coesite, subparallel, sometimes curving quartz subgrains and the truncation of one set of quartz subgrains by another are probable indicators of the earlier presence of coesite.

Petrologic evidence for cooling during decompression from depths in excess of 90 km requires that the exhumation of coesite-bearing terrains occurred while subduction/underflow continued to chill the overlying plate or that the eclogites were located near the top of an extensional structure's footwall. On the basis of radiometric data, the average rate of exhumation of the Dora-Maira coesite-bearing eclogite was ≈3.0 km Ma^{-1}, assuming that high-pressure metamorphism occurred ≈38 Ma. Exhumation from depths >90 km demands extensional tectonic thinning in addition to sedimentary erosion. Extensional features above the Dora-Maira eclogites have been described by Ballèvre et al. (1990); we predict that similar structures will be found above other UHPM terrains. More field work is required to reveal whether such extensional features exist in other terrains and to determine the age, magnitude, and sense of displacement along these structures.

Acknowledgments

B. R. H. was supported by the Department of Energy and the National Science Foundation. S. M. P. thanks the National Science Foundation for partial support of this research through grant EAR–9105741 and the Swiss Federal Institute of Technology (ETH) for providing a Guest Professorship during the 1991–1992 academic year. Gary Ernst, J. G. Liou, John Platt, Dave Rubie, and Xiaomin Wang provided reviews. Discussions in China with T. A. Carswell are also gratefully acknowledged.

References

Ames, L., Tilton, G. R., and Zhou, G. 1993. Timing of collision of the Sino-Korean and Yangtse cratons: U-Pb zircon dating of coesite-bearing eclogites. *Geology, 21*, 339–342.

Avigad, D. 1992. Exhumation of coesite-bearing rocks in the Dora-Maira massif (western Alps, Italy). *Geology, 20*, 947–950.

Ballèvre, M., Pinardon, J.-L., and Kienast, J. R. 1989. Reversal of Fe-Mg partitioning between garnet and staurolite in eclogite-facies metapelites from the Champtoceaux Nappe (Brittany, France). *Journal of Petrology, 30*, 1321–1349.

Ballèvre, M., Lagabrille, Y., and Merle, O. 1990. Tertiary ductile normal faulting as a consequence of lithospheric stacking in the western Alps. *Mem. Soc. geol. France, 156*, 227–236.

Barnicoat, A. C., and Frey, N. 1986. High-pressure metamorphism of the Zermat-Saas ophiolite zone, Switzerland. *Jour. Geol. Soc. London, 143*, 607–618.

Becker, H. 1993. Garnet peridotite and eclogite Sm-Nd mineral ages from the Lepontine dome (Swiss Alps): New evidence for Eocene high pressure metamorphism in the Central Alps. *Geology, 21,* 599–602.

Blake, M.C.J., and Jayko, A. S. 1990. Uplift of very high pressure rocks in the western Alps: evidence for structural attenuation along low-angle faults. *Mem. Soc. geol. France, 156,* 237–246.

Bohlen, S. R., and Boettcher, A. L. 1982. The quartz-coesite transformation: a pressure determination and the effects of other components. *J. Geophys. Res., 87,* 7073–7078.

Bundy, F. R. 1980. The P, T phase reaction diagrams for elemental carbon. *J. Geophys. Res., 85,* 6930–6936.

Carlson, W. D. 1983. Aragonite-calcite nucleation kinetics: an application and extension of Avrami transformation theory. *Jour. Geol., 91,* 57–71.

Carlson, W. D., and Rosenfeld, J. L. 1981. Optical determination of topotactic aragonite-calcite growth kinetics: Metamorphic implications. *Jour. Geol., 89,* 615–638.

Chopin, C. 1984. Coesite and pure pyrope in high-grade blueschists of the western Alps: A first record and some consequences. *Contrib. Mineral. Petrol., 86,* 107–118.

Chopin, C. 1986. Phase relationships of ellenbergite, a new high-pressure Mg-Al-Ti-silicate in pyrope-coesite-quartzite from the Western Alps. *Geol. Soc. Amer. Mem., 164,* 31–42.

Chopin, C., Henry, C., and Michard, A. 1991. Geology and petrology of the coesite-bearing terrain, Dora-Maira massif, Western Alps. *Eur. Jour. Mineral., 3,* 263–291.

Dachille, F., Zeto, R. J., and Roy, R. 1963. Coesite and stishovite: Stepwise reversal transformations. *Science,* 991–993.

Davis, G. A. 1988. Rapid upward transport of mid-crustal mylonitic gneisses in the footwall of a Miocene detachment fault, Whipple Mountains, southeastern California. *Geol. Rundsch., 77,* 191–209.

Davy, P., and Gillet, P. 1986. The stacking of thrust slices in collision zones and its thermal consequences. *Tectonics, 5,* 913–929.

Dewey, J. F., Helman, M. L., Turco, E., Hutton, D.W.H., and Knott, S. D. 1989. Kinematics of the western Mediterranean. *Geological Society Special Publication, 45,* 265–283.

Draper, G., and Bone, R. 1981. Denudation rates, thermal evolution and preservation of blueschist terrains. *Jour. Geol., 89,* 601–613.

Enami, M., and Zang, Q. 1990. Quartz pseudomorph after coesite in eclogites from Shandong Province, east China. *Am. Mineralogist, 75,* 381–386.

England, P., and Molnar, P. 1990. Surface uplift, uplift of rocks, and exhumation of rocks. *Geology, 18,* 1173–1177.

England, P. C., and Thompson, A. B. 1984. Pressure-temperature-time paths of regional metamorphism. Heat transfer during the evolutions of regions of thickened continental crust. *J. Petrol., 25,* 894–928.

Ernst, W. G. 1988. Tectonic history of subduction zones inferred from retrograde blueschist P-T paths. *Geology, 16,* 1081–1084.

Ernst, W. G., Zhou, G., Liou, J. G., Eide, E., and Wang, X. 1991. High-pressure and superhigh-pressure metamorphic terranes in the Qinling-Dabie Mountain Belt, central China; Early- to Mid-Phanerozoic accretion of the western paleo-Pacific rim. *Pacific Science Association Information Bulletin, 43,* 6–14.

Fossen, H., and Rykkelid, E. 1992. Postcollisional extension of the Caledonide orogen in Scandinavia: structural expressions and tectonic significance. *Geology, 20*, 737–740.

Gebauer, D., Grünenfelder, M., Tilton, G. R., and Trommsdorff, V. 1991. The geodynamic evolution of the gar-peridotite/eclogite association of Alpe Arami (central Alps) from early Proterozoic to Oligocene HP-metamorphism. *Terra Abstracts, 6*, 5.

Gillet, P., Choukroune, P., Ballèvre, M., and Davy, P. 1986. Thickening history of the western Alps. *Earth Planet. Sci. Lett., 78*, 44–52.

Gillet, P., Ingrin, J., and Chopin, C. 1984. Coesite in subducted continental crust P-T history deduced from an elastic model. *Earth Planet. Sci. Lett., 70*, 426–436.

Hacker, B. R., Kirby, S. H., and Bohlen, S. R. 1992. Metamorphic petrology: Calcite → aragonite experiments. *Science, 258*, 110–112.

Hacker, B. R., and Wang, Q. C. 1995. Cooling history of ultrahigh pressure rocks, Dabie Mountains, China. *Tectonics* (in press).

Hacker, B. R., Yin, A., Christie, J. M., and Snoke, A. W. 1990. Differential stress, strain rate, and temperatures of mylonitization in the Ruby Mountains: implications for the rate and duration of uplift. *J. Geophys. Res., 95*, 8569–8580.

Henry, C. 1990. L'Unite a coesite du massif Dora-Maira dans son cadre petrologique et structural (Alpes occidentales, Italie) Ph.D., Universite de Paris VI.

Hirajima, T., Ishiwatari, A., Cong, B., Zhang, R., Banno, S., and Nozaka, T. 1990. Coesite from Mengzhong eclogite at Donghai county, northeastern Jiangsu province, China. *Mineral. Mag., 54*, 579–583.

Hsü, K. J. 1989. Time and place in Alpine orogenesis – the Fermor lecture. *Geol. Soc. Amer. Spec. Pap., 45*, 421–443.

Hunziker, J. C., Desmons, J., and Martinotti, G. 1989. Alpine thermal evolution in the central and the western Alps. *Geol. Soc. Lond. Spec. Pub., 45*, 353–367.

Ingrin, J., and Gillet, P. 1986. TEM investigation of the crystal microstructures in a quartz-coesite assemblage of the western Alps. *Physics and Chemistry of Minerals, 13*, 325–330.

Kollé, J. J., and Blacic, J. D. 1982. Deformation of single-crystal clinopyroxenes: 1. Mechanical twinning in diopside and hedenbergite. *J. Geophys. Res., 87*, 4019–4034.

Kunzler, R. H., and Goodell, H. G. 1970. The aragonite-calcite transformation: A problem in the kinetics of a solid-solid reaction. *Am. J. Sci., 1269*, 360–391.

Li, S., and Wang, T. 1991. *Geochemistry of granitoids in Tongbaishan–Dabieshan, central China*. China University of Geosciences Press.

Massonne, H.-J. 1990. Geothermobarometrie mit Al-Gehalten in Talk. *Eur. Jour. Mineral., 2*, 169.

Massonne, H.-J., and Schreyer, W. 1989. Stability field of the high-pressure assemblage talc + phengite and two new phengite barometers. *Eur. Jour. Mineral., 1*, 391–410.

Molnar, P., and England, P. C. 1990. Temperature, heat flux, and frictional stress near major thrust fault. *J. Geophys. Res., 95*, 4833–4856.

Molnar, P., and Lyon-Caen, H. 1989. Fault plane solutions of earthquakes and active tectonic of the Tibetan plateau and its margins. *Geophys. J. Intern., 99*, 123–153.

Monie, P., and Chopin, C. 1991. [40]Ar/[39]Ar dating in coesite-bearing and associated units of the Dora-Maira massif, Western Alps. *Eur. Jour. Mineral., 3*, 239–262.

Murr, L. E. 1975. *Interfacial Phenomena in Metals and Alloys*. London: Addison-Wesley.

Paquette, J. R., Chopin, C., and Peucat, J. J. 1989. U-Pb zircon, Rb-Sr and Sm-Nd geochronology of high- to very-high-pressure meta-acidic rocks from the western Alps. *Contrib. Mineral. Petrol., 101*, 280–289.

Peacock, S. M. 1990. Fluid processes in subduction zones. *Science, 248*, 329–337.

Peacock, S. M. 1992. Blueschist-facies metamorphism, shear heating, and *P-T-t* paths in subduction shear zones. *J. Geophys. Res., 97*, 17,693–17,707.

Peacock, S. M. 1993. Metamorphism, dehydration, and the importance of the blueschist → eclogite transition in subducting oceanic crust. *Geol. Soc. Amer. Bull., 105*, 684–694.

Platt, J. P. 1992. Comment on "Exhumation of high-pressure metamorphic rocks" by K. Hsü. *Geology, 20*, 186–187.

Platt, J. P. 1986. Dynamics of orogenic wedges and the uplift of high-pressure metamorphic rocks. *Geol. Soc. Amer. Bull., 97*, 1037–1053.

Pollack, H. N., and Chapman, D. S. 1977. On the regional variation of heat flow, geotherms and the lithospheric thickness. *Tectonophysics, 38*, 279–296.

Porter, D. A., and Easterling, K. E. 1981. *Phase transformations in metals and alloys.* London: Chapman and Hall.

Reinecke, T. 1991. Very-high-pressure metamorphism and uplift of coesite-bearing metasediments from the Zermatt-Saas zone, Western Alps. *Eur. Jour. Mineral., 3*, 7–17.

Rossman, G. R., Berna, A., and Langer, K. 1989. The hydrous component of pyrope from the Dora-Maira massif, western Alps. *Eur. Jour. Mineral., 1*, 151–154.

Rubie, D. C. 1983. Reaction-enhanced ductility: The role of solid-solid univariant reactions in deformation in the Sesia Zone, Western Alps. *Tectonophysics, 96*, 331–352.

Rubie, D. C. 1984. A thermal-tectonic model for high-pressure metamorphism and deformation in the Sesia zone, western Alps. *J. Geol., 92*, 21–36.

Rubie, D. C. 1990. Role of kinetics in the formation and preservation of eclogites. *Eclogite Facies Rocks.* Glasgow: Blackie, 111–140.

Rubie, D. C., and Thompson, A. B. 1985. Kinetics of metamorphic reactions at elevated temperatures and pressures: an appraisal of available experimental data. *Advances in Physical Geochemistry.* New York: Springer-Verlag, 27–79.

Scaillet, S., Feraud, G., Lagabrielle, Y., Ballèvre, M., and Ruffet, G. 1990. $^{40}Ar/^{39}Ar$ laser-probe dating by step heating and spot fusion of phengites from the Dora-Maire Nappe of the western Alps, Italy. *Geology, 18*, 741–744.

Schertl, H.-P., Schreyer, W., and Chopin, C. 1991. The pyrope-coesite rocks and their country rocks at Parigi, Dora-Maira Massif, Western Alps: detailed petrography, mineral chemistry and PT-path. *Contrib. Mineral. Petrol., 108*, 1–21.

Schreyer, W. 1977. Whiteschists: their compositions and pressure-temperature regimes based on experimental, field, and petrographic evidence. *Tectonophysics, 43*, 127–144.

Schreyer, W. 1988. Experimental studies on metamorphism of crustal rocks under mantle pressures. *Mineral. Mag., 52*, 1–26.

Sharp, Z. D., Essene, E. J., and Hunziker, J. C. 1993. Stable isotope geochemistry and phase equilibria of coesite-bearing whiteschists, Dora-Maira massif, western Alps. *Contrib. Mineral. Petrol., 114*, 1–12.

Smith, D. C. 1984. Coesite in clinopyroxene in the Caledonides and its implications for geodynamics. *Nature, 310*, 641–644.

Smith, D. C. 1988. A review of the peculiar mineralogy of the "Norwegian-eclogite province," with crystal-chemical, petrological, geochemical and gfeodynamical notes and an extensive bibliography. *Eclogites and eclogite-facies rocks.* Amsterdam: Elsevier, 1–206.

Spencer, J. E., and Reynolds, S. J. 1991. Tectonics of mid-Tertiary extension along a transect through west-central Arizona. *Tectonophysics, 10,* 1204–1221.

Tilton, G. R., Schreyer, W., and Schertl, H.-P. 1991. Pb-Sr-Nd isotopic behavior of deeply subducted crustal rocks from the Dora-Maira Massif, Western Alps-II: what is the age of the ultrahigh-pressure metamorphism? *Contributions to Mineralogy and Petrology, 108,* 22–33.

van der Molen, I., and van Roermund, H.L.M. 1986. The pressure path of solid inclusions in minerals: the retention of coesite inclusions during uplift. *Lithos, 19,* 317–324.

Walcott, R. I. 1987. Geodetic strain and the deformational history of the North Island of New Zealand during the late Cainozoic. *Phil. Trans. R. Soc. Lond., A321,* 163–181.

Wang, X., Jing, Y., Liou, J. G., Pan, G., Liang, W., Xia, M., and Maruyama, S. 1990. Field occurrences and petrology of eclogites from the Dabie Mountains, Anhui, central China. *Lithos, 25,* 119–131.

Wang, X., and Liou, J. G. 1991. Regional ultrahigh-pressure coesite-bearing eclogitic terrane in central China: evidence from country rocks, gneiss, marble, and metapelite. *Geology, 19,* 933–936.

Wang, X., and Liou, J. G. 1993. Ultrahigh-pressure metamorphism of carbonates in the Dabie Mountains, central China. *J. Metamorph. Geol., 11,* 575–588.

Wang, X., Liou, J. G., and Mao, H. K. 1989. Coesite-bearing eclogites from the Dabie Mountains in central China. *Geology, 17,* 1085–1088.

Wang, X., Liou, J. G., and Maruyama, S. 1992. Coesite-bearing eclogites from the Dabie Mountains, central China: Petrogenesis, P-T paths and implications to tectonics. *J. Geol., 100,* 231–250.

Wheeler, J. 1991. Structural evolution of a subduction continental sliver: the northern Dora-Maira massif, Italian Alps. *J. Geol. Soc. London, 148,* 1101–1113.

Wendt, A. S., D'Arco, P., Goffe, B., and Oberhansli, R. 1993. Radial cracks around a-quartz inclusions in almandine: constraints on the metamorphic history of the Oman mountains. *Earth Planet. Sci. Lett., 114,* 449–461.

Xu, S., Okay, A. I., Ji, S., Sengor, A.M.C., Su, W., Liu, Y., and Jiang, L. 1992. Diamond from the Dabie Shan metamorphic rocks and its implication for tectonic setting. *Science, 256,* 80–82.

6

The Role of Serpentinite Melanges in the Unroofing of UHPM Rocks: An Example from the Western Alps of Italy

M. C. BLAKE, JR., D. E. MOORE, AND A. S. JAYKO

Abstract

The ultrahigh pressure metamorphic rocks (UHPM) of the Dora-Maira continental massif are overlain by a stack of oceanic nappes. Metamorphic grade appears to increase downward but with marked discontinuities between each of the nappes, suggesting that section has been removed along the bounding faults. This apparent omission of section is greatest in the lowest oceanic unit where a serpentinite melange containing blocks and slabs of eclogite, metamorphosed at 12–19 kbar, lies on the UHPM rocks. We suggest that this serpentinite melange represents a highly attenuated upper mantle section that structurally overlay the UHP rocks during subduction. Similar serpentinite melanges are known from other high-pressure (HP) and UHPM areas and may have a similar origin.

Introduction

Earlier, we presented evidence that the ultrahigh pressure metamorphic (UHPM) rocks of the continental Dora-Maira massif of the Western Alps, Italy, were structurally overlain by gently dipping oceanic units that were tectonically thinned during or following Alpine subduction (Blake and Jayko, 1990). Subsequent authors (Philippot, 1990; Avigad, 1992) have reached similar conclusions. However, the timing of metamorphism and uplift remains controversial.

In this chapter, we present new pressure-temperature estimates for eclogitic rocks from several of the oceanic subunits, including blocks from the basal melange. These new data support our earlier suggestion that the serpentinite melange represents a greatly attenuated section of upper mantle that may have once formed the upper plate to the UHPM rocks of the Dora-Maira massif. Similar serpentinite melanges have been described from other UHPM areas

(Wang and others, 1989; Dobretsov, 1991), and these melanges may also represent remnants of once-thicker upper mantle sections.

Geologic Setting

For a description of many recent geologic and geophysical investigations in the Western Alps, the reader is referred to a special joint publication of the Geological Societies of France, Italy, and Switzerland (Roure, Heitzman, and Polino, eds., 1990). UHPM rocks were first discovered here in the Dora-Maira massif (Chopin, 1984), and have been the object of many recent studies (see Chopin and others, this volume, for a list of references). Our work has concentrated on the ophiolitic rocks that lie on top of the Dora-Maira continental fragment. These dismembered ophiolitic rocks and their sedimentary cover are all that remain of the small Tethyan ocean that once lay between the European and African continents. Beginning in the Early Cretaceous (ca. 120 Ma), the ocean was subducted southeastward beneath the African plate. When the oceanic rocks had been consumed, the European continental plate entered the subduction zone and, based on the UHPM mineral assemblages, must have reached depths as great as 100 km before collision and overthrusting of the deeply subducted rocks back onto the European margin.

The Mont Viso–Dora-Maira transect contains one of the best preserved and exposed records of this subduction and collision in the Alps. Figures 6.1 and 6.2 show the generalized geology and structure as well as the location of the analyzed samples that are described in the following section. As can be seen from the map and sections, the oceanic units form a stack of gently west-dipping thrust sheets (nappes) above the continental Dora-Maira massif and its sedimentary cover. Fault-shear indicators observed in the field and determined from oriented samples (Fig. 6.2) indicate that the nappes were emplaced by a westward (down-dip), normal sense of movement (Blake and Jayko, 1990). Similar kinematic directions were observed by Phillipot (1990), Chopin and others (1991), and Avigad (1992) along this same transect; however, Chopin and others relate these structures to late thrusting during retrograde greenschist-facies metamorphism rather than low-angle normal faulting. In addition, Ballèvre and others (1990) studied the Mont Viso area and concluded that the fault separating the Median Oceanic unit from their Lower Oceanic unit was a steep normal fault that required about 15 km down-to-the-west movement in order to account for the difference in metamorphic grade on opposite sides of the fault. Our mapping indicates that this fault dips moderately to the west and is subparallel to the other structures (see section A-A', Fig. 6.2).

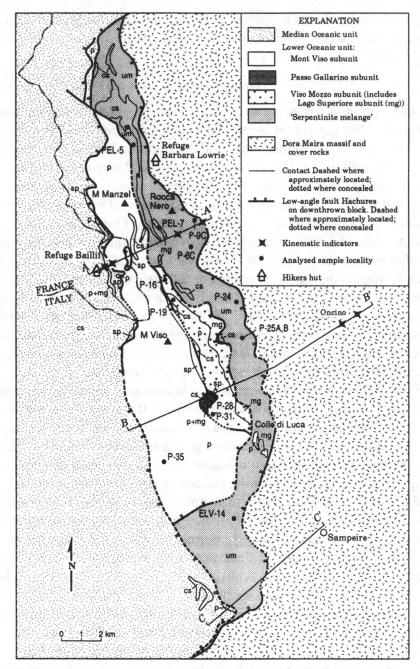

Figure 6.1. Simplified geologic map showing location of analyzed samples from the Mont Viso–Dora-Maira area (modified from Blake and Jayko, 1990). Symbols on map: cs = calcschist; p = prasinite; mg = metagabbro; sp = serpentinite; um = ultramafic rock.

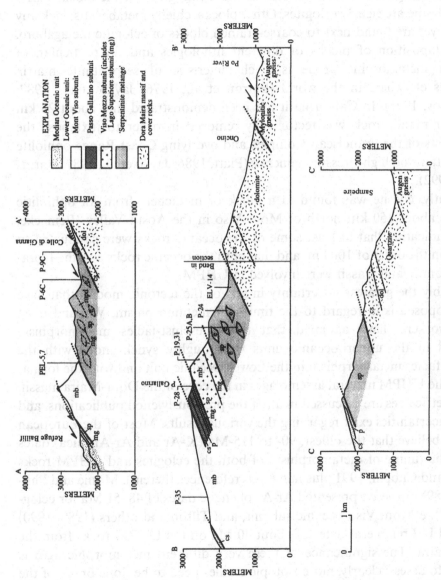

Figure 6.2. Structure sections from the Mont Viso–Dora-Maira area showing structural position of analyzed samples. Arrows on cross section indicate localities where the transport direction is known from field or thin section criteria.

The serpentinite matrix melange deserves further mention. All European workers have described this as ophiolite or serpentinite. In fact, it consists of blocks and slabs of metasedimentary and metaigneous rocks in a schistose matrix of antigorite serpentine. The metamorphic grade of the blocks varies greatly. Some are clearly eclogites. Other blocks, chiefly metabasalts, lack any garnets, yet are found next to coarse-grained blocks of eclogitic metagabbro. The juxtaposition of blocks of different lithologies and, in particular, of different metamorphic grade is a characteristic of serpentinite matrix melanges elsewhere in the world (Barron et al., 1976; Jayko et al., 1987; Dobretsov, 1991). In California it has been demonstrated that about 25 km of upper mantle rock was tectonically removed in order to juxtapose the blueschists of the Franciscan Complex and overlying Coast Range ophiolite that contains no high-pressure minerals (Platt, 1986; Jayko et al., 1987; Harms et al., 1992).

Recently, coesite was found in a block of metachert from an ophiolitic melange about 150 km north of Mont Viso in the Aosta Valley (Reinecke, 1991), indicating that at least some of the oceanic rocks were subducted to depths on the order of 100 km, and that both the oceanic rocks and the Dora-Maira continental massif were involved in UHPM.

Probably the greatest uncertainty in any of the tectonic models that have been proposed is in regard to the timing of metamorphism. We, and most other workers, have assumed that the blueschist-facies metamorphism recorded in the upper oceanic units was roughly synchronous with the eclogite-facies metamorphism in the Lower Oceanic unit and with the formation of the UHPM mineral assemblages in the core of the Dora-Maira massif. Radiometric ages are discussed in all of the previously cited publications, and many uncertainties exist regarding the various results. Most of the European workers believe that the oldest, 90- to 115-Ma (K-Ar and Ar-Ar mica) dates record the timing of metamorphism of both the eclogites and UHPM rocks (Monié and Chopin, 1991 plus numerous references therein). Monié and Phillipot (1989), however, presented Ar-Ar phengite dates of 48–51 Ma for eclogite from the Mont Viso oceanic subunit, and Tilton and others (1989; 1990) obtained U-Pb mineral dates of about 40 Ma on the UHPM rocks from the Dora-Maira. The significance of these very different metamorphic ages is difficult to assess; clearly, more isotopic studies need to be done on all of the units. In the meantime, we assume that we are dealing with one, possibly protracted, episode of subduction and metamorphism.

Lower Oceanic Unit

Mineral Assemblages

Thirteen garnet + clinopyroxene-bearing samples from the Lower Oceanic unit (Figs. 6.1, 6.2) were selected for microprobe analysis. Each structural subunit is represented by at least one sample, with the largest number (6) coming from the basal serpentinite melange (Table 6.1). Although the textures vary widely from sample to sample (brief rock descriptions are included in an Appendix), the eclogite-stage mineral assemblages are very similar (Table 6.1). Clinopyroxene and garnet are usually the principal minerals present, accompanied by small to moderate amounts of zoisite, rutile, apatite, quartz, and white mica (paragonite and/or phengite).

The garnets in different samples vary in size from less than 0.2 mm in diameter to porphyroblasts that are 5 mm across. With the exception of the grossular-rich garnets of P-6C (Fig. 6.3; Table 6.2), the garnets are all almandine-rich, with low spessartine contents. The pyrope components of the garnet rims range from about 5% to 43%. Moving from the cores to the rims, the pyrope contents increase at the expense of the almandine and/or grossular components. The garnet compositions are consistent with those reported by Lombardo and others (1978) and Kienast (1983). Almost all the clinopyroxenes are apple-green omphacites containing 20–45% jadeite molecule (Fig. 6.4; Table 6.3); the exceptions are the sodic augites of samples P-6C and P-24 from the basal serpentinite melange unit. Augitic as well as omphacitic eclogite-stage pyroxene compositions have also been reported by Lombardo and others (1978). Some of the pyroxenes have bluish-green, aegirine-augite cores.

Well-formed, fresh zoisite (Fig. 6.5A) was identified in a few samples; in most cases, only traces of zoisite were found in aggregates of clinozoisite + epidote that have pseudomorphically replaced the zoisite crystals (Fig. 6.5B). The zoisite contains little Fe (Table 6.4). Quartz generally occurs as a minor groundmass mineral, but in some samples it is found only as inclusions in garnet or zoisite (Table 6.1). Similarly, the Fe(Ti)-oxides, where present, occur principally as inclusions in other eclogite-stage minerals. Flakes of paragonite and phengite are scattered in the groundmass; paragonite also is interleaved with zoisite, in the association zoisite + paragonite + quartz. Both the paragonite and the phengite exhibit only minor K-Na exchange. A few samples from the basal serpentinite melange contain a nearly colorless, Mg-rich chlorite (Table 6.1). Amphiboles belonging to the eclogite-facies assemblages are very scarce in these samples: Inclusions of winchite (amphibole nomenclature

Table 6.1. *Eclogite-stage assemblages*

Subunit	Sample No.	Ga	Cpx	Rut	Zo	Qz	WhM	Mg-Chl	Apat	Fe(Ti)-Ox	Amph
Mont Viso	P-35	x	x	+	x		+ Ph				
Passo Gallarino	PEL-5	x	x	+	x	+	+ Pg		+		
	PEL-7	x	x	+	+	+	+ Ph		x		
	P-28	x	x	x	+	+			x	[+]	[+] Wi
Viso Mozzo	P-31	x	x	+	x				+		?
	P-19	x	x	+	x	+	+ Pg/Ph		+	[?]	+ Act
Lago Sup.	P-16	x	x	+	x	[+]			+		
Basal Serp.	P-6C	x	x	+	x			x	+	[+]	
	P-24	+	x	+	x	+	+ Ph	x	+		
	P-25A	x	x	+	[+]	+			+		
Melange	P-25B	x	x	x	[+]	[+]	[+]		+		
	P-9C	x	x	x	[+]	[?]	[+]				
	ELV-14	x	x	x	x	[+]	+ Pg		+		

Note: x = major constituent; + = minor constituent; [ ] = occurs only as inclusions in eclogite-stage mineral; Ph = phengite; Pg = paragonite; Wi = winchite; Act = actinolite.

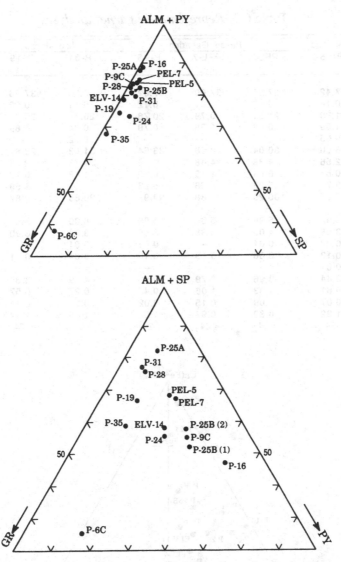

Figure 6.3. Representative garnet rim compositions from the analyzed samples of the Mont Viso–Dora-Maira area. For P–25B, the compositions of garnets formed during the second, blueschist stage are also included.

is after Leake, 1978) are contained in the garnets of P-28, and actinolite is a minor constituent of P-19, where it is principally found in quartz-rich patches. Sample P-31 contains the outlines of amphibole porphyroblasts that have been pseudomorphically replaced by retrograde minerals.

Retrograde metamorphism accompanying uplift of the Lower Oceanic unit is generally considered to encompass an early blueschist-facies event and a later greenschist-facies stage (e.g., Lombardo et al., 1978; Kienast, 1983; Nisio

Table 6.2. *Representative garnet analyses*

Sample No.	Mont Viso P-35	Passo Gallarino PEL-5	PEL-7	P-28	Viso Mozzo P-31	P-19	Lago Sup. P-16
SiO_2	37.42	37.81	37.60	37.02	37.20	37.56	39.57
TiO_2	0.14	0.10	0.04	0.12	0.11	0.02	0.02
Al_2O_3	21.33	21.51	20.79	20.78	20.71	20.76	22.12
Fe_2O_3	1.09	0.48	1.58	0.79	0.34	0.89	0.70
C_2O_3	0.05	--	--	--	0.02	--	0.06
FeO	26.00	30.04	28.58	33.64	31.03	28.87	21.44
MgO	2.56	4.38	4.48	1.76	1.13	2.39	11.25
MnO	0.54	0.69	1.13	0.13	1.48	0.79	0.68
CaO	10.74	5.28	5.68	5.83	7.83	8.59	4.03
Total	99.87	100.29	99.88	99.99	99.85	99.87	99.87
Si	5.92	5.96	5.96	5.96	6.00	5.99	5.98
Al	3.98	4.00	3.88	3.95	3.94	3.90	3.94
Ti	0.02	0.01	--	0.01	0.01	--	--
Fe^{3+}	0.13	0.06	0.19	0.10	0.04	0.11	0.09
Cr	0.01	--	--	--	--	--	0.01
Fe^{2+}	3.44	3.96	3.79	4.53	4.19	3.85	2.70
Mg	0.61	1.03	1.06	0.42	0.27	0.57	2.54
Mn	0.07	0.09	0.15	0.02	0.20	0.11	0.09
Ca	1.82	0.89	0.97	1.01	1.35	1.47	0.65
O	24	24	24	24	24	24	24

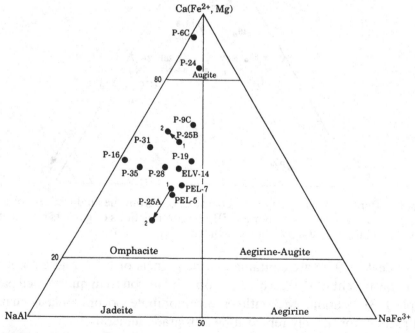

Figure 6.4. Representative groundmass clinopyroxene compositions from the analyzed samples of the Lower Oceanic unit. The recent IMA guidelines (Morimoto et al., 1988) for pyroxene nomenclature and the calculation of molecular proportions are adopted here. For P–25A and P–25B: 1 = eclogite stage; 2 = blueschist/eclogite stage.

Table 6.2 (*cont.*)

Sample No.	Basal serpentinite melange					
	P-6C	P-24	P-25A	P-25B	P-9C	ELV-14
SiO_2	38.88	38.03	37.02	38.99	38.91	38.52
TiO_2	0.22	0.10	0.12	0.03	0.08	0.05
Al_2O_3	20.90	22.43	20.37	21.75	21.14	21.73
Fe_2O_3	2.47	0.71	0.57	0.80	1.03	0.01
Cr_2O_3	0.10	0.15	--	--	0.02	--
FeO	10.76	22.38	36.01	23.59	24.58	26.11
MgO	3.90	5.89	1.89	8.15	7.80	5.55
MnO	1.09	2.15	0.57	0.60	0.25	0.44
CaO	21.76	8.21	3.45	6.18	6.13	7.55
Total	100.08	100.05	100.00	100.09	99.94	99.96
Si	5.94	5.89	6.00	5.98	6.00	6.00
Al	3.77	4.10	3.89	3.94	3.85	3.99
Ti	0.03	0.01	0.01	--	0.01	0.01
Fe^{3+}	0.29	0.08	0.07	0.09	0.13	--
Cr	0.01	0.02	--	--	--	--
Fe^{2+}	1.37	2.90	4.88	3.03	3.17	3.40
Mg	0.89	1.36	0.46	1.86	1.80	1.29
Mn	0.14	0.28	0.08	0.08	0.03	0.06
Ca	3.56	1.36	0.60	1.02	1.01	1.26
O	24	24	24	24	24	24

et al., 1987). Among these samples, however, the earlier retrograde metamorphic event encompasses both blueschist- and eclogite-facies assemblages. Glaucophane-forming reactions of the blueschist stage are well developed in the samples from the structurally higher subunits: Mont Viso, Passo Gallarino and Viso Mozzo. The glaucophane principally replaces omphacite, but it also forms rims around the eclogite-stage actinolite of sample P-19. Fe-rich clinozoisite (Table 6.5) probably formed at the same time as the glaucophane, because it partly replaces zoisite and is in turn rimmed by yellow epidote of the greenschist-facies stage. The Lago Superiore sample, however, shows no trace of sodic amphibole, and, of the six samples from the basal serpentinite melange, only P-25A contains a few small, idioblastic crystals of ferroglaucophane that appear to occur stably with the fine-grained, second-stage clinopyroxenes.

The structurally lower subunits – Lago Superiore and basal serpentinite melange – exhibit evidence for retrograde eclogite-facies rather than blueschist-facies assemblages. In P–25A, blue-green clinopyroxene porphyroblasts have recrystallized to very fine-grained groundmass pyroxenes that are enriched in jadeite and depleted in augite components compared to the older pyroxenes (Fig. 6.4). The younger pyroxenes are intergrown with quartz; the ferroglaucophane occurs along some pyroxene-quartz boundaries. Some garnets are partly replaced by quartz (Fig. 6.5D), but other, very small garnets in the quartz patches probably crystallized as part of the retrograde eclogite

Table 6.3. *Representative clinopyroxene analyses*

Sample No.	Mont Viso P-35	Passo Gallarino			Viso Mozzo		Lago Sup. P-16
		PEL-5	PEL-7	P-28	P-31	P-19	
SiO_2	55.52	55.26	55.60	55.27	55.29	54.79	56.29
TiO_2	0.09	0.08	0.02	0.05	0.06	0.01	0.04
Al_2O_3	10.49	9.12	8.88	9.02	8.07	6.97	11.47
Fe_2O_3	2.74	9.18	6.58	5.79	2.32	6.69	0.96
Cr_2O_3	0.04	--	--	--	0.01	--	--
FeO	2.65	1.65	2.44	3.36	5.73	3.61	2.53
MgO	8.24	6.38	7.62	7.19	7.99	7.90	8.77
MnO	0.03	0.01	0.05	--	0.06	0.02	0.03
CaO	12.78	9.85	11.52	12.08	14.22	13.63	12.67
Na_2O	7.06	8.73	7.70	7.44	6.02	6.52	7.10
Total	99.64	100.26	100.41	100.20	99.77	100.13	99.86
Si	1.98	1.98	1.99	1.99	2.01	1.99	1.99
Al^{IV}	0.02	0.02	0.01	0.01	--	0.01	0.01
Al^{VI}	0.42	0.37	0.36	0.37	0.35	0.29	0.47
Ti	--	--	--	--	--	--	--
Fe^{3+}	0.08	0.25	0.18	0.16	0.06	0.18	0.03
Cr	--	--	--	--	--	--	--
Fe^{2+}	0.08	0.05	0.08	0.10	0.17	0.11	0.07
Mg	0.44	0.34	0.41	0.38	0.43	0.43	0.46
Mn	--	--	--	--	--	--	--
Ca	0.49	0.38	0.44	0.47	0.55	0.53	0.48
Na	0.49	0.62	0.53	0.52	0.42	0.46	0.49
O	6	6	6	6	6	6	6

assemblage. These small garnets have a slightly smaller pyrope component than the older ones (Tables 6.2 and 6.5). In P-25B, some of the very large garnets have a narrow outer rim with slightly reduced pyrope contents (Fig. 6.3), whereas other garnets are partly replaced by Mg-chlorite. Clinopyroxene textures are especially complex in this sample, including zoned and partly recrystallized porphyroblasts and groundmass pyroxenes of various colors and sizes. Clinopyroxene textures in three other samples from the two lower subunits (P-16, P-9C, and P-24) are also suggestive of two stages of crystallization, but the overall reactions are not as obvious as in P-25A and P-25B.

Greenschist-facies reactions appear in nearly all samples, characterized by veins and patches of albite + chlorite + epidote + sphene + amphibole ± white mica. The greenschist amphibole is either a deep-greenish-brown mineral of variable composition or a pale-green actinolite that in some samples closely approaches tremolite in composition (Table 6.5; see the Appendix for occurrences). In previous studies (Nisio et al., 1987) a sodic-calcic amphibole was described as an early-formed greenschist mineral and actinolite as a later one. However, the ferro-barroisite of P-25A is the only sodic-calcic greenschist amphibole in the analyzed samples. Neither the ferro-barroisite nor any of the other greenish-brown amphiboles was found in the same sample as the greenschist-stage actinolites. The occurrence of these amphiboles thus

Table 6.3 (*cont.*)

Sample No.	Basal serpentinite melange					
	P-6C	P-24	P-25A	P-25B	P-9C	ELV-14
SiO_2	54.51	54.77	54.27	56.08	55.31	55.00
TiO_2	--	--	0.15	0.08	--	0.09
Al_2O_3	0.79	4.01	9.57	7.87	5.45	8.18
Fe_2O_3	1.28	3.43	7.54	1.52	6.66	6.52
Cr_2O_3	--	0.07	--	--	--	--
FeO	1.24	2.07	5.24	2.81	2.30	2.70
MgO	16.80	12.89	4.75	10.60	10.05	7.72
MnO	0.08	0.09	0.07	0.03	0.03	0.13
CaO	24.49	19.35	11.20	15.79	15.10	12.54
Na_2O	0.55	3.36	7.96	5.43	5.73	7.16
Total	99.74	100.04	100.75	100.31	100.63	100.05
Si	1.98	1.99	1.97	2.00	1.99	1.99
Al^{IV}	0.02	0.01	0.03	--	0.01	0.01
Al^{VI}	0.01	0.16	0.38	0.33	0.22	0.34
Ti	--	--	--	--	--	--
Fe^{3+}	0.04	0.09	0.21	0.04	0.18	0.18
Cr	--	--	--	--	--	--
Fe^{2+}	0.04	0.06	0.16	0.09	0.07	0.08
Mg	0.91	0.70	0.26	0.57	0.54	0.41
Mn	--	--	--	--	--	--
Ca	0.96	0.75	0.43	0.60	0.59	0.48
Na	0.04	0.24	0.56	0.37	0.40	0.50
O	6	6	6	6	6	6

appears to be a function of rock composition rather than differing metamorphic conditions.

The chlorite is an Fe-rich, green variety that replaces garnet and the colorless Mg-chlorite. In sample P-19, small crystals of crossite extend into patches of albite, suggesting that the crossite was stable at least during an early part of the greenschist-facies stage.

Pressure-Temperature Determinations

Temperature estimates of the eclogite and blueschist-eclogite event were obtained using the garnet-phengite geothermometer of Green and Hellman (1982) and, more extensively, the garnet-clinopyroxene geothermometer of Ellis and Green (1979) (Table 6.6; Fig. 6.6). Krogh (1988) reported that the Ellis and Green equation gives temperatures that are roughly 50° too high, and he proposed an alternative equation that incorporates a different type of correction for the Ca-content of garnet. Green and Adam (1991) tested the Ellis and Green geothermometer utilizing new experimental data. They concluded that the Ellis and Green equation does give overly high temperatures for high-grade metamorphic rocks such as granulites, but that it is preferable to the alternative garnet-clinopyroxene geothermometers for temperature esti-

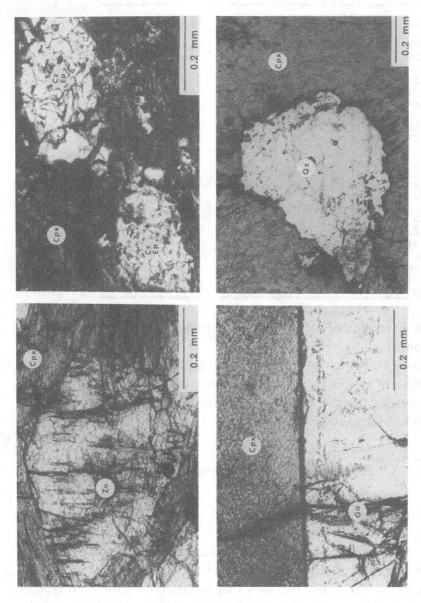

Figure 6.5. (A) Idioblastic zoisite (Zo) surrounded by clinopyroxene (Cpx) (P-28, Passo Gallarino subunit). (B) Aggregates of clinozoisite (Cz) rimmed by epidote (Ep), pseudomorphically replacing zoisite (PEL-5, Passo Gallarino subunit). (C) Zoned rim of a garnet porphyroblast (Ga) (P-25B, basal serpentinite melange). The maximum pyrope content of the crystal occurs just inside the line of inclusions. (D) Quartz aggregate (Qz) that may have replaced garnet during the retrograde, blueschist-eclogite stage of metamorphism (P-25A, basal serpentinite melange).

Table 6.4. *Compositions of other eclogite-stage minerals*

Mineral Sample No.	Ph P-35	Act P-19	Mg-Chl P-6C	Pg ELV-14	Zo P-35
SiO_2	53.96	55.07	29.38	46.31	38.93
TiO_2	0.07	0.06	0.03	0.03	0.06
Al_2O_3	21.56	2.11	19.33	39.25	33.05
Fe_2O_3	--	5.08	--	--	1.50
FeO	2.60	8.28	6.40	0.96	--
MgO	5.32	15.42	30.51	0.42	0.04
MnO	--	0.16	0.08	--	0.03
CaO	0.03	9.44	0.01	0.05	24.40
Na_2O	0.03	2.00	--	7.10	--
K_2O	11.56	0.04	--	0.99	NA
Total	95.13	97.66	85.74	95.11	98.01
Si	7.25	7.83	5.72	5.96	2.96
Al^{IV}	0.75	0.17	2.28	2.04	0.04
Al^{VI}	2.67	0.18	2.16	3.91	2.92
Ti	0.01	0.01	--	--	--
Fe^{3+}	--	0.54	--	--	0.09
Fe^{2+}	0.29	0.98	1.04	0.10	--
Mg	1.07	3.27	8.85	0.08	--
Mn	--	0.02	0.01	--	0.01
Ca	--	1.44	--	0.01	1.99
Na	0.01	0.55	--	1.77	--
K	1.98	0.01	--	0.16	NA
O	22	23	28	22	12.5

Note: NA = not analyzed.

mates below about 650–700°C. For the mineral pairs analyzed in this study, the Ellis and Green geothermometer yields temperatures that are more in keeping with the phase equilibria in Fig. 6.7 than does the Krogh geothermometer, consistent with the conclusion of Green and Adam (1991).

Minimum pressures of the eclogite-stage metamorphism are based on the coexistence of clinopyroxene + quartz, which was verified in several samples (Table 6.6). The pressures were determined using the isopleths for impure jadeitic pyroxenes in equilibrium with albite + quartz that were calculated by Maruyama and Liou (1987). Maximum pressures are based on the stability of paragonite, according to reactions (1) and (2) in Fig. 6.7. Either paragonite alone or the association zoisite + paragonite + quartz was identified in eight samples.

Temperatures in the range 450–550°C and pressures of 12–18 kb have generally been suggested for the prograde eclogite-facies metamorphism of this sequence (e.g., Kienast, 1983; Nisio et al., 1987). Temperature estimates for the individual units have only rarely been reported; the available published data are included in Table 6.6. Overall, the temperatures obtained in this study

Table 6.5. Compositions of retrograde minerals

| Mineral | Blueschist/eclogite stage | | | | | | Greenschist stage | | | |
| Sample | Gl | Cz | Ga | Cpx | Ph | Chl | Ep | Cpx | Act | Grn-Brn A |
No.	P-25A	ELV-14	P25-B	P-25A	PEL-7	P-19	PEL-7	P-9C	P-31	PEL-5
SiO_2	57.06	37.66	38.79	55.55	49.53	25.34	37.47	50.74	55.84	46.12
TiO_2	0.02	0.28	0.02	0.15	0.18	0.04	0.08	0.11	0.01	0.09
Al_2O_3	10.93	29.45	21.46	12.30	27.82	19.33	22.77	3.46	1.22	7.37
Fe_2O_3	3.27	5.02	0.50	7.40	NA	–	13.59	4.60	4.06	8.92
Cr_2O_3	NA	0.01	–	–	–	–	0.04	NA	–	0.01
FeO	12.03	–	26.23	2.47	3.95	29.06	–	9.32	8.28	14.84
MgO	7.37	0.07	7.13	4.54	2.70	12.98	0.02	9.69	16.57	8.04
MnO	–	0.06	1.21	0.02	0.02	0.41	0.27	0.17	0.15	0.35
CaO	0.17	24.63	4.87	8.08	0.02	0.13	23.47	19.01	11.20	9.05
Na_2O	7.17	NA	NA	9.84	0.47	NA	NA	2.07	0.77	2.86
K_2O	0.03	NA	NA	NA	10.53	NA	NA	NA	NA	NA
Total	98.05	97.18	100.21	100.35	95.22	87.29	97.71	99.17	98.10	97.65
Si	7.94	2.94	6.01	1.98	6.68	5.50	3.00	1.93	7.88	6.95
AlIV	0.06	0.06	3.92	0.02	1.32	2.50	–	0.07	0.12	1.05
AlVI	1.73	2.65	–	0.50	3.11	2.44	2.15	0.09	0.09	0.26
Ti	–	0.02	0.06	–	0.02	0.01	0.01	–	–	0.01
Fe^{3+}	0.34	0.30	–	0.20	NA	–	0.82	0.13	0.43	1.01
Cr.	NA	–	–	–	NA	–	–	NA	NA	NA
Fe^{2+}	1.40	–	3.40	0.07	0.45	5.27	–	0.30	0.98	1.87
Mg	1.53	0.01	1.64	0.24	0.54	4.20	–	0.55	3.49	1.81
Mn	–	–	0.16	–	–	0.07	0.02	0.01	0.02	0.04
Ca	0.03	2.06	0.81	0.31	–	0.03	2.01	0.77	1.69	1.46
Na	1.93	NA	NA	0.68	0.12	NA	–	0.15	0.21	0.83
K	–	NA	NA	NA	1.81	NA	NA	NA	NA	NA
O	23	12.5	24	6	22	28	12.5	6	23	23

Note: NA = not analyzed.

Table 6.6. *Pressure-temperature estimates*

Subunit	Sample No.	Temperatures (°C)			Pressures (kbar)	
		Ga-Cpx	Ga-Ph	Other sources	Minimum (Cpx+Qz)	Maximum (Pg)
Mont Viso	P-35	485-531 (6)[b]	480-491(2)			
Passo	PEL-5	400-430 (6)			9.5	10-12
Gallarino	PEL-7	459-513 (5)		425[c,d]	9-10	16-17
	P-28	381-434 (8)				
Viso	P-31	394-426 (7)			9.5	
Mozzo	P-19	447-485 (7)	452 (1)		9-9.5	14-16
Lago	P-16	560-623 (5)		506[c,d]	12.5	
Superiore				612[c]		
	P-6C	600-638 (4)				
	P-24	502-583 (4)			6	18-19
Basal	P-25A 1[a]	446-490 (7)			10	15-16
serpentinite	2	360-391 (14)			10	
melange	P-25B 1	589-635 (3)				19-20
	2	555-569 (3)				
	P-9C	513-526 (3)			8.5	18-19
	ELV-14	519-575 (13)				19-20

Note: [a]for P-25A and P-25B: 1 = eclogite stage, 2 = blueschist/eclogite stage; [b]( ) = Number of pairs included in the temperature estimate; [c]Kienast (1983); [d]Nisio et al. (1987).

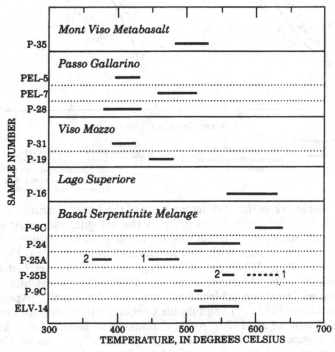

Figure 6.6. Temperature ranges of eclogite-stage and blueschist-eclogite-stage meta-morphism for the structural subunits of the Lower Oceanic unit, obtained from garnet-clinopyroxene and garnet-phengite geothermometers.

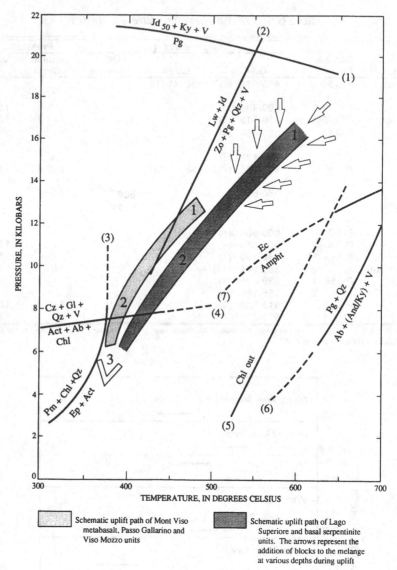

Figure 6.7. Possible *P-T* paths of different parts of the Lower Oceanic unit. The numbers 1, 2, and 3 in the uplift paths refer to the eclogite, blueschist-eclogite, and greenschist stages, respectively. Reactions: (1) upper pressure stability of paragonite, converting to a clinopyroxene with 50% jadeite molecule (Holland, 1979); (2) Heinrich and Althaus (1988); (3) Nitsch (1971); (4) greenschist- to blueschist-facies transition for basaltic rocks (Maruyama et al., 1986); (5) greenschist- to epidote-amphibolite-facies transition (chlorite-out reaction) for basalts (Apted and Liou, 1983); (6) Chatterjee (1972); (7) amphibolite to eclogite transition for quartz tholeiite compositions (Essene et al., 1970). Abbreviations: Jd = jadeite; Ky = kyanite; Pg = paragonite; V = vapor; Lws = lawsonite; Zo = zoisite; Qz = quartz; Cz = clinozoisite; Gl = glaucophane; Act = actinolite; Ab = albite; Chl = chlorite; Pm = pumpellyite; Ep = epidote; And = andalusite; Ec = eclogite; Ampht = amphibolite.

(Table 6.6; Fig. 6.6) are consistent with the previous determinations, but possible distinctions between some subunits can be made. The temperatures calculated for the Mont Viso, Passo Gallarino, and Viso Mozzo subunits are in the approximate range of 400–500°C (Fig. 6.6; Table 6.6), with the garnet-clinopyroxene and garnet-phengite geothermometers giving consistent results. The slight variations among samples within a given unit in this group may be related to the varying degrees of retrograde alteration affecting the garnets and clinopyroxenes. The temperature estimates for the Lago Superiore sample, however, are closer to 600°C than 500°C. Kienast (1983, p. 167) also reported a temperature of about 600°C for a sample from the Lago Superiore unit (Table 6.6). Temperatures of the eclogite-stage metamorphism in the blocks of the basal serpentinite unit cover a wide range, 450–650°C. For the samples in which two eclogite-facies assemblages were positively identified, the second assemblage crystallized at slightly lower temperatures than the first.

Discussion

Interpretations based on the temperature estimates should be made with caution, because of possible complications caused by the retrograde metamorphism. Nevertheless, the data suggest some possibilities that deserve further study. First, the contrasting temperatures obtained for separate eclogite blocks in the basal serpentinite are consistent with the hypothesis that this unit is a serpentinite melange containing fragments of rock from· several sources. In particular, the highest-temperature blocks may be remnants of the missing section that once separated the "cold" eclogite (Chopin et al., 1991) from the coesite-bearing rocks. Many more blocks from the basal serpentinite melange should be examined, to test this possibility.

The overlapping temperature ranges of the top three structural subunits in the Lower Oceanic unit – Mont Viso, Passo Gallarino, and Viso Mozzo – and the similarities among their retrograde histories support the idea that these units were closely associated throughout their history. The somewhat higher temperature range of the Lago Superiore sample compared to those of the overlying subunits suggests that the contact between the Lago Superiore and Viso Mozzo subunits could be a major structural break. The contrasting early retrograde histories of the subunits above and below this contact provide additional evidence for its being a major discontinuity. Whereas the top three subunits have a well-developed blueschist-facies overprint, the examined Lago Superiore and basal serpentinite samples have almost no traces of glaucophanic replacement. Instead, two samples were clearly subjected to a second, lower-temperature episode of eclogite-facies metamorphism, and the clino-pyroxene textures of some other samples are also best explained by a second or extended eclogite stage. Nisio and others (1987) also proposed contrasting histories for the Passo Gallarino and Lago Superiore subunits, based in part on differences in the size of antiphase domains in omphacite. According to their model, the Passo Gallarino subunit was rapidly uplifted to lower-temperature conditions subsequent to its initial, eclogitic metamorphism, whereas the Lago Superiore subunit was subjected to an extended period of eclogite-facies metamorphism.

Ridley (1984) proposed that the reaction

$$\text{glaucophane} + \text{epidote} = \text{omphacite} + \text{paragonite} + \text{quartz} + H_2O$$
$$\quad\text{(blueschist)} \qquad\qquad\qquad\qquad \text{(eclogite)}$$

has a steep negative slope in $P\text{-}T$ space and may be centered somewhere around 450°C for normal basaltic compositions. The eclogite assemblage occurs on the high-temperature side of the reaction. At a given temperature, the eclogite assemblage is favored by high Fe^{2+}/Mg ratios, low Fe^{3+}/Al ratios, and low activities of silica and H_2O (Ridley, 1984). Oh and others (1991) found that high Mn contents stabilize the eclogite assemblage to temperatures below 300°C.

The lack of blueschist-facies assemblages in the structurally lower subunits could be attributed to a variety of compositional factors rather than $P\text{-}T$ differences. This possibility must be investigated further. However, the presence of abundant Mg-chlorite in blocks from the basal serpentinite (Table 6.1) is an indication of high Mg contents in those blocks. If the development of the blueschist assemblages were a function of rock chemistry rather than $P\text{-}T$ conditions, the high Mg contents should have favored the blueschist instead of the eclogite assemblage. Similarly, the retrograde eclogite-facies assemblage of sample P-25A includes moderate amounts of quartz, which suggests that the activity of silica is not the controlling factor for blueschist versus eclogite occurrences in these samples. The regular distribution of blueschist assemblages in the structural sequence, despite the variety of rock types sampled, indicates that the blueschist and eclogite assemblages of the second metamorphic stage formed under different $P\text{-}T$ conditions. The development of glaucophane contrasts, for example, with the scattered occurrences of greenish-brown amphiboles and actinolite among the later, greenschist-facies assemblages. The minor ferroglaucophane in P-25A was a late-formed mineral of the second eclogite-facies assemblage in that block; its presence, therefore, is not inconsistent with the overall distribution of glaucophane in the sequence.

A highly schematic interpretation of the history of the Upper Oceanic unit is depicted in Fig. 6.7. The Mont Viso, Passo Gallarino, and Viso Mozzo subunits follow one uplift path, and the Lago Superiore subunit and basal serpentinite melange follow a second one. The temperatures of the eclogite assemblages are taken from Table 6.6. Pressures of the eclogite stage in the higher structural units are based on the maximum pressure estimate for sample PEL-5. Equivalent pressures for the lower structural units are based on the pressure maximum for sample P-25A; some blocks could have come from greater depths, however. Blocks probably also were added to the basal serpentinite melange at various levels during uplift; this is indicated in Fig. 6.7 by the small arrows. The late appearance of glaucophane during the blueschist-eclogite stage of P-25A suggests that the transition from eclogite to blueschist assemblages in that block occurred at about 360°C. If this transition temperature is roughly applicable to the other samples, then reaction (3) in Fig. 6.7 may be shifted to slightly lower temperatures for these rocks.

The final, greenschist-facies metamorphism appears to be identical in all five units. The various subunits of the Lower Oceanic unit may therefore have been tectonically juxtaposed prior to the greenschist-facies event. The later uplift history probably involved nearly isothermal decompression.

Conclusions

Estimated pressure-temperature conditions for ophiolitic rocks from the Mont Viso–Dora-Maira area support the idea that great thicknesses of over-burden have been tectonically removed along low-angle normal (detachment) faults. The basal serpentinite melange unit is probably a remnant of a once-thick (ca. 100 km?) upper mantle section that structurally overlay the UHPM rocks. The timing of UHPM and of subsequent uplift remains controversial. Available radiometric ages using different isotopic systems give very different results. We conclude, however, that both the continental and oceanic rocks were involved in the same subduction-related metamorphism and that both were largely uplifted by extensional processes.

Appendix: Sample Descriptions

P-35: Metabasalt; Mont Viso subunit. Small garnets, ≤0.25 mm in diameter, are scattered in a very fine-grained, faintly foliated, clinopyroxene-rich (≤0.05 mm long) groundmass. The sample is variably altered to blueschist (glaucophane) and greenschist (actinolite) facies assemblages.

PEL-5: Eclogitic metagabbro lens (or block) in calcschist; uppermost part of Passo Gallarino subunit. Small (≤0.2-mm diameter) garnets are dispersed in the clinopyroxene-rich matrix or concentrated in layers. Apple-green, groundmass clino-pyroxenes are ≤ 0.9 mm in length. The eclogite is heavily altered to blueschist and greenschist (magnesiohornblende, ferrohornblende, actinolitic hornblende) assemblages.

PEL-7: Coarse-grained eclogite; serpentinite of lowermost part of Passo Gallarino subunit. The rock is very garnet-rich, and individual garnets are as large as 5 mm in diameter. Interstitial, apple-green clinopyroxenes are ≤ 1.8 mm long. The eclogite has strong blueschist and greenschist (ferro-tschermakitic hornblende) overprints.

P-28: Foliated eclogite; serpentinite of lowermost part of Passo Gallarino subunit. Small, scattered or clustered, idioblastic garnets to 0.4 mm in diameter are distributed in a clinopyroxene-rich groundmass. The clinopyroxenes generally are fine-grained (<0.1 mm long) and grainy. Zoisite crystals occur in patches in the groundmass, and rutile is concentrated into narrow, discontinuous layers. Blueschist and greenschist (actinolite) overprints are relatively strong; the greenschist reactions are best developed along narrow veins oriented approximately perpendicular to the eclogite foliation.

P-31: Prasinite; Viso Mozzo subunit. The rock is dark and heavily altered, making eclogite-stage textures difficult to discern. Small garnets (≤ 0.25 mm diameter) are scattered in the groundmass or grouped into diffuse layers; zoisite-rich layers are also common. Partly altered clinopyroxene porphyroblasts can reach 5 mm in length; possible amphibole porphyroblasts have been replaced in turn by glaucophane and actinolite + chlorite.

P-19: Fine-grained eclogite; Viso Mozzo subunit. Garnets ≤ 0.6 mm in diameter are scattered in a clinopyroxene-rich matrix; the blocky, clean, apple-green clinopyroxenes are ≤ 0.25 mm long. Narrow, irregular concentrations of rutile, up to 3 mm in length, define a crude foliation. There are only minor blueschist and patchy greenschist (actinolitic hornblende) replacement reactions.

P-16: Foliated, prasinitic metagabbro; Lago Superiore subunit. The garnets are mostly ≤ 0.25 mm in diameter; some relict clinopyroxene porphyroblasts up to 1.25 mm long are preserved in a fine-grained (≤ 0.075 mm), clinopyroxene-rich groundmass. Well-preserved, elongate zoisite crystals are concentrated in layers; rutile forms thin, discontinuous streamers. No sodic amphiboles were found in the sample. Greenschist (actinolite) replacement occurs along veins that cut obliquely across the foliation and as a patchy replacement of the groundmass.

P-6C: Fine-grained chlorite-clinopyroxene-garnet block; basal serpentinite melange. Garnets ≤ 0.3 mm in diameter are either scattered in the Mg-chlorite + clinopyroxene (≤ 0.2 mm long) groundmass or grouped into irregular, discontinuous layers. The garnets have irregular, possibly corroded outlines, but the rock as a whole appears very fresh, with little evidence of either blueschist or greenschist reactions.

P-24: Foliated clinopyroxene-zoisite-chlorite schist block; basal serpentinite melange. Clinopyroxene and zoisite form monomineralic layers, and chlorite occurs in thin lenses that are up to 10 mm long. The clinopyroxenes are generally small and grainy, but a few, possibly older, crystals are 0.75 mm long. Scarce, skeletal (or pulled apart) garnets reach 3 mm in diameter. No sodic amphiboles were found, and greenschist-facies replacement is concentrated along a series of narrow, albite-rich veins without associated amphiboles.

P-25A: Eclogite block; basal serpentinite melange. Two stages of eclogite-facies crystallization are discernible in this block. Relict, bluish-green clinopyroxene porphyroblasts to 1.1 mm in length are partly broken down to fine-grained (0.1 mm long), apple-green clinopyroxene that is intergrown with quartz. Scattered, first-stage garnets are ≤1.5 mm in diameter. Tiny garnets (≤0.075 mm diameter) in some quartz + clinopyroxene patches may represent second-stage growth; other garnets may have been replaced by aggregates of quartz at that time. Minor, idioblastic ferroglaucophane may coexist with the groundmass clinopyroxenes. There is only modest greenschist (ferro-barroisite) replacement.

P-25B: Garnet-rich eclogite block; basal serpentinite melange. Eclogite-facies crystallization occurred in two stages. Sieved, first-stage garnets reach 3 mm in diameter; some have thin, second-stage rims whose inner boundaries are marked by a trail of tiny inclusions. The groundmass contains abundant clinopyroxene, apatite, rutile, and Mg-chlorite. Clinopyroxene relations are complex: porphyroblasts up to 4 mm in length have dark-green, grainy cores, a dull-green middle zone, and colorless rims; a few porphyroblasts appear to have been pulled apart. Patches of medium-grained (≤0.45 mm long), slightly bluish-green clinopyroxene occur in the groundmass, but most of the groundmass clinopyroxene is dark green and extremely fine-grained. No sodic amphiboles have been found, and greenschist-facies reactions are confined to a parallel array of narrow, albite + greenish-brown amphibole veins.

P-9C: Garnet-rich eclogite block; basal serpentinite melange. Individual garnets can reach 1.5 mm in diameter; garnets also occur in clusters of small (≤0.3 mm diameter) crystals. Rutile forms aggregates up to 1.25 mm wide. Clinopyroxene textures suggest two stages of growth. Relict clinopyroxene porphyroblasts (≤1.4 mm long) are pale green with darker green patches in the centers. The rims of the porphyroblasts are partly recrystallized to fine-grained (≤0.1 mm long), groundmass crystals. Some garnets adjacent to groundmass clinopyroxenes have a corroded appearance. No glaucophane was found; moderate greenschist-facies (hastingsite) reactions occur along veins and as a patchy replacement of groundmass pyroxenes.

ELV-14: Eclogite block or slab; basal serpentinite melange. Large, sieved garnets are up to 2.25 mm in diameter; garnet also occurs as clusters of moderate-sized crystals. The rock contains a dense groundmass of clinopyroxene + zoisite + rutile. The groundmass clinopyroxenes are generally 0.1 mm long. Idioblastic zoisite (≤0.8 mm long) is in places intergrown with paragonite and minor quartz. No traces of glaucophane were found; greenschist (magnesiohornblende) reactions are concentrated along a series of parallel veins and around garnet rims.

References

Apted, M. J., and Liou, J. G. 1983. Phase relations among greenschist, epidote-amphibolite, and amphibolite in a basaltic system. *Am. J. Sci.*, *283-A*, 328–354.

Avigad, D. 1992. Exhumation of coesite-bearing rocks in the Dora-Maira massif (Western Alps, Italy). *Geology*, *20*, 947–950.

Ballèvre, M., Lagabrielle, Y., and Merle, O. 1990. Tertiary ductile normal faulting as a consequence of lithospheric stacking in the western Alps. In Roure, F., Heitzmann, P., and Polino, R. (eds.), *Deep structure of the Alps. Mém. Soc. géol. Fr., Paris*, 156; *Mém. Soc. géol. suisse, Zürich*, 1; *Vol. spec. Soc. Geol. It., Roma*, 1, 227–236.

Barron, B. J., Scheibner, E., and Slanske, E. 1976. A dismembered ophiolite suite at Port Macquarie, New South Wales. *Records of the Geological Survey of New South Wales*, *18*, 69–102.

Blake, M. C., Jr., and Jayko, A. S. 1990. Uplift of very high-pressure rocks in the western Alps: Evidence for structural attenuation along low-angle-faults. In

Roure, F., Heitzmann, P., and Polino, R. (eds.), *Deep structure of the Alps.*
    Mém. Soc. géol. Fr., Paris, 156; *Mém. Soc. géol. suisse, Zürich,* 1; *Vol. spec. Soc.*
    Geol. It., Roma, 1, 237–246.

Chatterjee, N. D. 1972. The upper stability limit of the assemblage paragonite +
    quartz and its natural occurrences. *Contrib. Mineral. Petrol.,* 43, 288–303.

Chopin, C. 1984. Coesite and pure pyrope in high-grade blueschists of the Western
    Alps: A first record and some consequences. *Contrib. Mineral. Petrol.,* 86,
    107–118.

Chopin, C., Henry, C., and Michard, A. 1991. Geology and petrology of the
    coesite-bearing terrain, Dora-Maira Massif, Western Alps. *Eur. J. Mineral.,* 3,
    263–291.

Dobretsov, N. L. 1991. Blueschists and eclogites: A possible plate tectonic mecha-
    nism for their emplacement from the upper mantle. *Tectonophysics,* 186,
    253–268.

Ellis, D. J., and Green, D. H. 1979. An experimental study of the effect of Ca upon
    garnet-clinopyroxene exchange equilibria. *Contrib. Mineral. Petrol.,* 71, 13–22.

Essene, E. J., Hensen, B. J., and Green, D. H. 1970. Experimental study of amphibo-
    lite and eclogite stability. *Physics Earth Planet. Interiors,* 3, 378–384.

Green, T. H., and Adam, J. 1991. Assessment of the garnet-clinopyroxene Fe-Mg
    exchange thermometer using new experimental data. *J. Metamorph. Geol.,* 9,
    341–347.

Green, T. H., and Hellman, P. L. 1982. Fe-Mg partitioning between coexisting gar-
    net and phengite at high pressure, and comments on a garnet-phengite geother-
    mometer. *Lithos,* 15, 253–266.

Harms, T. A., Jayko, A. S., and Blake, M. C., Jr. 1992. Kinematic evidence for exten-
    sional unroofing of the Franciscan Complex along the Coast Range fault,
    northern Diablo Range, California. *Tectonics,* 11, 228–241.

Heinrich, W., and Althaus, E. 1988. Experimental determination of the reactions 4
    Lawsonite + 1 Albite = 1 Paragonite + 2 Zoisite + 2 Quartz + 6 H_2O and 4
    Lawsonite + 1 Jadeite = 1 Paragonite + 2 Zoisite + 1 Quartz + 6 H_2O. *Neues*
    Jahrbuch für Mineralogie Monatshefte, 11, 516–528.

Holland, T. J. B. 1979. Experimental determination of the reaction paragonite =
    jadeite + kyanite + H_2O, and internally consistent thermodynamic data for
    part of the system Na_2O-Al_2O_3-SiO_2-H_2O, with applications to eclogites and
    blueschists. *Contrib. Mineral. Petrol.,* 68, 293–301.

Jayko, A. S., Blake, M. C., Jr., and Harms, T. 1987. Attenuation of the Coast Range
    ophiolite by extensional faulting and nature of the Coast Range "thrust," Cali-
    fornia. *Tectonics,* 6, 475–488.

Kienast, J. R. 1983. *Le Metamorphisme de haute pression et basse temperature (eclog-*
    ites et schistes bleus): Données nouvelles sur la petrologie des roches de la croûte
    oceanique subductée et des sediments associés. Thèse d' Etat, Université Pierre et
    Marie Curie, 532 pp.

Krogh, E. J. 1988. The garnet-clinopyroxene Fe-Mg geothermometer – a reinterpre-
    tation of existing experimental data. *Contrib. Mineral. Petrol.,* 99, 44–48.

Leake, B. E. 1978. Nomenclature of amphiboles. *Am. Mineral.,* 63, 1023–1052.

Lombardo, B., Nervo, R., Compagnoni, R., Messiga, B., Kienast, J. R., Mevel, C.,
    Fiora, L., Piccardo, G. B., and Lanza, R. 1978. Osservazioni preliminari sulle
    ofioliti metamorfiche del Monviso (Alpi Occidentali). *Rendiconti, Societá Ital-*
    iana di Mineralogia e Petrologia, 34, 253–305.

Maruyama, S., Cho, M., and Liou, J. G. 1986. Experimental investigations of
    blueschist-greenschist transition equilibria: Pressure dependence of Al_2O_3 con-

tents in sodic amphiboles – a new geobarometer. *Geol. Soc. Am. Mem.*, *164*, 1–16.

Maruyama, S., and Liou, J. G. 1987. Clinopyroxene – a mineral telescoped through the processes of blueschist facies metamorphism. *J. Metamorph. Geol.*, *5*, 529–552.

Monié, P., and Chopin, C. 1991. $^{40}Ar/^{39}Ar$ dating in coesite-bearing and associated units of the Dora-Maira Massif, Western Alps. *Eur. J. Mineral.*, *3*, 239–262.

Monié, P., and Philippot, P. 1989. Mis en évidence de l'âge éocène moyen du métamorphisme de haute-pression dans la nappe ophiolitique du Monviso (Alpes Occidentales) par la méthode $^{39}Ar/^{40}Ar$. *C.R. Acad. Sci. Paris*, *309*, 245–251.

Morimoto, N., Fabries, J., Ferguson, A. K., Ginzburg, I. V., Ross, M., Seifert, F. A., Zussman, J., Aoki, K., and Gottardi, G. 1988. Nomenclature of pyroxenes. *Am. Mineralogist*, *73*, 1123–1133.

Nisio, P., Lardeaux, J.-M., and Boudeulle, M. 1987. Evolutions tectonometamorphiques contrastées des éclogites dans le massif du Viso; conséquences de la fragmentation de la croûte océanique lors de l'orogenèse alpine. *C.R. Acad. Sci. Paris*, *304*, Série II, 355–360.

Nitsch, K.-H. 1971. Stabilitätsbeziehungen von Prehnit- und Pumpellyit-haltigen Paragenesen. *Contrib. Mineral. Petrol.*, *30*, 240–260.

Oh, C. W., Liou, J. G., and Maruyama, S. 1991. Low-temperature eclogites and eclogitic schists in Mn-rich metabasites in Ward Creek, California; Mn and Fe effects on the transition between blueschist and eclogite. *J. Petrol.*, *32*, 275–301.

Philippot, P. 1990. Opposite vergence of nappes and crustal extension in the French-Italian western Alps. *Tectonics*, *9*, 1143–1164.

Platt, J. P. 1986. Dynamics of orogenic wedges and the uplift of high-pressure metamorphic rocks. *Geol. Soc. Am. Bull.*, *97*, 1037–1053.

Reinecke, T. 1991. Very-high pressure metamorphism and uplift of coesite-bearing metasediments from the Zermatt-Saas zone, western Alps. *Eur. J. Mineral.*, *3*, 7–17.

Ridley, J. 1984. Evidence of a temperature-dependent 'blueschist' to 'eclogite' transformation in high-pressure metamorphism of metabasic rocks. *J. Petrol.*, *25*, 852–870.

Roure, F., Heitzmann, P., and Polino, R., eds. 1990. *Deep structure of the Alps. Mém. soc. géol. Fr., Paris*, 156; *Mém. Soc. géol. suisse, Zürich*, 1; *Vol. spec. Soc. Geol. It., Roma*, 1, 350 pp.

Tilton, G. R., Schreyer, W., and Schertl, H. P. 1989. Pb-Sr-Nd isotopic behavior of deeply subducted crustal rocks from the Dora-Maira massif, western Alps, Italy. *Geochim. Cosmochim. Acta*, *53*, 1391–1400.

Tilton, G. R., Schreyer, W., and Schertl, H. P. 1990. Pb-Sr-Nd isotopic behavior of deeply subducted crustal rocks from the Dora-Maira massif, western Alps, Italy: The age of ultrahigh-pressure metamorphism. *EOS, Trans. Am. Geophys. Union*, *71*, 1708.

Wang, X., Liou, J. G., and Mao, H. K. 1989. Coesite-bearing eclogite from the Dabie Mountains in central China. *Geology*, *17*, 1085–1088.

7

Ultra-High-Pressure Metamorphic Rocks in the Western Alps

ROBERTO COMPAGNONI, TAKAO HIRAJIMA,
AND CHRISTIAN CHOPIN

Abstract

The focus of this paper is on the UHP rocks of the western Alps and particularly on the southern Dora-Maira massif where demonstrated coherency and a wide range of lithologies preserve UHP mineral species. The coesite-bearing Brossasco-Isasca Unit (BIU) of the southern Dora-Maira massif (DMM) is a coherent fragment of Variscan continental crust involved in the Alpine collisional orogenic belt. Two complexes were recognized: (1) a "polymetamorphic complex," consisting of paraschist with minor eclogite, marble, and fine-grained phengite-rich gneiss. Some paraschists contain relics of Variscan amphibolite-facies rocks; (2) a "monometamorphic complex," consisting of orthogneiss with local augen fabric, which contains relics of undeformed metagranitoids and pyrope-bearing whiteschist layers. The primary intrusive relations between "polymetamorphic" and "monometamorphic" complexes are indicated by the local occurrence, all along the main contact, of relics of thermally metamorphosed paraschists. The Variscan age of the granitoid emplacement into the pre-Alpine amphibolite-facies basement is constrained by a 303 ± 1 Ma U/Pb zircon age. Therefore, before the Alpine orogeny the BIU consisted of a Variscan amphibolite-facies basement intruded by late-Variscan granitoids. The peak metamorphic conditions within the BIU are constrained by the occurrence of pyrope + coesite instead of talc + kyanite and of the divariant assemblage jadeite + pyrope + glaucophane and are estimated to be P = 33 ± 3 kbar and T = $750 \pm 30^{\circ}$ C. The age of the peak metamorphism is controversial, with some workers reporting ages of 96–121 Ma and another group ages are of 41–28 Ma, using various techniques. The early-Alpine UHP conditions suggest that partial melting may have occurred simultaneously with the metamorphic climax, but the field and textural evidence indicate that this must have been a minor process.

Introduction

Since Chopin's (1984) discovery of coesite from "pyrope quartzite" in the southern Dora-Maira massif (DMM) of the Western Alps, ultrahigh pressure (UHP) metamorphic rocks with diamond, coesite, and/or coesite pseudomorphs have been reported from several lithotypes of presumed crustal origin in four other metamorphic belts: the Norwegian Caledonides (Smith, 1984, 1988; Smith and Lappin, 1989), the Variscides of the Saxonian Erzgebirge (Schmädicke, 1991), the Urals and Tien-Shan Mountains (Sobolev and Shatsky, 1990; Tagiri and Bakirov, 1990); Dabie Mountains and Su-Lu region in China (Okay et al., 1989; Wang et al.,1989; Yang and Smith, 1989; Enami and Zang, 1990; Hirajima et al., 1990; Wang and Liou, 1991; Xu et al., 1992; Wang et al., 1993). These records suggest that the ultrahigh pressure metamorphism (UHPM) is not an exception in continental collision belts.

In the Western Alps, coesite has also been found in metasedimentary rocks capping metamorphic ophiolites from the internal Piemonte Zone (Reinecke, 1991). This new occurrence makes the metamorphic belt of the Western Alps one of the best areas to study the subduction-related metamorphism, because both continental and oceanic lithotypes with all mineral assemblages from blueschist to coesite-eclogite facies are represented. To illustrate the UHPM in the Western Alps, the coesite-bearing unit of southern DMM is preferred for its demonstrated coherency and the wide range of lithologies, which contain well preserved evidence of the UHPM, new mineral species, and minerals with unusual composition and grain size. The southern DMM is, therefore, the focus of this chapter on the UHPM in the Western Alps.

Geology

The coesite-bearing unit (in the following named **Brossasco-Isasca Unit**, BIU) occurs in the southern **Dora-Maira massif** (DMM), which extends from Val di Susa as far as Val Maira to the south (Fig. 7.1). To the east, the DMM is bounded by the Quaternary deposits of the Po plain and to the west is in contact with the Mesozoic calcschists ("Schistes lustres") and metaophiolites of the Piemonte zone (Figs. 7.1 and 7.2). The DMM constitutes, along with Gran Paradiso and Monte Rosa, the Internal Crystalline Massifs of the Penninic Domain of the Western Alps (Fig. 7.1). They consist of Paleozoic rocks attributed to the passive margin of the European plate (e.g, Debelmas and Lemoine, 1970; Dal Piaz et al., 1972; Dal Piaz and Lombardo, 1986), considered as microcontinental fragments within the Mesozoic Thetian (or Tethyan) ocean (e.g., Pasquarè, 1975; Trümpy, 1982; Platt, 1986), or derived from the

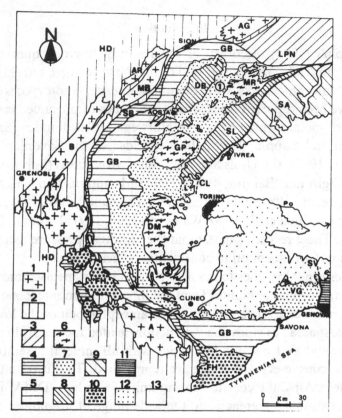

Figure 7.1. Tectonic sketch map of the western Alps. (1–2) *Helvetic–Dauphinois domain* (HD). (1) Basement of the "External Crystalline Massifs": Argentera (A), Pelvoux (P), Belledonne (B), Mont Blanc (MB), Aiguilles Rouges (AR), Aar- Gotthard (AG); (2) cover. (3–7) *Penninic domain* : (3) Lower Penninic nappes of the Ossola-Tessin region (LPN). (4) Subbriançonnais units and Sion-Courmayeur zone (SB). (5) St. Bernhard nappe, including the crystalline basement, the Permo-Carboniferous and the Briançonnais Mesozoic cover. (6) Upper Penninic nappes of the "Internal Crystalline Massifs": Monte Rosa (MR), Gran Paradiso (GP), and Dora-Maira (DM). (7) Piemonte zone. Calcschists ("Schistés lustrés") and metaophiolites derived from the Piemonte-Liguria oceanic basin; VG = Voltri Group metaophiolites. (8) *Austroalpine domain.* Dent Blanche nappe (DB) and Sesia Zone (SL). (9) *Southern Alps (SA),* including the Ivrea Zone. (10) Helminthoid Flysch nappes (FH). (11) Apenninic units. (12) *Late orogenic sedimentary sequences of Monferrato and Langhe.* (13) *Postorogenic molasse of the Po Plain.* CL = Canavese Line; SV = Sestri-Voltaggio line. Location of the coesite occurrences in the Western Alps: 1 (in circle): Lago di Cignana, upper Valtournenche, Val d'Aosta; 2 (in circle): Brossasco-Isasca unit of southern Dora-Maira massif. In the inset is the area of Fig.7.2.

margin of the Adria plate (Laubscher and Bernoulli, 1982; Desmons, 1986; Caron et al., 1987; Polino et al., 1990).

The southern DMM consists of a pile of thrust sheets, which are characterized by significantly different Early-Alpine metamorphic recrystallizations

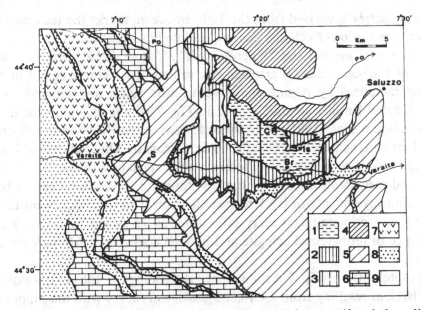

Figure 7.2. Tectonic sketch map of southern Dora-Maira massif and the adjoining Piemont Zone (modified from Monié & Chopin, 1991). (1–6) *Dora-Maira Massif.* (1) Pre-Alpine crystalline basement with middle-*T* eclogite-facies Alpine overprint (coesite-bearing Brossasco-Isasca Unit); (2) Pre-Alpine crystalline basement with low-*T* eclogite-facies Alpine overprint; (3) Pre-Alpine crystalline basement and Permo-Carboniferous + Permo-Triassic cover with low-*T* eclogite-facies Alpine overprint in the northern part; (4) Carboniferous (?) graphite-rich unit ("Pinerolo unit" of Vialon, 1966) with epidote-blueschist facies Alpine overprint; (5) Upper Paleozoic + Lower Triassic units with epidote-blueschist facies Alpine overprint; (6) Mesozoic cover series, including epicontinental calcschists ("Schistes lustrés"), with low-grade blueschist facies Alpine overprint. (7,8) *Piemont Zone.* (7) Monviso composite ophiolite unit with low-*T* eclogite facies Alpine overprint; (8) oceanic metasediments ("Schistes lustrés") with low-grade blueschist facies Alpine overprint. (9) *Postorogenic sediments.* Locality symbols: Br = Brossasco; CR = Case Ramello; Is = Isasca; S = Sampeyre. In the inset is the area of Fig. 7.3.

(Chopin et al., 1991; Henry et al., 1993; Michard et al., Chapter 4, this volume) (Fig. 7.2). The tectonic units are separated by low-angle tectonic contacts and the main regional foliation, which was formed mostly during the Mesoalpine greenschist facies event (see Petrography, below), is roughly parallel to the main tectonic contacts (see Michard et al., Chapter 4, this volume). To the south, the BIU is overlain by a crystalline tectonic unit (2 in Fig. 7.2), consisting of a polymetamorphic basement intruded by late-Variscan granitoids, which underwent Alpine low-*T* eclogite-facies recrystallization (near 15 kbar and ca. 550°C,: Chopin et al., 1991). This unit is, in turn, by Upper Paleozoic + Lower Triassic units (5 in Fig. 7.2) with Alpine epidote-blueschist facies metamorphic overprint (ca. 10 kbar and ca. 500°C: Henry, 1990). A narrow band of metaophiolites and their Mesozoic metasedimentary cover ("schistes lustrés"), derived from the Tethyan ocean, marks the tectonic contact between the low-T eclogite and the epidote blueschist facies units. To the north, the BIU overlies the "Pinerolo unit" (Vialon, 1966), which consists mainly of Carboniferous (?) graphite-rich schists and metaclastics (4 in Fig. 7.2), characterized by Alpine epidote-blueschist facies metamorphic overprint. A small lenslike tectonic unit (2 in Fig. 7.3), not distinguished in the map of Chopin et al. (1991), is interposed between the BIU and the Pinerolo units. This unit, named "San Chiaffredo unit" from the church north of Isasca (see Fig. 7.3), is composed of basement rocks, including paraschist, augengneiss and metagranitoid, similar to those found in the coesite-bearing BIU, but lacking any evidence of UHP recrystallization. This unit underwent Early-Alpine low-T eclogite-facies recrystallization, as suggested by the very rare occurrence of retrogressed eclogite and by the chemistry of Alpine garnet in metapelite.

The sketch-map of Fig. 7.3 summarizes the geology of the central portion of the BIU, as resulting from a 1:10,000 geological survey with the support of the study of about 800 rock thin sections. Two main lithologic complexes were recognized in the BIU: a monometamorphic and a polymetamorphic complex, respectively. These two complexes roughly correspond to the "gneiss and metagranite" complex and the "varied formation," respectively, of Chopin et al. (1991). The main difference concerns the recognition of a new tectonic unit (named "San Chiaffredo unit"), which straddles the boundary between BIU and the underlying "Pinerolo Unit" (see later on) as drawn by Chopin et al. (1991). The **monometamorphic complex** (3d in Fig. 7.3) consists of augengneiss grading to medium- to fine-grained mylonitic orthogneiss, which locally preserves relics of the granitoid protolith (crosses in Fig. 7.3) and contains layers of the pyrope-bearing whiteschist (black lenses in Fig. 7.3). The orthogneiss derives from Alpine deformation and metamorphic recrystallization of Late-Variscan, locally porphyritic granitoid (Biino and Compagnoni, 1992). This age of the igneous protolith is suggested by a zircon U/Pb radiometric age of 303 ± 1 Ma, obtained by Paquette (in Monié and Chopin, 1991) on a nearly undeformed porphyritic granitoid, cropping out near Brossasco (Fig. 7.3). The **polymetamorphic complex** consists of paraschist (3a in Fig. 7.3) with marble (3b in Fig. 7.3) and eclogite (7 in Fig. 7.3) intercalations, and of a medium- to fine-grained phengite-rich gneiss (3c in Fig. 7.3) with remnants of Pre-Alpine augengneiss and metagranite and very minor marble and metabasics. Especially interesting are the paraschists, which still preserve mineralogical and/or textural relics of a Pre-Alpine amphibolite-facies regional metamorphism (see The Polymetamorphic Complex, first two subsections, below).

The original intrusive relationships between the mono- and polymetamorphic complexes are proved by the preservation, in several localities along

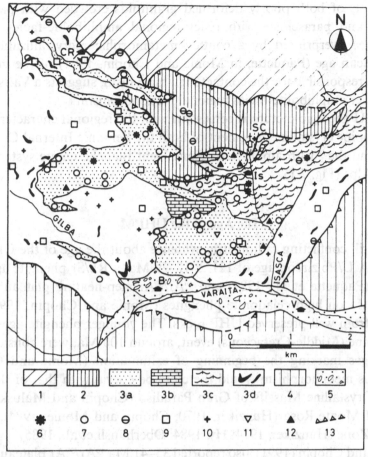

Figure 7.3. Simplified geologic map of the central part of the coesite-bearing "Brossasco-Isasca Unit." (1) Graphite-rich schists and metaclastics of the epidote-blueschist facies "Pinerolo Unit" (undifferentiated). (2) Lower tectonic unit ("San Chiaffredo unit") with basement rocks overprinted by low-T eclogite facies (undifferentiated). (3) *Coesite-bearing Brossasco-Isasca unit* (BIU). *Polymetamorphic complex.* (3a) Paraschists with middle-T eclogite (open circles) and marbles (asterisks); (3b) Main marble intercalations; (3c) Fine-grained phengite-rich gneiss with remnants of augengneiss and metagranite; *Monometamorphic complex.* (3d) Augengneiss grading to medium- to fine-grained orthogneiss with Late-Variscan metagranitoid relics (crosses) and pyrope-bearing whiteschist layers (black lenses). (4) Upper tectonic unit with basement rocks, overprinted by low-T eclogite facies (undifferentiated). (5) Alluvial deposits; (6) Small marble lens; (7) Fresh eclogite; (8) Retrogressed eclogite; (9) kyanite + jadeite occurrence; (10) relict Variscan granitoid; (11) relict regionally metamorphosed Variscan paraschist; (12) relict thermally metamorphosed Variscan paraschist; (13) thrust contact (ornaments on the upper unit). Locality symbols: B = Brossasco; CR = Case Ramello; Is = Isasca; SC = San Chiaffredo church; V = Venasca.

the contact, of both poorly deformed granitoids crowded with paraschist xenoliths and paraschists with relict regional amphibolite-facies mineral assemblages overprinted by a contact metamorphic recrystallization. The Late-Variscan age (Paquette, in Monié and Chopin, 1991) of the granitoid intrusion, responsible for the contact metamorphism, suggests a Variscan age for the older amphibolite-facies regional metamorphism.

This BIU geologic picture is consistent with the regional characteristics of the Pre-Alpine crystalline basement of both the Penninic Internal Crystalline Massifs and the Austroalpine Sesia Zone-Dent Blanche nappe system of the Western Alps (Fig. 7.1).

The Age of the UHPM

There is still conflicting radiometric evidence about the age of the UHPM in the BIU. A U/Pb zircon age of 121 ± 12–29 Ma, a Rb/Sr phengite age of 96 ± 4 Ma (Paquette et al., 1989) and $^{40}Ar/^{39}Ar$ step-heating plateau ages of 110–100 Ma on high celadonite phengites (Monié and Chopin, 1991) were obtained from well-preserved UHP rocks. The younger phengite ages of this Early-Alpine (Middle-Cretaceous) event, around 100 Ma, were considered as cooling ages marking the beginning of exhumation (Monié and Chopin, 1991). This metamorphism would therefore be coeval with that of the other Internal Crystalline Massifs of Gran Paradiso (Chopin and Maluski, 1978, 1980) and Monte Rosa (Hunziker, 1970; Chopin and Monié, 1984), and of the Sesia Zone (Hunziker, 1974; Hy, 1984; Oberhänsli et al., 1985).

Monié and Chopin (1991) also reported 35–41 Ma $^{40}Ar/^{39}Ar$ plateau (Meso-alpine) ages for low-celadonite phengites, separated from pervasively retro-gressed gneiss. They interpreted them as cooling ages of the Mesoalpine greenschist-facies event, extensively developed in the Western Alps (Hunziker, 1974; Hunziker et al., 1992), which is coeval with the main episode of nappe imbrication. $^{40}Ar/^{39}Ar$ biotite ages of 28–35 Ma by Monié and Chopin (1991) were also related to the closing stages of this medium- to low-P tectono-metamorphic evolution. Fine-grained white micas (muscovites), carefully sep-arated from a pervasively recrystallized greenschist-facies gneiss, yielded a $^{40}Ar/^{39}Ar$ plateau age of 31.3 ± 0.5 Ma, interpreted as dating the cooling after the regional greenschist-facies event (Hammerschmidt et al., 1992).

In contrast, Tilton et al. (1991) reported much younger ages of 38 ± 1.4 Ma (U/Pb) on zircon, 30–34 Ma (U/Pb) on ellenbergerite and monazite and about 38 Ma (Sm/Nd) on pyrope from the same UHP samples. The 38-Ma age is considered by Tilton et al. (1991) as dating the UHP event. Such data are in agreement with ion-microprobe data (SHRIMP II) by Gebauer et al.

(1993) performed on zircons from the BIU, which suggest "metamorphic or even magmatic domains" recrystallized or newly formed at 35 Ma.

Petrography

As a rule, most BIU lithotypes were pervasively deformed and recrystallized to greenschist facies mineral assemblages during the meso-Alpine event. In such rocks, evidence for the earlier UHP recrystallization is lacking or very difficult to recognize because it is scattered and/or preserved as rare relics in the cores of zoned minerals such as phengite (Chopin et al., 1991; Fig. 7.9). However, lithotypes locally occur that not only contain very fresh Early-Alpine UHP minerals, but also preserve Pre-Alpine fabric and part of the mineralogy of the Pre-Alpine protolith. Since the preservation of such relics is extraordinary, especially if the severe P (ca. 30 kbar) and T (ca. 750°C) conditions experienced by the unit are taken into consideration, this section deals primarily with (1) poorly deformed rocks, which preserve Pre-Alpine fabric; and (2) those rocks that best preserve UHP mineral assemblages.

The Monometamorphic Complex

Relics of Metagranitoids

Small relics of undeformed, massive granitoids with equant texture, ranging in composition from biotite granite to biotite tonalite, are locally exposed within the monometamorphic complex.

Biino and Compagnoni (1992) described the mineral transformations of a coronite metagranite exposed near Brossasco. The rock consists of small, undeformed parts of a porphyritic granite (from a few tens to a few hundreds of square meters), separated by anastomosing mylonitic shear zones. The undeformed metagranite grades into the surrounding orthogneiss, whose prevailing greenschist facies mineral assemblage indicates a continuous tectono-metamorphic recrystallization during the Alpine retrogressive evolution. The best preserved metagranite exhibits pseudomorphic and coronitic reactions. The igneous plagioclase was pseudomorphically replaced by almost pure granoblastic jadeite (up to jd_{96}, Fig. 7.4), zoisite and quartz, but close to the former igneous biotite a less jadeitic pyroxene (jd_{84-88}, Fig. 7.4) developed. Jadeite is preserved mainly in the pseudomorphs which developed after igneous plagioclase included in the igneous K-feldspar.

The igneous orthoclase was converted to a dactylitic intergrowth of high-celadonite phengite + quartz only around the igneous biotite (Fig. 7.9–3,

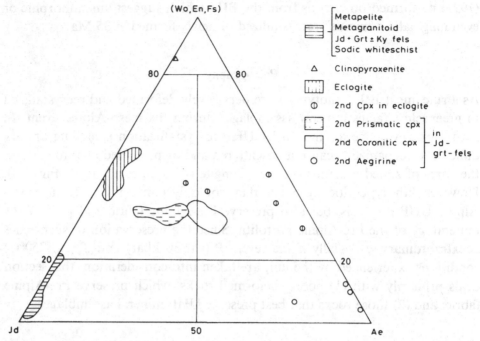

Figure 7.4. Compositional variation of UHP and secondary (2nd) clinopyroxenes in representative lithotypes of the Brossasco-Isasca Unit. Jd + Ky fels is short for the quartz/coesite-jadeite-kyanite-pyrope fels, and Jd + Alm fels is short for the quartz/coesite-jadeite-phengite-almandine fels in the text. Clinopyroxene nomenclature follows Morimoto et al. (1988). The main sources are Chopin et al. (1991), Schertl et al. (1991), Kienast et al. (1991), Compagnoni and Hirajima (1993), and Hirajima and Compagnoni (1993), and unpublished data.

below; see Biino and Compagnoni, 1992; fig. 12). The dactylitic phengite is zoned and displays a progressive decrease in the celadonitic substitution from ca. 3.40 in the core to 3.28 Si atoms pfu (for O = 11 basis, see Fig. 7.7) in the rim. This mica zoning was interpreted by Biino and Compagnoni (1992) as the evidence for a progressive decrease, in a closed system, of the pressure of the water (see Massonne and Schreyer, 1989), released from the transformation of biotite to coronitic garnet.

The igneous biotite is homoaxially replaced by high celadonite phengite and surrounded by a thin rim of coronitic garnet (Fig. 7.9–4, below). At the biotite-plagioclase contact, a second corona of intergrown dactylitic phengite + quartz develops on the plagioclase side. The coronitic garnet exhibits a strong asymmetrical zoning with high almandine + pyrope contents on the biotite side and high grossular content on the plagioclase side (Fig. 7.5 and see also Biino and Compagnoni, 1992). A narrow domain of very small (less than 0.1 mm across) idioblastic grossular-rich garnets (up to grs_{60-70}, Fig. 7.5) is developed within the plagioclase site close to igneous biotite.

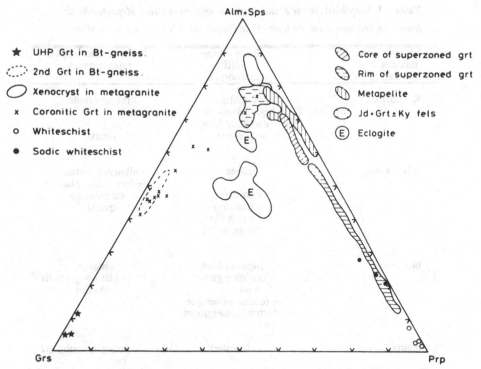

Figure 7.5. Compositional variation of UHP and secondary (2nd) garnets in representative lithotypes of the Brossasco-Isasca Unit. Abbreviations and data sources are the same as in Fig. 7.4.

A second type of garnet, up to ca. 1 cm in diameter, is frequently found in metagranite. As suggested by its chemical composition (alm_{80-90} prp_{5-15} grs_5, Fig. 7.5), this garnet is most likely a xenocryst, detached from the country-rock paraschist and incorporated by the granitoid magma during emplacement.

Coesite was never found in the BIU metagranitoids. However, the presence of polygonal, granoblastic quartz aggregates, pseudomorphic after primary igneous quartz is considered as the evidence of the former occurrence of coesite (Biino and Compagnoni, 1992). This interpretation is supported by the discovery of small, polycrystalline quartz aggregates (see Smith, 1988), included in a garnet xenocryst (Turello, personal communication). The minerals and the Alpine metamorphic assemblages are summarized in Table 7.1.

Most rocks of the monometamorphic complex were pervasively deformed and recrystallized to Meso-Alpine greenschist-facies mineral assemblages. The end product of this retrogression is an orthogneiss consisting of quartz, albite, clinozoisite/epidote, low-celadonite phengite, biotite and garnet, minor K-feldspar and chlorite, and accessory titanite, apatite, tourmaline, and zircon.

Table 7.1. *Simplified Alpine Metamorphic Evolution of Relict Metagranitoids from The Brossasco-Isasca Unit of the Dora-Maira Massif, Western Alps*

Late-Variscan igneous mineralogy	Early-Alpine UHP metamorphism	Meso- to Late-Alpine retrogression
K-feldspar	K-feldspar high-celadonite dactylitic Mus. + (coesite)	microperthite quartz
(plagioclase)	jadeite + zoisite + (coesite) + Ca-rich Grt [close to Bt]	albite/oligoclase + low-celad. Mus clinozoisite quartz
biotite	high-cel.Mus + coronitic garnet + rutile + (coesite)-phengite dactylitic intergrowth	chlorite red Bt --> green Bt titanite
quartz	(coesite)	polyg. granobl. qtz
(Ti-[Fe]-rich phase)	rutile	ilmenite-->titanite
apatite zircon tourmaline garnet [xenocryst]		

(+H2O) appears between the first and second columns at the K-feldspar row.

N.B. In parentheses are reported the minerals, whose former occurrence is unambiguously inferred on textural and mineralogical ground. Mineral symbols according to Kretz (1983).

In this orthogneiss, the presence of a celadonite-rich core in large phengite flakes (Chopin, 1984) as well as of the probable former occurrence of grossular-rich garnet (grs_{85}) + rutile (now titanite, Chopin et al., 1991) is practically the only evidence for former UHPM. Locally, an oligoclase rim (an_{17}) developed around albite (Fig. 7.9–8) suggests a moderate T increase during the climax of the meso-Alpine metamorphic event.

Relicts of Marginal Igneous Facies

At the contact between primary granitoid and basement metapelite, undeformed relics were found, where the igneous rock becomes crowded with partially assimilated paraschist xenoliths. The best example was found at Val Gilba (solid triangle, north of "GILBA" in Fig. 7.3).

The igneous portion of this marginal igneous facies consists of a polygonal granoblastic quartz matrix (after coesite), along with pseudomorphs after primary euhedral plagioclase and cordierite (Fig. 7.9.5). Plagioclase was replaced by jadeite (usually retrogressed to albite/oligoclase + white mica) + zoisite and cordierite mainly by garnet + kyanite (mostly altered to white mica). Locally, euhedral crystals of the former igneous biotite also occur, which are reequilibrated and partly replaced by phengite and rimmed with coronitic garnet.

The Pyrope-bearing Whiteschist

The type locality of the whiteschist is the famous outcrop of Case Ramello, Val Po (Fig. 7.3), where it was originally found by Vialon (1966) and described as a "pyrope-bearing nodular pegmatite." Chopin (1984) discovered coesite in pyrope, as the first example in metamorphic rocks derived from continental crust. This rock was described by Chopin (1984) as a "pyrope quartzite" and by Chopin (1987) as a "pyrope-phengite quartzite." Schertl et al. (1991) described the same rock as a "pyrope-quartzite" and differentiated a coarse-grained pyrope quartzite and a fine-grained pyrope quartzite, in order to clarify that there are two types of pyrope crystals (containing different mineral inclusions) formed by different reactions (megablasts: chlorite + kyanite + talc = pyrope + V; small pyropes: talc + kyanite = pyrope + coesite + V). However, the name "whiteschist" (Schreyer, 1973) is here preferred to emphasize both its peculiar chemical composition and its mineral assemblage, which is indicative of eclogite-facies conditions (Schreyer, 1977).

The whiteschist occurs within the orthogneiss of the monometamorphic unit (Fig. 7.3) as layers that range from a few centimeters to several meters in thickness and from a few meters to several tens of meters in length. The layers often appear as hinges of rounded folds with thinned or suppressed limbs or as lenslike boudins in the orthogneiss. The contact with the surrounding orthogneiss is either sharp or transitional, with interposition of a few decimeters to several meters of a gneiss, enriched in large phengite flakes but still locally preserving relics of the igneous K-feldspar. In the case of the Case Ramello locality, the contact between the pyrope quartzite and the biotite-phengite country rock is characterized by the interposition of phyllitic a quartzite a few decimeters thick (see Schertl et al., 1991, fig. 1). This phyllitic quartzite mainly consists of K-white mica (Si = 3.05–3.60), chlorite (0.85 < X_{Mg} < 0.95) and quartz.

The whiteschist usually consists of quartz/coesite, phengite, kyanite, pyrope, talc, rare jadeite and accessory rutile, zircon, and monazite. However, mineral mode and grain size are extremely variable, even within a single outcrop. Some whiteschists do not contain pyrope at all or may consist mainly of pyrope with interstitial quartz and/or phengite and/or talc. The pyrope grain size ranges from a few millimeters to about 20 cm. The pyrope megablast, originally an idioblastic trapezohedron, is now frequently rounded and broken. Fresh pyrope is light pink in colour, but unlike that found by Schertl et al. (1991), many megablasts are slightly zoned and show a deeper pinkish core and a paler pinkish rim, suggesting a difference in the Fe-content.

Pyropes are generally crowded with mineral inclusions, some of which are not found in the matrix. The commonest and most abundant inclusion is kyanite, whose modal amount within garnet locally exceeds 50 vol%. Rutile, zircon, and more rare monazite are ubiquitous accessory phases. Talc, Mg-chlorite, and K-deficient phlogopite or vermiculite (cf. Schertl et al., 1991) usually also occur, whereas phengite is very rare.

Coesite, or its pseudomorphic multicrystalline and/or polycrystalline quartz aggregates (Chopin, 1984; Smith, 1984, 1988), is included mostly in the small poikiloblastic pyrope or in the marginal part of pyrope megablast, as pointed out by Chopin (1984) and Schertl et al. (1991). Some centimeter-sized poikiloblastic pyropes contain a foliation defined by elongated coesite grains (see also Michard et al., Chapter 4, this volume). These fresh grains are transformed to quartz at their margins and occur as inclusions both in the garnet core and at its marginal embayments open to the rock matrix. These observations are accounted for when one is considering the contrasting elastic properties of coesite and its surrounding medium (Gillet et al., 1984).

Other less common phases included in pyrope megablasts are blood-red magnesiodumortierite (Schertl et al., 1991; Ferraris et al., 1994), purple ellenbergerite (Chopin, 1986, Chopin et al., 1994), and its phosphatian varieties (Chopin et al., 1986; Chopin and Langer, 1988), REE-rich epidote, apatite, wagnerite [$Mg_2PO_4(F,OH)$, cf. Chopin and Klaska, 1986], and the new mineral bearthite [$Ca_2Al(PO_4)_2(OH)$; see Chopin et al., 1993], glaucophane (possibly Li-bearing: Chopin, 1986), jadeite, dravitic tourmaline (Schertl et al., 1991), arsenopyrite and a series of phases, the prograde or retrograde status of which still has to be ascertained: gedrite, magnesio staurolite, corundum, enstatite, and sapphirine (Chopin et al., 1991).

Two typical pyrope-bearing whiteschist layers contained a few extraordinarily zoned (superzoned) garnets, from ca. 2 to 3 cm in diameter. These garnets are composed of a dark reddish core crowded with inclusions and a pinkish rim with sporadic inclusions. The garnet composition ranges from ca. $alm_{70} prp_{25} grs_5$ in the inner core, through ca. $alm_{60} prp_{35} grs_5$ at the core-rim boundary, to ca. $alm_{14} prp_{84} grs_2$ in the outer rim (Fig. 7.5). The inclusions in the core are kyanite, chloritoid, staurolite and paragonite, and those in the rim are kyanite, talc, and Mg-chlorite. It is noteworthy that the almandine core of the superzoned garnet displays the same composition and zoning as the most magnesian garnets from metapelite.

A detailed chemical study of the superzoned garnet, combined with geologic evidence of the gradual transition between the whiteschist and its surrounding orthogneiss, led Compagnoni and Hirajima (1992, 1993) to conclude that the BIU whiteschists are not Mg-rich metapelites in an evaporitic environment as claimed by Chopin (1987) and Schertl et al. (1991), but metasomatic rocks formed during Early-Alpine prograde UHPM at the expense of Variscan granitoids along active ductile shear zones in the presence of a hydrous fluid phase.

Schertl et al. (1991) distinguished two main stages of pyrope retrogression: an older one characterized by the assemblage phlogopite + kyanite + quartz, and a younger one with the assemblage chlorite + kyanite + muscovite + quartz.

Quartz / Coesite-Jadeite-Kyanite-Pyrope Fels

This lithotype occurs in the whiteschist as layers from a few centimeters to ca. 20 cm thick and several meters long. It was originally described as a "weathered bluish rock devoid of mica" by Chopin (1984), and later reported as a "jadeite-kyanite quartzite" by Chopin (1987) and a "garnet-jadeite-quartzite or garnet-jadeite-quartz bands" by Schertl et al. (1991). The best preserved fels (Schreyer et al., 1987) occurs in the whiteschist of Case Ramello, but identical ones are also found in some whiteschists of Val Varaita. The fels is medium-grained and massive, but some preferred orientation of the main mineral components may be observed locally under the microscope.

The peak metamorphic assemblage consists of coesite, but now quartz, associated with pyrope-rich garnet (prp_{70-80} alm_{20-30} grs_{1-5}; Fig. 7.5); jadeite ($jd_{95}di_4ae_1$, Fig. 7.4 and see also Schertl et al., 1991), kyanite, minor phengite, and talc (Schertl et al., 1991) and accessory apatite, rutile and zircon. The garnet includes not only coesite and paragonite (Schertl et al., 1991) but also phengite, talc, jadeite, kyanite, rutile, zircon, and apatite. Inclusions of colorless glaucophane and possibly of wagnerite and clusters of very small polycrystalline inclusions with negative garnet crystal shape, most likely former fluid inclusions, now consisting of intergrown white mica and Mg-chlorite, have also been found. These inclusions are similar to that found by Schertl et al. (1991) in the fine-grained pyropes.

Rectangular pseudomorphs, consisting of prismatic kyanite + quartz aggregates, distinctively occur in the quartz-rich matrix. They may have replaced a sheet silicate during the prograde stage.

A second glaucophane generation (Fig. 7.6), not armored in garnet, is

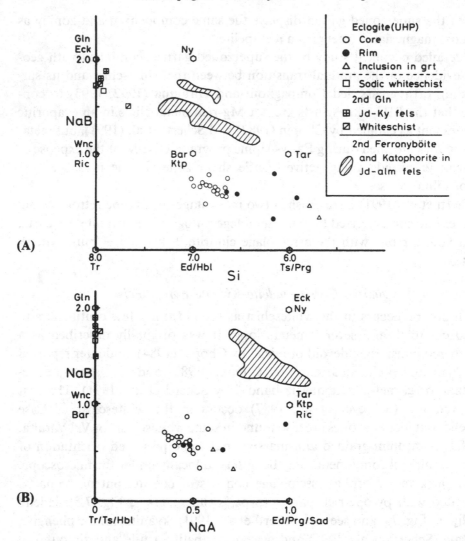

Figure 7.6. Si vs. Na(B) diagram (A) and Na(B) vs. Na(A) diagram (B) for UHP and secondary (2nd) amphiboles in representative lithotypes of the Brossasco-Isasca Unit. Amphibole nomenclature follows Leake (1978) and Smith (1988). Sad = Sadanagaite, Ric = Richterite. Other abbreviations follow Kretz (1983). The main sources are Chopin et al. (1991), Schertl et al. (1991), Kienast et al. (1991), Hirajima and Compagnoni (1993), and unpublished data.

locally recognized in the rock matrix and appears to have developed at the expense of jadeite during a very early retrogression stage, through the reaction:

$$\text{jadeite} + \text{kyanite} + \text{garnet} + H_2O \longrightarrow \text{glaucophane} + \text{paragonite} \quad (7.1)$$

Jadeite also breaks down to fibrous albite, which grows perpendicular to the grain boundary or the crack walls, ± paragonite ± hematite (Chopin, 1984;

Schreyer, 1985, fig. 15). Green phlogopite was also described as a garnet retrograde product by Schertl et al. (1991). Garnet alters to chloritoid (Chopin, 1984; if not mistaken for taramitic amphibole!), magnetite and green biotite (Schertl et al., 1991), especially where it is in contact with phengite.

The completely retrogressed fels is a fine-grained gneiss consisting of quartz, white micas, albite/oligoclase (Fig. 7.9.8) and minor zoisite / clinozoisite.

Quartz / Coesite-Jadeite-Phengite-Almandine Fels

Within both the mono- and the polymetamorphic complexes, decimeter to meter thick layers of leucocratic rocks locally occur, which most likely derive from UHP recrystallization of granitoid dikes (Chopin et al., 1991). They are very similar to the above mentioned fels (see previous section), but are devoid of kyanite. This fels consists of quartz/(coesite), jadeite (jd_{95}, Fig. 7.4), almandine garnet (alm_{70-80} grs_{5-20} prp_{10-20}, Fig. 7.5), phengite (Si = 3.25–3.41, Fig. 7.7) and accessory apatite, rutile and zircon.

The fels usually exhibits a complex polyphase retrogression. Jadeite breaks down first to granoblastic albite, then to sodic plagioclase (an_{8-14}), and finally to a fibrous albite (+ paragonite), identical to that of the quartz / coesite-jadeite-kyanite-pyrope-rich almandine fels (see previous section). The breakdown of jadeite to albite is accompanied by the growth of a Na-pyroxene, progressively poorer in the jadeite-component from omphacite (jd_{55}) to aegirine (jd_0, Fig. 7.4), and zoned amphiboles ranging in composition from ferronyboite (Fig. 7.6) (Hirajima and Compagnoni, 1993) to taramite (Chopin et al., 1991). Typically, the aegirine (ae_{85-95} jd_0 di_{5-15}, Fig. 7.4) homoaxially develops around the jadeite as a thin corona at the original interface with the quartz / (coesite) grains. Among retrogression phases, minor Mg-rich chlorite and zoisite were also reported by Chopin et al. (1991).

Sodic Whiteschist

A peculiar variety of whiteschist is a Mg-Al-Na-rich rock collected by Kienast et al. (1991) as a loose block near Case Ramello. It is coarse grained and contains the UHP peak assemblage: pyrope garnet (prp_{76} alm_{22} grs_2, Fig. 7.5), jadeite (jd_{82} di_{18}, Fig. 7.4), glaucophane, phengite (Si = 3.6, Fig. 7.7), quartz after coesite, and rutile. Glaucophane exhibits a chemical composition close to the pure Mg (= 2.94)-Al (= 1.82) end member, but is slight deficient in total B site cations (Ca = 0.12, Na = 1.67, see Kienast et al., 1991).

This lithotype of unusual composition and mineralogy is very important because it indicates that, at least in a Mg-Al-Na-rich system (NMASH), pure glaucophane is stable under UHP conditions, as suggested by Holland's (1988) thermodynamic calculations, but certainly close to its upper-pressure

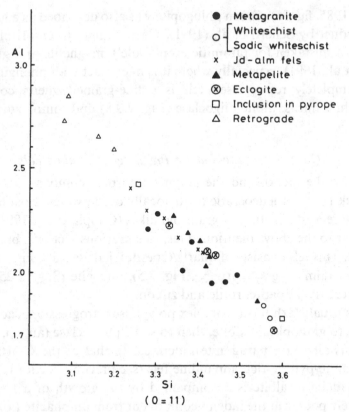

Figure 7.7. Si vs. Al diagram (for O = 11 basis) for UHP and retrograde phengites from representative lithotypes of the Brossasco-Isasca Unit. Abbreviations and data sources as in Figure 7.4.

stability limit, given the concurrent occurrence of the alternative pair talc + jadeite (see also Chopin et al., 1991; Fig. 5).

The Polymetamorphic Complex

Relict Regionally Metamorphosed Variscan Paraschist

Although remnants of Variscan minerals, especially garnet, are not uncommon all over the polymetamorphic complex, a well preserved Pre-Alpine regionally metamorphosed paraschist was discovered by Biino and Compagnoni (1991), approximately in the core of the polymetamorphic complex (Fig. 7.3). The rock consists of two portions: (1) a very dark graphite-rich, metamorphic-looking portion with garnet several millimeters across; and (2) a lighter, igneous-looking portion, where abundant quartz is recognizable.

Under the microscope, the *darker portions* consist of garnet porphyroblasts, wrapped around by a poorly defined foliation (Fig. 7.9.2, below). The pre-UHP mineralogy consisted of biotite, garnet, prismatic and fibrolitic silliman-ite (Fig. 7.9.1), minor feldspars, quartz, and most likely cordierite, now com-pletely replaced by a fine grained dark aggregate mainly consisting of garnet + kyanite.

Variscan garnet, typically almandine with small amounts of pyrope and grossular, is polyphase and contains a folded S_i (discordant to S_e), defined by alignments of very fine grained graphite inclusions toward the rim (Fig. 7.9.2), and by former fibrolitic sillimanite, biotite, ilmenite and quartz in the core. Sillimanite is replaced by kyanite aggregates, plagioclase by jadeite + zoisite, ilmenite by rutile and coronitic garnet, and orthoclase is partly altered to dactylitic phengite + quartz intergrowths.

The *lighter portions*, most likely original anatectic pockets, consist of K-feldspar and pseudomorphs after euhedral plagioclase and cordierite (transformed to albite + zoisite and garnet + kyanite, respectively), enclosed in a polygonal granoblastic quartz aggregate.

In conclusion, on the basis of the relict primary metamorphic assemblage (garnet, plagioclase, K-feldspar, quartz, sillimanite, cordierite, ilmenite and graphite of the darker portions) and the occurrence of the metatects (lighter portions), low- to medium-P and high-T amphibolite-facies conditions are inferred for the Pre-Alpine, most likely Variscan, regional metamorphism of the polymetamorphic complex (Fig. 7.8).

Relict Contact Metamorphosed Variscan Paraschists

As already reported in the section "Geology," the paraschist relicts, discovered in the polymetamorphic complex in close contact with the monometamorphic complex, display a thermal recrystallization clearly overprinting the Variscan regional metamorphic assemblages.

The less deformed specimens contain only Alpine pseudomorphic to coro-nitic topochemical reactions, so that the Pre-Alpine contact metamorphic minerals and their UHP transformations may be easily recognized.

In the thermally metamorphosed paraschists, which partly preserve the Pre-Alpine regional fabric and mineralogy, the Variscan foliation is largely obliter-ated by the random recrystallization of contact biotite and porphyroblastic andalusite. Variscan garnet usually exhibits a corroded habit and is sur-rounded by a dark aggregate, mainly consisting of garnet and minor kyanite (Fig. 7.9.2), interpreted as pseudomorphic after Pre-Alpine cordierite.

Regional sillimanite and contact andalusite are replaced during UHPM by kyanite aggregates with highly variable grain size. In some samples, the origi-nal andalusite zoning is still recognizable, owing to different color and grain

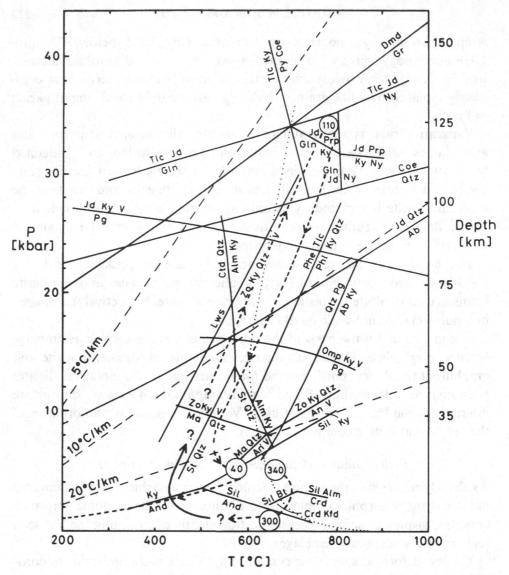

Figure 7.8. Petrogenetic grid showing the *P-T-t* path (short-dashed line) recorded in the UHP rocks of the Brossasco-Isasca Unit. Prograde path mainly follows Schertl et al. (1991) with supplementary newly found relic minerals of Cl-bearing hornblende inclusion in eclogitic garnet and Pre-Alpine almandine garnet in paraschist. The climax stage is defined by the existence of the Jd-Prp-Gln assemblage. An early retrograde stage is defined by the Fe-Mg distribution between omphacite-garnet (Hirajima and Compagnoni, 1993). The reactions graphite (Gr) = diamond (Dia) (Kennedy and Kennedy, 1976); quartz (Qtz) = coesite (Coe) (Mirwald and Massonne, 1980); glaucophane (Gln) = jadeite (Jd) + talc (Tlc); and the stability field of nyböite (Carman and Gilbert, 1983). Pg = Jd + Ky, Pg + Qtz = Ky + Ab, Ab = Jd + Qtz and Lws = Zo + Ky (Holland, 1979) are shown. Margarite stability field from Perkins et al. (1980). Long-dashed lines: geothermal gradients. Dotted line: onset of granite melting under water saturated conditions (Huang and Wyllie, 1975). Other curves are obtained with the GeO-Calc software (Brown et al., 1988) and the database of Berman (1988). Mineral abbreviations follow mainly Kretz (1983). Inset numbers in cir-

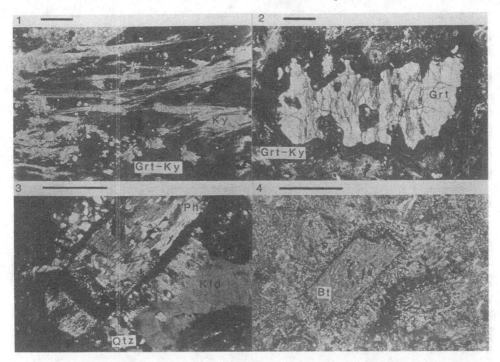

Figure 7.9. Photomicrographs. 7.9.1. *Relict of a regionally metamorphosed amphibolite facies paraschist* (DM 492). Variscan prismatic and fibrolitic sillimanite (light gray) is pseudomorphically replaced by Early-Alpine kyanite aggregates (Ky). The dark matrix, mainly consisting of fine-grained garnet and kyanite (Grt-Ky), most likely derives from primary cordierite. Plane polarized light. Scale bar is 1.0 mm.
7.9.2. *Relict of a regionally metamorphosed paraschist* (DM495). A corroded Variscan garnet (Grt), with a S_i (defined by very fine grained graphite alignments) perpendicular to S_e, is surrounded by a dark garnet-kyanite aggregate (Grt-Ky) pseudomorphic after primary cordierite. Plane-polarized light. Scale bar is 1.0 mm.
7.9.3. *Metagranite* (DM 989). The former igneous biotite is partly replaced by high-celadonite phengite (Ph: light gray to white) and surrounded by a thin garnet corona (black). Note the outer dactylitic intergrowth of phengite + quartz developed at the expense of the igneous K-feldspar (Kfd: gray, middle left and upper right) around biotite, and the granoblastic polygonal quartz aggregate (Qtz), derived from inversion of Early-Alpine coesite. Crossed polarizers. Scale bar is 0.5 mm.
7.9.4. *Metagranite* (DM 1103). Euhedral relicts of former igneous biotite (Bt) are included in a former primary plagioclase, now consisting of Middle-Alpine albite + low-celadonite phengite retrogressive from Early-Alpine jadeite + zoisite. Note the continuous garnet corona around biotite and the outer rim of very fine grained idioblastic grossular-rich garnet enclosed in albite (from former jadeite) devoid of zoisite. Crossed polarizers. Scale bar is 0.5 mm.

Caption for fig. 7.8 (cont.) cles are ages in Ma of the metamorphic events recorded in the Brossasco-Isasca Unit of Dora-Maira massif (see section "The Age of the UHPM") and follow the traditional radiometric data (see Hunziker et al., 1992). The question marks indicate a possible trajectory, connecting the (340–300 Ma) to the Alpine (110 to 40 Ma) *P-T* path of the BIU.

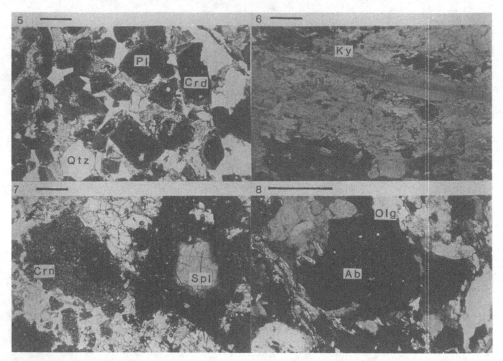

Figure 7.9.5. *Marginal granitoid facies* (DM 1116). Euhedral pseudomorphs after igneous plagioclase (Pl: dark gray), cordierite (Crd: almost black with discontinuous garnet corona) and biotite (dotted lath-shaped crystals) are enclosed in a matrix of quartz (white), which under crossed polarizers appears as a granoblastic polygonal aggregate (Qtz, cf. photomicrograph 7.9.3). Plane polarized light. Scale bar is 1.0 mm.

Figure 7.9.6. *Thermally metamorphosed Variscan paraschist* (DM 1143). Two porphyroblasts of Late-Variscan contact metamorphic andalusite (the upper one cut parallel and the lower one perpendicular to the z-axis, respectively) are pseudomorphically replaced by Early-Alpine kyanite aggregates (Ky). Note the original andalusite zoning, still recognizable in the upper porphyroblast, in spite of its replacement by kyanite. Plane polarized light. Scale bar is 1.0 mm.

Figure 7.9.7. *Aluminous dolomitic marble* (DM 1171). A relict Variscan greenish spinel (Spl: right) is progressively replaced by Early-Alpine very fine grained corundum aggregates (Crn). In the left part of the picture are evident coarser-grained corundum aggregates pseudomorphic after spinel. Plane-polarized light. Scale bar is 1.0 mm.

Figure 7.9.8. *Medium-grained orthogneiss from the monometamorphic complex* (DM 1015). Rim of oligoclase (Olg) developed around retrogressive albite (Ab) during the thermal peak of the Middle-Alpine metamorphic event. Crossed polarizers. Scale bar is 0.5 mm.

size of the kyanite pseudomorph (Fig. 7.9.6). Biotite is replaced by coronitic garnet + phengite, but locally biotite flakes, included in kyanite pseudomorphs after poikiloblastic andalusite, do not show apparent Alpine reequilibration. Plagioclase is pseudomorphically replaced by jadeite, kyanite, and minor zoisite. In some coarser grained quartz-rich paraschists, a kyanite +

quartz + jadeite aggregate with euhedral shape occurs, interpreted as derived from a Pre-Alpine sodic plagioclase.

In the rock matrix, quartz is always polygonal granoblastic, but in most inclusions in garnet it occurs as polycrystalline or multicrystalline aggregates with wavy extinctions, evidence of former coesite. In some sections, the polygonal granoblastic quartz aggregates appear to derive from recrystallization of a "columnar" or "palisade" quartz, similar to that described in the matrix of the pyrope-bearing whiteschist and considered as a typical product of coesite inversion (Chopin, 1984; Chopin et al., 1991, fig. 4; Schertl et al., 1991, fig. 3c; Michard et al., Chapter 4, this volume).

The Pre-Alpine muscovite flakes, usually developed around former fibrolitic sillimanite, now inverted to kyanite, are locally crossed by alignments of very fine grained kyanite aggregates, which appear to have developed along a system of conjugate kink planes. In some thermally metamorphosed peraluminous paraschists, corundum + rutile + kyanite aggregates occur, which are surrounded by an inner thin continuous mantle of kyanite and an outer wider corona of garnet. They are interpreted as UHP pseudomorphs after Pre-Alpine spinel + ilmenite aggregates, mantled by sillimanite.

The inferred Variscan regional and contact metamorphic minerals of metapelites and their Alpine transformations are summarized in Table 7.2.

Paraschists Pervasively Deformed and Recrystallized During UHPM

With increased rock deformation, the grain size of the UHP minerals progressively coarsens, the Pre-Alpine fabric gradually disappears, and the rock is converted to a kyanite-garnet-quartz / (coesite) + jadeite + high-Si phengite schist with minor zoisite and accessory rutile, apatite (up to 1 cm long) and local bearthite. Because of its susceptibility to retrogression (Chopin et al., 1991), fresh jadeite is rare in the matrix of metapelite, but is more commonly preserved as inclusions in garnet.

Kyanite and garnet locally develop as centimeter-sized porphyroblasts. Garnet is usually crowded with mineral inclusions, some of which (chloritoid and staurolite) are never found in the matrix. Porphyroblastic garnet, up to several centimeters in diameter, is usually almandine. Some garnets show a chemical zoning with Fe-rich core (alm_{80-85}) and Mg-rich rim (up to prp_{40}, Fig. 7.5). The core is crowded with inclusions of chlorite, chloritoid, kyanite, paragonite, quartz, and staurolite. The rim includes coesite (Chopin et al., 1991). Chloritoid, staurolite, paragonite, chlorite and quartz are incompatible with the climax mineral assemblage, as suggested by their absence from the rock matrix. The magnesian composition (up to ca. 50 mol% of the Mg-chloritoid end member) of chloritoid suggests it formed during prograde

Table 7.2 *Simplified Metamorphic Evolution of Relict Regionally and Thermally Metamorphosed Variscan Paraschists from the Polymetamorphic Complex of the Brossasco-Isasca Unit of Dora-Maira Massif, Western Alps.*

Regional Variscan metamorphism	Late-Variscan contact metamorphism [locally developed]	Early-Alpine UHP metamorphism	Meso-Alpine retrogression
garnet	(cordierite)		Bt + Chl
	(cordierite)	Grt + Ky	seric + Chl
biotite	biotite	Grt + Phe	Bt + Chl
K-feldspar	K-feldspar	K-feldspar (+H$_2$O) Phe + Qtz	microperthite
(plagioclase)	(plagioclase)	Jd + Qtz + Zo Jd + Qtz + Ky	Ab/Ol + CZo + Wm
(sillimanite)		kyanite	sericite
	(andalusite)	kyanite	sericite
(quartz)	(quartz)	(coesite)	quartz
muscovite	muscovite		
	(spinel?)	Grt + Cm	sericite
(ilmenite)	(Ilmenite)	Grt + Rut	Tit + Chl
graphite apatite monazite zircon tourmaline		epidote	epidote

Note: For explanation of parentheses, see note to Table 7.1

UHPM. On the contrary, staurolite is most likely a Pre-Alpine amphibolite-facies relict, as suggested by the not unusual chemical composition, the common corroded habit within white-mica aggregates, and its replacement by chloritoid (Chopin et al., 1991). The prograde nature of garnet zoning in metapelite is also supported by the Fe/Mg ratios of coexisting chloritoid and garnet, which suggest that the following continuous prograde reaction occurred:

$$\text{Fe-rich garnet} + \text{chloritoid} \longrightarrow \text{Mg-rich garnet} + \text{kyanite} + \text{water} \quad (7.2)$$

The final product of the Alpine paraschist retrogression is a two mica+albite + chlorite + clinozoisite/epidote micaschist with accessory titanite.

Marbles

Several lenses of usually dolomitic, silicate-bearing marbles, from a few decimeters to several tens of meters thick and from a few meters to several tens of meters long, occur in the paraschists of the polymetamorphic complex (Fig. 7.3). Millimetric to decimetric interlayers of boudinaged Ca-silicate fels or eclogite frequently occur within the marbles. The marble fabric varies from granoblastic to mylonitic or ultramylonitic through mortar, and carbonates may be either equant or elongated with preferred dimensional orientation.

The common mineral phases occurring in marbles, together with Fe-Mg-calcite and dolomite, are clinopyroxene, phengite, talc, quartz, garnet, zoisite, clinozoisite, Fe-rich epidote, microcline, and Mg-chlorite. For lack of unequivocal textural evidence, it is hard to group the mineral phases of marbles according to the metamorphic stages recognized in other BIU lithotypes. However, calcite/(aragonite), dolomite, quartz, omphacite, phengite, garnet, zoisite, and possibly talc, diopside, and olivine are minerals stable or developed at, or close to, the UHP peak. A wide spectrum of colorless to dark bluish-green secondary amphiboles occur, whose composition is clearly related to both chemistry of the parent mineral and to timing of retrogression. Ubiquitous accessories are fluor-apatite, zircon, and titanite, whereas graphite is only locally abundant as submillimetric millimetric deformed flakes. A zoned tourmaline is very rare. In a decimeter-thick lens of a banded dolomitic marble, forsterite, rimmed with a humite-group mineral, and diopside are also found.

The clinopyroxene ranges in composition from diopside to omphacite, most likely as a consequence of differences either in the primary bulk-rock chemistry or in the time of development during post-UHP evolution. In fact, several clinopyroxenes with a spongy appearance are diablastic intergrowths of Ca-rich pyroxene + albitic plagioclase, i. e., the decompressional products of former omphacite.

In spite of the usual, pervasive, polyphase deformation, a relict portion was discovered in the core of the largest marble lens. It is a dolomitic marble, which partially preserves relicts of a bluish-green spinel (most likely a Pre-Alpine contact metamorphic mineral), largely replaced by corundum aggregates (Fig. 7.9.7), and of ilmenite transformed to rutile. The corundum-dolomite-rutile peak metamorphic assemblage is locally surrounded by randomly-oriented margarite flakes and replaced by colorless Mg-chlorite (sheridanite). Chopin et al. (1991, p. 267, note) also reported in a dolomitic marble the occurrence of "millimeters-sized aluminous pockets (with corundum and late margarite) within calcite veinlets" dubiously interpreted as "former fracture filling."

The common presence of zoisite in the marble and the apparent stable coexistence of quartz with calcite suggest the presence of a relatively low X_{CO2} in the fluid phase and moderate temperatures during the whole metamorphic history.

A continuous reequilibration in marble, favored by the pervasive deformation, is suggested by the alteration of forsterite to both antigorite and lizardite, and by the local development of stilpnomelane at the expense of ferromagnesian silicates.

Eclogites

Eclogites, usually fine- or very fine-grained, occur all over the polymetamorphic complex as boudins from a few decimeters to ca. 10 meters long and from a few centimeters to a few meters thick, averaging a few decimeters in thickness. Most eclogites are characterized by a very strong mineral lineation and evident foliation, which result from a polyphase deformation including at least one transpositional event with millimeter-scale isoclinal folds and a later crenulation defined by centimeter-scale open folds. Both deformations took place under UHP conditions, as suggested by post-tectonic syn-eclogitic recrystallization of all UHP minerals.

Some eclogites exhibit a brecciated fabric, characterized by a very irregular network of millimeter- to centimeter-thick veins of coarse-grained prismatic omphacite with minor quartz and apatite, cross cutting a normal fine-grained foliated eclogite. Omphacite in the quartz-rich parts of these veins is usually idioblastic.

Eclogite is mostly banded or layered with different modal proportions and grain size of minerals. It exhibits highly variable modal amounts and compositions of minerals. Besides a bimineralic eclogite with omphacite and garnet, which typically occurs as darker lenses, pale-gray to pale-green quartz- and/or phengite-bearing eclogite is more widespread than kyanite- and/or zoisite-bearing eclogite. Many eclogites also contain a zoned porphyroblastic amphibole. Accessory phases are ubiquitous rutile, minor apatite, rare and locally large and corroded zircon, sulfides, and a radioactive zoisite. Some eclogites also contain large graphite flakes.

The almost ubiquitous quartz is locally so abundant that, in the presence of phengite, the eclogite makes a transition to a true "eclogitic micaschist" (i.e. a micaschist with garnet–omphacite assemblage) similar to that commonly occurring in the Sesia Zone of the Western Alps (see, e.g., Compagnoni, 1977). However, the presence of kyanite is consistent with the higher-T type of the BIU eclogites, unlike the lower-T Sesia eclogites, which are devoid of kyanite and locally contain climax paragonite.

In the BIU eclogites, quartz occurs mainly in coarse-grained aggregates, which appear to be derived from the deformation and transposition of former metamorphic veins. Such aggregates are devoid of garnet, but contain omphacite and minor zoisite and/or kyanite with subordinate amounts of accessory apatite and rutile. Quartz is coarse-grained, polygonal granoblastic, but the former occurrence of coesite is suggested by the common presence of fine-grained, polycrystalline quartz aggregates included in omphacite, kyanite, and zoisite.

Omphacite is commonly homogeneous and low or very low in aegirine content (Fig. 7.4, cf. Chopin, 1984; Kienast et al., 1991). However, some eclogites contain zoned omphacite with a sharp, irregular contact between a jadeite-poorer core (jd_{40-45}) and a jadeite-richer rim (jd_{45-55}). An aegirine-free sodic augite (ca. jd_{15}) was also reported by Kienast et al. (1991) from a clinopyroxenite (Fig. 7.4).

Eclogitic garnet is usually fine-grained and relatively homogeneous in each specimen. Its grossular content ranges from ca. 20 to 30 mol% (Fig. 7.5). The Fe/Mg ratio is usually homogeneous in each specimen, but varies from specimen to specimen (alm_{40-70} prp_{20-40}). However, some eclogites contain zoned garnets with Fe-rich cores. In one case, two garnet generations were observed: plurimillimetric poikiloblastic garnets with chloritoid inclusions and small submillimetric idioblastic garnets with very few inclusions. In rare eclogites, roundish aggregates of small garnets appear to be derived from UHP recrystallization of former garnet phenoblasts. In some phengite bearing eclogites, garnet locally contains sagenitic rutile or clusters of very small drop-like rutile most likely derived from recrystallization of former sagenite. This relict texture, quite common in the eclogitized rocks of the Western Alps, suggests the former presence of a pre-Alpine Ti-rich biotite, which exsolved Ti as rutile during transformation to garnet.

Phengite, locally abundant, defines the rock foliation. It appears in equilibrium with the eclogitic assemblage, but in most cases it is recrystallized post-tectonically. In kyanite-free eclogite, its Si content is up to Si = 3.58 pfu (Fig. 7.7).

Zoisite usually occurs as poikilitic porphyroblasts, elongated parallel to the rock foliation. It postdates the growth of both garnet and omphacite, but is in equilibrium with them. This is also supported by the finding within zoisite porphyroblasts of polycrystalline quartz inclusions after primary coesite. Zoisite predates the development of the porphyroblastic amphibole.

Kyanite is a minor component of some pale-gray to pale-green quartz-bearing eclogites. It occurs mainly in the quartz-rich portions and is usually altered to a paragonite aggregate (Kienast et al., 1991) and/or to a later coronitic margarite (Chopin et al., 1991).

The zoned porphyroblastic amphibole, which frequently contains worm-like quartz inclusions, develops mainly at the expense of omphacite and clearly overgrows the eclogitic foliation. The light pinkish purple core of porphyroblastic amphibole is intermediate in composition between barroisite and edenite (Si = 7.0–6.7, Na_B = 0.6–0.7 for 0 = 23 basis, Fig. 7.6), whereas the greenish rim is taramitic or pargasitic poorer in Si than the core.

A bluish-green pargasitic amphibole with a small amount of Cl is locally found as an inclusion in eclogitic garnet. Because the host eclogite is very fresh and does not contain retrograde amphibole from omphacite, the pargasitic amphibole is considered as a relict prograde Alpine mineral, armored within the garnet.

The BIU Evolution and Its *P-T-t* Trajectory

Pre-Alpine evolution

The oldest event recognized in the BIU is a regional, most likely Variscan (ca. 330–340 Ma) polyphase metamorphism, characterized by medium- to low-pressure amphibolite-facies conditions (Fig. 7.8). The presence of quartz-feldspars lens-like aggregates (metatects) in paraschists suggests that during the Variscan metamorphism temperatures higher than the wet granite solidus were reached.

The regional metamorphism was overprinted by a low-pressure and relatively high-temperature late-Variscan contact metamorphism, connected to the intrusion of ca. 300-Ma granitoids (Paquette, in Monié and Chopin, 1991; and Fig. 7.8). A moderate basement cooling, before and after the granitoid emplacement, is suggested by the significant thermal recrystallization of Variscan regional assemblages in paraschists and by the shallow character of the intrusion, respectively.

Alpine Evolution

Prograde history. Information on the Alpine prograde metamorphic evolution is obtained from the mineral inclusions in the early Alpine porphyroblastic garnet of metapelite. The presence of relict Variscan staurolite transforming to magnesian chloritoid in the garnet cores and the occurrence of kyanite in the garnet rims suggest that during prograde Alpine metamorphism the rock passed first through the stability field of chloritoid + quartz and later entered into that of almandine + kyanite. This steep trajectory is also constrained by the evidence that zoisite (± Ky ± Qtz) instead of lawsonite was the stable phase up to climax conditions, although lozenge-shaped pseudomorphs after lawsonite (?) were described in a BIU meta-aplite by Chopin et al. (1991, fig.8).

A similar prograde *P-T* path (for $a_{H_2O} = 1$) was estimated by Schertl et al. (1991). An early point at $P = 16$ kbar and $T = 560°C$ was estimated from the composition of a phengite (Si $= 3.26$) armored in a kyanite inclusion in pyrope and from the occurrence of talc + kyanite (Schreyer, 1988), assuming an initial Mg-chlorite + quartz mineralogy in the whiteschist protolith (Schreyer, 1977). The higher-*P* part of the path was estimated from the lack of Mg-chloritoid (+ talc) within pyrope, suggesting *P-T* conditions in the range of the alternative Mg-chlorite + kyanite pair (Chopin and Schreyer, 1983), commonly found in pyrope megablasts.

A further constraint is inferred by Schertl et al. (1991) from the occurrence of Ti-ellenbergerite inclusions in pyrope, whose lower-pressure stability is defined by the reaction:

$$\text{chlorite + kyanite + talc + rutile + water} \longrightarrow \text{ellenbergerite} \quad (7.3)$$

at $a_{H_2O} = 1$. Because this reaction curve intersects the curve:

$$\text{talc + Mg-chloritoid} \longrightarrow \text{Mg-chlorite + kyanite + water} \quad (7.4)$$

at its lower-temperature end, and the reaction:

$$\text{chlorite + kyanite + talc} \longrightarrow \text{pyrope + water} \quad (7.5)$$

at its higher-temperature end, the last part of the prograde *P-T* path is considered by Schertl et al. (1991) to pass through ca. 29 kbar and ca. 720°C. However, the potential petrologic constraints given by ellenbergerite (Chopin et al., 1988, 1992) are complicated by the oscillatory compositional zoning (e.g. Schertl et al., 1991, fig. 3a).

Climax conditions. The peak metamorphic conditions are quite well constrained, assuming a model NMASH system and $a_{H_2O} = 1$, by the occurrence of pyrope + coesite instead of talc + kyanite and of the divariant assemblage jadeite + pyrope + glaucophane. From the above mentioned constraints, **$P = 33 \pm 3$ kbar and $T = 750 \pm 30°C$** may be estimated for the BIU climax.

These values are consistent with the estimates of both Chopin (1984, 1987: $P \geq 28$ kbar, $T \sim 700°C$, assuming a low a_{H_2O}) and Kienast et al. (1991: 27 kbar < P < 35 kbar; 680°C < T < 750°C; low a_{H_2O})

Chopin et al. (1991) reported the climax jadeite + talc + quartz (from former coesite) assemblage, which is incompatible with the divariant jadeite + pyrope + glaucophane assemblage in the model NMASH system. However, a small amount of Fe and Ca in the relevant phases should increase the variance, so that a higher number of phases could coexist.

A bit higher peak conditions ($P \sim 37$ kbar, $T \sim 800°C$) were obtained by Schertl et al. (1991) by using the phengite geobarometer of Massonne and Schreyer (1989). This pressure value appears to be consistent with the result of the new geothermobarometer of Massonne (1990), based on the Al-content in talc. This new geothermobarometer gives a high water activity ($a_{H_2O} \geq 0.8$) for the metamorphic peak, whereas lower values ($a_{H_2O} = 0.40–0.75$) were estimated by Sharp et al. (1993).

The decompressional path. The postclimax evolution of the BIU requires a significant decompression coupled with a moderate cooling; otherwise an adiabatic or semi-adiabatic decompression should have reached approximately the boundary between the granulite- and the amphibolite-facies, for which no evidence was found.

Two steps of the decompressional P-T trajectory are recorded in the alteration of kyanite, which is first replaced by paragonite and later by margarite. Such alterations suggest the crossing of the pressure-dependent reaction (Holland, 1979):

$$\text{jadeite} + \text{kyanite} + \text{water} \longrightarrow \text{paragonite} \qquad (7.6)$$

and of the thermodynamically calculated isopleth (Newton, 1986)

$$\text{omphacite} + \text{kyanite} + \text{water} \longrightarrow \text{paragonite} \qquad (7.7)$$

Useful information on the decompressional P-T path arises from the retrogression of pyrope, which is altered first to phlogopite + kyanite + quartz and later to chlorite + muscovite + kyanite + quartz (Schertl et al., 1991). As pointed out by Chopin (1984), the first alteration assemblage suggests the crossing of the reaction:

$$\text{talc} + \text{phengite} \longrightarrow \text{phlogopite} + \text{kyanite} + \text{quartz} \qquad (7.8)$$

experimentally determined by Massonne and Schreyer (1989). Because in the whiteschist the mineral assemblages of both reaction sides occur, talc + phengite in the whiteschist matrix and phlogopite + kyanite + quartz in the pseudomorphs after pyrope, Schertl et al. (1991) suggest that the decompressional P-T path lies within the stability field of talc + phengite, near its upper stability limit (Fig. 7.8). The presence of the right-side assemblage of reaction (7.8) is interpreted by Schertl et al. (1991) as due to a moderate shift of the reaction curve toward lower temperatures during pyrope alteration, as the consequence of a reduced water activity in the metasomatizing hydrous fluid.

A second retrogression stage, at pressures of about 7–9 kbar for temperatures of about 500–600°C, was derived by Schertl et al. (1991), applying Massonne and Schreyer's (1989) phengite geobarometer to the low-celadonite muscovite (Si = 3.09) developed together with chlorite, kyanite and quartz from pyrope retrogression. The retrogression reactions mentioned above indicate, for the post-climactic BIU evolution, a decompression from ca. 30 to less than 7–9 kbar coupled with a cooling from 700–800 °C to 600–500°C or even less.

The later "retrograde" evolution refers to the meso-Alpine (30–40 Ma) metamorphic event, responsible for the extensive and pervasive greenschist-facies retrogression of the BIU rocks. However, the local occurrence of an oligoclase (+ biotite) rim around albite suggests a moderate late-metamorphic heating, postdating the nappe emplacement and predating the final BIU exhumation to the present position (Fig. 7.8). This regional thermal event, which indicates high-grade greenschist-facies conditions transitional to the amphibolite-facies, was also reported from the central and northern DMM (Sandrone and Borghi, 1992; Sandrone et al., 1992), Monte Rosa (Bearth, 1958; Dal Piaz, 1966), Gran Paradiso (Compagnoni et al., 1974), and Sesia Zone (Compagnoni, 1977).

Discussion and Conclusions

The coesite-bearing BIU of the southern DMM is a coherent fragment of Variscan continental crust involved in the Alpine collisional orogenic belt. Two complexes were recognized: (1) a "polymetamorphic complex," consisting of paraschist with minor eclogite, marble, and fine-grained phengite-rich gneiss. Some paraschists contain relics of Variscan amphibolite-facies rocks; (2) a "monometamorphic complex," consisting of orthogneiss with local augen fabric, which contains relics of undeformed metagranitoids and pyrope-bearing whiteschist layers. The primary intrusive relations between "polymetamorphic" and "monometamorphic" complexes are indicated by the local occurrence, all along the main contact, of relics of thermally metamorphosed paraschists. The Variscan age of the granitoid emplacement into the Pre-Alpine amphibolite-facies basement is constrained by the 303 ± 1 Ma U/Pb zircon dating obtained by Paquette (in Monié and Chopin, 1991) on an undeformed metagranitoid relict exposed within the orthogneiss of the monometamorphic complex near Brossasco. Therefore, before the Alpine orogeny the BIU consisted of a Variscan amphibolite-facies basement intruded by late-Variscan granitoids.

Pre-Alpine geologic and metamorphic features of the BIU basement are very similar to those of other crystalline tectonic units, which make up the nappe imbrication of the southern DMM (Borghi et al., 1992). However, these tectonic slices exhibit different early-Alpine climax metamorphic conditions, ranging from middle-T to low-T eclogite-facies and epidote blueschist-facies conditions (Chopin et al., 1991). Consequently, in the nappe pile of the southern DMM, crystalline thrust sheets, now in contact or only a few kilometers apart, suffered significantly different climax metamorphic conditions and followed contrasting retrograde P-T paths (Chopin et al., 1991). This geotectonic setting requires a mechanism of exhumation able to juxtapose slices of crustal rocks from the same basement, that were subducted to significantly different depths (Michard et al., 1993).

The BIU early-Alpine climax conditions, though peaking at ca. 750°C, plot on a geotherm of about 7°C/km, typical of high dP/dT metamorphism. Holding in due consideration the significantly higher peak pressure (ca. 30 kbar), also the shape of the clockwise P-T trajectory inferred for the Alpine evolution of the BIU (Fig. 7.8) is similar to those of other crystalline tectonic sheets of the Western Alps (see e. g. Bocquet, 1974; Sandrone and Borghi, 1992; Borghi et al., 1992).

In spite of the Alpine polyphase deformation, also active during the retrogressive P-T-t path, some small portions of Pre-Alpine rocks escaped repeated and pervasive Alpine deformation and recrystallization. The texturally best preserved lithotypes include fresh granitoids and thermally metamorphosed paraschists, i.e., rocks originally massive and dry or strengthened by the random recrystallization of contact minerals, respectively. The preservation of undeformed volumes of rocks in a tectonic unit that underwent pressures corresponding to depths of about 120 km and temperatures in excess of 700 °C, most likely was concentrated mainly along a network of ductile shear zones (Compagnoni and Hirajima, 1993). Such shear zones distinctively developed in the orthogneiss and yielded the pyrope-bearing whiteschist. This lithotype, characterized by an extremely high Mg/Fe ratio and very low Ca and Na content, is considered to have originated at the expense of granitoids during prograde UHPM by a metasomatic process active along ductile shear zones in the presence of a hydrous fluid phase. This metasomatic interpretation is supported by the discovery in the whiteschist of superzoned garnets, whose wide compositional range (see section on pyrope-bearing whiteschist, above) requires a significant change in the bulk rock chemical composition during prograde evolution close to the UHPM peak (Compagnoni and Hirajima, 1993).

It is interesting to note that the widespread eclogite does not preserve Pre-Alpine minerals or textures, but commonly exhibits a strong mineral fabric, defined by lineation of omphacite and preferred orientation of phengite. This tectonic fabric confirms on one side the existence of a high strain during the

UHPM and, on the other side, the tendency of eclogite easily to deform and recrystallize under UHP conditions, even in the absence of the water released by the amphibole breakdown during the transition from amphibolite to eclogite. In agreement with lack of relict amphibolite in eclogite boudins, the climax conditions of the UHP event (Fig. 7.8) plot outside the stability field of the hornblende. Therefore, polyphase deformation shown by the BIU eclogite occurred in a metabasic rock already eclogitized.

The Early-Alpine UHP conditions inferred for the BIU suggest that partial melting may have occurred at the metamorphic climax (cf. Schreyer et al., 1987). In the biotite-phengite gneiss, including the whiteschist of Case Ramello, "light coloured and coarser bands or veins of a metatectic character" were interpreted by Schertl et al. (1991, p. 9) as "indications for the beginning of partial melting at some stage during the metamorphic history." Owing to the evidence that the postclimactic decompression is accompanied by a continuous progressive cooling, any partial melting in the BIU should have occurred very close to the metamorphic peak. Therefore, it is quite improbable that early fabrics, such as "metatectic pockets," could have survived in the biotite-phengite gneiss, i. e., in a rock that experienced polyphase and pervasive deformation and recrystallization under greenschist-facies conditions.

Quartz-rich leucocratic layers, common in quartz bearing eclogites, were dubitatively interpreted by Kienast et al. (1991) as either recrystallized trondhjemite or alternatively the first product of partial melting in eclogite. In our opinion, such layers are metamorphic veins (developed at, or close to, the UHP climax) similar to those veins commonly occurring in middle-T eclogite. This means that during the same metamorphic event, i. e., under the same P-T conditions, different water activities may have taken place in different places of the same tectonic unit.

Other fabrics in texturally well-preserved Pre-Alpine paraschists could be considered as possible evidence for partial melting or, at least, of some element (alkalies and water) loss during the UHPM. Examples include the needle-like kyanite + quartz or kyanite + quartz + jadeite pseudomorphs, possibly after Pre-Alpine or prograde early-Alpine paragonite and plagioclase, respectively, or the very fine grained kyanite alignments developed along kinks of Pre-Alpine muscovite. In conclusion, clear evidence of partial melting in the BIU is thus far missing, either because during UHPM the wet solidus was never crossed when water activity was lower than one, or because primary metatectic textures, if any, were mostly obliterated by later deformation. Therefore, even if melting occurred in the BIU, it must have been a minor process.

Acknowledgments

R.C. is grateful to the Italian C.N.R., Centro di Studio sulla Geodinamica delle Catene Collisionali, Torino and to the Kyoto University Foundation, T. H. to the Ministry of Education of Japan (Grant-in-Aid Research, No. 04740456) and C. C. to the French INSU-CNRS programme "Dynamique et bilan de la Terre" for financial support. The constructive comments and suggestions of Dr. Diane Moore, U.S.G.S. and Dr. H.-P. Schertl, Bochum, are greatly appreciated.

References

Bearth, P. 1958. Über einen Wechsel der Mineralfazies in denWurzelzone des Penninikums. *Schweizerische mineralogische und petrographische Mitteilungen, 38*, 363–373.

Berman, R. G. 1988. Internally-consistent thermodynamic data for minerals in the system $Na_2O-K_2O-CaO-MgO-FeO-Fe_2O_3-Al_2O_3-SiO_2-H_2O-CO_2$. *J. Petrol., 29*, 445–522.

Biino, G. G., and Compagnoni, R. 1991. Evidence of a polymetamorphic crystalline basement in the very-high pressure Brossasco-Isasca Complex, Dora-Maira Massif, Western Alps. EUG VI meeting, 24–28 March 1991, Strasbourg (France). *Terra Abst., 3*, 84.

Biino, G., and Compagnoni, R. 1992. Very high pressure metamorphism of the Brossasco coronite metagranite, southern Dora-Maira massif, western Alps. *Schweiz. min. und petr. Mitt., 72*, 347–362.

Bocquet, J. 1974. Etudes minéraloques et pétrologigues sur les métamorphism d'âge alpin dans les Alpes françaises. Ph.D., Univ. Grenoble, France, 489 p.

Bocquet, J., Delaloye, M., Hunziker, J. C., and Krummenacher, D. 1974. K-Ar and Rb-Sr dating of blue amphiboles, micas and associated minerals from the western Alps. *Contributions to Mineralogy and Petrology, 47*, 7–26.

Borghi, A., Compagnoni, R., and Sandrone, R. 1992. The *P-T-t* paths of the internal Penninic units: petrological constraints to the geodynamic evolution of the western Alps. *29th IGC, Kyoto, Japan, Abstract Vol., 2*, 593.

Brown, T. H., Berman, R. G., and Perkins, E. H. 1988. Ge0-Calc: Software package for calculation and display of pressure-temperature-composition phase diagrams using an IBM or compatible Personal Computer. *Computers and Geoscience, 14*, 279–289.

Carman, J. H., and Gilbert, M. C. 1983. Experimental studies on glaucophane stability. *American Journal Science, 283-A*, 414–437.

Caron, J. M., Polino, R., Pognante, U., Lombardo, B., Lardeaux, J. M., Lagabrielle, Y., Gosso, G., and Allenbach, B. 1987. Où sont les sutures majeures dans les Alpes internes? (Transversale Briançon-Torino). *Memorie della Società Geologica Italiana, 29*, 71–78.

Carswell, D. A. 1990. Eclogites and the eclogite facies: definitions and classifications. *Eclogite facies rocks*. Glasgow: Blackie, 1–13.

Chopin, C. 1983. Magnesiochloritoid, a key-mineral for the petrogenesis of high-grade pelitic blueschist. *Bulletin Minéralogie, 106*, 715–717.

Chopin, C. 1984. Coesite and pure pyrope in high-grade blueschists of the western Alps: A first record and some consequences. *Contrib. Mineral. Petrol., 86*, 107–118.

Chopin, C. 1986. Phase relationships of ellenbergerite, a new high-pressure Mg-Al-Ti-silicate in pyrope-coesite-quartzite from the Western Alps. *Geol. Soc. Amer. Mem., 164*, 31–42.

Chopin, C. 1987. Very-high-pressure metamorphism in the western Alps: implications for subduction of continental crust. *Philosophical Transaction of the Royal Society of London, Series A, 321, no. 1557*, 183–197.

Chopin, C., Brunet, F., Gebert, W., Medenbach, O., and Tillmanns, E. 1993. Bearthite, $Ca_2Al(PO_4)_2OH$, a new mineral from high pressure terranes of the western Alps. *Schweiz. min. und petr. Mitt., 73*, 1–9.

Chopin, C., Henry, C., and Michard, A. 1991. Geology and petrology of the coesite-bearing terrain, Dora-Maira massif, Western Alps. *Eur. Jour. Mineral., 3*, 263–291.

Chopin, C., and Klaska, R.1986. *Metamorphism of continental crust at mantle depths: new rock-forming minerals.* 14th I.M.A. Meeting, 13–18 July 1986, Stanford University, CA., U.S.A., Abstracts with Program, p. 77., Stanford University, CA., U.S.A.

Chopin, C., Klaska, R., Medenbach, O., and Dron, D. 1986. Ellenbergerite, a new high-pressure Mg-Al-(Ti, Zr)-silicate with a novel structure based on face-sharing octahedra. *Contrib. Mineral. Petrol., 92*, 316–321.

Chopin, C., and Langer, K. 1988. Fe^{2+}-Ti^{4+} charge transfer between face-sharing octahedra: polarized adsorption spectra and crystal chemistry of ellenbergerite. *Bull. Minéralogie, 111*, 17–27.

Chopin, C., and Maluski, M. 1978. Résultats préliminaires obtenus par la méthode $^{39}Ar/^{40}Ar$ sur des minéraux alpins du massif du Grand Paradis et de son enveloppe. *Bulletin de la Societé géologique de France, 20*, 745–749.

Chopin, C., and Maluski, H. 1980. ^{40}Ar-^{39}Ar dating of high-pressure metamorphic micas from the Gran Paradiso area (western Alps): evidence against the blocking temperature concept. *Contrib. Mineral. Petrol., 74*, 109–122.

Chopin, C., and Monié, P. 1984. A unique magnesiochloritoid-bearing, high-pressure assemblage from the Monte Rosa: A petrologic and ^{39}Ar-^{40}Ar study. *Contrib. Mineral. Petrol., 87*, 388–398.

Chopin, C., and Schreyer, W. 1983. Magnesiocarpholite and magnesiochloritoid: Two index minerals of pelitic blueschists and their preliminary phase relations in the model system MgO-Al_2O_3-SiO_2-H_2O. *Am. J. Sci., 283-A*, 72–96.

Chopin, C., Schreyer, W., and Baller, T. 1992. Ellenbergerite stability, a reappraisal. *Terra Nova Abst. Suppl., 4*, 10.

Chopin, C., Schreyer, W., and Baller, T. 1988. Synthesis and stability limits of ellenbergerite. *Terra Cognita, 8*, 59–60.

Compagnoni, R. 1977. The Sesia-Lanzo Zone: high pressure-low temperature metamorphism in the Austroalpine continental margin. *Rendiconti della Società Italiana di Mineralogia e Petrologia, 33*, 281–334.

Compagnoni, R., Elter, G., and Lombardo, B. 1974. Eterogeneità stratigrafica del complesso degli "gneiss minuti" nel massiccio cristallino del Gran Paradiso. *Memorie della Società Geologica Italiana, 13*, 227–239.

Compagnoni, R., and Hirajima, T. 1991. Geology and petrology of the Brossasco-Isasca Complex, southern Dora-Maira massif, Western Alps. *Terra Abstract, 3*, 84.

Compagnoni, R., and Hirajima, T. 1992. Occurrence and significance of superzoned garnet in the coesite-bearing whiteschist of southern Dora-Maira Massif, western Alps. *29th IGC, 1992, Kyoto (Japan), Abs. 2*, 599.

Compagnoni, R., and Hirajima, T. 1993. The occurrence of a superzoned garnet in the coesite-bearing unit of Dora-Maira massif and the origin of the whiteschists. *Fourth International Eclogite Conference, Fuscaldo (Calabria, Italy), Terra Abst., 4,* 5.

Dal Piaz, G. V. 1966. Gneiss ghiandoni, marmo ed anfiboliti antiche del ricoprimento Monte Rosa nell'alta Val d'As. *Bollettino della Scocietà Geologica Italiana, 85,* 103–132.

Dal Piaz, G. V., Hunziker, J. C., and Martinotti, G. 1972. La Zona Sesia-Lanzo e l'evoluzione tettonico-metamofica delle Alpi noroccidentali interne. *Società Geologica Italiana, Memorie, 11,* 433–460.

Dal Piaz, G. V., and Lombardo, B. 1986. Early Alpine eclogite metamorphism in the Pennidic Monte Rosa-Gran Paradiso basement nappes of the northwestern Alps. *Geol. Soc. Amer. Mem., 164,* 249–265.

Debelmas, J., and Lemoine, M. 1970. The western Alps: paleogeography and structure. *Earth Science Review, 6,* 221–256.

Desmons, J. 1986. The Alpine metamorphisms and their environments in the western Alps: unsolved problems. *Schweizerische Mineralogische und Petrographische Mitteilungen, 66,* 29–40.

Enami, M., and Zang, Q. 1990. Quartz pseudomorphs after coesite in eclogites from Shandong Province, east China. *Am. Mineralogist, 75,* 381–386.

Ferraris, G., Ivaldi, G., and Chopin, C. 1994. Magnesio dumortierite, a new mineral from very-high-pressure rocks (Western Alps): Part I, crystal structure. *Eur. J. Mineral,* in press.

Gebauer, D., Tilton, G. R., Schertl, H.-P., and Schreyer, W. 1993. Eocene/Oligocene ultrahigh-pressure (UHP)- metamorphism in the Dora-Maira massif (western Alps) and its geodynamic implications. *Fourth International Eclogite Conference (Calabria, Italy). Terra Abst., 4,* 10–11.

Gillet, P., Ingrin, J., and Chopin, C. 1984. Coesite in subducted continental crust: *P-T* history deduced from an elastic model. *Earth Planet. Sci. Lett., 70,* 426–436.

Hammerschmidt, K., Schertl, H.-P., Friedrichsen, H., and Schreyer, W. 1992. Überschuss-Argon während der retrograden Metamorphose der Ultrahochdruckgesteine des Dora-Maira Massivs, italienische Westalpen. *Berichte der Deutschen Mineralogischen Gesellschaft (Abstract).*

Henry, C. 1990. L'unité à coesite du massif Dora-Maira dans son cadre petrologique et structural (Alpes occidentales) Italie. PhD.,Université de Paris VI.

Henry, C., Michard, A., and Chopin, C. 1993. Geometry and structural evolution of ultra-high-pressure and high-pressure rocks from the Dora-Maira massif, Western Alps, Italy. *Journal of Structural Geology, 15,* 965–982.

Hirajima, T., and Compagnoni, R. 1993. Petrology of a jadeite-quartz/coesite-almandine-phengite fels with retrograde ferro-nyböite from the Brossasco-Isaca unit, southern Dora-Maira massif, western Alps, Italy. *Eur. Jour. Mineral., 5,* 943–955.

Hirajima, T., Ishiwatari, A., Cong, B., Zhang, R., Banno, S., and Nozaka, T. 1990. Coesite from Mengzhong eclogite at Donghai county, northeastern Jiangsu province, China. *Mineral. Mag., 54,* 579–583.

Holland, T. J. B. 1979. The experimental determination of the reaction paragonite = jadeite + kyanite + H_2O and internally consistent thermodynamic data for part of the system $Na_2O-Al_2O_3-SiO_2-H_2O$, with applications to eclogites and blueschists. *Contrib. Mineral. Petrol., 68,* 293–301.

Holland, T. J. B. 1988. Preliminary phase relations involving glaucophane and applications to high pressure petrology: new heat capacity and thermodynamic data. *Contrib. Mineral. Petrol., 99,* 134–142.

Holland, T. J. B., and Powell, R. 1990. An enlarged and updated internally consistent thermodynamic dataset with uncertainties and correlations: The system $K_2O-Na_2O-CaO-MgO-MnO-FeO-Fe_2O_3-Al_2O_3-TiO_2-SiO_2-C-H_2-O_2$. *Jour. Metamorph. Geol., 8,* 89–124.

Huang, W. L., and Wyllie, P. J. 1975. Melting reactions in the system $NaAlSi_3O_8$-$KAlSi_3O_8$-SiO_2 to 35 kilobars, dry and with excess water. *J. Geol., 83,* 737–748.

Hunziker, J. C. 1970. Polymetamorphism in the Monte Rosa, western Alps. *Eclogae gelogicae Helvetiae, 63,* 151–161.

Hunziker, J. C. 1974. Rb-Sr and K-Ar age determination and the Alpine tectonic history of the western Alps *Istituti di Geologia e Mineralogia, Univ. di Padova, Memorie, 31,* 54 pp.

Hunziker, J. C., Desmons, J., and Hurford, A. J. 1992. *Thirty-two years of geochronological work in the Central and Western Alps: a review on seven maps.* Mémoires de Géologie, Lausanne, *13,* 1–59.

Hy, C. 1984. Métamorphisme polyphasé et évolution tectonique dans la croûte continental éclogitisée: les séries granitiques et pélitiques du Monte Mucrone (Zone Sesia Lanzo, Alpes Italiennes). PhD. Univ. P. et M. Curie, Paris.

Kennedy, C. S., and Kennedy, G. C. 1976. The equilibrium boundary between graphite and diamond. *J. Geophys. Res., 81,* 2467–2470.

Kienast, J. R., Lombardo, B., Biino, G., and Pinardon, J. L. 1991. Petrology of very-high-pressure eclogitic rocks from the Brossasco-Isasca Complex, Dora-Maira Massif, Italian Western Alps. *Jour. Metamorph. Geol., 9,* 19–34.

Kretz, R. 1983. Symbols for rock-forming minerals. *Am. Mineralogist, 68,* 277–279.

Laubscher, H. P., and Bernoulli, D. 1982. History and deformation of the Alps. *Mountain building processes.* New York: Academic Press, 170–180.

Leake, B. E. 1978. Nomenclature of amphiboles. *Am. Mineralogist, 63,* 1023–1052.

Massonne, H.-J. 1990. Geothermobarometrie mit Al-Gehalten in Talk. *Miner. Ges., Beih. 2 Europ. J. Mineral, 2,* 169.

Massonne, H.-J., and Schreyer, W. 1989. Stability field of the high-pressure assemblage talc + phengite and two new phengite barometers. *Eur. Jour. Mineral., 1,* 391–410.

Michard, A., Chopin, C., and Henry, C. 1993. Compression versus extension in the exhumation of the Doria-Maira coesite-bearing unit, Western Alps, Italy. *Tectonophysics, 221,* 173–193.

Mirwald, P. W., and Massonne, H.-J. 1980. The low-high quartz and quartz-coesite transition to 40 kbar between 600° and 1600°C and some reconnaissance data on the effect of $NaAlO_2$ component on the low quartz-coesite transition. *J. Geophys. Res., 85,* 6983–6990.

Monié, P., and Chopin, C. 1991. $^{40}Ar/^{39}Ar$ dating in coesite-bearing and associated units of the Dora-Maira massif, Western Alps. *Eur. Jour. Mineral., 3,* 239–262.

Morimoto, N., Ferguson, A. K., Ginzburg, I. V., Ross, M., Seifert, F. A., Zussman, J., Aoki, K., and Gottardi, G. 1988. Nomenclature of pyroxenes. *Am. Mineralogist, 73,* 1123–1133.

Newton, R. C. 1986. Metamorphic temperatures and pressures of group B and C eclogites. *Geol. Soc. America Mem., 164,* 17–30.

Oberhänsli, R., Hunziker, J. C., Martinotti, G., and Stern, W. 1985. Geochemistry, geochronology and petrology of Monte Mucrone: an example of Eo-Alpine eclogitisation of Permian granitoids in the Sesia-Lanzo zone, western Alps, Italy. *Chemical Geology, 52,* 165–184.

Okay, A., Sengör, A.M.C., Su, W., Liu, Y., and Jiang, L. 1992. Diamond from the Dabie Shan metamorphic rocks and its implications for tectonic setting. *Science, 256,* 80–82.

Okay, A. I., Xu, S., and Sengör, A.M.C. 1989. Coesite from the Dabie Shan eclogites, central China. *Eur. Jour. Mineral., 1*, 595–598.

Paquette, J.-R., Chopin, C., and Peucat, J. J. 1989. U-Pb zircon, Rb-Sr and Sm-Nd geochronology of high- to very-high-pressure meta-acidic rocks from the western Alps. *Contrib. Mineral. Petrol., 101*, 280–289.

Pasquarè, G. 1975. Geological summary of the central Alps. *Quaderni de "La Ricerca Scientifica", C.N.R., Roma, 90*, 121–148.

Perkins, D. III, Westrum, E. F., Jr., and Essene, E. J. 1980. The thermodynamic properties and phase relations of some minerals in the system $CaO-Al_2O_3-SiO_2-H_2O$. *Geochem. Cosmochim. Acta, 44*, 61–84.

Platt, J. P. 1986. Dynamics of orogenic wedges and the uplift of high-pressure metamorphic rocks. *Geol. Soc. Amer. Bull., 97*, 1037–1053.

Polino, R., Dal Piaz, G. V., and Gosso, G. 1990. Tectonic erosion of the Adria margin and accretionary processes for the Cretaceous orogeny of the Alps. *Soc. Geol. de France, Mémoires, 156*, 345–367.

Reinecke, T. 1991. Very-high-pressure metamorphism and uplift of coesite-bearing metasediments from the Zermatt-Saas zone, Western Alps. *Eur. Jour. Mineral., 3*, 7–17.

Rubie, D. C. 1990. Role of kinetics in the formation and preservation of eclogites. *Eclogite Facies Rocks*. Glasgow: Blackie.

Sandrone, R., and Borghi, A. 1992. Zoned garnets in the northern Dora-Maira Massif and their contribution to a reconstruction of the regional metamorphic evolution. *Eur. Jour. Mineral., 4*, 465–474.

Sandrone, R., Cadoppi, P., Sacchi, R., and Vialon, P. 1993. The Dora-Maira Massif. *Pre-Mesozoic geology in the Alps*. Berlin: Springer, 315–323.

Schertl, H.-P., Schreyer, W., and Chopin, C. 1991. The pyrope-coesite rocks and their country rocks at Parigi, Dora-Maira Massif, Western Alps: detailed petrography, mineral chemistry and PT-path. *Contrib. Mineral. Petrol., 108*, 1–21.

Schmädicke, E. 1991. Quartz psuedomorphs after coesite in eclogites from the Saxonian Erzgebirge. *Eur. Jour. Mineral., 1*, 231–238.

Schreyer, W. 1973. Whiteschist: a high-pressure rock and its geologic significance. *J. Geol., 81*, 735–739.

Schreyer, W. 1977. Whiteschists: their compositions and pressure-temperature regimes based on experimental, field, and petrographic evidence. *Tectonophysics, 43*, 127–144.

Schreyer, W. 1985. Metamorphism of crustal rocks at mantle depths: High-pressure minerals and mineral assemblages in metapelites. *Fortschritte der Mineralogie, 63*, 227–261.

Schreyer, W. 1988. Experimental studies on metamorphism of crustal rocks under mantle pressures. *Mineral. Mag., 52*, 1–26.

Schreyer, W., Massonne, H.-J., and Chopin, C. 1987. Continental crust subducted to depths of near 100 km: implications for magma and fluid genesis in collision zones. *Magmatic process: physicochemical principles. Geochemical Society Special Publication, 1*, 155–163.

Sharp, Z. D., Essene, E. J., and Hunziker, J. C. 1993. Stable isotope geochemistry and phase equilibria of coesite-bearing whiteschists, Dora-Maira massif, western Alps. *Contrib. Mineral. Petrol., 114*, 1–12.

Smith, D. C. 1984. Coesite in clinopyroxene in the Caledonides and its implications for geodynamics. *Nature, 310*, 641–644.

Smith, D. C. 1988. A review of the peculiar mineralogy of the "Norwegian coesite-eclogite province," with crystal-chemical, petrological, geochemical and geodynamical notes and an extensive bibliography. *Eclogites and eclogite-facies rocks.* Amsterdam: Elsevier, 1–206.

Smith, D. C., and Lappin, M. A. 1989. Coesite in the Straumen kyanite-eclogite pod, Norway. *Terra Nova, 1*, 47–56.

Sobolev, N. V., and Shatsky, V. S. 1990. Diamond inclusions in garnets from metamorphic rocks. *Nature, 343*, 742–746.

Tagiri, M., and Bakirov, A. 1990. Quartz pseudomorph after coesite in garnet from a garnet-chloritoid-talc schist, Northern Tien-Shan, Kirghiz, USSR. *Proceedings of the Japan Academy, 66, ser. B*, 135–139.

Tilton, G. R., Schreyer, W., and Schertl, H.-P. 1991. Pb-Sr-Nd isotopic behavior of deeply subducted crustal rocks from the Dora-Maira Massif, Western Alps-II: what is the age of the ultrahigh-pressure metamorphism? *Contrib. Mineral. Petrol., 108*, 22–33.

Trümpy, R. 1982. Alpine paleogeography: a reappraisal. *Mountain building processes.* London: Academic Press, 149–156.

Vialon, P. 1966. Etude géologique du massif cristallin Dora-Maira, Alpes cottiennes internes, Italie. Ph.D., Univ. de Grenoble.

Wang, Q., Ishiwatari, A., Zhao, Z. T. H., Enami, M., Zhai, M., Li, J., and Cong, B. 1993. Coesite-bearing granulite retrograded from eclogite in Weihai, Eastern China. *Eur. Jour. Mineral., 5*, 141–151.

Wang, X., and Liou, J. G. 1991. Regional ultrahigh-pressure coesite-bearing eclogitic terrane in central China: evidence from country rocks, gneiss, marble, and metapelite. *Geology, 19*, 933–936.

Wang, X., Liou, J. G., and Mao, H. K. 1989. Coesite-bearing eclogites from the Dabie Mountains in central China. *Geology, 17*, 1085–1088.

Xu, S., Okay, A. I., Ji, S., Sengor, A. M. C., Su, W., Liu, Y. & Jiang, L. 1992. Diamond from the Dabie Shan metamorphic rocks and its implication for tectonic setting. *Science, 256*, 80–82.

Yang, J., and Smith, D. C. 1989. Evidence for a former sanidine-coesite-eclogite at Lansantou, Eastern China, and the recognition of the Chinese 'Su-Lu coesite-eclogite province', East China [abs.]. *Terra Abst., 1*, 26.

8

HP and UHP Eclogites and Garnet Peridotites in the Scandinavian Caledonides

E. J. KROGH AND D. A. CARSWELL

Abstract

A review of the high and UHP eclogites and garnet peridotites from the Scandinavian Caledonides concerning models of their formation provides a background for individual descriptions of the various localities. The structure of the Scandinavian Caledonides is dominated by an assembly of thrust sheets transported eastward and finally emplaced on the Baltoscandian Platform during the Scandian Orogen in the Middle to Late Silurian. The tectonic units can conveniently be divided into five major complexes, the autochthon (and parautochthon), the lower, middle, upper and uppermost allochthons. Eclogites and eclogitic rocks are found within several of the tectono-stratigraphic levels of the Scandinavian Caledonides. Eclogites within the autochthon/parautocthon include rocks of the Proterozoic Western Gneiss Region, Træna Island, and the Lofoten-Vesterälen area. Eclogites within the Middle Allochthon include the Bergen Arcs and in the Hardangervidda-Ryfylke Nappe complex. The eclogites within the Upper Allochthon include the Seve Nappes. Eclogites within the Uppermost Allochthon are from the Tromsø Nappe Complex. Critical mineral assemblages are discussed for each locality along with known data on the age of metamorphism, and in some cases fluid inclusion studies provide new insights. Garnet peridotites within the Scandinavian Caledonides are present within Proterozoic basement Gneisses of the Western Gneiss Region of Norway and the Sev Nappe Complex in northern Sweden. Thermobarometric and geochronological data on occurrences of relict UHP mineral assemblages are given for Mg-Cr type peridotite bodies at a number of localities, mostly in the coastal region of the Western Gneiss Region. These data were taken to indicate that these UHP assemblages are of mid-Proterozoic age (ca. 1700 Ma) and represent mineralogies which equilibrated in old, subcontinental mantle, slices of which were tectonically emplaced into high grade crustal gneisses of the Western Gneiss Region during the Caledo-

nian orogeny. Tectonic models for the HP/UHP metamorphism in the Scanda-
navian Caledonides are related to subduction, collision and exhumation; however, all of the models fall short of complete explanation. It does seem evident
that a continuing, sequential underthrusting in response to crustal shortening
is essential for the refrigeration, survival and rapid exhumation of UHP/HP
rocks. The extensional features leading to formation of the Devonian basins
may have played only a minor role in the final unroofing of the eclogite-
bearing units.

Introduction

Historical Review / Controversies

Eclogites (sensu stricto) and eclogitic rocks are widely distributed within the
Scandinavian Caledonides (Fig. 8.1). They are commonly associated with
migmatites and gneisses (e.g. Eskola, 1921; Backlund, 1936; Bryhni, 1966;
Lappin, 1966; Bryhni et al., 1969, 1977; Mysen and Heier, 1972; Krogh 1977a,
1980a,b; Lappin and Smith 1978; Andreasson et al., 1985; Carswell et al.,
1985; Griffin and Carswell, 1985; van Roermund, 1985; Jamtveit, 1987a),
marbles and other metasupracrustals (Pettersen, 1878; Hernes, 1954; Råheim, 1972; Landmark, 1973; Kullerud, 1987; Stølen 1988; Krogh et al., 1990),
anorthosites in layered igneous complexes (Kolderup, 1903; Griffin, 1972;
Austrheim and Griffin, 1985; Brastad, 1985; Austrheim, 1987) and with garnet peridotites (Eskola, 1921; O'Hara and Mercy, 1963; Carswell, 1974; Lappin, 1974; Medaris, 1980a,b; Griffin and Qvale, 1985). In addition, eclogites
or eclogitic domains have been locally developed in doleritic sills and dikes,
and in gabbros which cross-cut the surrounding country-rocks (Gjelsvik,
1952; Griffin and Heier, 1973; Griffin and Råheim, 1973; Bryhni et al., 1977;
Tørudbakken and Råheim, 1981; Cuthbert, 1985; Griffin et al., 1985; Mørk,
1985a,b, 1986; Kullerud et al., 1990).

Most eclogites occur as lenticular bodies within amphibolite to granulite
facies host rocks, commonly showing tectonic contact relationships. Rims of
the eclogite lenses tend to be strongly sheared, and retrogressed to lower pressures, compatible with those recorded in the host gneisses. These features are
responsible for the great controversy concerning the origin of the Norwegian
eclogites. Historically, this controversy can be summarized as the "igneous
model" of Eskola, and the "metamorphic model" of Backlund, as briefly
described below.

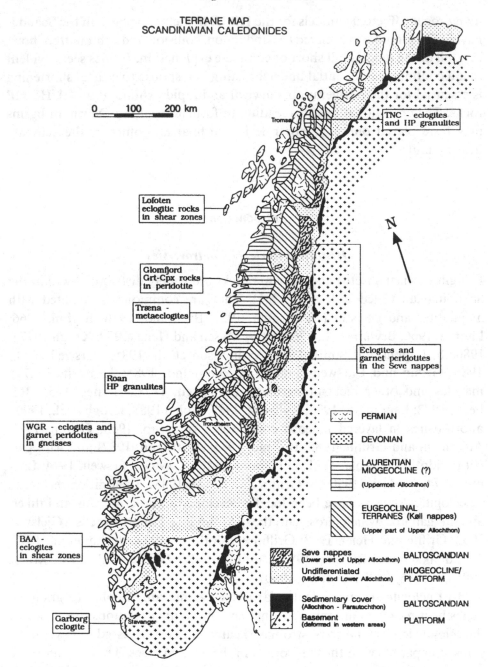

Figure 8.1. Tectonostratigraphic map of the Scandinavian Caledonides, showing the distribution of HP and UHP rocks. TNC–Tromsø Nappe Complex; WGR–Western Gneiss Region; BAA–Bergen Arcs anorthosites. (Modified after Dallmeyer and Gee, 1986; and Dallmeyer and Andresen, 1992.)

The Igneous Model of Eskola

As early as in 1862, Hiortdahl and Irgens pointed out an igneous origin for the eclogites in the Bergen region. They were followed by Kolderup (1903) who, according to Eskola (1920), provided full evidence that the eclogites are a differentiated variety derived from the same magma as the labradorite rocks (anorthosites/gabbros) of the Bergen Arcs. In 1920, Eskola stated that most eclogites have originated at temperatures above the melting temperatures of many other rocks (e.g. gneisses), and that all the Norwegian eclogites studied by him are of igneous origin. In Eskola's model, garnet and clinopyroxene start to crystallize from the "eclogite magma" at high temperature and pressure. During the course of crystallization and cooling, there may be a change in the facies of such rocks, and primary pale green amphibole starts to crystallize from the magma at the expense of garnet and clinopyroxene. This primary amphibole does not apparently belong to the eclogite facies proper, but to an independent facies, and the resulting rock should be called amphibole eclogite (Eskola, 1921). Another variety of eclogitic rock described by Eskola (1921) was termed "eclogite hornblende gabbro," of which his samples labeled "Romsdalshorn" (cf. Krogh, 1982) are representative. He proposed this name because the rock contains both the minerals of an eclogite and those of a hornblende gabbro. According to Eskola (1920, 1921), garnet and clinopyroxene were probably the first minerals to crystallize out in this rock. During cooling and crystallization of garnet, the remaining magma became enriched in Ca, and at the transition from eclogite to amphibolite facies, diopside started to crystallize along with hornblende and plagioclase. As common evidence for the igneous origin of Norwegian eclogites, Eskola referred to their inclusion-free garnets, which contrast with metamorphic garnets that are usually filled with quartz grains. In spite of this, he still mentioned "magmatic garnets" from the Romsdalshorn "eclogite-hornblende gabbro" containing inclusions of quartz and rutile (Eskola, 1921). Later Hernes (1954) and Krogh (1982) have shown that Eskola's eclogite-hornblende gabbro was wholly metamorphic.

A metamorphic origin for the Norwegian eclogites was thus dismissed by Eskola (1920, 1921). He stated that alteration from gabbro to eclogite had never been observed. As mentioned earlier, more recent studies have, however, shown that transitions from gabbroic to eclogitic rocks are not uncommon within the west Norwegian basal gneiss region.

The Metamorphic Model of Backlund

Contrary to the results of Eskola, Backlund (1936) found field and petrographical evidence for a wholly metamorphic origin of Caledonian eclogites from both eastern Greenland and Scandinavia. He described inclusions of hornblende in eclogite garnets, and concluded that these were not a result of retrogression but rather represented armored relicts. He also pointed out the general course for the transition from basalt to eclogite during prograde metamorphism:

basalt → uralite porphyry → greenstone → amphibolite → eclogite
amphibolite → eclogite

According to Backlund (1936) these prograde metamorphic reactions take place under increasing temperature at shallow to moderate depths. In his opinion the formation of eclogite results not from high hydrostatic pressure, but rather from high dynamic pressures.

Discussion

These two principally different models for the origin of eclogites, magmatic vs. metamorphic, have since the 1970s been redefined to the following fundamentally different interpretations for the origin of the eclogites within the Western Gneiss Region (WGR) of the Scandinavian Caledonides (as summarized by Cuthbert and Carswell, 1990; see also Figure 8.2 for locality names):

1. The "foreign model." The eclogites were formed at mantle depths and subsequently tectonically transported into gneisses, which lay at higher crustal levels. Eclogite margins were amphibolitized during emplacement into the gneisses. This model is not equivalent to Eskola's igneous model.
2. The "in situ model." The eclogites were formed in situ in their present host gneisses; differential retrogression has resulted in almost complete transformation of the gneisses to amphibolite (or granulite) facies assemblages, whereas kinetic factors have favored the preservation of eclogite facies assemblages in metabasic rocks.

Advocates of the "foreign model" have been Lappin (1966, 1977), O'Hara (1976), Lappin and Smith (1978, 1981), and Smith (1980, 1981). In a recent review paper Smith (1988) maintains the "foreign" model of "lithospheric interdigitation," and puts it into a new model: the "FIF" (foreign/in situ/foreign) model. This involves (Smith, 1988) "the tectonic introduction of the previously-eclogitized 'foreign' coesite-eclogite province into not-yet-eclogitized crust, a new lower-P eclogitization 'in situ', and finally another tectonic introduction of fragments of these rocks as 'foreign' slices into never-eclogitized crust."

The numerous advocates for a crustal in situ origin of the WGR eclogites include Gjelsvik (1952), Hernes (1954), Bryhni (1966), Bryhni et al. (1969,

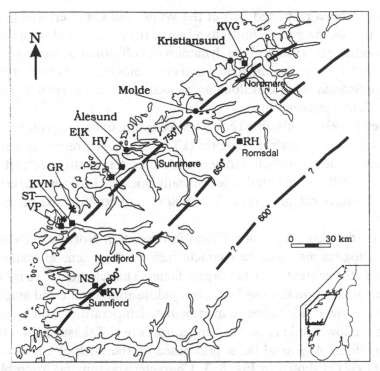

Figure 8.2. Locality map of eclogites in the WGR. * = Coesite eclogite; KV = Kvineset; NS = Naustdal; RH = Romsdalshorn; KVN = Kvalneset; VP = Verpeneset; GR = Grytting; ST = Straumen; EIK = Eiksunddal; HV = Hjørungavåg; KVG = Kvalvåg. (Isotherms from Griffin et al., 1985.)

1970, 1977), Skjerlie (1969), Råheim (1972), Mysen and Heier (1972), Griffin and Heier (1973), Griffin and Råheim (1973), Brueckner (1977), Krogh (1977a,b, 1980a,b, 1982), Krogh and Brunfelt (1981), Cuthbert et al. (1983), Griffin and Carswell (1985), Carswell et al. (1985), Brastad (1985), Griffin and Qvale (1985), Griffin et al. (1985), Mørk (1985a,b, 1986), Jamtveit (1987a,b), Chauvet et al. (1992).

A further discussion of the problems concerning the origin of the eclogites within the WGR will not be held here.

Previous Reviews

Since the 1980s, several reviews of eclogite facies metamorphism within the Scandinavian Caledonides have been published (Cuthbert et al., 1983; Griffin et al., 1985; Griffin, 1987; Smith, 1988; Cuthbert and Carswell, 1990). The first three of these reviews mainly concerned the relevance of the eclogite facies rocks to the orogenic evolution of the Scandinavian Caledonides. Smith's review mainly dealt with mineral assemblages and mineral chemistry of eclogites from a restricted part of the WGR, and Cuthbert and Carswell's work focused on the general understanding of the processes of eclogite formation in continental crust, and the dynamics of collisional orogenesis.

This chapter focuses on the spatial and chronologic distribution of eclogites within the Scandinavian Caledonides, as well as on the general P-T-t evolution of different eclogite suites.

Carswell (1990) has classified eclogites according to their equilibration temperatures. Low temperature eclogites (LT; T < 550°C) represent subducted oceanic crust and arc-trench sediments. Medium temperature eclogites (MT; 550C < T < 900°C) are found in tectonically thickened continental crust, and high temperature eclogites (HT; T > 900°C) are equilibrated in the upper mantle.

Based on detailed studies of eclogites from western Norway and elsewhere, the MT eclogites may also be characterized by their general metamorphic evolution in the context of metamorphic facies. The eclogite facies is bounded by blueschist facies on the low T side, by epidote amphibolite and amphibolite facies on the low P side at low to intermediate temperatures, and by granulite facies on the low P side at higher temperatures (e.g. Takasu, 1984; Oh et al., 1988; Oh, 1992). A general facies grid based on the work of Oh et al. (1988) and Oh (1992) is shown in Fig. 8. 3. Characteristic mineral assemblages of each of these metamorphic facies for a basaltic composition are given in Table 8.1.

During the course of tectonic thickening and subsequent decompression, a series of metamorphic evolutions can thus be expected, resulting in clockwise P-T-t paths. In the following, eclogites will be characterized from their general metamorphic evolution based on the modified grid shown in Fig. 8.3.

Spatial Distribution of Eclogites within the Scandinavian Caledonides

The structure of the Scandinavian Caledonides is dominated by an assembly of thrust sheets transported eastwards and finally emplaced on the Baltoscandian Platform during the Scandian Orogen in the Middle to Late Silurian. The tectonic units can conveniently be divided into five major complexes (Fig. 8.1), the autochthon (and parautochthon), the lower, middle, upper and uppermost allochthons (Roberts and Gee, 1985; Stephens and Gee, 1985; Stephens, 1988). A major regional unconformity locally separates marginal basin ophiolites from overlying upper Ordovician and younger Caledonian sequences (Thon, 1985; Dunning and Pedersen, 1988; Steltenpohl et al., 1990).

The **autochthon** and **parautochthon** consist of Precambrian (Middle Protero-

Table 8.1. *Characteristic Mineral Assemblages*
Defining Metamorphic Facies

Facies	Mineral assemblage
Greenschist (GS)	Act + Chl + Ep + Ab
Epidote-blueschist (BS)	Gl + Ep
Epidote-amphibolite (EA)	Ep + Hbl/Barr + Ab
Amphibolite (AM)	Hbl + Pl
Granulite (GR)	Grt + Cpx + Pl + Qtz (high-P)
	Cpx + Opx + Pl ± Grt (medium-P)
Eclogite (EC)	Grt + Omph + Qtz

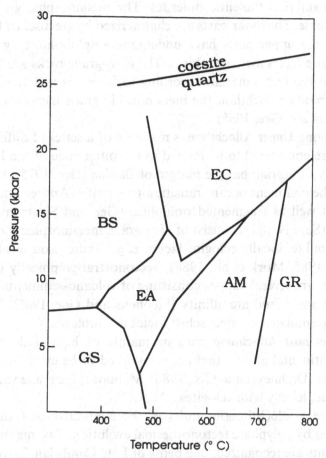

Figure 8.3. General facies grid for basaltic compositions (Modified from Oh et al., 1988, and Oh, 1992.) GS = greenschist facies; BS = blueschist facies; EA = epidote-amphibolite facies; AM = amphibolite facies; GR = granulite facies; EC = eclogite facies. Coesite–quartz transition from Bohlen and Boettcher (1982).

zoic and older) crystalline basement and a sedimentary cover of Vendian and/ or Lower Paleozoic age.

The **Lower Allochthon** consists of imbricated thrust slices of Precambrian crystalline rocks and Upper Proterozoic and/or Lower Paleozoic sedimentary cover. The metamorphic grade in lower allochthon varies from low to medium greenschist facies in the eastern and central parts of the orogen, up to amphibolite facies in the westernmost parts (Roberts and Gee, 1985; Stephens, 1988).

The **Middle Allochthon** consists mainly of Vendian and other Upper Proterozoic sandstone-dominated successions together with sheets of Precambrian crystalline rocks, all clearly related to the Baltoscandian miogeocline and outer shelf. The uppermost tectonic unit of the Middle Allochthon (Särv Nappes), derived from the outer part of the miogeocline, are extensively intruded by late Precambrian tholeiitic dolerites. The metamorphic grade shows an upward increase. The lower parts are charcterized by greenschist facies assemblages, and the upper parts have undergone amphibolite to granulite, and locally eclogite facies metamorphism. The high-grade rocks are "believed" to form part of the Baltic margin according to Stephens and Gee (1989). Also within the middle allochthon the metamorphic grade increases from east to west (Roberts and Gee, 1985).

The overlying **Upper Allochthon** is made up of a series of different tectonic units that are considered to be related to the outer edge of the Late Proterozoic to Early Cambrian passive margin of Baltica (Gee, 1975; Andreasson et al., 1979), the continent-ocean transition to Iapetus (Andreasson, 1986; Seve Nappes), as well as fragmented ophiolites (Gee and Sjøstrøm, 1984). The lower parts (Seve Nappes) consist of high grade metamorphic rocks (amphibolite to granulite, locally eclogite facies; e. g. Andreasson et al., 1985; van Roermund, 1985; Mørk et al., 1988). Tectonostratigraphically overlying the Seve Nappes are several nappes consisting of volcano-sedimentary sequences of ophiolite and island arc affinity (Stephens and Gee, 1985). These higher nappes are dominated by greenschist facies assemblages.

The **Uppermost Allochthon** consists mainly of high-grade metamorphic rocks of continental affinity that are considered to be exotic with respect to Baltoscandia (Dallmeyer and Gee, 1986). Metamorphic grade varies from low to high grade, locally with eclogites.

The different allochthonous units of the Scandinavian Caledonides are characterized by polyphase tectonothermal evolution. Two major Caledonian tectonic events are recognized, one being of Late Cambrian-Early Ordovician age (Finnmarkian orogeny; Sturt et al., 1978), and the other of Silurian age (Scandian orogeny; Gee and Wilson, 1974). Both orogenies are likely related to the closing of the Iapetus Ocean. In addition, older segments probably also contain records of Precambrian orogenic activity. The status of the Finn-markian orogeny is, however, uncertain. In the type area Seiland Igneous Province in Finnmark, northern Norway, the "Finnmarkian phase" has by recent dating appeared to be of Precambrian age (Krogh and Elvevold, 1990; Daly et al., 1991). However, a recent internal Sm-Nd isochron on a recrystal-lized metagabbro from the Seiland Igneous Province yielded 502 ± 28 Ma (Mørk and Stabel, 1990), and $^{40}Ar/^{39}Ar$ dating of hornblende by Dallmeyer (1988) from the same area records cooling ages of c. 490 Ma. Both these dates indicate that there was tectonic activity in this area during the Early Ordovician. In the Nordmannvik Nappe of the upper allochthonous unit in Troms, Lindstrøm and Andresen (1992) presented a RbSr isochron of 492 ± 5 Ma on a granulitic metadiorite, interpreted to represent a high-grade metamorphic event. The Nordmannvik Nappe is directely overlain by the Lyngen ophiolite complex.

Tectonostratigraphic Positions of Scandinavian Eclogites

Eclogites and eclogitic rocks are found within several of the tectono-stratigraphic levels of the Scandinavian Caledonides:

1a. Within autochthonous/parautochthonous Proterozoic basement gneisses of the Western Gneiss Region (WGR) between the Trondheimsfjord and Sognefjord areas
1b. Within Proterozoic basement gneisses on the island of Træna off the coast of Nordland
1c. Within the Proterozoic basement rocks in the Lofoten-Vesterålen area
2a. Within the Bergen Arcs, of uncertain tectonostratigraphic position, probably Middle Allochthon
2b. Within the middle allochthonous units overlying the WGR gneisses, Sætra nappe in the northern part of WGR, and the Jæren formation, Stavanger area
3. Within the upper allochthonous units (Seve Nappes) in northern Sweden
4. Within the Uppermost Allochthon in the Tromsø area (Tromsø Nappe Complex), northern Norway.

Eclogites Within the Autochthon/Parautochthon [1]

Eclogites Within the Western Gneiss Region [1a]

A great variety of country-rock eclogites occur within the large area desig-nated as the Western Gneiss Region (WGR) of western Norway (Fig. 8.1 and 8.2). Eclogites are recognized within this region all the way from the Sognefjord area in the south up to the Trondheimsfjord area in the north. The WGR eclogites commonly occur as pods and lenses from decimeters up to tens of meters in length, apparently representing boudinaged layers (Lappin 1966) within amphibolite to granulite facies gneisses. The numerous eclogites between Sunnfjord and Nordmøre have been thoroughly studied by several workers. There is a general trend across this area with regard to equilibration temperatures for the eclogites (Krogh, 1977a; Griffin et al. 1985). In the following description, the eclogites are classified according to their general metamorphic evolution with reference to Table 8.1 and the general facies grid modified after Oh et al. (1988) and Oh (1992), as shown in Fig. 8.3.

Mineralogic Evolution from Lower-P Metamorphic Protoliths

The EA-(BS-)EC-EA-GS Series. In the southeastern low-T part of the WGR (Sunnfjord), eclogites record a metamorphic evolution from an early epidote + amphibole-bearing assemblage found as inclusions in Mn-enriched garnet cores (Krogh, 1980a; Cuthbert, 1985). This can be assigned to the epidote-amphibolite facies (EA). It was succeeded, as exemplified by the Kvineset eclogite, by either (1) glaucophane + omphacite + phengite + quartz + rutile (blueschist/eclogite facies transition–BS/EG), or (2) barroisite + omphacite + phengite + quartz + rutile (epidote-amphibolite/eclogite facies transition, EA/EC) in garnet rims and in the matrix. The occurrence of glaucophane (1) or barroisite (2) is in this case a function of bulk chemistry (higher Na_2O/CaO in the former type). Cofacial amphibolites have barroisite + albite + epidote (EA). Thus the Kvineset eclogites seem to have equilibrated close to the EA-BS-EC triple point (KV, Fig. 8.4). Garnets from these eclogites typically display a prograde chemical zoning pattern (Krogh, 1980a; Cuthbert and Carswell, 1990, Fig. 8.4a). Retrograde features in the eclogites involve growth of barroisite and katophorite on glaucophane (1), or sodic hornblende on barroisite (2), and the breakdown of garnet to hornblende + plagioclase + magnetite, and later to epidote + chlorite + magnetite. Barroisitic amphibole in the cofacial amphibolite shows a rimward zoning to hornblende. Posteclogite stages thus have involved overprint in the EA and transitional EA/AM facies, and finally greenschist facies field. Host rocks are granitic gneisses with the general assemblage Ca-Mn-Fe garnet + sodic ferro-augite + ferro-hastingsite + albite + K-feldspar + phengite + biotite + quartz, equilibrated at ca. 500°C/10 kbar.

The EA-EC-AM-GS Series. Eclogites from Naustdal further west in Sunnfjord (Binns, 1967; Krogh, 1982) area have equilibrated at somewhat higher

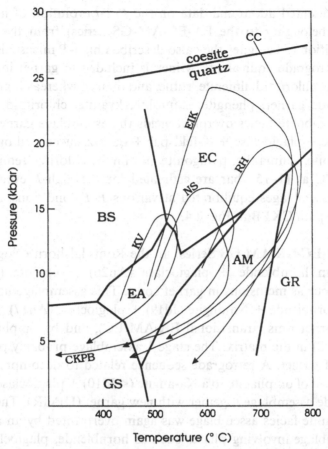

Figure 8.4. General *PT* evolution paths inferred for some WGR eclogites. KV = Kvineset (Krogh, 1980a); NS = Naustdal (Krogh, 1980a); RH = Romsdalshorn (Krogh, 1982); EIK = Eiksunddal (Jamtveit, 1987); CC = uplift trend for the highest *PT* eclogites in the WGR based on data from Cotkin et al. (1988), Lappin and Smith (1978, 1981), Smith (1984, 1988), Smith and Lappin (1989), and Fig. 8.5c; CKPB = uplift path for eclogite and HP micaschist from the vicinity of the Devonian Solund basin, after Chauvet et al. (1992).

temperatures, and thus show some differences in mineralogic evolution. An early epidote amphibolite facies episode was successively followed by eclogite, amphibolite and greenschist facies conditions (Fig. 8.4, NS). In the Naustdal eclogite, early mineral phases observed as inclusions in garnet cores are barroisite, paragonite, clinozoisite, and quartz (EA), succeeded by omphacite + paragonite + barroisite + quartz (EC) in garnet rims and the matrix. Garnets from this series show a prograde chemical zoning (Krogh, 1980a). The eclogite facies assemblage in turn breaks down to the lower pressure assemblage sodic augite or hornblende + plagioclase (AM). A further replacement of garnet by chlorite + magnetite is assigned to the greenschist facies. Recently, Chauvet et

al. (1992) presented additional data on the *P-T-t* evolution of high-*P* rocks
(apparently belonging to the EA-EC-AM-GS series) from the Sognefjord
area. In addition to eclogites they also described high-*P* micaschists (HPMS)
and their retrograde equivalents. Minerals included in garnet in the HPMS
are staurolite, chloritoid, ilmenite, rutile, and quartz, whereas high-*P* primary
phases include garnet, phengite, staurolite, kyanite, chloritoid, rutile, and
quartz. Amphibolite facies overprint yields the assemblage garnet + muscovite + biotite + plagioclase + K-feldspar + quartz, succeeded by the formation of ilmenite, sillimanite, paragonite and finally chlorite. Temperatures of
around 600°C at 11–15 kbar are estimated for the high-*P* event. Different
retrograde assemblages equilibrated at various *P-T* conditions, as indicated
by the *P-T-t* path CKPB in Fig. 8.4.

The EA/AM-EC-GR-AM-GS Series. In the Romsdalshorn eclogite (Krogh,
1982), brown hornblende + plagioclase (An26) + epidote (transitional
EA-AM) occur as inclusions in garnet cores. This assemblage has been succeeded by hornblende + Na-augite (Jd19) + plagioclase (An17) + epidote +
quartz in garnet rims (transitional EA-AM-EC), and by omphacite (Jd30)
+ quartz (EC) in the matrix. The stage 1 assemblage probably preceded the
formation of garnet. A retrograde sequence related to decompression yields
the breakdown of omphacite to a Na-augite (<Jd10) + plagioclase (An12–15)
± hornblende assemblage together with new garnet (HPGR). The secondary
high-*P* granulite facies assemblage was again overprinted by an amphibolite
facies assemblage involving phases such as hornblende, plagioclase, biotite,
magnetite, ilmenite, which in turn have been subject to a greenschist facies
overprint (chlorite) (Fig. 8.4, RH). A similar evolution has been recorded by
the orthopyroxene eclogites from Kvalneset, Nordfjord (Carswell et al., 1985),
and the kyanite-bearing Verpeneset (Nordfjord) eclogite (Bryhni et al., 1969;
Bryhni and Griffin, 1971; Bollingberg and Bryhni, 1972; Krogh, 1982). Garnets belonging to this series commonly show a prograde chemical zoning pattern (Bryhni and Griffin, 1971; Krogh, 1982; Carswell et al., 1985).

The (GR)-EC-GR-AM Series. Eclogites located along the northwestern edge
of the WGR have equilibrated at temperatures >700°C, and commonly display no, or slightly reverse (or retrograde), chemical zoning in the garnets
(e.g., Lappin and Smith, 1978; Krogh, 1980b; Carswell et al., 1985). Solid
inclusions in garnets are commonly the same as those occurring in the stable
eclogite facies matrix. Thus no evidence of a lower-grade protolith is found in
these eclogites. Maximum pressure estimates are difficult to evaluate for these
highest-*T* eclogites of the WGR, but the reported occurrence of coesite as

inclusions in clinopyroxenes from the Selje district, Nordfjord (Smith, 1984, 1988; Smith, and Lappin, 1989) indicates $P > 25$ kbar. Quartz is, however, the most common SiO_2-polymorph occurring as both inclusion and matrix phase throughout the WGR, and coesite seems to be only of local occurrence in the area around Selje/Stadtlandet. Lappin and Smith (1978, 1981) have also argued for even higher maximum pressures ($P = 30$–40 kbar) for the eclogites in this part of the WGR. Jamtveit (1987a) presented data from the Eiksunddal eclogite indicating maximum pressures >22–24 kbar at ca. 650°C. Retrogression is variably evolved in these highest T eclogites of the WGR, and includes assemblages belonging to the granulite facies, succeeded by amphibolite facies overprint. Decompression textures observed in eclogites in the Kristiansund area (Krogh, 1977a, 1980b), Hareidland (Mysen and Heier, 1972), and for orthopyroxene eclogites from Hjørungavåg and Kvalvåg (Carswell et al., 1985) involve the breakdown of high-P phases to orthopyroxene $\pm$ clinopyroxene + plagioclase $\pm$ hornblende (LPGR) assemblages. In the Kristiansund area, totally recrystallized post eclogite high-P granulite facies assemblages (Krogh, 1980b) record T-P conditions of c. 750°C, 9–13 kbar (Krogh, 1980b; Kohn and Spear, 1989, 1990). Jamtveit (1987a) presented a PT path (Fig. 8.4, EIK) where maximum pressures were estimated to >22–24 kbar at c. 650°C, succeeded by a period of increasing T during initial decompression to ca. 750°C/20 kbar. Then followed a period of rapid, nearly isothermal uplift into the sillimanite field. Similar posteclogite P-T paths (CC, Fig. 8.4) have been constructed for orthopyroxene-bearing eclogites from the Selje district (Lappin and Smith, 1978, 1981; Smith, 1984, 1988; Smith and Lappin, 1989) and for high-P meta-anorthosites from the Selje district (Cotkin et al., 1988).

Eclogite-like rocks in the Roan area, northern part of the WGR have been described in detail by Möller (1988, 1990) and Johansson and Möller (1986). High-P-T metamorphic assemblages include clinopyroxene + kyanite + garnet, equilibrated at 870 $\pm$ 50°C, 14.5 $\pm$ 2 kbar (Johansson and Möller, 1986). The thermal evolution during uplift, based on partial replacement of high-P minerals, indicate that the metamorphic peak was followed by uplift which involved a near-isothermal decompression into medium-pressure granulite and upper amphibolite facies conditions, with associated migmatization. This was followed by cooling at shallow levels.

Eclogites Derived from Igneous Protoliths

Eclogites derived from low-P magmatic protoliths are also common within the WGR. Gjelsvik (1952) described eclogitic mineral assemblages in reaction coronas around primary igneous phases in dolerites. The work on eclogitization of dolerites has been followed up by Griffin and Råheim (1973), Griffin and Heier (1973), Bryhni et al. (1977), Tørudbakken and Råheim (1981), and Cuthbert (1985). In a series of papers, Mørk and co-workers (Mørk, 1985a,b, 1986; Mørk and Mearns, 1986) presented detailed data on the progressive eclogitization of gabbros from the NW part of the WGR. Eclogitization is common in both olivine-gabbro and leuco-gabbro norite, and is partly confined to local shear zones. The olivine gabbros show all stages of transition from coronite to eclogite. Locally pristine gabbroic mineral assemblages and textures are well preserved, demonstrating a substantial degree of metastability similar to that described from the Bergen Arcs (see below). Isolated eclogite bodies within the enclosing gneiss probably represent disrupted parts of the eclogitized gabbro. Equilibrium conditions for the eclogite facies metamorphism are calculated as >700°C, 15–20 kbar, similar to estimates for the ordinary 'country-rock eclogites' of the NW part of the WGR. Post-eclogite depressurization yields symplectitization of omphacite, and local amphibolitization along fracture zones where granitic pegmatites were introduced, and along the margins toward the enclosing gneiss.

Fluid Inclusion Studies

In a fluid inclusion study of eclogites from the WGR, Andersen et al. (1989) were able to document a systematic evolution of the fluid phase interacting with the eclogites during the process of metamorphism. A general evolution from early higher density H_2O-N_2 dominated fluids to lower density H_2O-N_2 and pure N_2, and finally CO_2-dominated fluids was described. Model isochore calculations suggest that the earliest fluids are compatible with peak or early retrograde P-T conditions. The late pure CO_2 phase was introduced during the retrogression stage, and was partly mixed with the N_2-bearing fluid contained in preexisting inclusions in the eclogite minerals. H_2O was probably consumed in hydration reactions.

Age Constraints

The timing of eclogite facies metamorphism in the WGR has also been subject to controversies. Based on K-Ar dating of amphiboles (McDougall and Green, 1964), Krogh (1977a) proposed a Precambrian age for the eclogite formation. However, more recent Sm-Nd and U-Pb dating have yielded Caledonian ages. Krogh et al. (1974) and Gebauer et al. (1985) dated metamorphic zircons from eclogites from the WGR to 401 ± 20 Ma, and 446 + 54/–65 Ma, respectively. Griffin and Brueckner (1980, 1985) presented Sm-Nd garnet-omphacite ages on various eclogites in the range 407 ± 24 to 447 ± 20 Ma. Sm-Nd data on garnet-omphacite-whole-rock from the eclogitized metagabbros yield 400 ± 16 Ma (Mørk and Mearns, 1986). Tucker et al. (1987, 1992) presented U/Pb data from the WGR indicating metamorphic resetting at 401–395 Ma. They also concluded that the north-central WGR did not experience profound isotopic disturbance at any time between its postformational cooling (1687 to 1647 Ma) and resetting at ca. 400 Ma.

Different stages during post-eclogite facies cooling and depressurization can be dated by using various other isotope systems. Rb-Sr and K-Ar ages of hornblendes and micas from eclogites and gneisses commonly yield slightly lower ages, clustering around 385 Ma (see compilation of Kullerud et al., 1986). Lux (1985) and Berry and Lux (1991) presented $^{40}Ar/^{39}Ar$ data from the detachment zone and deeper levels. One amphibole from Nordfjord gave an isotope correlation age of 396 ± 5 Ma, whereas other analyzed amphibole, muscovite and biotite gave overlapping ages in the range from 404–410 Ma, with 2 sigma uncertainties <4 Ma, indicating very rapid uplift of the deep crustal rocks. Stiberg (1993) presented fission track data from apatites from Kristiansund yielding 248 Ma. This age is interpreted to represent the time when the rock cooled through the 120°C isotherm. Bimodal track length distribution for the Kristiansund sample shows that the rocks of this area have recorded a later thermal event, possibly related to Permian igneous activity. Dallmeyer et al. (1993) have dated the high-P peak metamorphism in the Roan area to 432 ± 6 Ma (Sm-Nd mineral age). Cooling ages of 400 ± 14 Ma and 393 ± 4 Ma have been recorded for the cooling through ca. 500°C and ca. 350°C. Fission track data from apatites from Roan yielded 315 Ma (Stiberg, 1993), again interpreted as the time when these rocks cooled through the 120°C isotherm. Möller (1988, 1990) concluded from the presence of late andalusite that at the time when these rocks passed the 500°C isotherm, they were already at a shallow (< ca. 3.7 kbar, or more likely ca. 2.5 kbar) crustal level. The average initial uplift rate for the Roan area was thus ca. 1.2 mm/year, and the average initial cooling rate ca. 12°C± / Ma.

Another exposition of the uplift-cooling-time path for the WGR is shown by various samples from the outer Nordfjord/Stadtlandet area that have been dated using different methods. Garnet-omphacite Sm-Nd isochron ages yield 423 ± 12 and 447 ± 20 Ma (Griffin and Brueckner, 1980, 1985). A nearby eclogite from Almklovdalen yields a garnet-omphacite-whole rock Sm-Nd isochron of 408 ± 8 Ma (Mearns, 1986). Hornblende and biotite from a retrograded eclogite yielded $^{40}Ar/^{39}Ar$ plateau ages of 410 ± 1 Ma and 374 ± 6 Ma, respectively (Lux, 1985). Rb-Sr biotite-whole rock ages of gneisses are 374 ± 2 Ma (Mearns, quoted in Kullerud et al., 1986), and 378 ± 22 Ma (Brueckner, 1972, recalculated by Kullerud et al., 1986). Using the commonly accepted closure temperatures for the various systems, the following cooling history for the outer Nordfjord/Stadtlandet area can be put forward (Cuthbert and Carswell, 1990; Cuthbert, 1991):

Peak metamorphic conditions (750 ± 50°C, 18 ± 2 kbar; Carswell et al., 1985) corresponding to the Sm-Nd ages. Cooling occurred through the 500°C isotherm at ca.410 Ma (^{40}Ar/^{39}Ar in hornblende, Harrison 1981), and through 300°C at ca. 375 Ma (K-Ar and Rb-Sr in biotite systems; Dodson, 1979). *P-T* estimates for retrograded assemblages are based on data from Cotkin et al. (1988) and Cuthbert (1991).

These data have been used to compile the pressure (depth)-time; pressure (depth)-temperature, and temperature-time curves presented in Fig. 8.5a–c. Importantly, Fig. 8.5a indicates initial near-isothermal decompression from eclogite facies conditions, with a fast mean uplift rate of ca. 4mm/year, decreasing substantially to ca. 0.2 mm/year in the later stages of the cooling and exhumation of these rocks.

Eclogites on Træna [1b]

On the island of Træna off the coast of Nordland Gustavson (1979 and personal communication, 1993) has reported retrograded eclogites occurring as pods within granitic gneisses. Their tectonostratigraphic position is uncertain, but most probably belongs to the autochthon/parautochthon. These eclogites have so far not been investigated in detail.

Eclogitic Rocks within the Proterozoic Lofoten-Vesterålen Area [1c]

The Precambrian crystalline rocks of the Lofoten-Vesterålen area has been regarded as a westerly extension of the Caledonian autochthonous/parautochthonous part of the Baltic Shield (e.g. Stephens et al., 1985), similar to the rocks of the Western Gneiss Region further south. The province consists of two generations of high-grade gneisses – a late Archaean migmatite complex and an overlying series of early Proterozoic supracrustals (Griffin et al., 1978; Jacobsen and Wasserburg, 1978). The whole area was subjected to low to intermediate pressure granulite facies metamorphism at ca. 1830 Ma (Griffin and Heier, 1969; Griffin et al., 1978; Krogh 1977c). Then followed a period of widespread igneous activity with intrusions of huge masses of gabbroic, mangeritic to charnockitic rocks crystallized at pressures of 8–10 kbar (Griffin et al., 1978; Ormaasen, 1977; Malm and Ormaasen, 1978). A thin sequence of late-Proterozoic metasediments (the Leknes group) metamorphosed at amphibolite facies conditions at ca. 1150 Ma occurs on Vestvågøy (Tull, 1978). An extensive retrogression of the late Archaean and early Proter-

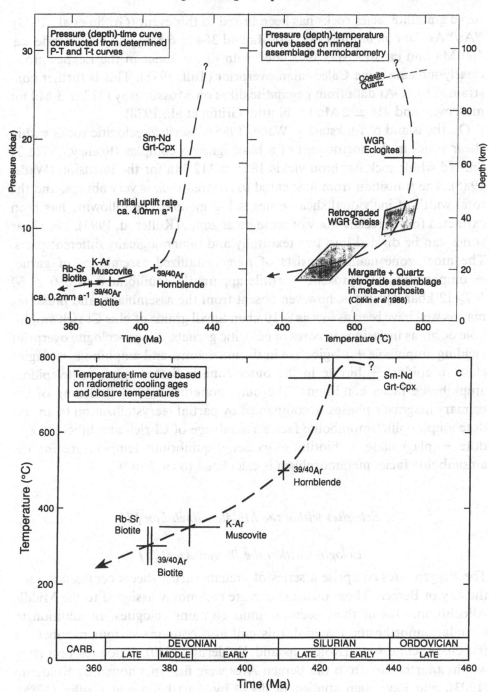

Figure 8.5(a–c). PTt uplift/cooling paths for high-pressure rocks in outer Nordfjord/Stadtlandet area of Western Gneiss Region, Norway.

ozoic granulite facies rocks has been linked to this event (Griffin et al., 1978). $^{40}Ar/^{39}Ar$ data on biotite (349 ± 6 Ma and 364 ± 6 Ma) and hornblende (ca. 460 Ma and ca. 615 Ma) associated with thrust zones in the Leknes group clearly indicate some Caledonian overprint (Tull, 1978). This is further constrained by K-Ar data from pegmatite dikes on Moskenesøy (317 ± 3 Ma for muscovite and 418 ± 2 Ma for biotite; Griffin et al., 1978).

On the island of Flakstadøy, Wade (1985) described eclogitic rocks within shear zones in the noritic part of a basic igneous complex (Romey, 1971). A Sm-Nd whole rock isochron yields 1803 ± 112 Ma for the intrusion (Wade, 1985). The transition from unsheared to sheared rock is very abrupt, and the total width of individual shear zones is 1–2 meters. The following has been extracted from a recent study of these shear zones (Kullerud, 1993). The shear zones can be divided into two texturally and mineralogically different parts. The inner zone mainly consists of a recrystallized assemblage of garnet + omphacitic clinopyroxene + rutile apparently equilibrated at 680 ± 60 °C/>12 kbar. Quartz is, however, absent from the assemblage, and pressures may as well have been as low as 8–10 kbar. Small grains of blue Cl-rich amphibole occur as inclusions in cores of eclogitic garnets. A post-eclogite overprint yielding amphibole + plagioclase in the inner zone, and amphibole + plagioclase + epidote + biotite in the outer zone indicates amphibolite/epidote amphibolite facies conditions. The outer zone consists of flattening of the primary magmatic phases, accompanied by partial recrystallization to an epidote amphibolite/amphibolite facies assemblage of Cl-rich amphibole + epidote + plagioclase + biotite + garnet. Equilibrium temperature for the amphibolite facies metamorphism is calculated to ca. 580 °C.

Eclogites within the Middle Allochthon [2]

Eclogites within the Bergen Arcs[2a]

The Bergen Arcs comprise a series of arcuate thrust sheets centered around the city of Bergen. These thrust sheets are commonly assigned to the Middle Allochthon. One of these tectonic units contains eclogites. In addition to abundant anorthositic material, this unit also contains various members of the charnockite suite (Kolderup and Kolderup, 1940). Eclogites occurring within anorthosites from the Bergen Arcs were first mentioned by Kolderup (1903), and have been studied in detail by Austrheim and Griffin (1985), Austrheim (1987), Austrheim and Mørk (1988), Andersen et al. (1990, 1991a,b), and Jamtveit et al. (1990).

The eclogite formation observed in the Bergen arcs was initially regarded as resulting from one continuous cycle caused by anorogenic cooling of the magmatic complex (Griffin, 1972; Griffin and Heier, 1973). A two stage model, with a Grenvillian granulite facies episode ($T = 700$–$800°C$, $P < 10$ kbar) succeeded by Caledonian eclogite (ca. 700 °C/18–21 kbar) and amphibolite facies (ca. 600 $\pm$ 75°C/4–14 kbar; cited in Andersen et al. 1991a) reworking, was put forward by Austrheim and Griffin (1985). Boundy et al. (1992) recently presented equlibration conditions for the eclogite-forming event of 670 $\pm$ 50°C/>14.6 kbar. It is clearly demonstrated that the granulite and eclogite facies metamorphic mineral assemblages are associated with different structural elements, and notably that the eclogites were developed from the dry granulite facies rocks along ductile shear zones and fractures where fluids have infiltrated. It is believed that deformation and/or fluid infiltration were necessary to trigger the eclogite forming reactions, as the granulite facies mineralogy has survived metastably in domains where fluids were absent. Fluid pressure must have been sufficiently high to cause hydraulic fracturing, and fluid transport was eased by the volume reduction caused by the granulite-eclogite transition. The metamorphic reactions also caused softening of the rocks, allowing shear zone formation. Boundy et al. (1992) presented a model for the temporal evolution of the eclogite shear zones, based on earlier writings (Austrheim and Griffin, 1985; Austrheim, 1987; Austrheim and Mørk, 1988; Jamtveit et al., 1990; Klaper, 1990) and their own investigations. Four sequential stages are recognized, including an initial phase with transformation of granulite to eclogite along fluid infiltrated fractures. In the second stage, discontinuous minor shear zones of well-foliated eclogite developed along these fractures, succeeded by the progression to anastomosing shear zones defining broader higher strain zones. In the fourth stage, these zones develop into major subparallel eclogite shear zones (30–100m thick) ranging several kilometers along strike. Jamtveit et al. (1990) calculated the composition of the infiltrating fluid to be H_2O-rich (XH_2O > = 0.75 at 700°C and 18–21 kbar) based on the possible coexistence of omphacite + paragonite + kyanite + quartz + dolomite in eclogitic shear zones. However, stable isotope data conclude that the carbonate in these rocks is a secondary mineral that formed from low temperature fluids that infiltrated the terrane during uplift (van Wyck et al., 1991).

Fluid Inclusion Studies

Fluid inclusion studies by Andersen et al. (1990, 1991a) have yielded evidence of different fluid regimes related to different metamorphic conditions. Fluids accompanying the granulite facies event were CO_2-rich, with less than 2.5 mol% N_2. The molar volumes are compatible with the *P-T* conditions of the

Proterozoic granulite facies metamorphism. Fluid inclusions in quartz from pegmatitic veins containing quartz + omphacite and quartz + white mica are N_2-rich, and the least disturbed inclusions have molar volumes that may agree with the Scandian high-P metamorphism. Younger low-density N_2 and N_2-H_2O fluid inclusions resulted from decrepitation and redistribution of early inclusions during the retrograde P-T evolution of the eclogites. Introduction of the H_2O-rich fluids were responsible for the amphibolitization of the eclogites and granulites. In a study of mineral-fluid-melt interactions in the high-P shear zones of the Bergen Arcs, Andersen et al. (1991b) described syntectonic extension veins consisting of albite, phengite, quartz, clinozoisite and minor calcite. Vein quartz contains two sets of early fluid inclusions: (1) aqueous brine (31–34 wt% NaCl), and (2) a CO_2-N_2 mixture with $XN_2 > = 5$ mol%. The two fluids are interpreted to have been present simultaneously and are thought to be immiscible at the P-T conditions of vein formation. Later generations of fluid inclusions comprise N_2-rich CO_2-N_2 mixtures and aqueous fluids with low salt contents, trapped at a late stage of the cooling/uplift history of the rocks. Andersen et al. (1991b) concluded that a fluid phase with high water activity was introduced into the rocks during the peak conditions of eclogite facies metamorphism, causing partial melting of the surrounding mafic to intermediate lithologies. The muscovite-granitic veins thus represent high-P anatectic melts migrating and forming in the shear zones.

Ages

The age of the granulite facies event has been determined by Austrheim and Råheim (1981) and Cohen et al. (1988). Sm-Nd ages from four coronitic granulite facies assemblages yielded consistent ages around 905 Ma (Cohen et al., 1988). Rb-Sr mineral ages also suggest a Grenvillian age for this event (best estimate 873 ± 51 Ma; Austrheim and Råheim, 1981; Austrheim et al., 1991). Cohen et al. (1988) also gave a mineral-whole rock age of 421 ± 69 Ma on a shear zone from Holsnøy. An Rb-Sr mineral isochron from the amphibolite facies minerals defines an age of 409 ± 8 Ma, and a U-Pb zircon age of 419 ± 9 Ma (lower intercept) from a post-granulite/pre- or syn-amphibolite facies trondhjemite dike constrains the second event to the Scandian phase of the Caledonian orogeny (Austrheim et al., 1991).

As there is a considerable time gap beween the granulite and eclogite facies events, and no mineralogic evidence for any intermediate episode, it is difficult to evaluate the pre-eclogite conditions during the initiation of the Caledonian high-P overprint. Cohen et al. (1988) proposed that the lower crust in the Bergen Arcs might have undergone a protracted cooling history that extended over 500 Ma or more, and that finally terminated by a rapid and near isother-

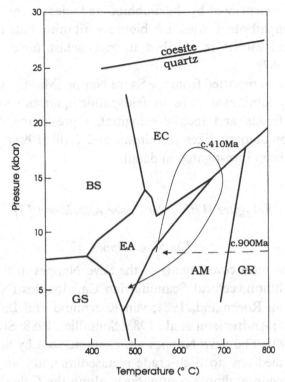

Figure 8.6. Schematic *P-T-t* path for the Bergen Arc eclogites (Austrheim and Griffin, 1985; Andersen et al., 1990, 1991a).

mal increase of pressure to produce the eclogite mineralogy of the shear zones and along fluid-filled cracks. Thus the Bergen Arc eclogites are examples of the granulite-eclogite facies transition. During the post-eclogite uplift stages the rocks entered the amphibolite facies field (Andersen et al., 1991a). A possible *P-T-t* evolution path for the Bergen Arc eclogites is shown in Fig. 8.6, based on the data given by Austrheim and Griffin (1985), Austrheim (1987), and Andersen et al. (1990, 1991a,b).

Eclogites Within the Hardangervidda-Ryfylke Nappe Complex [2b]

Rocks with relict eclogite facies mineralogy occur within the Jæren formation (Birkeland, 1981) of the Hardangervidda-Ryfylke Nappe Complex (Andresen and Færseth, 1982), in the area around Garborg in the southwesternmost part of the Scandinavian Caledonides (Hermans et al., 1987). The eclogites occur as lenses or layers in zoisite-rich amphibolites, probably representing retrograded eclogite. Other associated rocks are mainly garnet-quartz-mica schist. Inclusions in garnet are amphibole, zoisite, white mica, and rutile (EA facies). Eclogite mineral assemblage is garnet + omphacite + phengite + quartz +

zoisite + rutile, overprinted by the amphibolite facies assemblage plagioclase + blue-green amphibole + zoisite + biotite + titanite. Late chloritization of relict garnet and biotite is ascribed to greenschist facies metamorphism (Maijer et al., 1987).

Eclogites are also reported from the Sætra Nappe (Middle Allochthon) containing strongly laminated pure to feldspathic quartzite and amphibolite (locally garnetiferous and locally eclogitic), representing metamorphosed sandstone cut by diabase dikes (Robinson and Krill, 1991). These eclogites have so far not been investigated in detail.

Eclogites Within the Upper Allochthon [3]

The Seve Nappes

Eclogites are common constituents of the Seve Nappes in the lower part of the upper allochthon, central Scandinavian Caledonides (Nicholson, 1984; Stephens and van Roermund, 1984; van Roermund and Bakker, 1984; van Roermund, 1985; Andreasson et al., 1985; Santallier, 1988; Stølen, 1988; Kullerud et al., 1990). The Seve Nappes are characterized by an assemblage of predominantly medium- to high-grade metasedimentary and basic metaigneous rocks extending almost continuously along the Caledonian mountain chain in Scandinavia (Zachrisson, 1973). They are thought to have been derived from the transition zone between the continent Baltica and the ocean Iapetus (Gee, 1975; Stephens and Gee, 1989). The Seve Nappes are built up of several internal tectonic units, demonstrated or inferred to be thrust sheets (Trouw, 1973; Zachrisson and Stephens, 1984). The Seve Nappes has been subdivided into three major tectonic units; from the bottom to the top, these are the eastern, central, and western belts (Trouw, 1973; Zwart, 1974). The central belt reflects upper amphibolite to granulite facies, and the eastern and western belts have been subjected to lower to middle amphibolite facies (Williams and Zwart, 1977). Retrogression to medium- and low-grade conditions is prominent in the central and eastern belts (Williams and Zwart, 1977). The contact between the central and eastern belts is a thrust with development of blastomylonites and a drop in metamorphic grade, whereas the contact between the central and western belts is a metamorphic boundary, marked by the appearance of the mineral pair kyanite + potash feldspar and the onset of anatexis (Trouw, 1973). Eclogites are recognized in the central and eastern belts (van Roermund, 1985).

Eclogites of the eastern belt are hosted by feldspathic quartzite and psammitic schist in the Vaimok lens in southern Norrbotten (Andreasson et al., 1983; Santallier, 1988) and the Tjeliken lens in Jämtland (van Roermund and Bakker, 1984; van Roermund, 1985). In the central belt – the Tsäkkok lens in Norrbotten and the Ertsekey lens in Jämtland – eclogites occur within a series of marble, quartzite and quartz-feldspathic and pelitic schists (Kullerud, 1987; Kullerud et al., 1990; Stephens and van Roermund, 1984; Kullerud and Stephens, in preparation). So far, no evidence for high-P metamorphism has been found in the western belt.

Age constraints. Radiometric age datings on the Seve eclogites have yielded ages of 505 $\pm$ 18 and 503 $\pm$ 14 Ma (Sm-Nd internal isochrons, Mørk et al. 1988). Eclogitization here was succeeded by rapid uplift and cooling through the 500°C at 491 $\pm$ 8 Ma and ca. 400°C at ca.440 Ma (Dallmeyer and Gee, 1986). Dallmeyer and Stephens (1991) presented additional $^{40}Ar/^{39}Ar$ data for hornblendes and white mica from the eclogite-bearing units in the southern Norrbotten area. Hornblende cooling ages fall between ca. 470–460 Ma, and white mica records show cooling ages of ca. 468–448 Ma. These ages indicate that there is no simple model for the uplift history of the whole Seve Nappe Complex in the central part of the Scandinavian Caledonides during the early Paleozoic. Different areas and different tectonic segments appear to have attained the crustal level necessary for intracrystalline retention of argon in amphibole at different times. All the available data however, clearly demonstrate that the Seve eclogites were formed in an early Caledonian orogenic event.

Eclogites of the Central Belt of Jämtland

Eclogites of the EA-EC-GR-AM-EA-GS Series. Pre-eclogite facies minerals recorded as inclusions in garnets are epidote + Na Ca amphibole and phengite, belonging to the epidote-amphibolite facies. These minerals are succeeded by the eclogite facies assemblage garnet + omphacite + quartz + phengite $\pm$ zoisite equilibrated at 748 $\pm$31° C /18.6 $\pm$ 1.kbar. Post-eclogite alteration yielded (I) garnet + plagioclase + clinopyroxene + orthopyroxene + spinel (granulite facies), (II) clinopyroxene + hornblende + plagioclase $\pm$ scapolite (amphibolite facies), (III) blue-green amphibole + plagioclase + epidote $\pm$ garnet (epidote-amphibolite facies), and finally (IV) chlorite + actinolite + epidote (greenschist facies). A P-T path for the central belt (CB) eclogites based on the data from van Roermund (1985) is shown in Fig. 8.7.

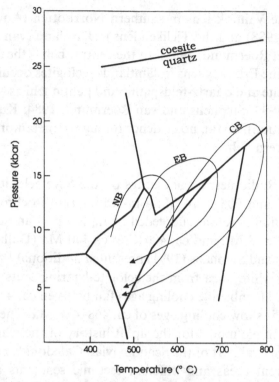

Figure 8.7. *P-T* paths for eclogites from the Seve Nappes. NB = Norrbotten (Kullerud et al., 1990; Kullerud and Stephens, in press); EB = eastern belt (van Roermund, 1985); CB = central belt (van Roermund, 1985).

Eclogite of the Eastern Belt of Jämtland

Eclogites of the EA-EC-AM-EA-GS facies series. Pre-eclogite mineral inclusions in garnet are phengite + Na-Ca amphibole + albite + epidote (epidote-amphibolite facies), succeeded by the eclogite facies assemblage garnet + omphacite + quartz ± phengite equilibrated at 564 ± 45° C, 15.4 ± 1 kbar (van Roermund, 1985). Post-eclogite breakdown assemblages include (I) garnet + clinopyroxene + plagioclase ± hornblende (amphibolite facies), (II) blue-green amphibole + plagioclase + garnet + biotite + epidote + sphene (epidote-amphibolite facies), and (III) chlorite + epidote + actinolite + albite (greenschist facies). A *P-T* path for the eastern belt (EB) eclogites is shown in Fig. 8.7.

Eclogites in Norrbotten

Eclogites of the EA-BS-EC-EA-AM-GS series. Kullerud et al. (1990) and
Kullerud and Stephens (in preparation) have described eclogitization of pillow
lavas from the Tsakkok Lens within the Seve Nappe sequence in Norrbotten.
Pre-eclogite assemblages recorded in garnets are (I) epidote + paragonite +
pargasite which may be designated to the epidote-amphibolite/amphibolite
facies transition. This was succeeded by the blueschist facies assemblage glaucophane + epidote + garnet + paragonite + Na-augite, and the eclogite
facies assemblage garnet + omphacite which equilibrated at 553 ± 66° C/ ca.
15 kbar. Post-eclogite stages include (I) barroisite (epidote-amphibolite
facies), (II) hornblende + plagioclase (amphibolite facies), and (III) chlorite
+ actinolite (greenschist facies). A *P-T* path for these eclogites (NB) is shown
in Fig. 8.7.

Eclogites within the Uppermost Allochthon [4]

Pods, lenses, and larger bodies of eclogite are intimately associated with metasupracrustals within the Tromsø Nappe Complex, being part of the Uppermost Allocthon of the Scandinavian Caledonides. The Tromsø Nappe Complex can broadly be subdivided into three major lithotectonic units, of which
only the uppermost unit (the Tromsdalstind sequence) contains eclogites
(Andresen et al., 1985; Krogh et al., 1990). The dominantly supracrustal
Tromsdalstind sequence shows strong lithologic variation, including marble,
calc-silicate rocks, garnet amphibolite, garnet-mica schist and gneiss (locally
kyanite-bearing), quartz-feldspathic gneiss, hornblende-biotite gneiss, biotite-
microcline gneiss, garnetiferous clinopyroxene gneiss (high-pressure granulite), ultrabasites and eclogite. The whole sequence shows a complex
polytectonic/metamorphic history (Krogh et al., 1990). The boundary
between the Tromsdalstind sequence and the underlying unit is tectonic
(Broks, 1985).

The Tromsø eclogites were first described by Pettersen (1878), and later by
Endell (1913). These earlier workers considered the eclogites and their retrograde alteration products to be metamorphosed gabbros that had intruded
supracrustal rocks. In a later description Landmark (1973) suggested an origin by metamorphism and metasomatism of Mg-rich calcareous sediments.

The eclogites commonly occur as smaller lenses and pods (seldom
exceeding 5 m in length) within intensely folded schistose calc-silicate rocks
and marbles and garnet mica gneiss and schist. Several "trains" of eclogite
pods indicate that they represent boudinaged layers. One larger body, the lay-

ered Tromsdalstind eclogite, has an exposed volume of ca. 0.4 km³. Locally, lathlike aggregates of garnet occur in this body, probably mimicking an original magmatic gabbro texture. Intimately interlayered with the eclogite are layers of trondhjemite/tonalite, exposing the high-*P* assemblage garnet + omphacite + sodic plagioclase + quartz ± kyanite.

P-T Evolution

The Tromsø eclogites typically belong to the AM-EC-GR-AM series, and were equilibrated at ca. 700° C/16–18 kbar (Krogh et al., 1990). Inclusions of hornblende and plagioclase, indicative for the amphibolite facies, are observed in eclogite garnet. Posteclogite decompression includes breakdown of omphacite to Na-augite + plagioclase in both eclogite and meta-trondhjemite/tonalite, and neocrystallization of garnet, defining the high-pressure granulite facies. These phases were overprinted by amphibolite facies assemblages including hornblende, plagioclase, and clinozoisite. A later deformational/prograde metamorphic episode locally involved total recrystallization in basic rocks to grossular-almandine-rich garnet + Na-augite + plagioclase ± hornblende + quartz (transitional high-*P* granulite facies/amphibolite facies; ca. 650° C/8–10 kbar). The new metamorphic fabric defines the regional foliation in the area, and was probably related to emplacement of the Caledonian nappes. This stage is succeeded by a new stage of deformation during decreasing temperatures (Fig. 8.8).

Age Constraints

Rb-Sr whole rock dating of a post-D1 / pre-D2 granite yields 433 ± 11 Ma, and K-Ar dating of secondary hornblende poikiloblasts in eclogite yields 437 ± 16 Ma (Krogh et al., 1990). Dallmeyer and Andresen (1992) reported well-defined ^{40}Ar / ^{39}Ar isotope-correlation ages on hornblendes from three samples from the Tromsø Nappe Complex of 432, 452, and 481 Ma, respectively. This would date the time when the rocks were cooled below the 500°C isotherm. Muscovite from three samples from the Tromsø Nappe Complex recorded plateau ages of 410, 418, and 427 Ma, thought to represent the cooling through 375°C. All these dates put a minimum age on the eclogite formation. Griffin and Brueckner (1985) gave Sm-Nd isotope data yielding an isochron age of 598 + 107 Ma on clinopyroxene, amphibole, and garnet from a retrograde and recrystallized eclogite. The status of this age is uncertain, but may represent the granulite facies overprint (D2/M2; Krogh et al.,1990). However, preliminary Sm-Nd data on garnet and whole-rock from an unretrograded eclogite and a recrystallized granulite equivalent yield ca. 480 and ca. 405 Ma, respectively (Andresen and Krogh, in progress). The youngest assem-

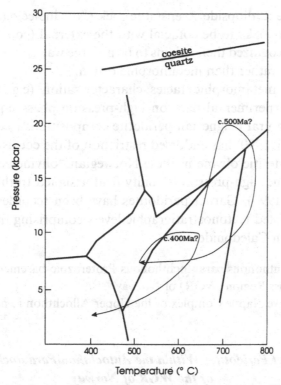

Figure 8.8. *P-T-t* path for the Tromsø Nappe Complex eclogites (Krogh et al., 1990; Andresen and Krogh, in preparation).

blage equilibrated at ca. 630°C/8–10 kbar. These results are thus partly in conflict with the $^{40}Ar/^{39}Ar$ data given by Dallmeyer and Andresen (1992), if intracrystalline retention of argon in amphiboles is effective only below 500°C. Further dating is thus necessary to better constrain the timing of the different events within the Tromsø Nappe Complex. It nevertheless appears that the eclogites in the Tromsø Nappe Complex are significantly older than those of the WGR, possibly of similar age to those occurring in the Seve Nappes.

Spatial Distribution of Garnet Peridotites within the Scandinavian Caledonides

Review of Occurrences

The classic early treatise by Eskola (1921) on "The Eclogites of Norway" also included descriptions of "eclogite in olivine-rock." He recognized that peridotite bodies in Møre and Romsdal districts locally contain interbanded sequences of garnet + diopside + enstatite rock ("enstatite-eclogite") and garnet + olivine ± diopside ± enstatite rock ("olivine-eclogite"). He clearly considered these rocks to be cofacial with the external (country-rock) eclogites. While he recognized their origins to be in some way connected, he emphasizes an igneous rather than metamorphic origin.

More recent metamorphic facies characterization (e.g. O'Hara, 1967), backed up by experimental data on high-pressure phase equilibria in both synthetic and natural magnesian peridotite compositions (see review in Harley and Carswell, 1990) has endorsed restriction of the coexistence of pyropic garnet with forsteritic olivine in these Norwegian "olivine-eclogites" (garnet peridotites) to the high-pressure stability field assigned to the eclogite facies (e.g., Carswell, 1990). Garnet peridotites have been recorded within two of the previously listed tectonostratigraphic levels comprising the nappe pile of the Scandinavian Caledonides:

1a. Within autochthonous/parautochthonous Proterozoic basement gneisses of the Western Gneiss Region (WGR) of Norway
3. Within the Seve Nappe Complex of the Upper Allochthon in northern Sweden

Garnet Peridotites Within the Autochthon/Parautochthon of the WGR of Norway

Coarse-grained garnet-bearing peridotites of rather spectacular appearance have now been documented at several localities in the tract of the WGR exposed between Nordfjord and the Molde peninsula (Fig. 8.9).

Many papers on these rocks published over the last 30 years, such as those by O'Hara and Mercy (1963), Lappin (1966, 1974), Carswell (1968ab; 1973; 1986), Brueckner (1974), Carswell and Gibb (1980), and Medaris (1980b, 1984), have utilized various geochemical and mineralogic data to propose an upper mantle origin for the high-pressure assemblages in these peridotites. Most of these papers have invoked tectonic emplacement of the mantle-derived "alpine-type" peridotite bodies into the high-grade lower crustal gneisses as a result of crust-mantle interaction during lithospheric plate collision, as illustrated in the tectonic models of the Norwegian Caledonides presented by Cuthbert et al. (1983) and Andersen et al. (1991). Accordingly, such an interpretation is consistent with, although not conclusive evidence for, the petrogenetic "foreign model" for the country-rock eclogites. Indeed the latter is in many respects just a conjectured extension of the upper mantle derivation interpretation for the peridotitic rocks.

Carswell et al. (1983) drew attention, however, to the existence of two contrasting types of garnet-bearing peridotites in the WGR. On the basis of geochemical distinctions, these were termed Mg-Cr and Fe-Ti types.

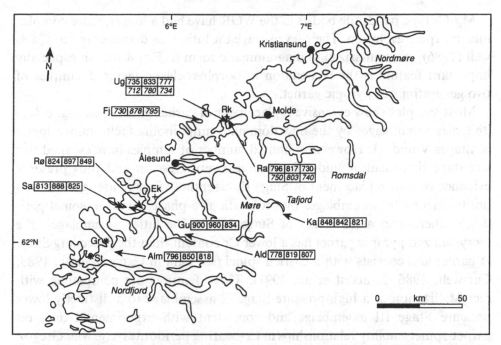

Figure 8.9. Locality map for Mg-Cr type garnet peridotites in the WGR, with indi-
cated mean equilibration temperatures at a nominal 30 kbar pressure based on the
data compiled by Medaris and Carswell (1990, table 1.1). Values indicated are for
Fe-Mg exchange between garnet-clinopyroxene (Powell, 1985), garnet-orthopyroxene
(Harley, 1984a), and garnet-olivine (O'Neill and Wood, 1979) pairs, respectively.
Roman numerals for clast core (Stage II) assemblages and italic numerals for recrys-
tallized (Stage III) assemblages. Mg-Cr type garnet peridotite localities (arrowed);
Ald = Aldalen; Alm = Almklovdalen; Fj = Fjørtoft; Gu = Gurskebotn; Ka = Kal-
skaret; Ra = Raudhaugene, Otrøy; Rø = Rødskar; Ug = Ugelvik, Otrøy. Fe-Ti type
garnet peridotite localites (crosses): Ek = Eiksunddal; Rk = Raknestangen, Otrøy.
Coesite-bearing eclogite localities (starred), after Smith (1988): Gr = Grytting;
St = Straumen

The Mg-Cr type peridotite bodies (tectonic lenses) have upper mantle geochemical signatures and contain relics of early high-pressure mineral assemblages indicative of an especially deep, subcrustal origin. By contrast, the
Fe-Ti type garnet peridotites occur as an integral, although minor, component of certain metamorphosed layered gabbro/peridotite complexes (as at
Eiksunddal on Hareidlandet) for which an origin from low-pressure, crustal
cumulates is indicated (Carswell et al., 1983; Mørk, 1985 a,b; Jamtveit,
1987a,b).

Geochronologic interpretation of the Fe-Ti type garnet peridotites is rather
equivocal because of frequent intergranular isotopic disequilibrium, but best
indications are of early Caledonian high-pressure metamorphism of Sveconorwegian (Grenvillian)-aged crustal intrusives (Jamtveit et al., 1991; Mørk
and Mearns, 1986).

Mg-Cr type peridotite bodies in the WGR have had a longer, more complex metamorphic evolution. Their seven-stage evolution, as documented by Carswell (1986), is summarized in diagrammatic form in Fig. 8.10. An especially important feature is the recognition in porphyroclastic textured samples of two generations of pyropic garnet.

Most samples show extensive replacement of earlier garnet-bearing, eclogite facies assemblages by the now dominant amphibolite facies mineralogies of Stages V and VI. However, a limited number of samples have escaped this late-stage deformation/fluid influx-induced retrogression and thus preserve evidence of part replacement of Stage II, essentially four-phase, garnet peridotite (lherzolite) assemblages by neoblastic five-phase garnet + spinel peridotite (lherzolite) assemblages of Stage III. In the latter assemblages, the recrystallized pyropic garnet has a lower Cr content than the preceding Stage II garnet and coexists with a Cr-rich spinel (see data in Carswell et al., 1983; Carswell, 1986; Jamtveit et al., 1991). These features are compatible with recrystallization of a high-pressure Stage II assemblage to a distinctly lower pressure Stage III assemblage and consistent with experimental data on garnet-spinel stability relationships in Cr-bearing peridotites (e.g. MacGregor, 1970; O'Neill, 1981; Nickel, 1986; Webb and Wood, 1986).

Carswell (1986) surmised that the development of the lower pressure (Stage III) neoblastic garnetiferous peridotite assemblages was related to the tectonic intercalation of lithospheric mantle into lower continental crustal gneisses during Caledonian plate collision. Subsequent Sm-Nd dating by Jamtveit et al. (1991) of selected garnet-bearing peridotite samples with good petrographic control, although again encountering repeated problems of isotopic disequilibrium, has provided important geochronologic constraints. These are most reasonably taken to indicate early Caledonian (ca. 440 Ma) recrystallization to Stage III assemblages of earlier, higher P-T (Stage II) Middle-Proterozoic (ca. 1700 Ma) Mg-Cr type garnet peridotite of upper mantle origin. Such an interpretation runs counter to some previous suggestions (e.g., Jamtveit, 1984; Griffin and Qvale, 1985; Griffin et al., 1985) that the garnet-bearing assemblages in these peridotite bodies may have been generated by Caledonian prograde, high-pressure metamorphism of original low-pressure spinel peridotites of possible ophiolite origin.

Mg-Cr peridotites are widely distributed across the WGR in Møre and Romsdal districts in the form of tectonically bounded lenses with dimensions varying from just a few meters to over a kilometer. However, high-pressure garnetiferous assemblages are only sporadically preserved at the localities shown in Fig. 8.9. Most of these are close to the coast, since the degree of retrogressive replacement by late Caledonian amphibolite facies assemblages

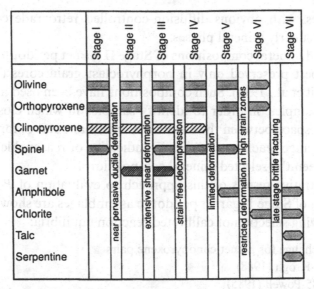

Figure 8.10. Summary of the seven-stage metamorphic evolution of Mg-Cr type peridotites in the WGR, based on Carswell (1986).

of Stages V and VI generally increases in an easterly direction across the WGR, as also with the country-rock eclogites.

Also indicated in Fig. 8.9 are comparative mean equilibration temperatures (calculated for a nominal 30-kbar pressure) for Stages II and III garnet peridotite assemblages at the various localities, based on the data compilation in Medaris and Carswell (1990). These data provide no obvious evidence for replication by the Stage II (porphyroclast core) assemblages of the regional trend of declining equilibration temperatures in a southeasterly direction across the WGR, as outlined by Krogh (1977a) and Griffin et al. (1985) for country-rock eclogite assemblages. This is consistent with geochronologic data which indicate a mid-Proterozoic age for the Stage II garnet peridotite assemblages (Jamtveit et al., 1991) in contrast to Caledonian mineral ages of ca 400–450 Ma (Krogh et al., 1974; Gebauer et al., 1985; Griffin and Brueckner, 1985) for the high pressure assemblages in the country-rock eclogites which display the regional temperature variation.

Experimental calibrations exist for several cation-exchange and net-transfer reaction equilibria applicable to garnet peridotite (lherzolite) assemblages, which should ideally enable reliable estimates to be made of both pressures and temperatures of equilibration and hence fix the conditions (and depths) for formation of the Stages II and III garnet peridotite assemblages during their evolutionary history. However, a major obstacle is the complexity of the latter and the fact that available samples invariably comprise disequilibrium assemblages of minerals belonging to two or more of the metamorphic evolutionary stages, with obvious diffusion-controlled retrograde compositional zoning in relict early mineral phases.

Unmodified mineral compositions of Stage II garnet peridotite assemblages are thus at best preserved only in porphyroclast grain cores (see Carswell, 1986; Jamtveit et al., 1991). Such compositions have been used in a large suite of analysed samples, in order to identify samples in which close correspondence in P-T space between the calculated univariant curves for a number of different exchange reaction equilibria is indicative of reasonable overall equilibrium between the selected mineral compositions.

Some illustrative results of this approach to evaluation of P-T conditions for formation of Stage II garnet peridotite assemblages are shown in Fig. 8.11 for the following selection of calibrated reaction equilibria:

(1) Fe-Mg exchange for garnet-clinopyroxene pairs
  (a) KR88-Krogh (1988)
  (b) POW85-Powell (1985)
(2) Fe-Mg exchange for garnet-olivine pairs OW79; O'Neill and Wood (1979, corrected 1980)
(3) Fe-Mg exchange for garnet-orthopyroxene pairs H84; Harley (1984a)
(4) Ca-Mg exchange for clinopyroxene-orthopyroxene pairsBK90-Brey and Köhler (1990)
(5) Garnet-orthopyroxene (Al) barometer based on Mg Tschermaks substitution in orthopyroxene
  (a) BK90(Al); Brey and Köhler (1990) for Cr-bearing natural peridotite compositions,
  (b) H84(Al) – Harley (1984b) for synthetic FMAS – and CFMAS – system compositions

Expectations were that calibrations 1a, 2, 4 and 5a would yield the most reliable results (cf., Carswell and Gibb, 1987; Brey and Köhler, 1990) In reality, (1a) almost invariably indicates appreciably lower temperatures than the other methods and the results from the alternative calibration (1b) of Powell (1985) for Fe-Mg exchange between garnet and clinopyroxene seem preferable. By contrast, calibration (3) of Harley (1984a) for garnet-orthopyroxene pairs mostly indicates unrealistically high temperatures, out of line with other mineralogic indicators. These problems with Fe-Mg exchange reaction equilibria may relate to unidentified variations in Fe^{3+}/Fe^{2+} ratios between the paired minerals. With such low-Fe mineral assemblages it is not possible to reliably calculate Fe^{3+} contents in minerals by charge balance stoichiometry and an assumption that all Fe is present as Fe^{2+} has had to be made.

The calculated reaction equilibria judged to yield the most reliable results for the individual samples featured in Figs. 8.11 and 8.12 are shown as solid

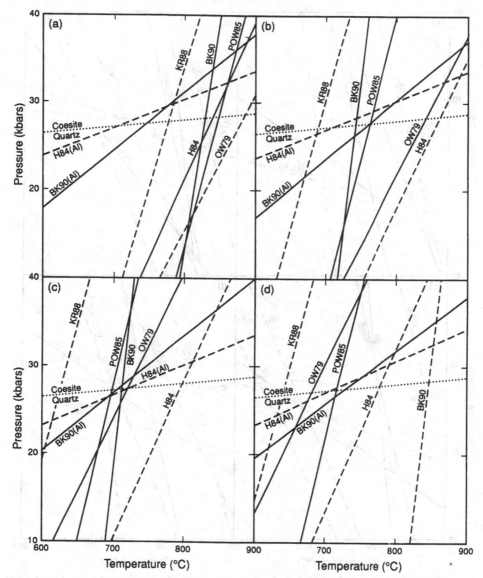

Figure 8.11. *P-T* evaluation of Proterozoic (Stage II) UHP garnet peridotite assemblages (clast cores) in selected WGR samples. (a) Sample T96 from Kalskaret, Tafjord; microprobe data in Jamtveit et al. (1991). (b) Sample SU-IK from Almklovdalen (Lien), Nordfjord; microprobe data from Medaris (1984). (c) Sample U157 from Ugelvik, Otrøy–unpublished microprobe data by Carswell (University of Sheffield). (d) Sample U45 from Raudhaugene, Otrøy; unpublished microprobe data by Carswell (University of Sheffield).

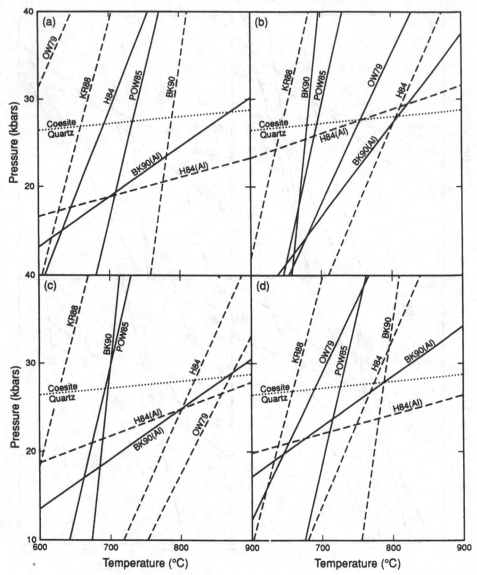

Figure 8.12. *P-T* evaluation of Caledonian garnet peridotite assemblages in selected WGR samples. (a) Sample T337 from Kalskaret, Tafjord. Neoblast (Stage III) assemblage in porphyroclastic Mg-Cr type garnet peridotite; microprobe data by Al-Samman (1985). (b) Sample U539 from Ugelvik, Otrøy. Recrystallized Stage III assemblage in Mg-Cr type garnet peridotite–microprobe data in Jamtveit et al. (1991). (c) Sample U157 from Ugelvik, Otrøy. Neoblast (Stage III) assemblage in porphyroclastic Mg-Cr type garnet peridotite; unpublished microprobe data by Carswell (University of Sheffield). (d) Sample U125 from Raknestangen, Otrøy. Fe-Ti type garnet peridotite interlayered with eclogite; microprobe data in Carswell (1983).

lines. These results indicate that the "best" *P-T* estimates for the earlier Stage II garnet peridotite assemblages (Fig. 8.11) lie close to, although mostly on the high-pressure side of, the coesite-quartz reaction curve (Bohlen and Boettcher, 1982). Mean "best" *P-T* estimates have been evaluated on the same basis for suites of eight samples each from the Otrøy (Ugelvik and Raud-haugene) and Kalskaret (Tafjord) localities, all showing reasonable concordance between the different reaction equilibria calculated for individual samples. Indicated *P/T* values are:

For Otrøy samples: 28 ± 3 kbar/725 ± 30 °C
For Kalskaret samples : 32 ± 2 kbar/825 ± 30 °C

As both sets of results lie predominantly within the coesite stability field, it is appropriate to classify the relict mid-Proterozoic upper mantle Stage II garnetiferous assemblages in the Mg-Cr type peridotite bodies as *UHP assemblages*.

By contrast, the recrystallized Caledonian-aged Stage III garnetiferous assemblages in the Mg-Cr type peridotite lenses (Fig. 8.12 a–c) have demonstrably equilibrated at much lower pressures within the quartz stability field – "best" *P/T* estimates being around 18 ± 2 kbar/680 ± 30°C. Broadly similar *P/T* values are indicated for a Caledonian-aged Fe-Ti type garnet peridotite assemblage of likely initial low-pressure crustal origin (Fig. 8.12d), as also for the high-pressure mineral assemblages in many country-rock eclogites in the WGR (e.g., Fig. 8.4).

A synoptic *P-T-t* path for the metamorphic evolutionary Stages II $\pm$ VII in the Mg-Cr type peridotite bodies is presented in Fig. 8.13, with indicated *P/T* estimates for Stage II and III assemblages based on the aforementioned data for samples from the localities on Otrøy. It is thought unlikely that these rocks followed a direct *P-T* path between Stage IIa and III on Fig. 8.13, in view of the long time-interval between these assemblages. It seems more likely, as argued by Jamtveit et al. (1991), that the IIa mid-Proterozoic mineral assemblages were held (with frozen-in mineral equilibria) at ambient sub-continental Moho conditions (IIb in Fig. 8.13) immediately prior to tectonic emplacement into the crustal gneiss terrane during the profound tectonism of the Caledonian orogeny. It is thus envisaged that the *P-T-t* paths for the mantle-derived Mg-Cr type peridotites converged, at the *P-T* conditions of their Stage III mineral assemblages, with the clockwise prograde paths for the development of the Caledonian high-pressure assemblages in the country-rock eclogites (Fig. 8.4) and in the Fe-Ti type garnet peridotites.

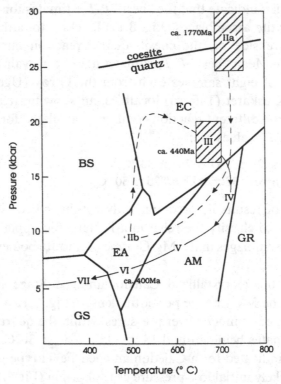

Figure 8.13. *P-T-t* path for metamorphic evolution of Mg-Cr type peridotites on Otrøy in the WGR of Norway, based on integration of isotopic age data (Jamtveit et al., 1991) with the revised *P-T* estimates detailed in this paper.

Garnet Peridotites in the Seve Nappe Complex of the Upper Allochthon in Northern Sweden

Petrographic descriptions and mineral composition data on garnet-bearing Mg-Cr type peridotites/websterites occurring as isolated lenses within two thrust sheets (referred to as the Ertsekey and Tjeliken tectonic lenses) of the Seve Nappe Complex in the county of Jämtland have been provided by van Roermund (1989). These particular tectonic units of dominant migmatitic gneisses display evidence of retrograde amphibolite facies assemblages super-imposed on earlier high-pressure assemblages for which *P/T* conditions for formation of associated metabasic eclogites were estimated by van Roermund (1989) as ca. 18 kbar/780°C (Ertsekey lens) and ca. 15 kbar/575°C (Tjeliken lens).

Garnet peridotites from the Ertsekey lens have a notably porphyroclastic microstructure with coarse-grained clasts of a four-phase (M1) garnet lherzol-ite assemblage mantled by a recrystallized five-phase (M2) garnet-spinel lher-zolite assemblage, sometimes with additional amphibole. Garnet-bearing peridotites from the Tjeliken lens have an equigranular tabular microstructure interpreted by van Roermund (1989) to reflect complete strain-induced recrystallization to an M2 assemblage.

The respective isotopic ages of these two generations of garnet-bearing peridotite remain to be determined, but it is tempting to speculate on a similar evolutionary history from subcontinental upper mantle material as proposed for comparable rocks in the WGR of Norway.

Mineralogic disequilibrium is similarly commonplace in the Swedish garnet-bearing peridotites reflecting the five-stage metamorphic growth history recognized by van Roermund (1989). Hence retrograde reaction textures and compositional zoning in minerals again cause problems with thermobarometric evaluation of the early high-pressure assemblages. The lack of provision of M1 garnet compositions by van Roermund (1989), presumably because of replacement by secondary kelyphite, prevents calculation of *P-T* conditions for formation of the relict M1 garnet peridotite assemblages in the Ertsekey lens. Evaluation of *P/T* equilibration conditions for the M2 garnet + spinel peridotite/websterite assemblages in two samples from the Ertsekey and Tjeliken lenses (by the same procedures as outlined for the garnet peridotite assemblages in the WGR) yields "best" estimates of 22 kbar/650°C and 20 kbar/680°C, respectively. These values are within error of the calculated "best" estimates for the Caledonian-aged, second generation garnet (+ spinel) peridotite assemblages in the WGR of Norway.

Reality or Otherwise of a Caledonian UHP Metamorphic Province in the WGR of Norway

In a previous section, Spatial Distribution of Garnet Peridotites, we documented thermobarometric and geochronological data on occurrences of relict UHP mineral assemblages preserved in Mg-Cr type peridotite bodies at a number of localities, mostly in the coastal region of the WGR just to the north and south of the town of Ålesund (Fig. 8.9). These data were taken to indicate that these UHP assemblages are of mid-Proterozoic age (ca. 1700 Ma) and represent mineralogies that equilibrated in old, subcontinental mantle, tectonic slices of which were tectonically emplaced into high-grade crustal gneisses of the WGR during the Caledonian orogeny.

General mineralogic constraints and thermobarometric evaluation indicate that the Caledonian-aged recrystallized garnet + Cr spinel assemblages in the mantle-derived Mg-Cr type peridotites, as well as the contemporaneous prograde high-pressure assemblages in the Fe-Ti type peridotites and in the majority of the country-rock eclogites, equilibrated at maximum pressures of around 20 kbar (see also Cuthbert et al., 1983; Cuthbert and Carswell, 1990).

Contradictory evidence for the existence of possible Caledonian UHP metamorphism in a "**Norwegian coesite-eclogite province**" (Smith, 1984, 1988; Smith and Lappin, 1989) comes from reports of relict coesite inclusions preserved within garnet or omphacite from two country-rock eclogite occurrences in the outer Nordfjord district (Figs. 8.2 and 8.9) and more tenuously of pseudomorphs of polycrystalline quartz after supposed coesite at five other, more geographically scattered, localities in the coastal part of the WGR south of Ålesund. However, the widespread lack of evidence to indicate that the encompassing quartzo-feldspathic gneisses ever contained jadeite, let alone coesite, raises doubts over the reality of a regional-scale UHP province. Thus, as Smith (1988) clearly excludes all the WGR gneisses from the UHP province, on the basis of their evidenced much lower- pressure assemblages, the regional extent of his supposed "coesite-eclogite province" is highly conjectural.

There are, as yet, no published age constraints on the UHP assemblages in the coesite-eclogite pods at Grytting and Straumen, but their locations are broadly consistent with the regional Caledonian P-T pattern recognized by Krogh (1977a) and Griffin et al. (1985) with highest P-T values, and hence supposed deepest level subduction, in the coastal region of Möre. However, at this stage the recognition of a Caledonian UHP province in this part of Norway remains highly speculative, and confirmation of its reality urgently requires isotopic age data and further supportive mineralogic evidence.

One possible alternative interpretation is that the coesite-bearing eclogites (like the mantle-derived garnetiferous peridotites) comprise older, deeper level UHP assemblages which were tectonically emplaced into the crustal gneisses during the Caledonian orogeny. As there is no obvious field association between the UHP coesite eclogites and the UHP garnet peridotite occurrences in the WGR (Fig. 8.9), it may, however, be necessary to consider other possible explanations of the apparent contradictory evidence that currently exists.

Profound crust–mantle interaction during Caledonian continental plate collision, resulting in tectonic intercalation of older UHP lithologies into a crustal terrane experiencing high-pressure metamorphism, would be consistent with at least the first part of the elaborate "FIF" (foreign/in situ/foreign) lithospheric interdigitation model of Smith (1988). This latter model emphasises the existence of an intimate mix of lithologies (a "gigantic tectonic melange" – Smith, 1980) with reputedly different metamorphic grades and histories. However, current evidence provides little justification for recognition of any substantial volume of the WGR as tectonically disrupted portions of the Caledonian-aged UHP (coesite-eclogite) province or terrane of deeply subducted crustal rocks.

Tectonic Models for the HP/UHP Metamorphism in the Scandinavian Caledonides

Various models concerning the subduction and exhumation of the eclogite-bearing WGR have been proposed (Krogh, 1977a; Lappin and Smith, 1978; Smith, 1980; Cuthbert et al., 1983; Jamtveit, 1987; Cuthbert and Carswell, 1990), and a discussion of these is given by Cuthbert and Carswell (1990). The model elaborated by Cuthbert and Carswell (1990), and two more recent models presented by Andersen and co-workers (Andersen and Jamtveit, 1990; Andersen et al., 1991), and Austrheim (1991) are briefly presented below.

The model presented by Cuthbert and Carswell (1990) is based on the general tectonic model for eclogite facies metamorphism in continental crust developed by Carswell and Cuthbert (1986), in which multiple subduction is achieved by continental collision, followed by telescoping of the underthrust plate along low-angle crustal fractures. The model predicts that each time new crust is subducted, it drives the uplift of previously subducted, eclogite-bearing crust, and that the age of eclogite facies metamorphism will tend to increase with tectonostratigraphic height in the orogen. This sequential stacking model may be broadly applicable to the Scandinavian Caledonides, as illustrated by Fig. 8.14 (adapted from Cuthbert and Carswell, 1990). Although the age of the TNC eclogites is still uncertain, this model fairly well illustrates the formation of the UHP/HP terranes in the Scandinavian Caledonides. Crustal extension or "collapse" is proposed as an effective mechanism to bring the HP/UHP rocks back to the surface.

Andersen and Jamtveit (1990) and Andersen et al. (1991) presented a new model for the uplift of the eclogite-bearing WGR involving extensional collapse of the Caledonide orogen. They stated that the eclogite-bearing lower crust is separated from the middle and upper crust by major detachment zones formed during extensional collapse of the orogen. Formation of the Devonian basins and the uplift of the deep crust are related to this extension. The footwall of the detachment zones comprises three structural and metamorphic zones (Fig. 8.15). The upper zone (zone 1) is characterized by down-to-the-west simple shear developed under retrograde greenschist-facies metamorphism. Zone 2 suffered inhomogeneous simple shear of the same polarity. Petrography and mineral chemistry data from the lower zone 3 show a record of initial eclogite facies metamorphism at $T = 600°C$ and $P > 16$ kbar, which was decompressed almost isothermally to amphibolite facies conditions at

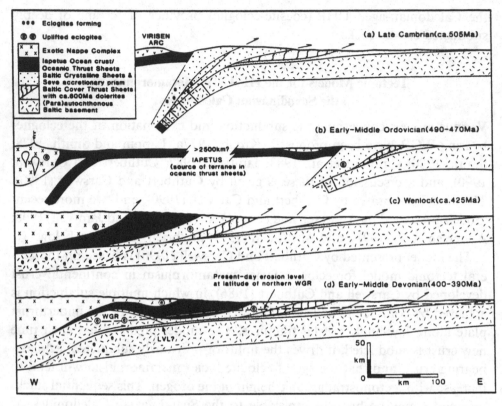

Figure 8.14. Model for the formation of UHP/HP rocks in the Scandinavian Caledonides. (Adapted from Cuthbert and Carswell, 1990.)

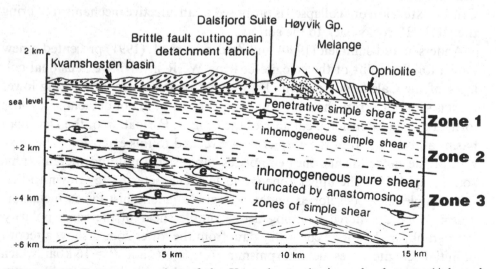

Figure 8.15. East–west section of the Kvamshesten basin and substrate. (Adapted from Andersen and Jamtveit, 1990.)

550°C and 10–12 kbar. Both the eclogite and amphibolite facies metamorphism developed in a regime of pure shear with vertical shortening. A schematic model for the crustal uplift and extension is shown in Fig. 8.16. In stage 1, the compressional construction of the orogenic welt is completed, and the orogen contains a structural anisotropy defined by the compressional thrust. Stage 2 of the model represents the main uplift stage, caused by removal of the thickened thermal boundary layer (TBL). At this stage vertical (σ 1) compressional stresses cause pure shear eclogite facies deformation in the lower crust. The isothermal decompression results in amphibolite facies overprinting of the eclogite facies fabric. Stage 3 of the model represents the main spreading event with reworking of structures formed in the pure shear regime by simple shear at greenschist facies. In their later model Andersen et al. (1991) suggested that the initial eclogite fabrics were developed by vertical nonrotational constrictional deformation and that the subhorizontal nonrotational fabrics at eclogite to amphibolite facies characterized the earliest part of the decompression of the deeply buried rocks. This model is also supported by the data given by Chauvet et al. (1992). Some critical remarks on the model of Andersen et al. (1991) must, however, be mentioned. First, the reality of the necessary uplift-promoting delamination of the thermal boundary layer is questionable. Second, exhumation along extensional normal faults (shear zones) does not account for observations of eclogite-bearing tectonostratigraphic units now at high structural levels in the Caledonian nappe pile, overlying lower grade (and younger?) rock units - as also seen in other major UHP/HP belts such as the Alps and the central European Variscides.

Austrheim (1991) presented a model for the dynamics of crustal root zones based on field observations from the deepest parts of the Scandinavian Caledonides (Bergen Arcs and WGR), and geophysical data available for more recent continental collision zones such as the Alps (Fig. 8.17). One of Austrheim's key points is the fluid phase locally infiltrating the granulite facies rocks of the root zone causing eclogitization along shear zones and fluid pathways. The eclogitization is associated with dramatic changes in rock properties such as density and rheology. A density fractionation may occur as dense, low-viscosity, eclogitic materials descend and lighter, unreacted, and partly eclogitized rocks ascend. The incomplete eclogitization and felsic nature of exposed eclogite-bearing terrains are interpreted as the relatively lighter fraction after density fractionation in orogenic root zones, although Austrheim's model rather fails to provide an adequate explanation for its rapid exhumation. The descending eclogitic material may accumulate temporarily at the crust–mantle boundary before descending to greater depths. Crust–mantle interaction and exchange of material down to depths of >90 km by solid flow is suggested.

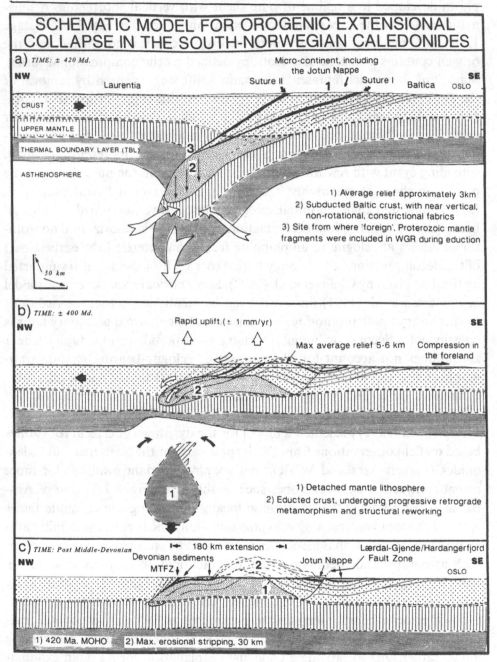

SCHEMATIC MODEL FOR OROGENIC EXTENSIONAL COLLAPSE IN THE SOUTH-NORWEGIAN CALEDONIDES

a) *TIME: ± 420 Md.*

NW

Micro-continent, including the Jotun Nappe

SE

Laurentia   Suture II   1   Suture I   Baltica   OSLO

CRUST
UPPER MANTLE
THERMAL BOUNDARY LAYER (TBL)
ASTHENOSPHERE

3
2

1) Average relief approximately 3km
2) Subducted Baltic crust, with near vertical, non-rotational, constrictional fabrics
3) Site from where 'foreign', Proterozoic mantle fragments were included in WGR during eduction

50 km

b) *TIME: ± 400 Md.*

NW

Rapid uplift (± 1 mm/yr)

Max average relief 5-6 km   Compression in the foreland

SE

2

1

1) Detached mantle lithosphere
2) Educted crust, undergoing progressive retrograde metamorphism and structural reworking

c) *TIME: Post Middle-Devonian*

NW

180 km extension

Devonian sediments
MTFZ

2

Jotun Nappe

Lærdal-Gjende/Hardangerfjord Fault Zone

SE

OSLO

1

1) 420 Ma. MOHO   2) Max. erosional stripping, 30 km

Figure 8.16. Schematic model for orogenic extensional collapse in the WGR. (Adapted from Andersen et al., 1991.)

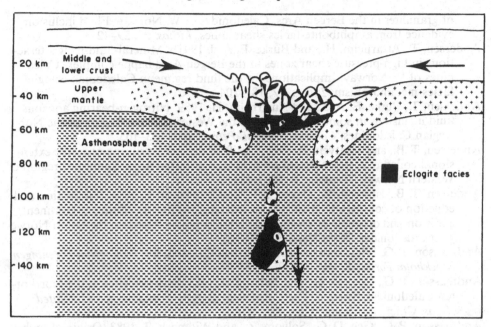

Figure 8.17. Eclogitization and density fractionation in the root zone of a continental collision belt. (Adapted from Austrheim, 1991.)

All of these three models have their advantages and weaknesses. However, we believe that continuing, sequential underthrusting in response to further crustal shortening (a key feature in both the Cuthbert et al., 1983, and the Cuthbert and Carswell, 1990, models) is essential for the refrigeration, survival, and rapid exhumation of UHP/HP rocks. The extensional features leading to formation of the Devonian basins could possibly be regarded just as relatively high level collapse structures which played only a minor role in the final unroofing of the eclogite-bearing units.

Acknowledgments

This chapter has been greatly improved by constructive reviews by T. B. Andersen and H. Austrheim.

References

Al-Samman, A. H. 1985. Mineralogy and geochemistry of ultramafic rocks in the west Norway basement gneiss trrain. Ph.D., University of Sheffield.

Andersen, T., Austrheim, H., and Burke, E. A. J. 1990. Fluid inclusions in granulites and eclogites from the Bergen Arcs, Caledonides of W. Norway. *Mineral. Mag.*, *55*, 145–158.

Andersen, T., Austrheim, H., and Burke, E. A. J. 1991a. Fluid induced retrogression of granulites in the Bergen Arcs, Caledonides of W. Norway: Fluid inclusion evidence from amphibolite-facies shear zones. *Lithos, 27*, 29–42.

Andersen, T., Austrheim, H., and Burke, E. A. J. 1991b. Mineral-fluid-melt interactions in high-pressure shear zones in the Bergen Arcs nappe complex, Caledonides of W. Norway: Implications for the fluid regime in Caledonian eclogite facies-metamorphism. *Lithos, 27*, 187–204.

Andersen, T., Burke, E. A. J., and Austrheim, H. 1989. Nitrogen-bearing aqueous fluid inclusions in some eclogites from the Western Gneiss Region of the Norwegian Caledonides. *Contrib. Mineral. Petrol., 103*, 153–165.

Andersen, T. B., and Jamtveit, B. 1990. Uplift of deep crust during orogenic extensional collapse: A model based on field studies in the Sogn–Sunnfjord region of Western Norway. *Tectonics, 9*, 1097–1111.

Andersen, T. B., Jamtveit, B., Dewey, J. F., and Swensson, E. 1991. Subduction and eduction of continental crust: Major mechanisms during continent-continent collision and orogenic extensional collapse, a model based on the South Norwegian Caledonides. *Terra Nova, 3*, 303–310.

Andreasson, P. G. 1986. Seve terranes, Swedish Caledonides. *Geologiska Foreningen Stockholm Förhandlingar, 108*, 261–263.

Andreasson, P. G., Gee., D. G., and Sukotjo, S. 1985. Seve eclogites in the Norrbotten Caledonides, Sweden. *The Caledonide Orogen–Scandinavia and Related Areas.* Chichester: John Wiley, 887–901.

Andreasson, P.-G., Gee, D. G., Solyom, Z., and Widmark, T. 1983. Origin of high-P metamorphism in the Scandes. High pressure metamorphism: indicator of subduction and crustal thickening. *Terra Cognita (abstract), 1*, 182.

Andreasson, P. G., Solyom, Z., and Roberts, D. 1979. Petrochemistry and tectonic significance of basic and alkaline-ultrabasic dykes in the Leksdal Nappe, northern Trondheim Region, Norway. *Norges Geologiske Undersøkelse, 348*, 47–42.

Andresen, A., Fareth, E., Berg, S., Kristensen, S. E., and Krogh, E. J. 1985. Review of Caledonian lithotectonic units in Troms, North Norway. *The Caledonide Orogen–Scandinavia and Related Areas.* New York: John Wiley, 569–578.

Andresen, A., and Færseth, R. 1982. An evolutionary model for the southwest Norwegian Caledonides. *Am. J. Sci., 282*, 765–782.

Austrheim, H. 1987. Eclogitization of lower crustal granulites by fluid migration through shear zones. *Earth Planet. Sci. Lett., 81*, 221–232.

Austrheim, H. 1991. Eclogite formation and dynamics of crustal roots under continental collision zones. *Terra Nova, 3*, 492-499.

Austrheim, H., and Mørk, M. B. E. 1988. The lower continental crust of the Caledonian mountain chain: evidence from former deep, crustal sections in western Norway. *Norges Geologiske Undersøkelse Special Publication, 3*, 102–113.

Austrheim, H., and Råheim, A. 1981. Age relationships within the high grade metamorphic rocks of the Bergen Arcs, western Norway. *Terra Cognita, 1*, 33.

Austrheim, H., Råheim, A., and Krogh, T. E. 1991. Age relationships and petrogenesis of the high grade metamorphic anorthosite-mangerite-charnockite suite of the Bergen Arcs, W. Norway. In: Austrheim, H., Fluid induced processes in the lower crust as evidenced by Caledonian eclogitization of Precambrian granulites, Bergen Arcs, western Norway. Ph.D., University of Oslo.

Backlund, H. G. 1936. Zur genetischen Deutung der Eklogite. *Geologische Rundschau, 28*, 47–61.

Berry, H. N., and Lux, D. R. 1991. Preliminary results from an Argon isotopic study of rapidly uplifted, high-pressure rocks in southwest Norway. *Geological Society of America Annual Meeting 1991, Abstracts with Program, 23*, A311.

Binns, R. A. 1967. Barroisite-bearing eclogite from Naustdal, Sogn og Fjordane, Norway. *J. Petrol., 8*, 349–371.

Birkeland, T. 1981. The geology of Jæren and adjacent districts. A contribution to the Caledonian nappe tectonics of Rogaland, southwestern Norway. *Norsk Geologisk Tidsskrift, 61*, 213–235.

Bohlen, S. R., and Boettcher, A. L. 1982. The quartz-coesite transformation: a pressure determination and the effects of other components. *J. Geophys. Res., 87*, 7073–7078.

Bollingberg, H. J., and Bryhni, I. 1972. Minor element zonation in an eclogite garnet. *Contrib. Mineral. Petrol., 36*, 113–122.

Boundy, T. M., Fountain, D. M. and Austrheim, H. 1992. Structural development and petrofabrics of eclogite facies shear zones, Bergen Arcs, western Norway: implications for deep crustal deformational processes. *Jour. Metamorph. Geol., 10*, 127–146.

Brastad, K. 1985. Relations between peridotites, anorthosites and eclogites in Bjørkedalen, Western Norway. *Caledonide Orogen*. New York: J. Wiley, 859–872.

Brey, G. P., Köhler, T., and Nickel, K. G. 1990. Geothermobarometry in four-phase lherzolites I. Experimental results from 10 to 60 kb. *J. Petrol., 31*, 1313–1352.

Broks, T. M. 1985. Petrologiske undersøkelser innen Tromsø dekkekompleks i området Tromsdalen-Ramfjord-Breivikeidet, Troms Ph.D., University of Tromsø.

Brueckner, H. K. 1972. Interpretation of Rb-Sr ages from Precambrian and Paleozoic rocks of Southern Norway. *Am. J. Sci., 272*, 334–358.

Brueckner, H. K. 1974. "Mantle" Rb/Sr and $^{87}Sr/^{86}Sr$ ratios for clinopyroxenes from Norwegian garnet peridotites and pyroxenenites. *Earth Planet. Sci. Lett., 24*, 26–32.

Brueckner, H. K. 1977. A crustal origin for eclogites and a mantle origin for garnet peridotites: Strontium isotopic evidence from clinopyroxenes. *Contrib. Mineral. Petrol., 60*, 1–15.

Bryhni, I. 1966. Reconnaissance studies of gneisses, ultrabasites, eclogites and anorthosites in Outer Nordfjord, Western Norway. *Norges Geologiske Undersøkelse, 241*, 1–68.

Bryhni, I., Bollingberg, H. J., and Graff, P. R. 1969. Eclogites in quartzo-felspathic gneisses of Nordfjord, West Norway. *Norsk Geologisk Tidsskrift, 49*, 193–225.

Bryhni, I., Green, D. H., Heier, K. S., and Fyfe, W. S. 1970. On the occurrence of eclogites in Western Norway. *Contrib. Mineral. Petrol., 26*, 12–19.

Bryhni, I., and Griffin, W. L. 1971. Zoning in eclogite garnets from Nordfjord, west Norway. *Contrib. Mineral. Petrol., 32*, 112–125.

Bryhni, I., Krogh, E. J., and Griffin, W. L. 1977. Crustal derivation of Norwegian eclogites: a review. *Neues Jahrb. Mineralogy Abh., 130*, 49–68.

Carswell, D. A. 1968a. Picritic magma-residual dunite relationships in garnet peridotite at Kalslaret near Tafjord, south Norway. *Contrib. Mineral. Petrol., 19*, 97–124.

Carswell, D. A. 1968b. Possible primary upper mantle peridotite in Norwegian basal gneiss. *Lithos, 1*, 322–355.

Carswell, D. A. 1973. Garnet pyroxenite lens within Ugelvik layered garnet peridotite. *Earth Planet. Sci. Lett., 20*, 347–352.

Carswell, D. A. 1974. Comparative equilibration temperatures and pressures of garnet lherzolites in Norwegian gneisses and in kimberlite. *Lithos, 7*, 113–121.

Carswell, D. A. 1986. The metamorphic evolution of Mg-Cr type Norwegian garnet peridotites. *Second International Eclogite Conference, Lithos*, 279–297.

Carswell, D. A. 1990. Eclogites and the eclogite facies: definitions and classifications. *Eclogite facies rocks.* Glasgow: Blackie, 1–13.

Carswell, D. A., and Cuthbert, S. J. 1986. Eclogite facies metamorphism in the lower continental crust. *The nature of the lower continental crust, Geological Society special publication, 25,* 193–209.

Carswell, D. A., and Gibb, F. G. F. 1980. The equilibration conditions and petrogenesis of European crustal garnet lherzolites. *Lithos, 13,* 19–29.

Carswell, D. A., and Gibb, F. G. F. 1987. Evaluation of mineral thermometers and barometers applicable to garnet lherzolite assemblages. *Contrib. Mineral. Petrol., 95,* 499–511.

Carswell, D. A., Harvey, M. A., and Al-Samman, A. 1983. The petrogenesis of contrasting Fe-Ti and Mg-Cr garent-peridotite types in the high grade gneiss complex of Western Norway. *Bulletin Mineralogie, 106,* 727–750.

Carswell, D. A., Krogh, E. J., and Griffin, W. L. 1985. Norwegian orthopyroxene eclogites: calculated equilibration conditions and petrogenetic implications. *The Caledonide Orogen.* New York: J. Wiley, 823–841.

Chauvet, A., Kienast, J. R., Pinardon, J. L., and Brunel, M. 1992. Petrological constraints and PT path of Devonian collapse tectonics within the Scandian mountain belt (Western Gneiss Region, Norway). *J. Geol. Soc. London, 159,* 383–400.

Cohen, A. S., O'Nions, R. K., Siegentaler, R., and Griffin, W. L. 1988. Chronology of the pressure-temperature history recorded by a granulite terrain. *Contrib. Mineral. Petrol., 98,* 303–311.

Cotkin, S. J., Valley, J. W., and Essene, E. J. 1988. Petrology of margarite-bearing meta-anorthosite from Seljeneset, Nordfjord, western Norway: Implications for the P-T history of the Western Gneiss Region during Caledonian uplift. *Lithos, 21,* 117–128.

Cuthbert, S. J. 1985. Petrology and tectonic settings of relatively low-temperature eclogites and related rocks in the Dalsfjord area, Sunnfjord. University of Sheffield.

Cuthbert, S. J. 1991. Evolution of the Devonian Hornelen Basin, west Norway: new constraints from petrological studies of metamorphic clasts, in Morton, *Geol. Soc. Lond. Spec. Pub., 57,* 343–360.

Cuthbert, S. J., and Carswell, D. A. 1990. Formation and exhumation of medium-temperature eclogites in the Scandinavian Caledonides. *Eclogite-facies rocks.* Glasgow: Blackie, 180–203.

Cuthbert, S. J., Harvey, M. A., and Carswell, D. A. 1983. A tectonic model for the metamorphic evolution of the Basal Gneiss Complex, Western South Norway. *Jour. Metamorph. Geol., 1,* 63–90.

Dallmeyer, R. D. 1988. Polyorogenic $^{40}Ar/^{39}Ar$ mineral age record within the Kalak Nappe Complex, Northern Scandinavian Caledonides. *J. Geol. Soc. London, 145,* 705–716.

Dallmeyer, R. D., and Andresen, A. 1992. Polyphase tectonothermal evolution of exotic Caledonian nappes in Troms, Norway: Evidence from $^{40}Ar/^{39}Ar$ mineral ages. *Lithos, 29,* 19–42.

Dallmeyer, R. D., and Gee, D. G. 1986. $^{40}Ar/^{39}Ar$ mineral dates from retrogressed eclogites within Baltoscandian miogeocline: Implications for a polyphase Caledonian orogenic evolution. *Geol. Soc. Amer. Bull., 97,* 26–34.

Dallmeyer, R. D., Johansson, L. and Möller, C. 1992. Chronology of Caledonian high-pressure granulite facies metamorphism, uplift and deformation within northern portions of the Western Gneiss Region, Norway. *Geol. Soc. Amer. Bull., 104,* 449–455.

Dallmeyer, R. D., and Stephens, M. B. 1991. Chronology of eclogite retrogression within the Seve Nappe Complex, Ravvejaure, Sweden: evidence from $^{40}Ar/^{39}Ar$ mineral ages. *Geologische Rundschau, 80*, 729–743.

Daly, J. S., Aitcheson, S. J., Cliff, R. A., and Rice, A. H. N. 1991. Geochronological evidence from discordant plutons for a late Proterozoic orogen in the Caledonides of Finnmark, northern Norway. *J. Geol. Soc. London, 148*, 29–40.

Dodson, M. H. 1979. Theory of cooling ages. *Lectures in Isotope Geology*. Berlin: Springer Verlag, 194–202.

Dunning, G. R., and Pedersen, R. B. 1988. U/Pb ages of ophiolites and arc-related plutons of the Norwegian Caledonides: Implications for the development of Iapetus. *Contrib. Mineral. Petrol., 96*, 123–23.

Endell, K. 1913. Ueber Granatamphibolite und Eklogite von Tromsø und von Tromsdalstind. *Centralblatt fur Mineralogie, Geologie und Paleontologie, 5*, 29–33.

Eskola, P. 1920. The mineral facies of rocks. *Norsk Geol. Tidsskrift, 6*, 43–194.

Eskola, P. 1921. On the eclogites of Norway. *Skrift. Videnskaps-selsk. Christiania Mat.-Naturv, Kl 18*, 1–118.

Gebauer, D., Lappin, M. A., Grünenfelder, M., and Wyttenbach, A. 1985. The age and origin of some Norwegian eclogites. A U-Pb zircon and REE study. *Chemical Geology, 52*, 227–247.

Gee, D. G. 1975. A tectonic model for the central part of the Scandinavian Caledonides. *Am. J. Sci., 275A*, 468–515.

Gee, D. G., and Sjøstrøm, H. 1984. Early Caledonian obduction of the Handøl Ophiolite. *Stockholm Universitet Geologisk Institut Med., 255*, 72.

Gee, D. G., and Wilson, M. R. 1974. The age of orogenic deformation in Swedish Caledonides. *Am. J. Sci., 274*, 1–9.

Gjelsvik, T. 1952. Metamorphosed dolerites in the gneiss area of Sunnmøre on the west coast of southern Norway. *Norsk Geologisk Tidsskrift, 30*, 33–134.

Griffin, W. L. 1972. Formation of eclogites and the coronas in anorthosite, Bergen Arcs, Norway. *Geol. Soc. Amer. Mem., 135*, 37–63.

Griffin, W. L. 1987. On the eclogites of Norway–65 years late. *Mineral. Mag., 51*, 333–343.

Griffin, W. L., Austrheim, H., Brastad, K., Bryhni, I., Krill, A. G., Krogh, E. J., Mørk, M.-B. E., Qvale, H., and Tørudbakken, B. 1985. High-pressure metamorphism in the Scandinavian Caledonides. *The Caledonide Orogen*. New York: J. Wiley, 783–801.

Griffin, W. L., and Brueckner, H. K. 1980. Caledonian Sm-Nd ages and a crustal origin for Norwegian eclogites. *Nature, 285*, 319–321.

Griffin, W. L., and Brueckner, H. K. 1985. REE, Rb-Sr and Sm-Nd studies of Norwegian eclogites. *Chemical Geology, 52*, 249–271.

Griffin, W. L., and Carswell, D. A. 1985. In situ metamorphism of Norwegian eclogites: an example. *The Caledonide Orogen*. New York: J. Wiley, 813–822.

Griffin, W. L., and Heier, K. S. 1969. Parageneses of garnet in granulite-facies rocks, Lofoten-Vesterålen, Norway. *Contrib. Mineral. Petrol., 23*, 89–116.

Griffin, W. L., and Heier, K. S. 1973. Petrological implications of some corona structures. *Lithos, 6*, 315–335.

Griffin, W. L., and Qvale, H. 1985. Superterrian eclogites and the crustal origin of garnet peridotites, Almklovdalen, Norway. *The Caledonide Orogen*. New York: J. Wiley, 804–812.

Griffin, W. L., and Råheim, A. 1973. Convergent metamorphism of eclogites and dolerites, Kristansund area, Norway. *Lithos, 6*, 21–40.

Griffin, W. L., Taylor, P. N., Hakkinen, J. W., Heier, K. S., Iden, I. K., Krogh, E. J., Malm, O. A., Olsen, K. I., Ormaasen, D. E., and Tveten, E. 1978. Archaean and Proterozoic crustal evolution in Lofoten-Vesterålen, N. Norway. *J. Geol. Soc. London, 135*, 629–647.

Gustavson, M. 1979. Et meta-eklogittførende gneiskompleks på ytre Helgeland. *Geolognytt, 12*, 12.

Harley, S. L. 1984a. An experimental study of the partitioning of Fe and Mg between garnet and orthopyroxene. *Contrib. Mineral. Petrol., 86*, 359–373.

Harley, S. L. 1984b. The solubility of alumina in orthopyroxene coexisting with garnet in $FeO-Al_2O_3-SiO_2$ and $CaO-FeO-MgO-Al_2O_3-SiO_2$. *J. Petrol., 25*, 665–696.

Harley, S. L., and Carswell, D. A. 1990. Experimental studies on the stability of eclogite facies mineral parageneses. *Ecologite Facies Rocks*. Glasgow: Blackie, 53–82.

Harrison, T. M. 1981. Diffusion of ^{40}Ar in hornblende. *Contrib. Mineral. Petrol., 78*, 324–331.

Hermans, G. A. E. M., Maijer, C., Jansen, J. B. H., and Tobi, A. C. 1987. The geology of the Jæren district. Precambrian basement rocks and precambrian rocks overthrust during the Caledonian orogeny. In Maijer and Padget, eds., The geology of southernmost Norway: An excursion guide. *Norges Geologiske Undersøkelse Special Publication, 1*, pp.98.

Hernes, I. 1954. Eclogite-amphibolite on the Molde Peninsula, southern Norway. *Norsk Geologisk Tidsskrift, 33*, 163–184.

Hiortdahl, T., and Irgens, M. 1862. Geologiske Undersøgelser i Bergen Omegn: Universitetsprogram for andet halvaar 1862.

Jacobsen, B., and Wasserburg, G. J. 1978. Interpretation of Nd, Sr, and Pb isotope data from Archaean migmatites in Lofoten-Vestrålen, Norway. *Earth Planet. Sci. Lett., 41*, 245–253.

Jamtveit, B. 1984. High-P metamorphism and deformation of the Gurskebotn garnet-peridotite, Sunnmøre, Western Norway. *Norsk Geologisk Tidsskrift, 2*, 97–110.

Jamtveit, B. 1987a. Metamorphic evolution of the Eiksunddal eclogite complex, Western Norway, and some tectonic implications. *Contrib. Mineral. Petrol., 95*, 82–99.

Jamtveit, B. 1987b. Magmatic and metamorphic controls on chemical variations within the Eiksunddal eclogite complex, Sunnmøre, western Norway. *Lithos, 20*, 369–389.

Jamtveit, B., Bucher, K., and Austrheim, H. 1990. Fluid controlled eclogitization of granulites in deep crustal shear zones, Bergen arcs, western Norway. *Contrib. Mineral. Petrol., 104*, 184–193.

Jamtveit, B., Carswell, D. A. and Mearns, E. W. 1991. Chronology of the high-pressure metamorphism of Norwegian garnet peridotites/pyroxenites. *Journal of Metamorphic Petrology, 9*, 125–139.

Johansson, L., and Möller, C. 1986. Formation of sapphirine during retrogression of a basic high-pressure granulite, Roan, Western Gneiss Region. *Contrib. Mineral. Petrol., 94*, 29–41.

Klaper, E. M. 1990. Reaction-enhanced formation of eclogite-facies shear zones in granulite-facies anorthosites. *Geol. Soc. Lond. Spec. Pub., 64*, 167–173.

Kohn, M. J., and Spear, F. S. 1989. Empirical calibration of geobarometers for the assemblage garnet + hornblende + plagioclase + quartz. *Am. Mineralogist, 74*, 77–84.

Kohn, M. J., and Spear, F. S. 1990. Two new geobarometers for garnet amphibolites, with applications to southeastern Vermont. *Am. Mineralogist, 75,* 89–96.

Kolderup, C. F. 1903. Die Labradorfelse des westliches Norwegens II. Die Labradorfelse und die mit denselben verwandten Gesteine in dem Bergengebiete. *Bergens Museums Aarbog, 12,* 1–129.

Kolderup, C. F., and Kolderup, N. H. 1940. Geology of the Bergen Arc system. *Bergens Museums Skrifter, 20,* 1–137.

Krogh, E. J. 1977a. Evidence for a Precambrian continent-continent collision in western Norway. *Nature, 267,* 17–19.

Krogh, E. J. 1977b. Crustal and in situ origin of Norwegian eclogites. A reply. *Nature, 269,* 730.

Krogh, E. J. 1977c. Origin and metamorphism of iron formations and associated rocks, Lofoten-Vesteralen, N. Norway. I. The Vestpolltind Fe-Mn deposit. *Lithos, 10,* 243–255.

Krogh, E. J. 1980a. Geochemistry and petrology of glaucophane-bearing eclogites and associated rocks from Sunnfjord, western Norway. *Lithos, 13,* 355–380.

Krogh, E. J. 1980b. Compatible P-T conditions for eclogites and surrounding gneisses in the Kristiansund area, western Norway. *Contrib. Mineral. Petrol., 75,* 387–393.

Krogh, E. J. 1982. Metamorphic evolution of Norwegian country-rock eclogites, as deduced from mineral inclusions and compositional zoning in garnets. *Lithos, 15,* 305–321.

Krogh, E. J. 1988. The garnet-clinopyroxene Fe-Mg geothermometer – a reinterpretation of existing experimental data. *Contrib. Mineral. Petrol., 75,* 387–393.

Krogh, E. J., Andresen, A., Bryhni, I., Broks, T. M., and Kristensen, S. E. 1990. Eclogites and polyphase P-T cycling in the Caledonian uppermost allochthon in Troms, northern Norway. *Journal of Metamorphic Geology, 8,* 289–309.

Krogh, E. J., and Brunfelt, A. O. 1981. REE, Cs, Rb, Sr and Ba in glaucophane-bearing eclogites and associated rocks, Sunnfjord, western Norway. *Chemical Geology, 332,* 295–305.

Krogh, E. J., and Elvevold, S. 1990. A Precambrian age for an early gabbro-monzonitic intrusive on the Øksfjord peninsula, Seiland Igneous Province, northern Norway. *Norsk Geologisk Tidsskrift, 70,* 267–273.

Krogh, T. E., Mysen, B. O., and Davis, G. L. 1974. A Palaeozoic age for the primary minerals of a Norwegian eclogite. *Carnegie Institution Washington Yearbook, 73,* 575–576.

Kullerud, K. 1987. Origin and tectonometamorphic evolution of the eclogites in the Tsäkkok Lens (Seve Nappes), southern Norrbotten Sweden. Ph.D., University of Oslo.

Kullerud, K. 1993. Shear zones and fluid induced metamorphism within the Flakstadøy Basic Complex, Lofoten, Norway. *Norsk Geologisk Tidsskrift* (in press).

Kullerud, K., and Stephens, M. B. 1993. Tectonometamorphic evolution of the eclogites of the Tsäkkok lens (Seve Nappes), southern Norrbotten Caledonides, Sweden. *Jour. Metamorph. Geol.* (in review).

Kullerud, K., Stephens, M. B., and Zachrisson, E. 1990. Pillow lavas as protoliths for eclogites: evidence from a late Precambrian-Cambrian continental margin, Seve Nappes, Scandinavian Caledonides. *Contrib. Mineral. Petrol., 105,* 1–10.

Kullerud, L., Tørudbakken, B. O., and Ilebekk, S. 1986. A compilation of radiometric age determinations from the Western Gneiss Region, South Norway. *Norges Geologiske Undersøkelse Bulletin, 406,* 17–42.

294   *E. J. Krogh and D. A. Carswell*

Landmark, K. 1973. Beskrivelse til de geologiske kart 'Tromsø' og 'Mälselv'. II. Kaledonske bergarter. *Tromsø Museum Skrifter, 15*, 1–263.

Lappin, M. A. 1966. The field relationships of basic and ultrabasic masses in the basal gneiss complex of Stadlandet and Almklovdalen, Nordfjord, southwestern Norway. *Norsk Geologisk Tidsskrift, 4*, 439–496.

Lappin, M. A. 1974. Eclogites from the Sunndal-Grubse ultramafic mass, Almklovdalen, Norway, and the T-P history of the Almklovdalen masses. *J. Petrol., 15*, 567–601.

Lappin, M. A. 1977. Crustal and in situ origin of Norwegian eclogites. *Nature, 269*, 730.

Lappin, M. A., and Smith, D. C. 1981. Carbonate, silicate and fluid relationships in eclogites, Selje district and environs, S. W. Norway. *Transactions Royal Society Edinburgh: Earth Sciences, 72*, 171–193.

Lappin, M. A., and Smith, D. C. 1978. Mantle-equilibrated orthopyroxene eclogite pods from the basal gneisses in the Selje district, western Norway. *J. Petrol., 19*, 530–584.

Lindstrom, M., and Andresen, A. 1992. Early Caledonian high-grade metamorphism within exotic terranes of the Troms Caledonides. *Norsk Geologisk Tidsskrift, 72*, 375–379.

Lux, D. R. 1985. K/Ar ages from the Basal Gneiss Region, Stadlandet area, Western Norway. *Norsk Geologisk Tidsskrift, 65*, 277–286.

MacGregor, I. D. 1970. The effect of CaO, Cr_2O_3, Fe_2O_3 and Al_2O_3 on the stability of spinel and garnet peridotite. *Phys. Earth Planet. Int., 3*, 372–377.

Maijer, C., Hermans, G.A.E.M., Tobi, A. C., and Jansen, J.B.H. 1987. Caledonides and westernmost Precambrian intrusions. In Maijer and Padget, eds., The geology of southernmost Norway: An excursion guide. *Norges Geologiske Undersøkelse Special Publication, 1*, 99.

Malm, O. A., and Ormaasen, D. E. 1978. Mangerite-charnockite intrusives in the Lofoten-Vesterälen area, north Norway: Petrography, chemistry and petrology. *Norges Geologiske Undersøkelse, 338*, 83–114.

McDougall, I., and Green, D. H. 1964. Excess radiogenic argon in pyroxenes and isotopic ages on minerals from Norwegian eclogites. *Norsk Geologisk Tidsskr, 44*, 183–196.

Mearns, E. W. 1986. Sm-Nd ages for Norwegian garnet peridotite. *Lithos, 19*, 269–278.

Mearns, E. W., and Lappin, M. A. 1982. A Sm-Nd isotopic study of "internal" and "external eclogites", garnet lherzolite and grey gneiss from Almklovdalen, Western Norway. *Terra Cognita, First International Eclogite Conference (abstract), 2*, 324–324.

Medaris, L. G. J. 1980a. Convergent metamorphism of eclogite and garnet-bearing ultramafic rocks in West Norway. *Nature, 283*, 470–472.

Medaris, L. G. J. 1980b. Petrogenesis of the Lien peridotite and associated eclogites, Almklovdalen, Western Norway. *Lithos, 13*, 339–353.

Medaris, L. G. J. 1984. A geothermobarometric investigation of garnet-peridotites in the Western Gneiss Region of Norway. *Contrib. Mineral. Petrol., 87*, 72–86.

Medaris, L. G., Jr., and Carswell, D. A. 1990. The petrogenesis of Mg-Cr garnet peridotites in European metamorphic belts. *Eclogite facies rocks*. Glasgow and London: Blackie, 260–290.

Möller, C. 1988. Geolgy and metamorphic evolution of the Roan area, Vestranden, Western Gneiss Region, Central Norwegian Caledonides. *Norges Geologiske Undersøkelse Bulletin, 413*, 1–31.

Möller, C. 1990. Metamorphic evolution of granulite facies supracrustals, Roan, northern Western Gneiss Region, Norwegian Caledonides. University of Lund.

Mørk, M. B. E. 1985a. A gabbro to eclogite transition on Flemsøy, Sunmøre, Western Norway. *Chemistry and Petrology of Eclogites. Chemical Geology.* 283–310.

Mørk, M. B. E. 1985b. Incomplete high P-T metamorphic transitions within the Kvamsøy pyroxenite complex, west Norway: a case study of disequilibrium. *Jour. Metamorph. Geol., 3*, 245–264.

Mørk, M. B. E. 1986. Coronite and eclogite formation in olivine gabbro (western Norway): reaction paths and garnet zoning. *Mineral. Mag., 50*, 417–426.

Mørk, M. B. E., Kullerud, K., and Stabel, A. 1988. Sm-Nd dating of Seve eclogites, Norrbotten, Sweden–evidence for early Caledonian (505 Ma) subduction. *Contrib. Mineral. Petrol., 99*, 344–351.

Mørk, M. B. E., and Mearns, E. W. 1986. Sm-Nd isotopic systematics of a gabbro-eclogite transition. *Lithos, Second International Eclogite Conference,* 255–267.

Mørk, M. B. E., and Stabel, A. 1990. Cambrian Sm-Nd dates for an ultramafic intrusion and for high grade metamorphism on the Øksfjord peninsula, Finnmark, North Norway. *Norsk Geologisk Tidsskrift, 70*, 275–291.

Mysen, B. O., and Heier, L. 1972. Petrogenesis of eclogites in high grade metamorphic gneisses, exemplified by the Hareidland eclogite, Western Norway. *Contrib. Mineral. Petrol., 36*, 73–94.

Nicholson, R. 1984. An eclogite from the Caledonides of southern Norrbotten. *Norsk Geologisk Tidsskrift, 64*, 165–169.

Nickel, K. G. 1986. Phase equilibria in the system SiO_2-MgO-Al_2O_3-CaO-Cr_2O_3 (SMACCR) and their bearing on spinel/garnet lherzolite relationships. *N. Jb. Mineral. Abh., 155*, 259–287.

O'Hara, M. J. 1967. Mineral facies in ultrabasic rocks. *Ultramafic and Related Rocks.* New York: John Wiley, 7–18.

O'Hara, M. J. 1976. Origin of the Norwegian eclogites. *Progress in Experimental Petrology, NERC 3*, 252.

O'Hara, M. J., and Mercy, E.L.P. 1963. Petrology and petrogenesis of some garnetiferous peridotites. *Royal Society of Edinburgh Transactions, 65*, 251–314.

O'Neill, H. S. C. 1981. The transition between spinel lherzolite and garnet lherzolite, and its use as a geobarometer. *Contrib. Mineral. Petrol., 77*, 185–194.

O'Neill, H. S. C., and Wood, B. J. 1980. An experimental study of Fe-Mg partitioning between garnet and olivine and its calibration as a geothermometer, corrections. *Contrib. Mineral. Petrol., 72*, 337.

O'Neill, S. C., and Wood, B. J. 1979. An experimental study of Fe-Mg partitioning between garnet and olivine and its calibration as a geothermometer. *Contrib. Mineral. Petrol., 70*, 59–70.

Oh, C. 1992. The petrogenetic relationship among high-P/T metamorphic facies including the eclogite and epidote-amphibolite facies in model basaltic system. *Journal Geological Society of Korea, 28*, 298–313.

Oh, C., Liou, J. G., and Krogh, E. J. 1988. A petrogenetic grid for the eclogite and related facies at high pressure metamorphism. *Geological Society of America, Abstracts with Programs, 20*, A344.

Ormaasen, D. E. 1977. Petrology of the Hopen mangerite-charnockite intrusion, Lofoten, north Norway. *Lithos, 10*, 291–310.

Pettersen, K. 1878. Det nordlige Sveriges og Norges geologi. *Archiv for Matematiske Naturvidenskaber, 3*, 1–38.

Powell, R. 1985. Regression diagnostic and robust regression in geothermometer/ geobarometer calibration: the garnet-clinopyroxene geothermometer revisited. *Jour. Metamorph. Geol., 3*, 231–243.

Råheim, A. 1972. Petrology of high grade metamorphic rocks of the Kristiansund area. *Norges Geologiske Undersøkelse, 279*, 1–75.

Roberts, D., and Gee, D. G. 1985. An introduction to the structure of the Scandinavian Caledonides. *The Caledonide Orogen-Scandinavia and Related Areas.* Chichester: John Wiley, 55–68.

Robinson, P., and Krill, A. G. 1991. Extension of Trollheim tectono-stratigraphy and evidence for late Scandian sinistral shear in the Molde and Brattväg areas, Western Gneiss Region, Norway. *Terra Nova, 4*, 26–27.

Romey, W. D. 1971. Basic igneous complex, mangerite and high grade gneisses of Flakstadøy, Lofoten, northern Norway. I. Field relations and speculations on origins. *Norsk Geologisk Tidsskrift, 51*, 33–61.

Santallier, D. S. 1988. Mineralogy and crystallization of the Seve eclogites in the Vuoggatjålme area, Swedish Caledonides of Norrbotten. *Geologiska Föreningen Stockholm Förhandlingar, 110*, 89–98.

Skjerlie, F. J. 1969. The Pre-Devonian rocks in the Askvoll-Gaular area and adjacent districts, western Norway. *Norges Geologiske Undersøkelse, 258*, 325–359.

Smith, D. C. 1980. A tectonic melange of foreign eclogites and ultramafites in the Basal Gneiss Region, West Norway. *Nature, 287*, 366–368.

Smith, D. C. 1981. A reappraisal of factual and mythical evidence concerning the metamorphic and tectonic evolution of eclogite-bearing terrain in the Caledonides (abstract). *Terra Cognita, 1*, 73–74.

Smith, D. C. 1984. Coesite in clinopyroxene in the Caledonides and its implications for geodynamics. *Nature, 310*, 641–644.

Smith, D. C. 1988. A review of the peculiar mineralogy of the "Norwegian coesite-eclogite province," with crystal-chemical, petrological, geochemical and geodynamical notes and an extensive bibliography. *Eclogites and eclogite-facies rocks.* Amsterdam: Elsvier, 1–206.

Smith, D. C., and Lappin, M. A. 1989. Coesite in the Straumen kyanite-eclogite pod, Norway. *Terra Research, 1*, 47–56.

Steltenpohl, M. G., Andresen, A., and Tull, J. F. 1990. Lithostratigraphic correlation of the Salangen (Ofoten) and Balsfjord (Troms) Groups: evidence for the post-Finnmarkian unconformity, North Norwegian Caledonides. *Norges Geologiske Undersøkelse Bulletin, 418*, 61–77.

Stephens, M. B. 1988. The Scandinavian Caledonides: a complexity of collisions. *Geology Today, Jan-Feb 1988*, 20–26.

Stephens, M. B., and Gee, D. G. 1985. A tectonic model for the evolution of the eugeoclinal terranes in the central Scandinavian Caledonides. *The Caledonide Orogen-Scandinavia and Related Areas.* Chichester, England, Wiley and Sons, 953–978.

Stephens, M. B., and Gee, D. G. 1989. Terranes and polyphase accretionary history in the Scandinavian Caledonides. *Geol. Soc. Amer. Spec. Pap., 230*, 17–30.

Stephens, M. B., Gustavson, M., Ramberg, I. B., and Zachrisson, E. 1985. The Caledonides of central-north Scandinavia – a tectonostratigraphic overview. *The Caledonide Orogen–Scandinavia and Related Areas.* Chichester: John Wiley, 135–162.

Stephens, M. B., and van Roermund, L. M. 1984. Occurrence of glaucophane and crossite in eclogites of the Seve Nappes, southern Norrbotten Caledonide, Sweden. *Norsk Geologisk Tidsskrift, 64*, 155–163.

Stiberg, J.-P. 1993. Apatite fission track analysis from the Western Gneiss Region, preliminary results: Constraints to the Cenozoic uplift of the Norwegian passive margin. *Geonytt, 1/93,* 46.

Stølen, L. K. 1988. Tectonostratigraphy and structure of the Staloluokta area, Padjelanta, southern Norrbotten Caledonides, Sweden. *Geologiska Föreningens Stockholm Förhandlingar, 110,* 341–349.

Sturt, B. A., Pringle, I. R., and Ramsay, D. M. 1978. The Finnmarkian phase of the Caledonian orogeny. *J. Geol. Soc. London, 13,* 381–391.

Takasu, A. 1984. Prograde and retrograde eclogites in the Sanbagawa metamorphic belt, Besshi district. *Japan Journal of Petrology, 25,* 619–643.

Thon, A. 1985. Late Ordovician and Early Silurian cover sequences to the west Norwegian ophiolite fragments: stratigraphy and structural evolution. *The Caledonide Orogen–Scandinavia and Related Areas.* Chichester: John Wiley. 395–406.

Torudbakken, B., and Råheim, A. 1981. An in situ metamorphosed eclogite near the Surnadal syncline and its implications for the metamorphic relationships in the Surnadal area. *Terra Cognita, 1,* 152.

Trouw, R. A. J. 1973. Structural geology of the Marsfjallen area, Caledonides of Västerbotten, Sweden. *Sveriges Geologiska Undersökning, C689,* 1–115.

Tucker, R. D., Krogh, T. E., and Råheim, A. 1992. Proterozoic evolution and age-province boundaries in the central part of the Western Gneiss Region, Norway: Results of U-Pb dating of accessory minerals from Trondheimsfjord to Geiranger. *Geological Association of Canada, Special Paper, 38,* 149–173.

Tucker, R. D., Råheim, A., Krogh, T. E., and Corfu, F. 1987. Uranium-lead zircon and titanite ages from the northern portion of the Western Gneiss Region, south-central Norway. *Earth Planet. Sci. Lett., 81,* 203–211.

Tull, J. F. 1978. Geology and structure of Vestvågöy, Lofoten, N. Norway. *Norges Geologiske Undersøkelse, 333,* 1–59.

van Roermund, H. 1985. Eclogites of the Seve nappe, central Scandinavian Caledonides. *The Caledonide Orogen – Scandinavia and related areas.* New York: Wiley & Sons, 873–886.

van Roermund, H. L. M. 1989. High-pressure ultramafic rocks from the allochthonous nappes of the Swedish Caledonides. *The Caledonide Geology of Scandinavia.* London: Graham and Trotman, 205–221.

van Roermund, H. R. M., and Bakker, E. 1984. A cross section through the Seve Nappe, central Scandinavian Caledonides, the geology of the Tangen-Inviken area. In van Roermund, H. L. M., On eclogites from the Seve Nappe, Jämtland, central Scandinavian Caledonides (Chapter 4), University of Utrecht.

van Wyck, N., Valley, J. W., and Austrheim, H. 1991. Carbon and oxygen ratios from carbonate-bearing eclogites and calc-silicates from the Bergen Arcs, Norway. *Geological Society of America Abstracts with Programs, 23,* A394.

Wade, S. J. R. 1985. Radiogenic isotope studies of crust-forming processes in the Lofoten-Vesterålen province of North Norway. University of Oxford.

Webb, S. A. C., and Wood, B. J. 1986. Spinel-pyroxene-garnet relationships and their dependence on Cr/Al ratio. *Contrib. Mineral. Petrol., 92,* 471–480.

Williams, P. F., and Zwart, H. J. 1977. A model for the development of the Seve-Köli Caledonian Nappe Complex. *Energetics of Geological Processes.* New York: Springer Verlag, 170–187.

Zachrisson, E. 1973. The westerly extension of Seve rocks within the Seve-Köli Nappe Complex in the Scandinavian Caledonides. *Geologiska Föreningen Stockholm Förhandlingar, 95,* 243–251.

Zachrisson, E., and Stephens, M. B. 1984. Mega-structures within the Seve Nappes,

southern Norrbotten Caledonides, Sweden. *16e Nordiska Geologiska Vintermö-
tet, Stockholm 9–13 januari 1984, Stockholm Universitet Geologisk Institut Med,
255,* 241.
Zwart, H. J. 1974. Structure and metamorphism in the Seve-Köli Nappe Complex
and its implications concerning the formation of metamorphic nappes. *Geologie
des domaines crystalline.* Societe Geologie Belgique, 129–144.

9

Microcoesites and Microdiamonds in Norway:
An Overview

DAVID C. SMITH

Abstract

The known occurrences in Norway of micron-sized coesite or diamond crystals and their overall geologic context, including possible petrologic and geodynamic models, are briefly reviewed in essay style. Some implications of problems concerning conventional and unconventional models for the creation or exhumation of UHPM rocks are briefly discussed.

Definite or deduced coesite occurs as microinclusions within garnet or clinopyroxene in several eclogites of the Norwegian *Coesite-Eclogite Province* (CEP) which has been described as "a fragmented infolded nappe" within the larger Caledonized terrane known as the *Western Gneiss Region* (WGR) in the western part of South Norway. These microcoesites confirmed the preexistence of UHPM conditions that had previously been deduced from other petrologic data for certain eclogite localities in the WGR.

Microdiamonds have recently been reported from two nearby localities of gneiss within the WGR situated further North than the definite microcoesite occurrences of the "Selje District," but well within the probable limits of the CEP. Unfortunately the petrogenetical significance of these microdiamonds is still uncertain, as several very different hypotheses can be proposed, such that it cannot yet be established whether or not they again confirm the preexistence of UHPM conditions. In any case they provide sufficient data for the definition of an embryonic *Diamond Gneiss Province* (DGP), and they also underline the importance of parts of the WGR, as well as of the discipline of *micromineralogy* (cf. *micropalaeontology*), for a better understanding of deep metamorphic processes during orogenic events.

Priorities for further research are rather clear: more detailed micropetrographical and microchemical mapping of eclogites and gneisses on the 0.001–1-mm scale in order to clarify the pressure, temperature, composition, temporal, and spatial relationships between the different generations of microinclusions, and more detailed petrologic, structural, and geochronologic mapping on the 0.01 to 10 meter and 0.1–100-km scales in order to distinguish between geological units of different *P-T-X-t-s* histories and convincingly to correlate those with the same histories, in particular to clarify the relationships between the CEP, the DGP, the GUPs (*Garnet-Ultrabasite Provinces*) and the various non-UHPM subterranes that constitute the WGR terrane.

Introduction to the Geological Context

The Scope of This Chapter

This chapter cannot practically nor usefully repeat the long recent mineralogical review of Smith (1988); thus the reader is referred to that work as the most comprehensive, even if still far from complete, review available to date; its bibliography, pp. 178–206, is rather exhaustive and includes many published abstracts and articles that have been ignored in various other reviews. Instead this chapter summarizes the essential data concerning *microcoesites and microdiamonds only* and does not deal in detail with other minerals, or with most of the various petrologic or geodynamic problems concerning the CEP or the whole WGR–in particular, other UHPM eclogites with low-Al orthopyroxene or magnesite but without microcoesite or microdiamond, and various other kinds of non-UHPM eclogites in Scandinavia that do not indicate pressures exceeding ~20 kbar. The "Garnet-Ultrabasite Province(s)" (GUP), which was considered by Smith (1988) to represent one, or more probably two, separate subterranes distinct from the CEP, is also set aside except for a few important comments on the geochronologic data therefrom. These other rock types were discussed in the cited review as well as in several others (e.g., Bryhni et al., 1977; Carswell et al., 1983, 1985; Cuthbert et al., 1983; Griffin et al., 1985; Carswell and Cuthbert, 1986; Griffin, 1987; Mørk and Krogh, 1987; Cuthbert and Carswell, 1990).

The microdiamonds have been described only very recently by Dobrzhinetskaya et al. (1993a). The small amount of available data is described and discussed here, as it is of great relevance to the whole question of UHPM in Norway which is still rejected or doubted by some people (Griffin, 1987; Cuthbert and Carswell, 1990).

Essential Field Relations

As described by Eskola (1921) most eclogites in the Sunnmöre county of western Southern Norway ("Vestlandet" region), which lies at the heart of what is now called the "Western Gneiss Region" (WGR, previously known as the

"Basal Gneiss Complex" (BGC) or other expressions, and stretching essentially from Bergen to Trondheim), occur as meter- to kilometer-sized "pods" ("lenses," "boudins," "inclusions," "truncated layers") within principally amphibolite-facies gneisses (or schists), but granulite- or greenschist-facies gneisses also occur. The *gneisses* are generally plagioclase + quartz + clinoamphibole amphibolites but commonly they also contain K-feldspar and/or biotite and/or garnet, and sometimes epidote, clinopyroxene, muscovite, kyanite or sillimanite can be found. Rarely they do not contain plagioclase and their facies attribution is unclear (e.g., Smith, 1982a). A polymetamorphic polydeformational history is evident at many localities but probably some Lower Paleozoic sediments were metamorphosed only once by the Caledonian Orogeny. In fact, the commonly used expression "the gneisses" is as inappropriate as the equivalent expression "the eclogites" because there exist several distinct kinds of both rock types. At least four orientations of fold axes of the gneisses were recorded in the Stadlandet Peninsula in the "Selje District" by Lappin (1962, 1966) and in the "Vartdalsfjorden District" by Smith (1976a) (Figs. 9.1–2); even if some of these could be explained by one continuous deformational event, the structural situation is quite complex. Lithological breaks in the gneissic country-rocks are commonplace but trying to distinguish tectonic contacts, which could be major and separate different provinces, from non-tectonic contacts is often very difficult. Small- and large-scale shear zones have been recognized, for example on the coast near Selje: "Elongate, vertical dipping lenticular shear zones, which may vary from 2 cm to more than 200 m in width, occur in the gneiss complex" (Lappin, 1966; p. 458); the dunites "occur in a complex shear zone which may have an axial extension for some hundred kilometers and may thus represent a fundamental tectonic feature of the Norwegian Caledonides" (Lappin, 1966; p. 488).

The *eclogite* parageneses are "fresh" only in the "cores" that have escaped the pervasive amphibolitization retrogression which transformed the pod "margins," and often also late fracture zones cutting the pod, into amphibolites showing all transitions from cryptocrystalline symplectitic intergrowths of principally plagioclase + diopside and/or clinoamphibole toward coarse plagioclase + clinoamphibole assemblages similar to those in the adjacent gneisses. Whereas the eclogite cores often conserve a strong compositional layering oblique to the foliation of the gneisses, the amphibolitized margins developed a strong foliation parallel to that of the gneisses. The eclogite layering is usually manifested by garnet-, clinopyroxene- or quartz-rich zones (? relict igneous layering; ? metamorphic segregation) and is sometimes strongly folded internally pre- or syn-eclogitization (Lappin, 1966; Smith, 1976a).

Occasionally ultrabasites + garnet + spinel, "mangerites," metaanorthos-

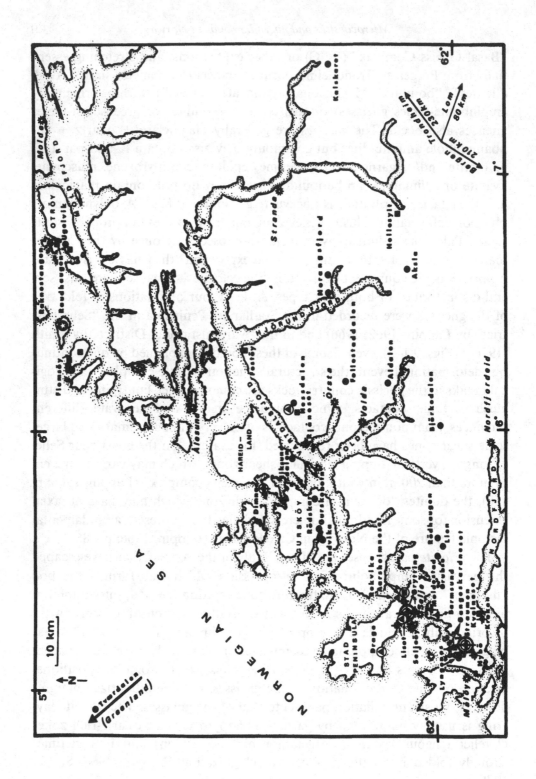

ites, metagranites, metagabbros, metadolerites, or marbles are associated with eclogites but the field relations are often ambiguous, tectonic juxtaposition usually, but not always, being the preferred interpretation. The garnet-ultrabasites (serpentinized dunite, lherzolite, harzburgite or wehrlite + garnet) often enclose layers, lenses, or "dykes" of garnet-websterite or garnet-pyroxenite which are usually referred to as *internal eclogites* as opposed to the *external eclogites* enclosed within the gneisses. Further detailed field relationships of eclogites and the other mentioned rock types of the WGR are provided in Brynhi (1966) and Lappin (1966) in particular, as well as in numerous other publications mentioned in the reviews cited above.

Essential Geochronology

The relative amounts of Caledonian vs. Pre-Caledonian metamorphism and deformation in the WGR, as well as in the rest of Norway, have been a subject of active discussion for over half a century, and probably still will be for a few more decades as new data frequently overturn former interpretations. The radiometric dates obtained by several methods applied to many different rock types from the WGR vary considerably with statistical peaks at ~1750 Ma ("Svecofennian" = "Middle Proterozoic"), ~1500 Ma, ~1000 Ma ("Sveconorwegian" = "Grenvillian" = "Late Proterozoic") and ~400 Ma ("Scandian-Acadian" = "Late Caledonian" = "Middle Paleozoic") (e.g. see the excellent compilation by Kullerud et al., 1986, and the review books edited by Gee and Sturt, 1985, and Harris and Fettes, 1988).

Various *gneissic rock types* ("augen gneiss," "grey gneiss," "banded gneiss," "schist," "quartzite") from the Selje District of Sunnmöre (Fig. 9.1) were examined by Lappin et al. (1979). They reported calculated igneous or metamorphic Rb-Sr model ages in the range 1855–642 Ma (up to 2755 Ma for

Figure 9.1. Map of that part of the WGR from Målöy (bottom left) to Molde (top right) (towns = open stars), showing all the known microcoesite (double circle for "definite coesite" and single circle for "deduced coesite") and microdiamond + "deduced coesite" occurrences (solid circle with central open star on Fjörtoft island, top center) and hence displaying embryonic data for the spatial distribution of the CEP and DGP. The "Selje," "Vartdalsfjorden" and "Moldefjorden" districts occur, respectively, in the bottom-left, center and top-right parts; these districts are particularly rich in known eclogites with exceptional mineralogy (crosses, solid squares, or solid triangles), and also in garnet-ultrabasites (solid circles) of the Mg-Cr GUP or Fe-Ti GUP, but unfortunately it is not yet possible to draw in the boundaries between the various tectonic provinces owing to a lack of sufficient petrologic/structural mapping. (Figure taken from Fig. 1.1 in Smith, 1988, where further details are available; figure is unmodified except for the addition of the Fjörtoft encircled star.)

"mangerite syenite" with garnet coronas); interestingly the youngest, probably most disturbed dates (840–642 Ma) were from garnet-mica schists collected very close to the famous Grytting UHPM eclogite locality (see below). They interpreted their data in terms of a 1760 ± 70 Ma *whole-rock* isochron "based upon thirteen apparently undisturbed data points from all four localities." Four of these garnet-mica schists gave a whole-rock isochron at 385 ± 70 Ma. Other Rb-Sr whole rock isochrons in the Selje District gave ages of 1556 Ma ("augen gneiss"), 968 Ma ("augen gneiss"), and 904 Ma ("grey gneiss") (Mearns, 1984).

Rb-Sr *mineral ages* from non-eclogite rock types, mostly of pegmatite, show a strong concentration around 385 Ma (e.g. Kullerud et al., 1986). Concerning the Selje District, Brueckner (1972) published a two-mica + epidote + whole-rock Rb-Sr isochron at 383 + 12 Ma for a "schist," and J.-J. Peucat and D. C. Smith (unpub. data) obtained a two-mica + whole-rock Rb-Sr isochron at 380 ± 6 Ma for a two-mica two-feldspar vein cross-cutting the famous Aarsheimneset headland UHPM eclogite pod (see below). Lux (1985) described K-Ar and ^{40}Ar-^{39}Ar data from amphiboles and micas from gneisses and eclogites in the Selje District; he discussed in detail the problems of excess argon that gave rise to apparent ages between 5208 and 172 Ma and finally deduced that the hornblendes in the country-rock gneisses had cooled through 500°C at ~410 Ma "shortly after eclogite formation" and that the biotites cooled through 300°C at ~370 Ma.

Lappin et al. (1979) also recorded U-Pb zircon ages of 1520 ± 10 Ma (upper intercept) and ~400 Ma (lower intercept) for a "mangerite syenite" in the Selje District; three "banded gneisses" at Grytting also indicated a lower intercept at ~400 Ma with upper intercepts of various gneisses between ~1650 and 1700 Ma. Mearns (1984) noted a Sm-Nd garnet-whole-rock age of 410 Ma for a "schist" in the same district.

Whereas many of the earlier dates from Rb-Sr, K-Ar and ^{40}Ar-^{39}Ar data should be treated with great caution, if not rejected, it is thus fairly clear that one (or more) substantial Late Caledonian metamorphic event(s) *overprinted* many, but not necessarily all, preexisting Svecofennian and Sveconorwegian gneiss units, and may have also metamorphosed for the first time some Lower Paleozoic and Precambrian igneous and sedimentary rocks. There are thus *several different types of gneisses* [not only in terms of protolith, degree of metamorphism and of deformation, presence or not of eclogite pods, but also in terms of age(s)], even in very small areas, such that there remains an enormous amount of petrologic/structural/geochronologic mapping to be done before a viable geologic map of the WGR can be produced.

Concerning the age of *eclogite formation* based on radioisotopic data, there are strong indications from studies at several eclogite localities within the WGR and including the Selje District (Sm-Nd: Griffin and Brueckner, 1980, 1985; Mearns and Lappin, 1982a,b; U-Pb lower intercepts: Krogh et al., 1973; Gebauer et al., 1985) of at least one *major metamorphic eclogite-forming event* at ~450–400 Ma (e.g., 447 ± 20 Ma [Sm-Nd] and 398 Ma [^{206}Pb-^{238}U] for the Grytting orthopyroxene-eclogite; and 423 ± 30 Ma for the Ulstein-Dimnöy eclogite, both described below). Gebauer et al. (1985) interpreted a primary igneous gabbroic protolith age at ~1500 Ma (upper intercept) for certain eclogites but a basaltic protolith for others (with no indication of the igneous age). Earlier K-Ar Precambrian dates of McDougall and Green (1964) from the same orthopyroxene-eclogite at Grytting were disbelieved by those authors, and all the following ones, because of expected excess argon anomalies. It is relevant to note that ages of eclogite formation at ~500 Ma or ~600 Ma are recorded from other parts of the Scandinavian Caledonides (Bergen Arcs, North Norway, and North Sweden, see below) such that *several distinct periods* of major eclogite-generating orogenic events are recognized in Scandinavia as a whole; however, since there is so far no evidence of UHPM at these other localities, they are not reviewed here.

Mineral ages from eclogites are relatively rare; McDougall and Green (1964) recorded Rb-Sr ages of 393 and 374 Ma (recalculated by Kullerud et al., 1986) from phlogopite from the same orthopyroxene-eclogite at Grytting, and Griffin and Brueckner (1985) found Rb-Sr ages of 487, 434, 405, 403, 398, and 267 Ma elsewhere in the WGR. Many of these ages, mostly based on micas, fall in the range 410–390 Ma which might well represent the eclogite formation or their symplectitic retrogression. So far there is no strong evidence to permit a clear discrimination in formation age between those eclogites assigned to the CEP and those that are not (or not yet); the data allow the possibility of UHPM at ~440 Ma and HPM at ~410 Ma.

A very important point is raised, but not adequately solved, in the discussion by Kullerud et al. (1986, p. 27) on the different blocking T of different isotopic systems and on false dates derived by isotopic disturbance. They argue, with reason, that only the oldest Rb-Sr whole-rock ages are reliable for a given orogenic event as the younger dates are most probably disturbed ones. However, they express surprise that even the oldest ages, i. e., ~1750 Ma for the "major crust-forming event," are retained in view of the "extensive" Caledonian eclogite facies metamorphism at "750–800°C, 15–20 kbar" which should have reset the Rb-Sr system. They explained this anomalous "resistance" in terms of "dry conditions" despite the fact that petrologic data exist that indicate that the conditions in the gneisses were not dry. An alternative solution is that those gneisses that do retain the ~1750 Ma original age repre-

sent precisely those country-rock gneisses that did *not* experience the Caledonian eclogite facies metamorphism but were juxtaposed at a late stage among
other gneiss units (subterranes, provinces) that did (cf. the FIF model of
Smith, 1988, discussed below).

Mention must be made briefly of a major development in the geochronology of the GUP(s). Jacobsen and Wasserburg (1980) suggested a ^{143}Nd-^{144}Nd
age of 1477 $\pm$ 7 Ma for a garnet-websterite layer within a garnet-ultrabasite
mass at Almklovdalen in the Selje District (Fig. 9.1) but these data were widely
rejected at the time. Mearns (1986) and Jamtveit et al. (1991) calculated Sm-
Nd ages of ~1703, 1303, 1163, 1040, 599, 561, 511, 437, 412 and 292 Ma in
some garnet-ultrabasites and "internal" eclogites from several localities in the
WGR. After eliminating certain dates on the basis of inferred isotopic read-
justment and disequilibrium, and eliminating certain P estimates on other
grounds, they interpreted their data from the *Mg-Cr type* of garnet-ultrabasite
in terms of an early UHPM event ("average" $P = 35 \pm 3$ kbar, $T = 850° \pm$
50°C) at 1703 $\pm$ 29 Ma and a later HPM event (P ~22 kbar [despite some
data at 40 kbar], T ~700°C) at 437 $\pm$ 58 Ma or 511 $\pm$ 18 Ma (respectively
for the Raudhaugene and Fjörtoft localities), i. e. *two distinct eclogite-facies
equilibrations* with a considerable time gap, the first not being Caledonian and
the second not being merely an age of tectonic introduction but of a new
eclogite-facies metamorphism. The *Fe-Ti type* of garnet-ultrabasite was constrained only by the 412 $\pm$ 12 Ma age from Eiksunddalen, and a 599 $\pm$ 42
Ma date from a "super-terrian," "dyke-like" internal eclogite body cross-
cutting the Gurskebotn garnet-ultrabasite, i. e., there are no Precambrian
dates. These data contribute toward defining a distinction between the Caledonian history of two GUPs, as suggested by Smith (1988, p. 58), rather than
toward a concept whereby the two different types of garnet-ultrabasite had
only different protoliths but a similar, if not identical, metamorphism as
deduced by Carswell et al. (1983), Griffin et al. (1985) and others.

Whereas the cited 437 and 412 ages overlap with the 450–400 Ma ages for
the external eclogites, the 511 and 599 Ma ages clearly do not. Jamtveit et al.
(1991, p. 135) argued that "the consistency between the P-T estimates" for the
Caledonian event in the garnet-ultrabasites and "the P-T conditions prevailing during the formation of the country-rock eclogites suggests that the Mg-
Cr type peridotites were *already in place* during the maximum T conditions
of the Caledonian high-P metamorphism." If, on the contrary, the real age at
Raudhaugene is on the upper side of its uncertainty bracket, which just overlaps with that for Fjörtoft at 494 Ma, then the Caledonian eclogite facies
metamorphism in the Mg-Cr GUP would have been quite distinct from that
in the CEP such that these two provinces would have come into contact *only*

for the final Late Caledonian symplectitic retrogression of all the mentioned rock types, at around 385 Ma. This is precisely what was suggested by Lappin (1966, pp. 477, 485, 491) who deduced from structural data that the Almklovdalen garnet-ultrabasites were the last rock type to be tectonically introduced, i. e., after the metaanorthosites which in turn had arrived after the external eclogites. The present writer favors the model of an older Caledonian UHPM metamorphism for the "Mg-Cr GUP" but can accept that the "Fe-Ti GUP" was most probably metamorphosed along with the external eclogites (at least at "stage A," if not also "stage I," see below).

Finally it is emphasized that similar P-T-t conditions in any two rock units outcropping nearby to each other unfortunately do *not* prove their coeval development side by side, since in a complex orogen it is quite possible to have similar P-T-t conditions at different places, i. e., different P-T-t-s conditions, prior to tectonic juxtaposition to permit the final petrologic event to be in common (e.g., the symplectitic retrogression).

Essential Bulk-Rock Chemistry and Protolith Hypotheses

Bulk-rock major-element chemistry of eclogites from several districts of the WGR is available (e.g., Binns, 1967; Smith, 1968; Bryhni et al., 1969; Green, 1969; Mysen and Heier, 1972; Smith, 1976a; Kechid, 1984; see Griffin, 1987, for more recent references). Concerning the Selje District (Fig. 9.2) it may be noted that whereas the cloud of compositions plotted on any chemical diagram centers around various "mean" gabbroic/doleritic/basaltic compositions except for a paucity in K_2O, a periphery of extreme compositions (e.g. 0.00–11.00 wt% Na_2O; 4.49–23.05 wt% Al_2O_3) (Smith, 1976a; Kechid, 1984; see Smith, 1988, table 1.1 for a summary) make comparison with typical igneous rocks meaningless as the protolith compositions must have been drastically modified; various metamorphic processes have been suggested for this (e.g., metasomatism, exsolution, carbonation, partial melting, mechanical segregation) (Smith, 1968, 1976a, 1988 pp. 12–23). The compositional variation is not confined to between-pods fluctuations since very strong within-pod zonations are common; these were described as "enrichment trends" since they corresponded to the enrichment of the modal abundance of a specific mineral: clinopyroxene, orthopyroxene or kyanite.

Bulk-rock analysis of several pairs of fresh and retrogressed eclogite from the same layer in nine different pods (Smith, 1976a, 1988) showed that the consistent differences were the additions of K^+, O^{2-}, and OH^- during amphibolitization, i. e., an influx of a potassium-rich oxidizing aqueous fluid, presumably derived from the adjacent gneisses.

Figure 9.2. Aerial photograph [by the author] of the SE coast of Stadlandet Peninsula in the Selje District of the WGR. Special eclogite localities: D = Drage, G = Grytting (or Grytingvaag), L = Liset (or Liseter), all with deduced and/or definite coesite. Note also the houses, port and beach of Selje town (lower center-left), the island of Selja (or Seljeöya) (center-left) whose cloister was a Middle Age capital of Christendom, the bay of Hoddevik (top with white beach) which excellently displays three orientations of folding in the gneisses, and the road (bottom right, between lakes) to Sandvik which lies in the center of the 3–4 km-wide "shear zone" mapped by Lappin (1966), which encompasses the Grytting and Aarsheimneset headland localities.

Amongst the ten arguments in favor of *crustal protoliths* assembled by the present author in Agrinier et al. (1985) was the significant content in certain eclogite pods of light elements: H (abundant hydrous minerals), B (tourmaline), C (carbonates), F (Al-rich sphene and amphiboles), and/or Na. In contrast to several eclogite occurrences *outside of the CEP* where definite low-P crustal protoliths can easily be identified by means of such evidence as relict ophitic texture with plagioclase still existing in eclogitized sheared gabbros (e.g. on Flemsøya: Mørk, 1985) or eclogitized shear zones cross-cutting non-eclogitized granulite-facies anorthositic rocks (e.g. on Holsnöy: Austrheim and Griffin, 1985; Austrheim, 1987, 1990, 1991), not a single eclogite attributed to the CEP up to now displays any such textural evidence. Distinct basaltic and gabbroic protoliths were favored on the basis of zircon morphology by Gebauer et al. (1985).

It is important to explain that there has long been a strong tendency among geologists working in Norway to believe that the eclogites in the WGR were simply metamorphosed dolerites since metadolerites in the same region show well-developed garnet coronas typical of incipient eclogitization (e.g., Gjelsvik, 1951, 1952, 1953; Hernes, 1953; Kolderup, 1960; Bryhni, 1966; Bryhni et al., 1969; Griffin and Råhem, 1973; Bryhni and Andréasson, 1985). Some of these metadolerites were reexamined in detail by Smith (1976a) who assembled chemical analyses of the *Sunnmøre Metadolerite Suite* (SMS). This showed trends well compatible with low-P olivine and plagioclase fractionation but there was no similarity with the within-pod or between-pods trends of the *Sunnmøre Eclogite Suite* (SES) which is predominantly composed of analyses from the CEP (Smith, 1976a, Plates 70–73; Smith, 1988, Fig. 1.3) such that Smith (1976a) argued against the SMS being the protolith of the SES. This did not exclude the possibility that some other kind(s) of crustal basic igneous rock constituted the protolith(s), but low-P igneous fractionation *alone* in the SMS could not explain the contrasting trends observed in the SES, and an extensive modification of the SMS compositions would have had to have been rather fortuitous in order to transform the trend of one suite into the trends of the other (nearly perpendicular when SiO_2 is plotted).

Suggestions of mantle protoliths are mentioned below.

First Records of UHPM in Norway and in the World

UHPM conditions in Norway were first suspected during the 1960s in the form of interpretations of mantle pressures for the crystallization of certain eclogites in the Selje District of the Norwegian WGR (Lappin, 1962, 1966; O'Hara and Mercy, 1963; Smith, 1968). The term UHPM was not of course in vogue at the time and specific *P-T* conditions were not presented; it is clear however from these texts that all these authors were thinking of *P* way above 20 kbar by analogy with the petrology of certain garnet-ultrabasites which occur nearby and for which *P* in the range 30–45 kbar was postulated (e.g., O'Hara and Mercy, 1963; Carswell 1973, 1974; Lappin, 1974).

The only other known pre-1975 record of suspected UHPM conditions elsewhere in the world (excluding kimberlite and other diatremes, and meteorite impact craters) was the discovery of radial cracks around quartz inclusions in garnet and other minerals at Shubino Village in the USSR and in the Münchberger Gneiss Massif, Germany (Chesnokov and Popov, 1965). These authors favored the former presence of coesite, rather than other hypotheses involving only the relative compressibility and/or expansivity of *alpha* or *beta* quartz, but again no specific *P-T* numbers were presented and by discussing the possibility of catalysis they hinted at possible metastable growth of the coesite.

Specific *P-T* conditions for eclogite genesis in the range 30–45 kbar at 700–850°C (for the amphibole-free "initial" stage I) and of 15–28 Kbar at 700–850°C [for the "early amphibole"-bearing "early" stage E (later called stage A for amphibole)] were presented by Smith (1976a) and Lappin and Smith (1978) (Fig.9.3). These values were based upon diverse mineralogical/petrologic criteria, principally the very low Al-contents of orthopyroxene coexisting with garnet and the coexistence of magnesite + diopside instead of the lower-*P* equivalent enstatite + dolomite. Furthermore the works of the present author first established the existence of *two distinct eclogite-facies metamorphic stages ("initial" and "early")* in the SES (Smith, 1968, 1971, 1976a, 1976b; Lappin and Smith, 1978, 1981) of which at least the initial stage was definitely under UHPM conditions.

A number of other crystal-chemical compositions or mineral associations, sometimes rare or unique, discovered in the same set of eclogites (the SES) indicate at least > 15 kbar and are at least compatible with > 30 kbar, as in the Italian Dora-Maira CEP which also displays various critical associations, such as talc + kyanite, talc + phengite, and new minerals like ellenbergerite and bearthite (e.g. Chopin, 1984, 1986; Chopin and Klaska, 1985; Chopin et al., 1993), for example:

1. Coexisting *olivine + garnet* (Eskola, 1921; Smith, 1976a; Lappin and Smith, 1978)
2. *Mid-solvus omphacite* and APD's (Smith, 1976a; Smith et al., 1980; Carpenter and Smith, 1981; Rossi et al., 1983a,b; Kechid, 1984)
3. The new mineral *nyböite* (Ungaretti et al., 1980, 1981; Smith and Lappin, 1982; Kechid and Smith, 1982; Kechid, 1984)

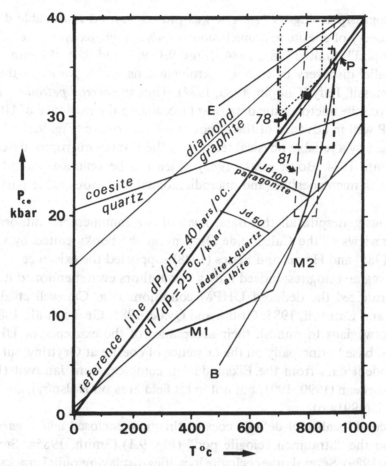

Figure 9.3. *P-T* diagram showing the UHPM estimates of Lappin and Smith (1978 and 1981) (lightly-dashed rhombohedra labeled 78 or 81) and the "best estimate" for stage I of Smith (1988) (solid square and thick dashed rectangle at 32 + 5 kbar, 800° + 100°C). [From Fig. 1.6 in Smith (1988), where further details are available (unmodified).]

4. The Si content of *phengite* (Lappin and Smith, 1981; Smith, unpub. data)
5. Coexisting *jadeite + quartz* instead of albite (Kechid, 1984)
6. Coexisting *jadeite + kyanite* instead of paragonite (Kechid, 1984)
7. Diverse probable *exsolution phenomena* within silicates and oxides (Lappin and Smith, 1978; see list in Smith, 1988, pp. 34–35).

Specific *P-T* calculations were often rather vague because of a lack of appropriate experimental or thermodynamic calibrations, but these data indicated or supported very high pressures. UHPM was therefore recognized and supported by a long series of distinct criteria by at least one school of mineralogists/petrologists *before* it was proved beyond reasonable doubt by the discovery of coesite microinclusions *within clinopyroxene* in the "Grytting eclogite pod" (Smith, 1983, 1984) (Figs. 9.1, 9.2, and 9.4), the same year as the parallel discovery of coesite microinclusions *within garnet* in the Dora-Maira massif, Italy (Chopin, 1983, 1984). Thus the *overall petrologic context*, apart from the microcoesite data, was favorable to the existence of UHPM in the CEP well in advance of the discovery of the coesite (Fig. 9.3) which was confirmed by a reliable physical technique: the Raman microprobe (see Boyer and Smith, 1984; Boyer et al., 1985) which has become a standard way of confirming microcoesites (and microdiamonds) in various other parts of the world.

Curiously, despite all this data, *none* of the innumerable authors of the major reviews of the Caledonide Orogeny in the books edited by Gee and Sturt (1985) and Harris and Fettes (1988) supported the existence of UHPM in Norwegian eclogites; indeed only a few authors even mentioned it, and all these criticized the deduced UHPM conditions (e.g. Carswell et al., 1985; Griffin and Carswell, 1985; Griffin and Qvale, 1985; Griffin et al., 1985). The first Norwegians to publish their acceptance of the existence of UHPM in Norway, based principally on the existence of coesite at Grytting but also on independent data from the Eiksunddalen complex, were Jamtveit (1987, p. 95), Austrheim (1990, 1991, but not in his field area on Holsnöy), and Andersen et al. (1991a,b).

A second locality of definite coesite, this time enclosed *within garnet*, was found in the "Straumen eclogite pod" (Fig. 9.1) (Smith, 1985a; Smith and Lappin, 1989). Several other eclogite localities displaying radial cracks around "polycrystalline" or "multicrystalline" quartz microinclusions in garnet, clinopyroxene or kyanite were also detected. This led to the recognition of the Norwegian "Coesite-Eclogite Province" (CEP) (Smith, 1985b, 1988, 1992) as *a group of UHPM eclogites* previously belonging to a single UHPM subterrane ("nappe," "slice", "slab", "block", "unit") but folded, fragmented, and dispersed by orogenic events within the WGR.

The Sequence of Metamorphic Stages in the Petrologic Evolution of Those Eclogites in the CEP

The CEP in Vestlandet was defined not only on the basis of seven localities with definite or deduced microcoesite (all described below) but also on the basis of a large number of other eclogite pods deduced, but not proved, to be coeval, i. e., eclogites containing one or more of the following indicator minerals: olivine, low-Al orthopyroxene with garnet, phlogopite, tourmaline, jade-

Figure 9.4. Field photograph [by the author] at the famous Grytting locality. The light-colored sub-triangular block (center left, labelled GN) is a gneissic sliver which thickens westward (to the top-left, toward the sea, outcrop broken on the western side). The other two sides of the block are geologic contacts. In front lies the "pegmatitic" (Eskola, 1921) magnesite-orthopyroxene-eclogite pod (Lappin and Smith, 1978), eroded downward for more than 1 m (probably assisted by overcollecting and perhaps dynamiting of this spectacular rock type). Behind and to the right lies the dolomite-coesite-eclogite (Smith, 1984) into which the white wooden post was drilled to delimit the area within which it is forbidden to collect any rocks (to protect the remaining orthopyroxene-eclogite). Sample C51 came from the freshest core which is not central but which occurs close to the gneissic sliver (lighter patch labeled CS). Thus the two contrasting orthopyroxene-lineage and kyanite-lineage pods are separated by only a gneissic sliver which thins to less than 5 cm wide (bottom center) before widening again eastward (below grass, bottom right). Coesite has now been found over a wide area further to the left and back such that the coesite-eclogite pod was most probably much larger before having been dissected by innumerable fractures that are now strongly amphibolitized.

ite, kyanite, phengite, zoisite, magnesite, dolomite (see Smith, 1988, p. 127 and pp. 148–154 for more details on localities). Other eclogites in which P really never exceeded ~20 kbar might contain one or more of these minerals and the distinction between these and those others that really did reach ~30 kbar UHPM conditions is an obvious and serious problem when no further indications are present. It is thus evident that the geographic boundaries of the CEP cannot easily be drawn on maps.

Those eclogites showing SiO_2 exsolution from a supersilicic clinopyroxene in the Rendeelven area of Liverpool Land, East Greenland (Smith and Cheeney, 1980; Smith 1982e), were tentatively included in the same CEP as this area is rather close to the Norwegian WGR on pre-Atlantic reconstructions. In the same way the garnet-ultrabasites from the Tværdalen area of Liverpool Land (Smith and Cheeney, 1981) were assimilated with the GUPs in the WGR.

The different *P-T-t* stages in the evolution of the CEP have been codified in order to reduce confusion in petrographic descriptions where it is essential to distinguish different generations of the same mineral type; the latest version is provided in Smith (1988, 1992). It is important to note that the first three stages **Z** (protolith), **Y** (eclogitization) and **X** (pre-exsolution) are poorly defined since the principal eclogite-facies event of stage **I** ("initial" eclogite) thoroughly recrystallized most of· the rocks involved. Stage **A** (amphibole-eclogite) caused the growth of non-symplectitic "early amphibole" (Al^{VI}-rich, Al^{IV}-poor) at the expense of garnet and pyroxene to form a new amphibole-garnet-clinopyroxene "amphibole-eclogite" paragenesis (Smith, 1968, 1971, 1976a; Lappin and Smith, 1978); no plagioclase was formed, and thus this transformation was quite distinct from the later amphibolitization transformation, both petrologically and also texturally as "seas" of early amphibole surround "islands" of remaining garnet, pyroxene, and other stage I minerals. The last stage **S** ("symplectitic") is the easiest to recognize as it is represented by numerous retrogressive mineral reactions, most of which involve the first growth of plagioclase or of other Al^{VI}-poor, Al^{IV}-rich phases including "late amphibole," and most of which occur in a symplectitic (or kelyphitic) textural fashion (cf. Eskola, 1921). Note that the discrimination of the octahedral or tetrahedral coordination of Al, for example by means of the factors $f_{Al} = Al^{VI}/(Al^{VI} + Al^{IV})$ and $delta_{Al} = (Al^{VI} - Al^{IV})$ per 24 cations, is considered as a critical feature of minerals stable or not stable in eclogites (Smith, 1982a; 1988, p. 59, pp. 89–91) and it can be used for distinguishing different *P-T* generations of the same mineral group, at least relatively if not absolutely (Smith et al., 1986b; Ungaretti et al., 1986). Although stage S began with amphibolite or granulite-facies parageneses, it continued through lower-*T* facies during gradual exhumation as testified by, for example, the development of epidote, chlorite, oxychlorite, pumpellyite, and serpentine.

Stage Z is represented only by a possible relict "early magnetite" (Pinet and Smith, 1985; Smith and Pinet, 1985) such that there is effectively zero petrographic evidence of the nature of the protoliths, in contrast to several non-UHPM localities where gabbroic textures are preserved (see above). Stage Y is possibly represented by microinclusions of anthophyllite, cummingtonite, Na-rich gedrite, tschermakite, Mg-rich staurolite, pure paragonite, K-eastonite, "Na-eastonite," and Al-rich titanite in garnet (Smith, 1988; pp. 35–36) ; however, it is far from certain which, if any, of these phases were definitely trapped during eclogitization since some or all may be of syn- or post-eclogitization age, having been formed by reactions due to fluid movement within the widely fractured garnet host. At the moment the writer favors attribution of the staurolite, titanite, and Al-poor amphiboles to stage Y, but the micas and Al^{IV}-rich amphiboles to stages A or S. Stage X is represented principally by probable exsolution phenomena within pyroxene (garnet from orthopyroxene and orthopyroxene from clinopyroxene: Lappin and Smith, 1978; and SiO_2 exsolution from supersilicic clinopyroxene: Smith and Cheeney, 1980; Smith 1982f, 1984), but other phenomena are also recorded (see Smith, 1988, pp. 34–35).

The crystal-chemical/petrographical/P-T distinction between stages I and A was one of the essential points in the works of Smith (1968, 1971, 1976a) followed by Lappin and Smith (1978, 1981), but this key point was almost entirely ignored by the "in situ school" (see below). It has turned out to represent the significant difference between coesite-eclogite (sub)facies and quartz-eclogite (sub)facies events (Smith, 1985b) in the CEP. Since many of the retrogressive reactions of stage S probably occurred at P ~10 kbar (with an uncertainty band of at least ± 5 kbar as for the other stages), Smith (1992) summarized the P-T-t evolution of the CEP by the following simplified formula:

> ~30 kbar = initial eclogite stage I = *coesite-eclogite (sub)facies*
>
> = (+ COESITE − QUARTZ − PLAGIOCLASE − CLINOAMPHIBOLE)
>
> ~20 kbar = early-amphibole stage A = *quartz-eclogite (sub)facies*
>
> = (− COESITE + QUARTZ − PLAGIOCLASE + CLINOAMPHIBOLE)
>
> ~10 kbar = symplectitic stage S = *amphibolite facies*
>
> = (− COESITE + QUARTZ + PLAGIOCLASE + CLINOAMPHIBOLE)

This simplification has a positive side in concisely distinguishing the three principal metamorphic facies involved, but on the negative side it implies a coeval growth of quartz and of clinoamphibole (of various compositions including nyböite and Mg-Al-taramite) that may not be correct.

A long list of mineral species or varieties crystallized during stage I, depending upon the sample locality and the bulk-rock chemical composition: coesite, garnet, clinopyroxene (diopside, omphacite or jadeite), low-Al orthopyroxene, olivine, uvite tourmaline, phlogopite, phengite, magnesite, dolomite, calcite, kyanite, "plate corundum," allanite, rutile, ilmenite, hematite; most of these minerals persisted stably into stage A but often with some change in chemical composition. In addition to quartz and clinoamphibole, several other minerals developed during the stage I to A transition and it is quite probable that they developed at *different moments* during a gradual slide in *P-T* conditions: paragonite after kyanite + jadeite, high-Al titanite after rutile, dravite tourmaline margins to uvite cores, Al-richer margins of ortho-pyroxene, rim dolomite on initial dolomite or magnesite, rim calcite on initial calcite, talc + magnetite veins cutting orthopyroxene, zoisite/clinozoisite around allanite, nyböite and Mg-Al-taramite and some more ordinary Al^{VI}-rich clinoamphiboles such as tremolite, winchite and barroisite, and most probably a resetting of all Fe/Mg distributions between coexisting phases (Smith, 1988, especially pp. 127–128 and Tables 1.2–1.4).

Since nyböite is abundant at an early stage in two different CEPs [in the "Nybö eclogite pod" (Ungaretti et al., 1980, 1981) and in the "Liset eclogite pod" (Kechid and Smith, 1982; Smith and Lappin, 1982; Kechid, 1984; Oberti et al., 1989) in the Norwegian CEP, and in the "Jianchang eclogite pod" in the Chinese "Su-Lu" CEP (Yang and Smith, 1989; Smith et al., 1990; Smith and Yang, 1991; Hirajima et al., 1992)], it was suggested by Smith et al. (1990) that the nyböite growth may have preceded the breakdown of coesite; if this was so, then the so-called amphibole-free stage I would exceptionally have contained this one amphibole species of coesite-eclogite (sub)facies vintage at certain suitable localities (i.e., Na-Al-rich). Although amphibole stability dur-ing metamorphism is clearly dependent on fluid composition, the crystal-chemical kind of amphibole stable (e.g. in terms of Al^{vi} and Al^{iv} proportions, especially for the more extreme compositions in multidimensional chemical space) is more dependent on *P* and *T* than on *X* in the opinion of the author (see the f_{Al} vs. V_{24a} *P/T* sequence glaucophane > nyböite > Mg-Al-taramite > pargasite in fig. 1.11 in Smith, 1988); in other words the *P-T* conditions deter-mine if a particular amphibole composition can be *physically stable* (i.e. no crystallographic collapse or explosion; e. g. nyböite if P is high enough and T is too hot for glaucophane), and then the local X conditions (e.g. Na, Al, Fe^{3+}, OH etc . . .) determine if it will or will not be *chemically stable*. Thus nyböite is considered as a valid indicator of at least HPM if not UHPM, regardless of fluid conditions (unfortunately nyböite is still sometimes referred to by its old IMA-discredited name "miyashiroite").

Another long list of other mineral species or varieties joined plagioclase, diopside and "late clinoamphibole" during stage S or later, depending upon local chemical compositions: aegirine, aegirine-augite, Ca-Mg-Fe-rich parag-onite, Na-Mg-Fe-rich margarite, pure margarite, "Na-eastonite," K-eastonite, sericite, preiswerkite, lisetite, "Ca-nepheline," Mg-rich kandite, "bead corundum," "blitz rutile," "late magnetite," Fe- or Mn-geikielite, Fe- or Mg-pyrophanite, Mg- or Mn-ilmenite, Mg-poor staurolite, Al-poor titanite, Cr-Zn-rich högbomite, spinel, epidote, chlorite, oxychlorite, pumpellyite, Fe-Al-taramite, Mg-ferri-winchite and many more ordinary clinoamphiboles such as hornblende, katophorite, pargasite, and tschermakite (Smith, 1988, 1992).

Some other minerals cannot convincingly be placed in the $Z > Y > X > I > A > S$ sequence (e.g. "initial corundum" blade/lath/plate/tablet inclusions in rutile, Mg-kandite, clinozoisite), whereas others may develop at more than one stage (e.g. Al-rich titanite as inclusions in garnet during stage Y and after rutile during stage A). A few accessory minerals (apatite, zircon, monazite, and sulfides) most probably were stable during all of the metamorphic stages.

Table 1.5 in Smith (1988) summarizes the distinguishing features of the two eclogite "lineages" (orthopyroxene- and kyanite-eclogite lineages) defined originally by Smith (1976a) and Lappin and Smith (1981). They refer to different sequences of mineral development (lines of descent). In the 1970s it was not clear if the different sequences were due to different origins or only to different bulk-rock chemical compositions. Now it seems certain that the latter is the only factor since both lineages appear to share all of the petrologic and deformational stages as well as their geographic distribution, and in particular 450–400 Ma ages and deduced coesite occurs in both.

Given that the Caledonian 450–400 Ma event represents the major, ~20 kbar, quartz-eclogite (sub)facies amphibole-eclogite recrystallization (stage A), then the initial UHPM coesite-eclogite (sub)facies (stage I) must either have happened only just before (i.e. within the same uncertainty band of about ± 25 Ma and hence be undetectable), or have happened much earlier with all geochronologic data thereon having been eliminated by the strong stage A event (see Smith, 1982c, and Smith, 1988, p. 31–32, for a discussion on whether these ages apply to the eclogites in their present structural position and whether they apply regionally to the entire WGR gneisses or not).

Petrologic/Geodynamic Modeling

The Norwegian Eclogite Controversy

Certain other research schools did not believe the original UHPM 30–45 kbar P estimates of Smith (1976a) and Lappin and Smith (1978) and there has been much heated discussion at international meetings [e.g., Griffin et al. (1981) vs. Smith (1981b); and Cuthbert and Carswell (1982) vs. Smith (1982c)], as well as in numerous publications. The expression "the Norwegian eclogite controversy" has been widely applied but effectively this began in 1966 with the publication of the rival hypotheses of Bryhni (1966) and Lappin (1966) who both provided similar lengthy petrologic/structural descriptions of the eclogites, garnet-ultrabasites, metaanorthosites, metagabbros, etc., and gneisses from nearby districts of the WGR. These two papers should be treated as "classical" works to constitute necessary reading as is the case of the much earlier work of Eskola (1921) which was a most valuable contribution to science. Whereas Bryhni (1966) argued in favor of the metamorphism of basaltic/ gabbroic rocks into eclogites in situ within their host gneisses and hence at *the same P-T-t-s conditions*, Lappin (1966) argued for the solid tectonic introduction of *foreign* eclogites previously equilibrated elsewhere *at higher P* conditions than those experienced by the enclosing "country-rock gneisses." This controversy thus also became known as the "in situ vs. foreign" dispute [e.g., Krogh (1977a,b) vs. Lappin (1977); Medaris (1980a,b) vs. Smith (1980a)]; this expression is, however, unfortunate since the essential discussion was over the *P* difference, the tectonic difference being a *consequence* of the *P* difference.

It is also important to note that this controversy was also confused in its early days by a third aspect: the *crustal or mantle* origin of the eclogite protoliths. Although a mantle origin was favored in the early days by O'Hara and Mercy (1963), Lappin (1962, 1966), and Lappin and Smith (1978), a crustal origin was included in the options by Smith (1976a, 1980a) and Lappin and Smith (1978) and was firmly preferred in Smith (1984) and Agrinier et al. (1985), despite the stable isotope data of the latter authors being compatible with a mantle origin (see below).

A fourth important aspect was the dispute over the "dry" or high-X_{CO_2} conditions said to be necessary for eclogite formation (e.g., Bryhni et al., 1970, 1977; Green and Mysen, 1972), whereas it was argued that the conditions in Norway were in fact "wet" and/or low-X_{CO_2} (e.g., Smith, 1971, 1976b; Lappin and Smith, 1978, 1981).

A Single Regional In Situ Metamorphism

As noted above, several reviews of Norwegian eclogites have been published (e.g. Bryhni et al., 1977; Cuthbert et al., 1983; Carswell et al., 1985; Carswell and Cuthbert, 1986; Griffin et al., 1985; Griffin, 1987; Mørk and Krogh, 1987; Cuthbert and Carswell, 1990), but they *all* present the same in situ model as the only viable model, in particular, by not allowing P_{max} in the eclogites to exceed ~20 Kbar, and by not admitting any difference in *P* between eclogites and gneisses and hence also not accepting the existence of two distinct eclogite-facies metamorphic events. These works thus argue for a *single regional in situ* Caledonian eclogite-forming event (except Bryhni et al., 1977, who favored a Precambrian collision following Krogh, 1977) when the entire WGR (part of "Baltica") was subducted under "Laurentia" and created coeval cofacial mineral parageneses in the eclogites, garnet-ultrabasites, metaanorthosites, metagabbros, etc., and gneisses, despite much published evidence that there were two distinct eclogite events (stages I and A), that at least the initial one was not regional, since other rock types did not experience the same stage I UHPM eclogite-forming event, that evidence for such high *P* as ~20 kbar in the gneisses is limited to very few localities, etc. (see section 1.6.1 in Smith, 1988, where 23 contentious points were analyzed in detail since it is in such details where lie the facts that make or break a model).

Given that, for the tenants of this petrologic model, the eclogites and gneisses were metamorphosed together, their complementary geodynamic (or "tectono-thermal") model [the references just above plus Medaris and Wang (1986) and Jamtveit (1987) but less Bryhni et al. (1977)] was relatively straightforward and consisted essentially of successive continental underthrusting (from the East) in order to place sufficient buoyant material below the WGR such that it rose by isostasy followed by erosion or "tectonic stripping" to take away the excess overburden. Cuthbert and Carswell (1990) added the possibility of extensional tectonics to aid the transport toward the surface of the dense "HP rocks," following Platt (1976). The present author has always dismissed this basic geodynamic model and its variants (e.g., Smith, 1980a, 1981b, 1984, 1988, 1991), but *not* because of the underthrusting or erosion or extensional tectonics aspects, but because P_{max} was never allowed to exceed ~20 kbar such that the model was not dealing with reality.

The Ultrareactivity Model

The same petrologic model involving in situ metamorphism, underthrusting and erosion but with P_{max} at ~30 kbar has never been suggested in the literature (except for the "*ultrametastability*" model, see below). Although the present author has several reasons to be reticent, it can be put forward here just as a model to contemplate in the form of the ~30 kbar "*ultrareactivity*" model (cf., Smith, 1993: mineral reactivity to an extreme degree) as follows: the westernmost part of the WGR was *all* subducted to depths appropriate for ~30 kbar UHPM metamorphism and all rock types recrystallized accordingly (i.e. no metastability), and then when this rock unit eventually returned to amphibolite-facies *P-T* conditions the rock types recrystallized again, but

almost entirely leaving little or no UHPM or HPM relics in the gneisses and
UHPM or HPM relics only in some eclogite cores.

The Foreign/In Situ/Foreign (FIF) Model

The first geodynamic model that *reconciled* both the foreign and in situ stories
was presented in section 1.6.2 of Smith (1988): the "foreign/in situ/foreign"
(FIF) model. This involved:

(1) The *first tectonic introduction* of eclogite slabs already equilibrated at ~30 kbar
    (the foreign CEP sub-terrane with stage I parageneses) into non-eclogitized inho-
    mogeneous continental crust.
(2) The subsequent subduction of this mixed crust to ~20kbar conditions which cre-
    ated new *in situ* eclogites from various basic rocks and which modified the crystal-
    chemistry and petrography of the CEP eclogites (the stage A readjustments, see
    above).
(3) A *second tectonic introduction* of parts of this unit into other never-eclogitized
    gneisses having ~10 kbar parageneses.

A variant of this FIF model, mentioned by Smith (1988, p. 131) and
adopted in Smith (1992), was that events (1) and (2) may have been contemporaneous, i. e. that the CEP subterrane was tectonically introduced into the
subducted crust *while* it was already undergoing 20 kbar metamorphism
rather than earlier at a higher level in the crust (i.e. P in the CEP rocks did
not descend below the P of stage A until after stage A). This still seems the
best overall interpretation today such that it is re-expressed in the conclusions
below along with some specific ages indicated.

It is interesting to note that Carswell et al. (1993) and Cong et al. (1993)
have deduced a very similar lithospheric interdigitation model involving the
existence of definite and deduced coesite, of UHPM in eclogites in a specific
distinct structural "block," of the lack of UHPM in adjacent rock units and,
hence, of the "tectonic juxtapositioning of rocks with different P-T histories"
in the Dabieshan CEP in China. Likewise a similar geodynamic scenario was
also suggested for the Su-Lu CEP in China by Smith and Yang (1991).

The Ultrametastability Model

Based largely on the work of Austrheim (1987, 1990) in the Holsnöy District
of the Bergen Arcs Anorthosites (BAA) who had argued for the metastable
preservation of "dry" undeformed granulite facies parageneses under eclogite
facies conditions, Andersen et al. (1991a,b) suggested that dense eclogites

formed at depths exceeding 100 km could be *educted* toward the surface by larger lighter masses of non-eclogitized continental crust, previously sucked downwards by the mass of the subjacent Thermal Boundary Layer (TBL) by asthenospheric convection but subsequently decoupled therefrom. Whereas the "in situ school" had previously argued strongly for a maximum P of ~20 kbar for the eclogites and gneisses and other rock types (numerous references above), in a few printed lines Andersen et al. (1991b) accepted *both* the in situ model and the ~30 kbar UHPM estimates of the present writer and coauthors. Thus without clearly spelling it out, they extended the in situ model to ~30 kbar, but not in the ultrareactivity manner as discussed above.

The author disagrees with the implicit assumption that most of the rock units in the WGR were subducted to ~30 kbar as this necessitates that all the eclogites and gneisses, not to mention the garnet-ultrabasites, metaanortho-sites, and metagabbros, etc., passed through stage I, although not all of them reacted according to their model: "It is quite obvious from petrographic observations that major parts of the orthogneisses, gabbros, and dolerites within the WGR never recrystallized to assemblages compatible with the *P-T* conditions inferred from the eclogites; they retained their low-density (-pressure) assemblage during crustal thickening" (Andersen et al., 1991b, p. 305), i.e, "to 100 km or more" (p. 303).

Smith (1993) was obliged to reject the idea that plagioclase can survive for more than a few hours at P conditions much above the maximum P stability for this Al^{iv}-rich phase as the crystal structure should rapidly collapse (the "ultrametastability" argument of Smith, 1993). Thus the geodynamic model of Andersen et al. (1991b) is also considered untenable, but not because of the eduction aspect, nor other structural aspects, which in fact constitute the main part of their paper. Note that this rejection does not mean that the writer also rejects the metastability aspect of Austrheim's (1987, 1990, 1991) model (see also Boundy et al., 1992), because that applied only to the situation in the BAA where there is no evidence of UHPM (P = 15–18 kbar) such that only a few kilobars of metastability are required, and this is considered here to lie within the limits of credibility. Andersen et al. (1991a,b) extrapolated the BAA model to the entire WGR, despite a great number of significant petrologic and structural differences, and added well over 10 kbar of extra metastability, an extreme notion which merits the prefix "ultra." Incidentally Andersen et al. (1991b) also made a number of useful criticisms of the pre-viously well-entrenched petrologic or geodynamic models published by Cuthbert et al. (1983) and Griffin (1987) such that it is evident that consider-able evolution of thinking among geologists based in Norway is at last under way.

Baltica, Laurentia or a Microcontinent

Gilotti et al. (1991), Gilotti and Hull (1991, 1993), Gilotti (1993), and Gilotti and Brueckner (1993) argued against the supposed westward-dipping subduction of WGR-bearing Baltica under Laurentia. This radical new approach was based partly on a reassessment of structural data in the "Central Norwegian Hinterland" (the Fosen Peninsula, N. W. of Trondheim) and partly on the discovery of a new large eclogite-bearing "province" in the Caledonides in the Skærfjorden Area of N. E. Greenland (around and North of Danmarkshavn, around latitude 77° N.). Although preliminary data suggested Early Proterozoic to Archean eclogite facies metamorphism (Gilotti et al., 1991), recent Sm-Nd data indicates Caledonian ages between 434 and 346 Ma (Gilotti and Brueckner, 1993) which led the latter authors to argue that this HP metamorphic event could not have happened at an inconceivably high structural level in the upper plate of the collision. Gilotti and her coauthors found it extremely difficult to consider that the new province once belonged to Baltica rather than to Laurentia. Gilotti and Hull (1991) boldly stated "The current orthodoxy of Baltica, including the WGR and Vestranden, underthrusting Laurentia must be abandoned" and suggested, that "Vestranden, and by implication the WGR, probably represent microcontinents or perhaps fragments of Laurentia."

It should be noted that the eclogite- and ultramafic-bearing Skaerfjorden province is much farther north than the eclogite- and ultramafic-bearing subterrane in Liverpool Land (east of Scoresby Sund, around latitude 71° N). This was considered to represent part of the WGR, and possibly also of the CEP and GUP provinces, detached by the opening of the Atlantic Ocean (Smith and Cheeney, 1980, 1981; Smith, 1988). There are so far no indications of UHPM in the "Skærfjorden Eclogite Province" (SEP) and no suggestions that it is directly related to the WGR. The SEP thus represents yet another *non-UHPM eclogite province* in the Caledonides distinct from the WGR, as is the case of all the following subterranes: the "Tromsö Nappe Complex" (TNC) in northern Norway (e.g., Krogh et al., 1982; Andresen et al., 1985); the "Seve Thrust Complex" (STC) in Sweden (e.g. Van Roermund, 1981; Dallmeyer and Gee, 1986; Mørk et al., 1988); and the "Bergen Arcs Anorthosites" (BAA) in S. W. Norway (e.g. Austrheim, 1987, 1990; Cohen et al., 1988) [which all indicate older ages of respectively ~600, ~500 and ~500 Ma]; another new eclogite province further south in eastern Greenland in the "Nagssugtoqidian mobile belt" (NAG) (Messiga et al., 1990); other eclogite or blueschist provinces in Vest Spitsbergen (e.g. Gee, 1966; Ohta et al., 1986); and units bearing garnet-pyroxene rocks in the "Vestranden Gneiss Complex" (VGC) (e.g. Johansson and Möller, 1986) and around Glomfjorden (e.g. Sörensen, 1955) [which both occur geographically between the WGR and the TNC in central Norway].

Whether the WGR belonged to Baltica or to Laurentia or was a *microcontinent* in between them is now a wide open question after a couple of decades of monotonous conventional wisdom; furthermore, the Vestlandet CEP and/ or GUP(s) may well have belonged to a *different continent* from that of the non-UHPM parts of the WGR. A new era of paleogeographic reconstructions may thus be expected as researchers attempt to fit together no less than at least 15 distinct eclogite- and/or garnet-pyroxene- and/or garnet-ultrabasite-bearing provinces (along with their coeval gneisses where appropriate, and additional, non-coeval, gneiss units) in the North Atlantic Caledonides [ignoring here such rocks occurring as far south as Scotland (e.g. Sanders, 1982), Ireland (e.g. Sanders et al., 1987) or Newfoundland (e.g. Church, 1968)]. By comparison with the Alpine Orogen which juxtaposed numerous subterranes from the African Plate, the European Plate and from ophiolite suites in between (often showing different ages of peak metamorphism), it would seem statistically unlikely that the mentioned 15 provinces in the Caledonide Orogen, or even only those existing today in mainland Norway, *all* originally belonged to the same lithospheric plate.

Definite Microcoesite Localities

The Grytting Dolomite-Eclogite Pod

The Grytting dolomite-coesite eclogite pod has a rather uninteresting mineralogic composition, except for the coesite. It occurs on the rocky coast at Grytting near Selje as a largely retrogressed amphibolitized pod with a relatively fresh eclogite paragenesis conserved only in the core where the first sample C51 was collected (Fig. 9.4). This sample, described in Smith (1984), is composed principally of millimeter-sized garnet and clinopyroxene with minor coesite, dolomite, rutile, ilmenite, zircon, and sulphide, all of presumed stage I age (or earlier for at least the zircon). Minor interstitial "early clinoamphibole" of stage A occurs. Traces of interstitial zoisite/clinozoisite also occur but cannot be convincingly attributed to stage I or A. Plagioclase + "late amphibole" symplectites of stage S replace garnet and clinopyroxene at grain boundaries and make up about 5 vol% of the rock.

Several definite coesite crystals up to 100 μm in length occur in each thin section near the center of 100–300-μm long SiO_2 inclusions *within clinopyroxene only*; this was a new phenomenon since the coesites discovered in Italy by Chopin (1984) occurred within garnet. The margins of the inclusions are always composed of strained polycrystalline quartz. "Polycrystalline," "multi-crystalline," and "monocrystalline" quartz, as defined in Smith (1984), occur in other inclusions in clinopyroxene and occasionally also within garnet. Most SiO_2 inclusions are the source of some cracks that radiate outward, often reaching the boundary of the host grain, but usually in an irregular fashion. Rarely concentric fractures may be observed (fig. 1a in Smith, 1984). It is most probable that these inclusions were all originally composed of coesite alone. Larger grains of quartz occur in the matrix and were most probably also origi-nally composed of coesite, but the greater possibility of intergranular rather than intragranular movement would have avoided the necessity of creating fractures due to the volume expansion of the coesite to quartz transition (Smith, 1984; cf. Chopin, 1984). Photomicrographs have been published in Boyer and Smith (1984), Boyer et al. (1985) and Smith (1984, 1988).

Stable isotope data on carbonates from several eclogites plot in the low $\delta^{13}C$ and low $\delta^{18}O$ field of "primary igneous carbonatites" (Agrinier et al., 1985), but the authors argued that these data were also explicable by extended sub-duction of, and CO_2 escape from, oceanic or continental crust carbonates; the subduction must have been considerable, but this conveniently fitted well with the presence of definite coesite in the Grytting sample C51 which gave very low isotopic values. A similar trend of both isotope factors decreasing with increasing subduction, but to a lesser extent, is recorded by Spencer and Spencer-Cervato (1993) for an eclogitized basaltic slab in the Himalayas.

It was formerly supposed that sample C51 represented a small 2-m-long pod separated by slivers of country-rock gneiss from a number of other similar-looking pods outcropping on the northern side, but it was difficult in the field to distinguish amphibolitized eclogite from amphibolite gneiss, all being largely covered by microvegetation. Subsequent field work (Smith, unpub. data) revealed the presence of definite coesite in several of the cores of relatively fresh eclogite in those other nearby pods outcropping within about 20 m from sample C51. It now seems more likely that in fact there is only one large coesite-eclogite pod that has been retrogressed not only at its outer margins but also along numerous fractures that divide it up into several apparently separate pods. Such retrogression macrotextures are very common in Norway (cf. Eskola, 1921; Lappin, 1966; Bryhni, 1966).

A very important feature of the Grytting locality is that this dolomite-coesite eclogite pod is separated by only a zone of gneissic material less than 10 cm wide from the better known magnesite-orthopyroxene eclogite pod with spectacular "pegmatitic" orthopyroxene described by Eskola (1921) and vis-ited by every eclogitologist in Norway. The two pods are completely different in almost every aspect of crystal-chemistry, rock chemistry and petrography.

Interestingly this pod was one of the critical localities from which Smith (1976a) and Lappin and Smith (1978) estimated *P* in the range 30–45 kbar based on orthopyroxene/garnet and clinopyroxene/magnesite relationships. Thus UHPM conditions were first deduced from one pod and then confirmed by the adjacent one many years later. Since the beautiful orthopyroxene pod has been considerably damaged and depleted by rock hunters, it was classed as a national protected site in the early 1980s. Fortuitously one of the posts marking the limits of the site was drilled into C51's pod on the outer side (Fig. 9.4) such that the first locality of coesite in Norway is already protected, but for a different reason!

The Straumen Eclogite Pod

The Straumen coesite-eclogite pod, found by Lappin (1966), occurs on the hillside East of Sörpollen fjord overlooking the Nybö eclogite pod, famous for being the type locality of nyböite (Ungaretti et al., 1980, 1981) and also of mid-solvus omphacite-jadeite (Smith et al., 1980; Carpenter and Smith, 1981; Rossi et al., 1983a,b). Nyböite may in fact be the only clinoamphibole stable in the coesite-eclogite (sub)facies (see above), but it requires a Na-rich bulk-rock composition which is neither the case at Grytting nor at Straumen. In contrast to the Grytting dolomite-coesite-eclogite pod, that at Straumen contains coesite within polycrystalline quartz in SiO_2 inclusions *within garnet only*. Poly-, multi-, and mono-crystalline quartz inclusions again occur, sometimes also within clinopyroxene or kyanite. Otherwise the microcoesite/microquartz textural relationships are essentially identical to those at Grytting although the coesites at Straumen tend to be better preserved (see the photomicrographs published in Smith, 1988, and Smith and Lappin, 1989).

The significance of the clinopyroxene vs. garnet host is probably related to the $Ab/(Ab + An)$ proportion of the plagioclase in the gabbroic protolith of the eclogite, less *Ab* promoting garnet growth at the plagioclase site to trap the excess SiO_2 within garnet and more *Ab* promoting clinopyroxene, as argued by Smith and Lappin (1989, p. 52).

The Straumen eclogite pod is remarkable for three additional reasons. First, sample 62/62b is totally fresh with no trace of retrogression. Second, it has a great number of mineral phases in apparent equilibrium: garnet, clinopyroxene, dolomite, calcite, quartz, kyanite, phengite, zoisite/clinozoisite, early clinoamphibole, and rutile. All minerals except quartz and early amphibole are attributed to stage I, but they persisted during stage A when the coesite transformed to quartz and the early amphibole grew and narrow rims of new dolomite surrounded stage I dolomite (as explained in detail in Lappin and Smith,

1981). Third, it is the only known eclogite with calcite of stage I vintage, and this phase is surrounded by new calcite during stage A.

Deduced Microcoesite Localities

The Drage Eclogite Pod

This pod, first recorded by Lappin (1966) on the western coast of Stadlandet North of Selje (Fig. 9.2), contains the same nine stage A mineral phases as that at Straumen (except for calcite), but not all occur in the same thin section. Sample 168 contains excellent polycrystalline quartz textures in SiO_2 inclusions in garnet (Smith, 1988), sometimes adjacent to a clinopyroxene grain in a two-phase microinclusion. One of the textures was considered as the type example of polycrystalline quartz after coesite (plate 1.3a in Smith, 1988) and may be similarly considered as the type example of the former three-phase assemblage: garnet + clinopyroxene + coesite.

The Liset Eclogite Pod

This extraordinary kyanite-lineage pod, discovered by Smith (1968) on the coast between Grytting and Drage (Fig. 2) and noted for the first known occurrence worldwide of F-bearing high-Al titanite (Smith, 1976a, 1977, 1980b, 1981a; Smith and Lappin, 1982; Oberti et al., 1985, 1991), was the object of the entire Ph.D. thesis of Kechid (1984). That work, as well as Kechid and Smith (1985), Smith (1988), and continued studies by the present author revealed data unknown elsewhere, such as evidence of:

(1) *Two* tectonic introductions.
(2) *Syn-eclogite deformation and boudinage* of garnet-rich + kyanite + zoisite layers within clinopyroxene-rich + quartz layers.
(3) *Possibly lower T* than the other kyanite eclogites in the district.
(4) *Extreme oxidation in a static fashion* that promoted the symplectitization of stage I or stage A grains while perfectly conserving their original grain boundaries in pseudomorph fashion.

Point (4) significantly indicates that the exhumation process did not cause any deformation-induced recrystallization within the pod.

However, this pod is even more interesting because of the unusually large number of exceptional mineral compositions that occur, in addition to the Al-rich titanite. These include the so far unique minerals lisetite and K-poor "Ca-nepheline" (Rossi et al., 1986, 1989; Smith et al., 1986), Fe-Mn-geikielite and Fe-Mg-pyrophanite (Pinet and Smith, 1985; Smith and Pinet, 1985), as well as "plate corundum" and Mg-rich kandite inclusions within rutile inclusions within kyanite (Smith, 1988), pure paragonite and Mg-rich staurolite inclusions within garnet (as in the Aarsheimneset headland eclogite pod), mid-solvus omphacite-jadeite (as in the Nybö eclogite pod), jadeite + quartz, "bead corundum" + plagioclase after paragonite laths themselves after jadeite + kyanite, abundant allanite with zoisite/clinozoisite rims, nyböite and Mg-Al-taramite (as in the Nybö eclogite pod) (Kechid and Smith, 1982; Oberti et al., 1989), Fe-Al-taramite (Smith and Lappin, 1982; Ungaretti et al., 1985), Ca-Fe-Mg-rich paragonite after corundum + Ab-rich plagioclase, Na-Fe-Mg-rich margarite after corundum + An-rich plagioclase, "Na-eastonite" after garnet (Tlili et al., 1989), preiswerkite (Oberti et al., 1993; Smith and Kechid, 1983), "film hematite" lamellae in rutile, and Mg-Mn-ilmenite, all in addition to the more usual epidote, aegirine, aegirine-augite, hematite, "blitz rutile," and of course plagioclase and other "late clinoamphiboles." The aegirine or aegirine-augite always occurs around quartz whether occurring in the matrix or as inclusions in other phases and has been deduced to be the product of Na^+ migrating from the breakdown of jadeite, Fe^{2+} migrating from the breakdown of garnet, and oxidation which is endemic in this pod during stage S. Many of the mineralogic features of this pod are primarily due to the extreme Na-Al-F-rich, K-Mg-poor local bulk-rock compositions rather than to extreme *P-T* conditions; for this reason, several of them have been found in the Jianchang eclogite pod in China which has similar chemical compositions and deduced *P-T* history (Smith et al., 1990). Most are thus of stage S vintage and are thus irrelevant to UHPM but this is not the case for all, in particular the deduced coesite, F-bearing nyböite and jadeite (all also at Jianchang).

Deduced coesite in this pod is represented by typical polycrystalline quartz enclosed within garnet showing radial cracks (plates 1.3e,f in Smith, 1988). A unique feature is the additional presence of a narrow rim of aegirine around the polycrystalline quartz but this fits perfectly with the petrologic evolution of this pod. Strange "spherules" composed of albitic plagioclase, magnetite, and clinoamphibole occurring within pseudomorphed symplectitized clinopyroxene grains were very tentatively attributed to the former presence of coesite (plate 1.12c in Smith, 1988).

The Aarsheimneset Headland Eclogite Pod

This pod, described by Lappin (1966), occurs on the small headland south of Sandvik on the east coast of Stadlandet (Fig. 9.2). It is famous as the locality of the coarse lamellae of garnet occurring within large, ~10-cm-thick, ortho-pyroxenes and deduced to represent exsolution from a previous Al-richer orthopyroxene (Smith, 1976a; Lappin and Smith, 1978). Lappin and Smith (1978) also recorded evidence of orthopyroxene exsolution from clinopyro-xene here. These data contributed to the recognition of the pre-exsolution stage X. A review of the data and discussion over whether this really was an exsolution phenomenon rather than one of parallel growth and over the *P-T-t* deductions is given in Smith (1988, p. 34, 68 and 93). This is one of the two orthopyroxene-lineage pods amongst all those pods exhibiting definite or deduced coesites and for this reason it is important as a link connecting the evolution of the two lineages, as suggested by Lappin and Smith (1978, 1981).

Inclusions of pure paragonite, Na-rich gedrite, tschermakite, "Na-eastonite," K-eastonite, and Mg-rich staurolite occur in garnet and have been tentatively attributed to stage Y (see above). The orthopyroxene coexisting with garnet is low in Al (0.30–0.49 wt % Al_2O_3) which contributed to the estimation of extremely high P beyond the scale of all available calibrations, but several other pods have yet lower Al contents (Lappin and Smith, 1978). Phlogopite occurs as does "early amphibole" and cummingtonite/actinolite pairs. Mg-ferri-winchite occurs rarely in symplectites with aegirine and albite in a cross-cutting vein deduced to be a retrogressed previous omphacite + quartz vein (Smith, 1982d). Superb exsolution textures among the Fe-Ti-oxides are displayed and indicate more than one petrologic event (Pinet and Smith, 1985; Smith and Pinet, 1985).

Excellent polycrystalline and multicrystalline quartz textures in SiO_2 inclu-sions within garnet are displayed as well as some two-phase inclusions, as at Drage, one photomicrograph (plate 1.3d in Smith, 1988) having been selected as the type example of multicrystalline quartz. This is one of the eclogite pods in which other workers completely independent of the author's research team have also observed deduced coesite textures in their own rock samples (D.Waters, personal communication).

The Ulstein-Dimnöy Eclogite Pod

This pod is famous for being one of the largest in the WGR, some 6 km long (Mysen and Heier, 1971, 1972). It lies further north in the Vartdalsfjorden District (Fig. 9.1). It is generally strongly retrogressed and has a rather simple mineralogy. Sample A2 from Dimna contains only garnet, clinopyroxene, dolomite, quartz, early amphibole and rutile, and much fine symplectite. A neighboring sample (A3) contains zoisite and biotite + plagioclase symplec-tite after phengite. The dolomite gave different isotopic values of $\delta^{13}C$ and

$\partial^{18}O$ compared to the Grytting eclogite pod (Agrinier et al., 1985). In Lappin and Smith (1981) all the principal mineral phases were attributed to stage A since there was no specific evidence available then for the higher *P* of stage I.

Polycrystalline, multicrystalline, and monocrystalline quartz occur in SiO_2 inclusions in garnet which shows some radial fractures (Plates 1.3 g,h in Smith, 1988) and thus provide evidence for UHPM in the form of deduced coesite, as in the other pods mentioned above.

This locality was the first, and still is the best, for evidence of former omphacite in the adjacent gneisses, now replaced by plagioclase + diopside symplectites (Mysen and Heier, 1972). This has been widely cited as evidence of compatible *P-T* conditions in the eclogites and in the *country-rock gneisses*, in particular abundant migmatitic ones (innumerable references, e. g. Griffin, 1987), but this interpretation has been highly controversial. Smith (1980a, 1988, p. 118) argued for equivalent *P* (~20 kbar of stage A) in the eclogites [a Late Caledonian age] and *only* in those *"garnet-biotite-gneisses"* [age unknown] adjacent to the eclogite body but not in the migmatitic *"dioritic gneisses"* [a Precambrian Rb-Sr age] which constitute the country-rocks and into which the combined basic eclogite + "acid eclogite" sheath ("garnet-biotite-gneiss") were tectonically introduced as a single but composite foreign body. It has now become more probable that even stage I pertained in this gneissic sheath (see discussion below concerning the Fjörtoft locality). However, more detailed field work is necessary as the writer's preliminary observations suggest that the geologic relationships are more complex than indicated in the original map of a single eclogite within a single sheath of garnet-biotite gneiss.

The Hessdalen Eclogite / Garnet-Ultrabasite / Metaanorthosite Complex

This interesting complex found by Smith (1976), which includes quartz-bearing orthopyroxene-eclogites which are uncommon elsewhere, shares many similarities with the eclogite/garnet-ultrabasite complex at Eiksunddalen (Schmitt, 1960, 1964) on the opposite side of Vartdalsfjorden (Fig. 9.1). These complexes were the object of the theses of respectively Erambert (1985) and Jamtveit (1986). In sample C254, Smith (1984) found multicrystalline quartz enclosed within clinopyroxene showing many short radial cracks, and also excellent needles of quartz exsolved from a previous supersilicic clinopyroxene, and deduced the former presence of coesite (see also plate 1.2d in Smith, 1988). This sample also contains low-Al orthopyroxene, phlogopite, and "early clinoamphibole" as is typical of orthopyroxene-lineage eclogites (cf. at Aarsheimneset headland, see above). Unfortunately sample C254 came from a loose block, and Erambert (1985) did not record similar textures from her new rock collection. However, she did confirm the low-Al nature of the orthopyroxenes coexisting with garnet and supported the very high P estimates previously deduced in the Selje District.

The Fjörtoft Garnet-Biotite-Gneisses

Quartz inclusions in garnet surrounded by radial fractures that could have been produced by the coesite-to-quartz transition are recorded by Dobrzhinetskaya et al. (1993a) from the island of Fjörtoft (at the top of Fig. 9.1). Significantly, these textures have not been recorded in eclogite, which is said to outcrop nearby, but in kyanite and/or hornblende-bearing *garnet-biotite gneiss*.

In his conclusions, Smith (1988, p. 139) declared that the CEP "excludes most, if not all, of the intervening country-rock gneisses." In other words, stage I coesite-eclogite (sub)facies assemblages were not expected to have occurred in the bulk of the gneisses but could have occurred in *small* volumes of gneissic material that had been introduced along with and adjacent to some CEP eclogites, in contrast to larger volumes of gneiss which had experienced only stage A. The former presence of stage A quartz-eclogite (sub)facies assemblages had already been recognized in certain gneisses, significantly *"garnet-biotite-gneiss"* at Ulstein-Dimnöy, locally *kyanite-bearing*, presumably formerly subducted argillaceous sediments, but cofacial in the form of quartz-phengite-"acid" eclogite at least at the ~20 kbar P conditions of stage A in the Ulstein-Dimnöy eclogite (Smith, 1980a, 1988, pp. 118, 126). Since the Ulstein-Dimnöy eclogite also contains "deduced coesite" (see above), the probability that the adjacent garnet-biotite-gneiss had also passed previously through stage I increased such that the present author considered that if any gneisses had been subjected to UHPM conditions, then garnet-biotite-gneiss, especially kyanite-bearing, was the most likely candidate. This opinion is now reinforced by the data from Fjörtoft, although it cannot be *proved* that the SiO$_2$ inclusion textures were due only to the previous presence of coesite rather than to a P-T dependent compression/expansion phenomenon. The problem of deducing former coesite or not when the evidence available consists only of unsymmetrically strained multicrystalline quartz and uneven radial fractures, instead of more convincing radially orientated polycrystalline quartz and regular radial fractures, was addressed by Smith (1984) and Smith and Lappin (1989). Effectively there is no solution and one must rely upon other petrologic features. If none are available, then the risks are either (1) of accepting "deduced coesite" and hence possibly incorrectly deducing a nonexistent event, or (2) of rejecting it and hence possibly missing a key geological event.

Following the data from Ulstein-Dimnöy and the new discoveries on Fjörtoft island, the present author now considers that *greater volumes of gneiss* did pass through stage I such that not only mineralogic relics of stage A, but also of stage I, should occur in various parts of the WGR and are only awaiting their discovery (Fig. 9.5). The CEP's definition is thus enlarged here to include at least some substantial parts of the gneisses; the mentioned garnet-biotite-gneisses on Fjörtoft island, and in particular the kyanite-bearing ones, are thus tentatively attributed to the CEP. Hence the infolded sub-terrane of the CEP, which was previously recognized in the southern (Selje District) and central (Vartdalsfjorden District) coastal parts of Fig. 9.1 and suspected in the northern part (Moldefjorden District) on the basis of the presence of some orthopyroxene-eclogites there, is now more firmly recognized in the northern part. Note that these three districts all also display garnet-ultrabasite localities of both of the Mg-Cr and Fe-Ti GUPs (Fig. 9.1).

Microdiamond Localities

Since P_{max} conditions in the CEP cannot be constrained at less than about 50 kbar, owing to the absence of the beginning of pyroxene solution in garnet (i.e. some octahedral Si), then the presence of diamond in the CEP would not be surprising. Smith (1988, pp. 92–93) suggested that the chances of finding diamond in Norway were low but "not nil"; this was deduced on the basis of the frequent presence of carbonates (e.g. Lappin and Smith, 1981) which suggested that fo_2 may have been too high, i. e., *not* because of P being insufficiently high. However, in different bulk rocks, this may not have been the case. Since microdiamonds have been discovered included in garnet or in zircon in UHPM rocks in Kazakhstan (e.g. Sobolev and Shatsky, 1987, 1990; Sobolev et al., 1991, 1994) and China (Xu et al., 1992; Okay et al., 1993), the chances of finding microdiamond in the Norwegian CEP have increased.

It is thus particularly relevant that Dobrzhinetskaya et al. (1993a) now report the finding of 20–50-μm-sized microdiamonds of two morphologic types in kyanite-garnet-biotite-gneisses and hornblende-garnet-biotite-gneisses close to eclogites on Fjörtoft island in Norway (Fig. 9.1) from *the same garnet-biotite gneiss* locality where the deduced microcoesite textures were observed (see the preceding section). Superficially, attribution to the UHPM CEP would seem rather obvious. However, there are real possibilities (1) that the occurrence of deduced coesite microtextures and microdiamonds

```
                                    Pmax   stages   critical
                                    kBar   passed   minerals
                                    ----   ------   --------
...............................  )
...............................  )                 PL   amphibolite-facies
...............................  ) ~ 10     S       QZ   country-rock
...............................  )                 CA   gneisses
...............................  )
###############################  foreign relation
.........o*********************
...........o******************** ) ~ 30    I>A>S    CS   coesite-eclogite
..................o************* )                       (sub)facies
.......................o*******  )                       eclogites
...........................o**** )
..........................o      in situ relation
...............................  )
...............................  )                      coesite-eclogite
...............................  ) ~ 30    I>A>S    CS   (sub)facies
...............................  )                      gneisses
...............................  )
..........................#      foreign relation
...............................  )
++++++++++++++++++.    #........ )                 PL   amphibolite-facies
...................   #........  ) ~ 10     S       QZ   country-rock
..................   #........   )                 CA   gneisses
.................#               foreign relation
...............................  )
...............................  )                      quartz-eclogite
...............................  ) ~ 20     A>S      QZ  (sub)facies
...............................  )                 CA   gneisses
.................o               in situ relation
...............o*******
.............o*********           ) ~ 20    A>S      QZ  quartz-eclogite
..........o**************         )                      (sub)facies
........o*****************         )                 CA  eclogites
###############################  foreign relation
...............................  )
...............................  )                 PL   amphibolite-facies
...............................  ) ~ 10     S       QZ   country-rock
...............................  )                 CA   gneisses
###############################  foreign relation
...............................  )
...............................  )                 PL   a different unit
...............................  ) ~ 10     S       QZ   of country-rock
...............................  )                 CA   gneisses
...............................  )


        ....   acid          ****   basic
        ....  = rock         ****  = rock
        ....   material      ****   material

   oooooo = non-tectonic contact since before stage I, e.g. igneous intrusion
   ++++++ = tectonic introduction of stage I rocks into stage A rocks
   ###### = tectonic introduction of stage I, A or S rocks into stage S rocks
```

Figure 9.5. Schematic field map corresponding to the updated FIF petrologic model displaying the various different kinds of geologic contacts that would exist, especially *between gneiss units* of very different *P-T-X-t-s* history, but which are often very difficult to recognize in the field, having been largely welded together. CA = clinoamphibole, CS = coesite, PL = plagioclase, QZ = quartz. Note that although attributed to stage S (equivalent to the symplectitic amphibolite-facies retrogression in the eclogites), the never-eclogitized country-rock gneisses would not display symplectitic textures. The UHPM CEP clearly applies to the *combined* acid + basic units attributed to the coesite-eclogite (sub)facies. The DGP and the two GUP provinces, as well as the "mangerites," metaanorthosites, metagranites, metagabbros, metadolerites, and marbles, are not included here for the sake of clarity, but they exist and may or may not constitute different tectonic provinces. The previous single regional coherent body model of the "in situ" school, whereby the whole WGR was subducted to ~20 kbar conditions, allowed only the existence of the two quartz-eclogite (sub)facies units (labelled ~20), i.e., neither the CEP nor the never-eclogitized gneisses (labeled ~30 and ~10 respectively) existed.

in the same rock unit is co-incidental, or (2) the radial fractures around quartz were not really due to preexisting coesite, or (3) the microdiamonds were not really due to UHPM (see below); hence such an attribution cannot be made with confidence at the present time.

Dobrzhinetskaya et al. (1993a) record that their microdiamonds were observed after thermochemical dissolution of rock samples under clean room conditions. RMP, XRD and SEM techiques were employed to confirm the diamond crystal structure and to discern an "inner mosaic texture" which was tentatively attributed to "anisotropic stress conditions" and hence not necessarily to the same *P-T* conditions as those that generated the deduced microcoesites. If, on the other hand, there were no problems concerning the equilibrium growth of the diamond structure in these rocks, then the *higher P part* of the 30–45 kbar estimates of Smith (1976a), and Lappin and Smith (1978) would have been closer to the truth (Fig. 9.3).

Further work is necessary to try to clarify the distributions of microdiamonds and microcoesites in the different rock types on Fjörtoft island. Such work might reveal the first direct link in Norway between microcoesite and microdiamond (already indicated in Kazakhstan and China), as well as establish more solidly the first case of deduced microcoesite within gneiss in Norway and the first case of microdiamond in Norway, all of which of course are of utmost importance to any geodynamic modeling. Since kyanite-bearing garnet-biotite-gneiss provides a favored target for finding new occurrences of microcoesites in Norway (for the reasons noted above) and since microdiamonds have been found in the same rock type, then new searches for microdiamonds could be focused on this rock type; this does not, however, exclude the possibility that other rock types contain these rather special microinclusions.

In other works Dobrzhinetskaya and Molchanova (1992) and Dobrzhinetskaya et al. (1993b, 1994) have presented important new data, very relevant to the present discussion, on the microdiamond occurrences in the Kokchetav massif in Kazakhstan. These were first recorded by Rozen et al. (1972, 1979) and Rozen (1973) and in several unpublished reports kept secret because of the strategic interest of a potentially economic ore deposit, and were later described in more detail by others (e.g. Letnikov, 1983; Sobolev and Shatsky 1987, 1990; Nadezhdina and Posukhova, 1990; Lavrova, 1991; Ekimova et al., 1992). In part following Letnikov (1983), Nadezhdina and Posukhova (1990), and Lavrova (1991), Dobrzhinetskaya et al. (1993b) raise some doubts about the UHPM nature of the crystallization of some, if not all, of the microdiamonds, as metastable growth at lower *P* in later shear zones is favored by the latter authors for at least one kind of microdiamond occurrence in the

Kokchetav massif. If this is the case in Kazakhstan, then clearly it could also be the case in Norway such that the microcoesite and microdiamond occurrences may well have been dissociated in their *P-T-t* histories, as has already been deduced for the *P-T-t* histories of the Norwegian CEP and Mg-Cr GUP provinces (see above).

It is the opinion of the present author that four strong points in the works of Dobrzhinetskaya et al. (1993b, 1994) are their statements (1) that the "Kumdy Kol diamond province" is situated "*within a tectonic melange zone*," (2) that the high concentrations of microdiamonds are associated with post-eclogite *shear zones* that cross-cut eclogites and gneisses of the Kokchetav Massif, (3) that microdiamonds *occur within many non-UHPM mineral phases* including diopside, phlogopite, biotite, quartz or calcite-chlorite-sericite aggregates (cf. Lavrova, 1991; Ekimova et al., 1992), and (4) that *overpressure* may have pertained during microdiamond growth; however, their hypothesis, following Letnikov (1983) and Nadezhdina and Posukhova (1990) suggesting the *metastable growth* of the microdiamonds at relatively low *P*, is considered here to have a less solid basis.

Apart from the uncomfortable contamination model, which it is effectively impossible to exclude with 100% certainty (cf. Bovenkerk et al., 1993, who describe the contamination of a diamond synthesis run by a natural diamond) unless the microdiamonds are observed *enclosed within a key paragenetic mineral* or unless the microdiamonds possess *specific distinguishing properties*, four fundamentally different petrologic/geodynamic models for the existence of microdiamonds in Norway, Kazakhstan, China and elsewhere can be formulated and summarized as follows, along with several variants:

1. The UHPM model:
    (1a) The microdiamonds (within eclogites, but possibly also within gneisses) *grew stably* within their equilibrium stability field and *survived metastably* while their UHPM *host rocks* reacted *partially* leaving typical traces of mineralogical evidence of UHPM (cf. Sobolev and Shatsky, 1987, 1990, and Sobolev et al., 1991, 1994, for Kazakhstan).
    (1b) The microdiamonds (within gneisses, but possibly also within eclogites) *grew stably* within their equilibrium stability field and *survived metastably* while their UHPM *host rocks* reacted *totally* to annihilate all evidence of UHPM [this is a mind-boggling, but not inconceivable, combination of the ultrametastability model (for the diamonds) and the ultrareactivity model (for the gneiss) of Smith, 1993].
2. The overpressure + local UHPM model: The microdiamonds *grew stably* owing to some form of *overpressure* which *locally* increased P-$_{effective}$ (or at least the principal stress) in their eclogite or gneissic host rocks such that it exceeded

the lower *P* limit of the equilibrium transformation of graphite into diamond, whereas the surrounding rocks were at P_{load} *less than this limit* (cf. the ultra-overpressure model of Smith, 1993).

3. The metastability model: The microdiamonds *grew metastably* below their equilibrium *P* during a (3a) pre-, (3b) syn-, or (3c) post-peak metamorphic stage and hence do not indicate former UHPM in their eclogite or gneissic host rocks [cf. Dobrzhinetskaya et al., 1993b, 1994, for Kazakhstan (model 3c)].

4. The premetamorphic relicts model: The microdiamonds are *relicts from a premetamorphic stage* which survived subduction and/or plate collision and substantial metamorphism (not necessarily with *P* as high as UHPM), such as:

(4a) Residual diamonds from *sedimentary protoliths* of some gneisses [cf. *alluvial diamonds* recorded in Norway by Velain (1891), and Wegmann (1929), and elsewhere by many others, or *carbonados* in gravels described by Trueb and Butterman (1969)], or

(4b) Residual diamonds from *igneous protoliths* of some eclogites (e.g., UHPM nodules in *kimberlite* or *lamproite*), or

(4c) Residual diamonds from *metamorphic protoliths*, i. e., formed during a previous metamorphic event, or

(4d) Residual extraterrestrial microdiamonds from *meteorite falls* or *meteoritic impact craters*.

Some of these models are less credible than others: for example, microdiamonds, which are apparently rather abundant in interstellar dust, have only nanometric sizes as well as several distinguishing isotopic and other chemical features in meteorites (see the review by Anders and Zinner, 1993); despite these critical points model 4d is not eliminated from this text since the mentioned distinguishing features may be modified by metamorphism, and since an attempt is made here to provide a comprehensive list of possibilities.

As research work is pursued, the increasing data set will serve to narrow down the choice among the possible models, but the single real one at each locality may remain elusive for some time. At the moment the interpretation of the origin of the microdiamond occurrences at Fjörtoft is still wide open, and it is too early to formulate solid conclusions; models 1b and 2 are presently favored by the writer (i.e., stable diamond formation during the UHPM metamorphism) but models 4a and 4c are certainly not excluded. Models 3 and 4d are considered the least credible, whereas models 1a and 4b are inappropriate for the gneiss host rocks.

In the meantime it nevertheless seems appropriate to consider the two nearby microdiamond localities as landmarks of an embryonic "Diamond Gneiss Province" (DGP) constituting *yet another subterrane* within the WGR distinct from other gneisses in which microdiamond probably does not occur.

Note that at this time the DGP is intended as a petrographic rather than petrogenetic term and is thus independent of the origin, UHPM or not, of its microdiamonds.

Geodynamic Models for the Creation and Exhumation of UHPM Terranes

Concerning the WGR eclogites in particular, some published suggestions of geodynamic mechanisms to explain their return to the surface (e.g. successive underthrusting, eduction) have been briefly mentioned above. Concerning the exhumation of HPM or UHPM rocks in general, Platt (1993) presents a useful, serious review of possible concepts and processes including expulsion, extrusion, strike-slip faulting, underthrusting, underplating, frontal accretion, corner flow, buoyancy, diapirism, extensional tectonics and erosion. His first conclusion is unequivocal: "There is no single mechanism that can explain the exhumation of all high-pressure terrains." However, from the tone of his paper, it seems that even a combination of some of the mentioned processes is inadequate, at least in many natural examples of presently exposed HPM or UHPM rocks. Thus a global conclusion is that the essential geodynamic process(es) of exhumation is (are) still unknown.

It is, however, relevant to note that the review did not include any mention of possible overpressure ($P_{effective} > P_{load}$), presumably because J. P. Platt is one of many respected structural geologists who cannot entertain such a heresy. Likewise the review failed to mention the extensive suction and convection or "advection" model of Perchuk et al. (1992) [referred to as the credibility-stretching "ultrasuction" and "ultraconvection" models in Smith (1993), by which coesite-eclogites and enveloping granitic gneisses could be brought up from some 200-km depth by "huge rising diapirs causing circulation"]. Again the review did not discuss models involving variable terrestrial dimensions and speed of rotation [e.g., the "rhythmic tectonism" or "helicyclic tectonics" of Wezel (1988) discussed in Smith (1989)]. Clearly J. P. Platt was constraining his review to lie within the narrow limits of the now traditional Plate Tectonic Theory. The validity of his approach may be judged in relation to the comments "Is it reasonable to consider that the geodynamic process(es) responsible for the generation and preservation of coesite in eclogites in crustal metamorphic environments are not part of the data bank of presently-accepted processes." (Smith, 1985b); and "Perhaps they are all wrong such that we should deduce that the real major petrological/geodynamic processes involved do not lie within the realm of already-suggested models" (Smith, 1993).

Based upon a wide range of phenomena (including the apparent lack of overburden appropriate for the formation of blueschists or UHPM rocks; inverted pressure gradients deduced in layered metamorphic sequences; over-pressure or underpressure recorded in boreholes and above tunnels; different mineral assemblages observed in the hinges and limbs of folds; rock fractures, brecciation and shear zones; overpressure between opposing pistons deter-mined in the laboratory), some old ideas on overpressure (e.g. Sobolev, 1960; Dobretsov, 1964; de Roever, 1967; Coleman, 1972) have frequently been resur-rected, reviewed and reformulated by Smith (e.g. 1983; 1984; 1985b; 1988, pp. 133–136; 1989) in terms of expressions such as: "stress focalization," "dynamic excess pressure," and "small continuous pressure gradients" which lead to "macroscale local or regional dynamically-stabilized non-lithostatic tectonic overpressure," and hence to *"inhomogeneously-distributed isobars."* Rock viscosity and internal and external friction were considered to be critical factors whose significance has been underestimated, and Smith (1984) con-cluded, "We are thus not constrained to assume that eclogites were in general subjected to homogeneous lithostatic pressure."

Furthermore a discussion of *intracrystalline overpressure and underpressure* measured in situ in solid and fluid inclusions in minerals by spectroscopy or deduced by mineral/fluid equilibria led Smith (1988, pp. 136–138) to conclude that "one can no longer assume that the P_{ce} inside mineral inclusions or the $P_{hydrostatic}$ inside fluid inclusions within natural minerals are/were at the same P_{ce} as at the outer margin of their host crystal, either today in the field or laboratory, or in the past at various points along the *P-T* trajectory of their host rock" [$P_{ce} = P_{crystal-effective}$] such that *P-T* estimation for rocks based upon inclusions in their minerals is a risky enterprise. Another confirmation of this is provided by Schrauder and Navon (1993) who deduce 50-kbar intracrystal-line overpressure today inside a solid CO_2 inclusion within a diamond.

Another argument, simple but strong, is added here: Atmospheric wind and ocean currents ("hydrospheric wind") are due principally to the combination of three physical processes:

1. Terrestrial, lunar and solar *gravitation*
2. Terrestrial *rotation* on axes involving the Earth, Moon and Sun
3. *Heat exchange* between and along different terrestrial layers

The same three processes apply to the lithosphere and the asthenosphere such that "lithospheric wind" and "asthenospheric wind" should exist also. These internal winds in the crust and upper mantle would of course be very, very slow and very, very strong by comparison with the external winds in the air and the sea because of the great differences in densities and viscosities involved; when someone eventually attempts to calculate them [not on the standard basis of varying densities, viscosities and temperatures with $P_{effective}$ constrained as being proportional to depth (i.e. $= P_{load}$), but with $P_{effective}$ allowed to vary as well] they might turn out to be of an appropriate order of magnitude for creating some 10 kbar or more of overpressure in natural rocks, sufficient to trigger certain mineral reactions which subsequently might remain metastable. Also such winds, along with wavefronts, anticyclones and other manifestations of Chaos Theory, may well provide a powerful new mechanism for the rapid exhumation of dense rock units. In this connection, is it just a coincidence that many UHPM or HPM rocks occur in sheared zones displaying *vertical or subvertical lineations/foliations*: e. g. at the Grytting and Aarsheimneset CEP localities (Lappin, 1966); the Tafjord Mg-Cr GUP locality (O'Hara and Mercy, 1963); and the Almklovdalen Mg-Cr GUP locality (Lappin, 1966)? Many field relations are thus compatible with the idea of simple subvertical solid intrusion/extrusion of "foreign" tectonic slices, although this flies in the face of conventional rock mechanics and plate tectonics. A source of overpressure, such as wind in the mantle, could be all that is necessary to exhume rapidly, certain slabs of rock. Of course overpressure alone may not be the key mechanism for exhumation, but it would certainly *modify* the behavior of rocks and hence of other geodynamic mechanisms, as well as *require* less depth of retrieval in the first place.

The concept of overpressure seems to be well accepted by some mechanical engineers, engineering geologists, and solid-state chemists and physicists, but the subject is still asphyxiatingly taboo among most structural geologists and petrologists, at least in "Western" countries (usually it is not merely dismissed, it is simply not even considered). Those textbooks on petrology that do mention overpressure usually simply *declare* that it cannot exceed more than a few tens or hundreds of bars (even modern books like the monumental work of Spear, 1993, pp. 9–10) and/or merely cite refutations of greater overpressure by, for example Rutland (1965) and Brace et al. (1970), whereas challenges to those refutations are bypassed without comment. The rejection of significant overpressure, and of the concomitant necessary direct relationship between $P_{effective}$ and depth, gravity, and average density (as conventionally plotted on *P-T-depth* diagrams, has, for the reasons assembled above, long been considered by the present author as constituting the greatest geologic misconception of this century [cf. Smith (1989): "An important result might be that horizontal stress has more influence than vertical stress such that depth would not be the critical factor for eclogite genesis."] In the former USSR there exists a considerable literature which for several decades has challenged the conventional relationship between $P_{effective}$ and depth, but the articles are often inaccessible and are rarely cited in the West; one example is Milanovsky (1983) who reviews ideas on the Earth's expansion and pulsation that affect the magnitude of the gravitational force, and hence of $P_{effective}$, at a specified depth at a specific time.

It is thus of great relevance to record here that another respected structural geologist (Mancktelow, 1993) has just published detailed calculations confirming the high probability of the existence of significant overpressure in Nature which contributes toward a stronger justification of some of the above-mentioned deductions/predictions/assertions. It is hoped that this will serve to stimulate deeper reflections on the part of previous skeptics and provoke reevaluations of many geodynamic processes. The main theme of Mancktelow's paper is that during "layer-parallel shortening," the difference in competence between adjacent rock layers can lead to deviatoric stresses of up to 8 kbar in magnitude at $T < = 550°C$ which "may" explain the origin of "low-T" higher-P eclogite pods surrounded by lower-P gneisses. A few selected citations demonstrate the importance of his work: "The basic argument that variation in deviatoric stress states related to the heterogeneity of the lithosphere should produce variation in the effective pressure has never been refuted"; "the prevalent assumption that the metamorphic pressure reflects depth alone cannot always be justified"; "isolated strong layers can therefore be subjected to much higher stress conditions than that acting on the bulk material"; "the stress magnitude may also be strongly scale-dependent"; "overpressure within these limb regions . . ." [of folds]; ". . . tectonic underpressure . . ." Furthermore, on a related theme, *regional tectonic overpressure in areas of extrusion*, Mancktelow also makes some comments: "More regional nonlithostatic gradients in pressure may also accompany the extrusion of continental material between more rigid indenters, whether the extrusion is upwards in the case of flower structures in transpressive zones, or sideways during regional-scale continental escape."

Mancktelow (1993) argues strongly that for $T \geq 550°C$ his model does not work because of rock weakening due to power-law creep in clinopyroxene. Two consequences of this phenomenon that are of immense importance to eclogite genesis need comment. On the one hand, if 8 kbar is sufficient to cover the gap between nonoverlapping P estimates for a particular low-T eclogite and its host gneiss, then the tectonic model of a foreign introduction (initially based upon the existence of the P gap) is no longer necessary and an in situ model becomes possible. On the other hand if such P differences are not possible at $T \geq 550°C$ then this *in situ overpressure model* would not be applicable to HPM and UHPM eclogites in the Norwegian WGR since most, if not all, of them indicate $T = 750 \pm 150°C$ for eclogite crystallization. In any case, the apparent P gap between UHPM eclogites in the WGR and those gneisses showing no indications thereof is much greater than 8 kbar. However, this should not detract from the essential points that, first, *the taboo has at last been broken* on a solid structural geologic basis, and second, that this might be only *the first step* since other researchers now incited to come to grips with this question may find other ways of allowing significant overpressure at higher T (cf. the way many scientific revolutions have developed in the past).

The overpressure model could be taken further. For example if lithospheric winds can create relatively high and low P domains on a regional scale, and competence differences can create higher P within eclogite pods and lower P within the enclosing gneisses (i.e., two distinct mechanisms superimposed), then one could conceivably explain most if not all P-T differences and thus remove the need for any foreign introductions, i. e., all the rocks were metamorphosed in situ. However the author does not go this far for the following reasons: (1) The magnitude of the overpressure due to lithospheric wind has not yet been estimated and application of the competence model (or another one) to higher T is not yet acquired. (2) Identical P-T-X-t conditions for two adjacent geologic units does not prove the absence of tectonic juxtaposition. (3) On the basis of lineations/foliations, mylonites, shear zones, faults, tectonized contacts, differing mineralogies, differing bulk-rock compositions and differing ages being widespread in the WGR, he is disposed to favor the hypothesis of considerable tectonic movements, regardless of the existence of overpressure or not. Thus overpressure is considered to be a *significant but not an essential factor*; hence the single FIF model is maintained but with two options left open: either no overpressure, which means over 100-km depth for UHPM, or with overpressure and hence less depth.

Finally it seems important once again to underline two structural correlations that have been noted at many worldwide eclogite localities. First, many eclogites occur *at or close to shear zones, "mega-shear-zones," "breccias," or major structural boundaries* (innumerable references) such that these zones of orientated stress and movement almost certainly played a critical role in the *transport* of the eclogites in some cases, but perhaps also in the *creation* of the eclogites in other cases (cf. Smith et al., 1982; Smith, 1983, 1984, 1988, pp. 134–135; Austrheim, 1987, 1990, 1991), whether or not one believes in overpressure. Second, many eclogites occur in rock units that are recognized to be *stacked slices of differing P-T-X-t-s origin* (variously referred to as "foreign introduction," "exotic block," "tectonic melange," "lithospheric interdigitation," "lithospheric melange," "tectonic juxtaposition," "tectonic mix," "nappe stacking," etc.), e.g., numerous localities and references including the Selje District, Norway (Smith, 1980a); and more recently: Dora-Maira massif,

Italy (Chopin et al., 1991); Cima Lunga and Adula Nappes, Switzerland (Trommsdorff, 1991); Kokchetav Complex, Kazakhstan (Shatsky et al., 1991); and Dabie Shan, China (Cong et al., 1993; Okay, 1993). Massive tectonism has thus become "à la mode" such that continued suggestions that very large crustal metamorphic complexes such as the Norwegian WGR (well over 10,000 km^2) were "largely a coherent body during eclogite facies metamorphism" (Cuthbert and Carswell, 1990, p. 201) seem somewhat inopportune.

Conclusions

1. The arguments in favor of UHPM (~30 kbar or more) in the Norwegian CEP based on the presence of definite coesites (Smith, 1984; Smith and Lappin, 1989) and supported by evidence of deduced coesites and of other petrologic criteria (such as the original data on low-Al orthopyroxene coexisting with garnet and on carbonate/silicate relationships, e.g., 30–45 kbar: Lappin and Smith, 1978) remain powerful such that the Norwegian CEP constitutes a key locality, if not the *type locality* on the basis of historical precedence, for UHPM in the world (excluding nodules in kimberlite and other diatremes, and meteorite impact craters).

2. The discovery of *microdiamonds* in the WGR by Dobrzhinetskaya et al. (1993a), coming precisely a decade after the discovery of microcoesites, provides a second revolution quite unsuspected before the first ideas of UHPM in Norway were formulated in the 1960s and specified in the 1970s. However, it does not necessarily confirm the existence of UHPM as other models can be proposed (as listed above; e.g., the hypothesis that the microdiamonds are premetamorphic relicts). It is quite possible that in Norway the microdiamond distribution (DGP) is distinct from that of the microcoesite distribution (CEP). In the absence of sufficient data the present author risks the speculation that they are in fact coincident and were coeval at $P_{effective}$ within the equilibrium stability field of diamond, and that several new localities of microcoesite and/or microdiamond will be found in Norway, in both eclogites and gneisses.

3. The long-standing "Norwegian eclogite controversy" is almost extinct from the point of view of disputing differing P-T conditions for different tectonic provinces (subterranes, nappes, slices, slabs, blocks, units) as many workers now recognize such differences. It was already extinct a decade ago concerning the mantle vs. crustal protoliths debate and the "dry" vs. "wet" debate, since apparently no one maintains the idea of mantle protoliths for the external eclogites and the presence of H_2O-rich fluids during eclogite metamorphism is now widely accepted. The remaining aspect, "foreign" vs.

"in situ" metamorphism, used to depend merely on the existence or nonexist-ence of P contrasts between eclogites and their adjacent gneisses, and this is bound to evolve each time new P-T data is presented; this aspect now depends in general also on the possibility of *in situ overpressure* within the more compe-tent eclogite pods but the medium-to-high T of the UHPM metamorphism of the CEP precludes this possibility according to Mancktelow (1993).

4. If one sets aside the possibility of overpressure, then the FIF model of a Late Caledonian, in situ, stage A, eclogitization of mixed crust situated in time between two foreign tectonic introduction events, as devised by Smith (1988) and improved in Smith (1992) and here, remains the only available *petrologic model* to reconcile the previous rival "foreign" and "in situ" models of eclogite generation and emplacement in the WGR. It is improved here by the recognition that at least some gneisses experienced the UHPM stage I event prior to their foreign introduction along with the UHPM eclogites as part of the CEP. It will no doubt be further improved in the light of future data and ideas, but it should constitute the starting point for any new studies unless a different and better model comes along. The available geochronologic constraints allow the updated FIF model to be summarized as follows:

~1750 Ma: intrusion and/or low-P metamorphism of *gneiss-forming rocks*;

~1500 Ma: intrusion and/or low-P metamorphism of *gneiss-forming rocks*;

~1000 Ma: intrusion and/or low-P metamorphism of *gneiss-forming rocks* (these three events not necessarily in the same place, nor even in the same continent).

~1700 Ma: UHPM of peridotites (probably of mantle derivation) to create the *Mg-Cr GUP* elsewhere (its first eclogite facies event).

? 600 Ma: HPM or UHPM of peridotites and/or layered basic complexes which cre-ated the *Fe-Ti GUP* elsewhere.

~500 Ma: HPM or UHPM of the *Mg-Cr GUP* elsewhere (its second eclogite facies event).

~440 Ma: UHPM of gabbros/basalts, and of probably adjacent gneisses (coesite-bearing), which created the *CEP* (stage I) elsewhere (its first eclogite facies event).

?440 Ma: UHPM of gneisses (coesite + diamond-bearing) which created the *DGP*, either *elsewhere*, or cofacial and coeval with the *CEP*.

≥410 Ma: *the first foreign introduction*: tectonic interdigitation of the *CEP* and *DGP* (? and *Fe-Ti GUP*) into already-subducted already-mixed acid + basic noneclogi-tized crust of various ages.

~410 Ma: large-scale HPM at around 20 kbar of the entire mixed crust which: (1) created *new HPM gneisses* (omphacite and/or phengite-bearing); (2) created quartz-eclogite (sub)facies in situ *new HPM eclogites* (*their first and only* eclogite facies event = stage A); (3) recrystallized in situ the former foreign *CEP* (its second eclogite facies event = stage A) and *DGP* (? and *Fe-Ti GUP*); (4) reset Fe/Mg geothermometers and most isotope systems, partially or totally;

≥385 Ma: the *second foreign introduction*: tectonic interdigitation of the stage A mixed terrane with the **Mg-Cr GUP** and with some *never-eclogitized gneisses* of the three Precambrian ages and with some *nonmetamorphosed* Lower Paleozoic sediments and volcanics (all of which bypassed the ~410 Ma HPM event).

~385 Ma: regional metamorphism at around 10 kbar which formed stage S retrogressive symplectites in all rock types that had already been through stage A (*CEP, DGP, GUPs*, stage I and A *gneisses*) and created *new low-P gneisses*.

< 385 Ma: gradual exhumation, rapid enough to preserve certain UHPM phases but slow enough to allow growth of epidote, serpentine, pumpellyite, etc.

5. The previous standard *geodynamic model* for the *creation* of eclogite facies parageneses in the WGR by subducting Baltica under Laurentia has been seriously questioned by Gilotti and Hull (1991, 1993) and thus needs detailed reevaluation; their suggestion that the WGR did not constitute part of Baltica but of Laurentia or of an intervening microcontinent is attractive; indeed the CEP and other tectonic provinces within the WGR may well have originated in different (micro)continents. The process of subduction is not necessarily in question, although the depth required to acquire a specific pressure is debatable in terms of possible overpressure, but some models do not involve subduction (e.g. lithospheric advection; Perchuk et al., 1992). Major continental collision seems to be an inescapable ingredient, but it may well represent only a final stage following the involvement of island arcs and/or microcontinents in order to create so many (around fifteen) different eclogite, garnet-pyroxenite, and garnet-ultrabasite provinces in the North Atlantic Caledonides.

6. No adequate geodynamic model for the exhumation of the UHPM and HPM rocks is yet available, but active tectonism must have played an important part, both in creating the lithospheric interdigitation of different *P-T-X-t-s* provinces and in facilitating their upward movement. Successive underthrusting followed by erosion, with or without extensional tectonics, were probably major processes, but in addition some form of overpressure is as least as credible as some other models (e.g., eduction by ultrametastable continental crust: Andersen et al., 1991a,b).

7. Substantial regional or local overpressure (i.e. $P_{effective}$ exceeding P_{load} by at least 10 kbar) is thus still considered a viable option and is in fact reinforced by the work of Mancktelow (1993). Concerning petrologic modeling, the existence of overpressure would not negate the FIF model which is based upon apparently much greater P differences between UHPM rocks and nevereclogitized gneisses; rather it complements it by reducing the depths that would be required by the conventional relationship between depth and pressure. Concerning geodynamical modeling, the existence of overpressure would

provide an excellent mechanism for promoting subvertical transport as is indicated by some field relationships of certain UHPM provinces.

Future Research Orientations

The discovery of deduced microcoesite within gneisses as well as of microdiamonds (in the same geologic unit) should provide a strong impetus for new research on the WGR; a decline in research activity since about 1989 is noticeable and is due to the fact that most of the key personalities in the Norwegian eclogite saga have been preoccupied with other matters. It is clearly essential to establish the *distribution* of microcoesite and microdiamond occurrences in order to clarify whether the presently-known ones are just exceptional cases or are part of regional distributions. It is quite possible that by the end of the century there could be 30 or more new microcoesite and/or microdiamond occurrences in which case one could hope to be able to discern whether or not they are confined to particular geographic areas and geologic units, which would permit better definitions of the CEP and DGP, and to see if they correlate or not with each other or with the GUPs.

Two principal rock types need urgent attention: the eclogites themselves and the enclosing gneisses, especially the latter which have received less attention in the past, partly because they are less spectacular in the field. It is essential to hunt for microcoesites, microdiamonds and other microinclusions in the gneisses in order to try to distinguish among different P-T generations of gneisses, as has been done with the eclogites of the CEP. There are certainly many different kinds of gneisses on other criteria (e.g. geochemistry, petrography, deformation, age, etc., as described above) but there are few data available on P-T histories.

Three principal scales of study are required. First the 10^0 -10^3 μm (1 μm-1 mm) scale is necessary to discover and map on the crystal scale the *various microinclusions* already known, especially the enigmatic possible eclogitization stage Y ones and the possible post-stage X exsolution lamellae, and perhaps others yet to be discovered in this mineralogically rich terrane. They need identifying and distinguishing which means using essentially the electron, ion and Raman microprobes. Second, the 10^4-10^7 μm (10 mm–10 m) scale is necessary to map in detail the petrologic and structural relations of the contacts between eclogites and their adjacent gneisses, and also between gneiss units that may have experienced different P-T-X-t-s histories before having been welded together. Structural mapping without the petrology or petrologic mapping without the structures would be considerably less useful. Third, petrologic/structural mapping on the 10^8–10^{11} μm (100 m–100 km) scale is required in order to be able to correlate between districts and to produce the first ever geologic map depicting the limits of the different tectonic provinces that compose the WGR.

Although geochemical, geophysical, experimental, thermodynamic and geodynamic modeling and manipulations all contribute to a better understanding of any geologic problem, they seem of less urgency now in the WGR than the traditional "naturalistic" approach of "detect, observe, and describe" (including crystal-chemical characterizations). Geochronologic studies are of course also important, but they need a very careful selection of samples to analyze (collected from petrologically and structurally mapped localities) in order properly to understand which events are being dated.

In brief, the emphasis should be: Back to the field, and more work with maps, microscopes, and microprobes, rather than with microcomputers!

Acknowledgments

The writer is grateful to Larissa F. Dobrzhinetskaya and Brian A. Sturt for fruitful discussions concerning microdiamonds, and to Larissa F. Dobrzhinetskaya, Haakon Austrheim, and Robert G. Coleman for making several constructive suggestions after reviewing the submitted manuscript. Thanks are of course also due to the editors for their invitation to prepare this chapter.

References

Agrinier, P., Javoy, M., Smith, D. C. and Pineau, F. 1985. Carbon and oxygen isotopes in eclogites, amphibolites, veins and marbles from the Western Gneiss Region, Norway. In D. C. Smith and P. Vidal (eds), *Isotope Geochemistry and Geochronology of Eclogites, Isotope GeoScience*, 145–162.

Anders, E., and Zinner, E. 1993. Interstellar grains in primitive meteorites: diamond, silicon carbide, and graphite. *Meteoritics, 28*, 490–514.

Andersen, T. B., Jamtveit, B., Dewey, J. F., and Swensson, E. 1991a. Orogenic extensional collapse and eduction of high-P terranes in S. W. Norwegian Caledonides. *Terra Abstracts suppl. 4 to Terra Nova, 3*, 1.

Andersen, T. B., Jamtveit, B., Dewey, J. F., and Swensson, E. 1991b. Subduction and eduction of continental crust: Major mechanisms during continent-continent collision and orogenic extensional collapse, a model based on the South Norwegian Caledonides. *Terra Nova, 3*, 303–310.

Andresen, A., Fareth, E., Berg, S., Kristensen, S. E., and Krogh, E. J. 1985. Review of Caledonian lithotectonic units in Troms, North Norway. *The Caledonide Orogen–Scandinavia and Related Areas*. New York: John Wiley, 569–578.

Austrheim, H. 1987. Eclogitization of lower crustal granulites by fluid migration through shear zones. *Earth Planet. Sci. Lett., 81*, 221–232.

Austrheim, H. 1990. The granulite-eclogite facies transition: a comparison of experimental work and a natural occurrence in the Bergen Arcs, western Norway: Third Int. Eclogite Conf., Spec. Issue (ed. M.Okrusch). *Lithos, 25,* 163–169.

Austrheim, H. 1991. Eclogite formation and dynamics of crustal roots under continental collision zones. *Terra Nova, 3,* 492–499.

Austrheim, H., and Griffin, W. L. 1985. Shear deformation and eclogite formation within granulite-facies anorthosites of the Bergen Arcs, Western Norway. *Chemistry and Petrology of Eclogites. Chemical Geology,* 267–281.

Binns, R. A. 1967. Barroisite-bearing eclogite from Naustdal, Sogn og Fjordane, Norway. *J. Petrol., 8,* 349–371.

Boundy, T. M., Fountain, D. M. and Austrheim, H. 1992. Structural development and petrofabrics of eclogite facies shear zones, Bergen Arcs, western Norway: implications for deep crustal deformational processes. *Jour. Metamorph. Geol., 10,* 127–146.

Bovenkerk, H. P., Bundy, F. P., Chrenko, R. M., Codella, P. J., Strong, H. M., and Wentorf Jr., R. H. 1993. Errors in diamond synthesis. *Nature, 365,* 19.

Boyer, H., Smith, D. C., Chopin, C., and Lasnier, B. 1985. Raman microprobe (RMP) determinations of natural and synthetic coesite. *Physics and Chemistry of Minerals, 12,* 45–48.

Brace, W., F., Ernst, W. G., and Kallberg, R. W. 1970. An experimental study of tectonic overpressure in Franciscan rocks. *Geol. Soc. Amer. Bull., 81,* 1325–1338.

Brueckner, H. K. 1972. Interpretation of Rb-Sr ages from Precambrian and Paleozoic rocks of Southern Norway. *Am. J. Sci., 272,* 334–358.

Bryhni, I. 1966. Reconnaissance studies of gneisses, ultrabasites, eclogites and anorthosites in Outer Nordfjord, Western Norway. *Norges Geologiske Undersøkelse, 241,* 1–68.

Bryhni, I., and Andréasson, P. G. 1985. Metamorphism in the Scandinavian Caledonides. *The Caledonide Orogen.* New York: J. Wiley, 763–781.

Bryhni, I., Bollingberg, H. J., and Graff, P. R. 1969. Eclogites in quartzo-feldspathic gneisses of Nordfjord, West Norway. *Norsk Geologisk Tidsskrift, 49,* 193–225.

Bryhni, I., Green, D. H., Heier, K. S., and Fyfe, W. S. 1970. On the occurrence of eclogites in Western Norway. *Contrib. Mineral. Petrol., 26,* 12–19.

Bryhni, I., Krogh, E. J., and Griffin, W. L. 1977. Crustal derivation of Norwegian eclogites: a review. *Neues Jahrb. Mineralogy Abh., 130,* 49–68.

Carpenter, M., and Smith, D. C. 1981. Solid-solution and cation ordering limits in high temperature sodic pyroxenes from the Nybö eclogite pod, Norway. *Mineral Mag., 44,* 37–44.

Carswell, D. A. 1973. Garnet pyroxenite lens within Ugelvik layered garnet peridotite. *Earth Planet. Sci. Lett., 20,* 347–352.

Carswell, D. A. 1974. Comparative equilibration temperatures and pressures of garnet lherzolites in Norwegian gneisses and in kimberlite. *Lithos, 7,* 113–121.

Carswell, D. A., and Cuthbert, S. J. 1986. Eclogite facies metamorphism in the lower continental crust. *The nature of the lower continental crust, Geological Society special publication, 25,* 193–209.

Carswell, D. A., Harvey, M. A., and Al-Samman, A. 1983. The petrogenesis of contrasting Fe-Ti and Mg-Cr garnet-peridotite types in the high grade gneiss complex of Western Norway. *Bulletin Mineralogie, 106,* 727–750.

Carswell, D. A., Krogh, E. J., and Griffin, W. L. 1985. Norwegian orthopyroxene eclogites: calculated equilibration conditions and petrogenetic implications. *The Caledonide Orogen.* New York: J. Wiley, 823–841.

Carswell, D. A., Wilson, R. H., Cong, B., Zhai, M. and Zhao, Z. 1993. Areal extent of the ultra-high pressure metamorphism of eclogites and gneisses in Dabie-shan, Central China. *Terra Abst. sup. 4 to Terra Nova, 5,* 5.

Chesnokov, B. V., and Popov, V. A. 1965. Increasing of volume of quartz grains in eclogites of the South Urals. *Dokl. Akad. Nauk SSSR, 162,* 176–178.

Chopin, C. 1983. High pressure facies series in pelitic rocks: a review. *Terra Cognita, 3,* 183.

Chopin, C. 1984. Coesite and pure pyrope in high-grade blueschists of the western Alps: A first record and some consequences. *Contributions to Mineralogy and Petrology, 86,* 107–118.

Chopin, C. 1986. Phase relationships of ellenbergerite, a new high-pressure Mg-Al-Ti-silicate in pyrope-coesite-quartzite from the Western Alps. *Geol. Soc. Amer. Mem., 164,* 31–42.

Chopin, C., Brunet, F., Gebert, W., Medenbach, O., and Tillmanns, E. 1993. Bear-thite, $Ca_2Al(PO_4)_2OH$, a new mineral from high pressure terranes of the west-ern Alps. *Schweiz. min. und petr. Mitt., 73,* 1–9.

Chopin, C., Henry, C., and Michard, A. 1991. Geology and petrology of the coesite-bearing terrain, Dora-Maira massif, Western Alps. *Eur. Jour. Mineral., 3,* 263–291.

Chopin, C., Klaska, R., Medenbach, O., and Dron, D. 1986. Ellenbergerite, a new high-pressure Mg-Al-(Ti, Zr)- silicate with a novel structure based on face-sharing octahedra. *Contrib. Mineral. Petrol., 92,* 316–321.

Church, W. R. 1968. Metamorphic rocks of Burlington Peninsula and adjoining areas of Newfoundland and their bearing on continental drift in the North Atlantic. *A A P G, Memoir, 12,* 212–233.

Cohen, A. S., O'Nions, R. K., Siegentaler, R., and Griffin, W. L. 1988. Chronology of the pressure-temperature history recorded by a granulite terrain. *Contrib. Mineral. Petrol., 98,* 303–311.

Coleman, R. G. 1972. Blueschist metamorphism and plate tectonics. *24th Interna-tional Geological Congress,* Sec. 2, 19–26.

Cong Bolin, Zhai Mingguo, Zhao Zhongyan, Carswell, D. A., and Wilson, R. H. 1993. A typical area of UHP coesite-bearing eclogite and related rocks in Shu-anghe village, Dabie Mountains, Central China: its petrology, metamorphic mineralogy and history, and evidence for collision and exhumation. *Terra Nova Abstracts suppl. 4, 5,* 3.

Cuthbert, S. J., and Carswell, D. A. 1982. Petrology and tectonic setting of eclogites and related rocks from the Dalsfjord Area, Sunnfjord, West Norway. *Terra Cog-nita, 2,* 315–316.

Cuthbert, S. J., and Carswell, D. A. 1990. Formation and exhumation of medium-temperature eclogites in the Scandinavian Caledonides. *Eclogite-facies rocks.* Glasgow: Blackie.

Cuthbert, S. J., Harvey, M. A., and Carswell, D. A. 1983. A tectonic model for the metamorphic evolution of the Basal Gneiss Complex, Western South Norway. *Jour. Metamorph. Geol., 1,* 63–90.

Dallmeyer, R. D., and Gee, D. G. 1986. $^{40}Ar/^{39}Ar$ mineral dates from retrogressed eclogites within Baltoscandian miogeocline: Implications for a polyphase Caledonian orogenic evolution. *Geological Society of America Bulletin, 97,* 26–34.

De Roever, W. P. 1967. Overdruk van tektonische oorsprong of diepe metamofose: Koninkl. Nederl. Akad. Wetensch. Versl. Gew. Vergad. Afd. Natuurk, v. 76, p.

69–74 (Overpressure of tectonic origin or deep metamorphism). *Metamorphism and plate tectonic regimes: Benchmark Papers Geology.* Benchmark Papers. 348–352.

Dobretsov, N. L. 1964. The jadeite rocks as indicators of high pressure in the Earth's crust. *22nd International geological Congress (India), 16,* 177–196.

Dobrzhinetskaya, L. F., and Molchanova, T. V. 1992. Deformation history of eclogites and diamond-bearing metamorphic rocks, Northern Kazakhstan, USSR. *Abstracts, 29th Int. Geol. Congr., Kyoto, Japan, 2,* 448.

Dobrzhinetskaya, L. F., Molchanova, T. V., Sheshkel, G. G., and Podkuiko, Y. A. 1994. Geology and structure of diamond-bearing rocks of the Kokchetav Massif (Kazakhstan). *Tectonophysics 233,* 293–313.

Dobrzhinetskaya, L. F., Posukhova, T., Tronnes, R., Korneliussen, A., and Sturt, B. 1993a. A microdiamond from eclogite-gneiss area of Norway. *Terra Abstracts, 5,* 8.

Dobrzhinetskaya, L. F., Sheshkel, G. G., and Podkuiko, Y. A. 1993b. The structural distribution of crustal microdiamonds within eclogite-gneiss formation (Northern Kazakhstan). *Terra Abstracts suppl. 4 to Terra Nova, 5,* 8.

Ekimova, T. E., Lavrova, L. D., and Petrova, M. A. 1992. Diamond inclusions in rock-forming minerals of the metamorphic rocks. *Doklady Acad. Nauk., 322,* 366–368.

Erambert, M. 1985. Etude pétrologique de lentilles éclogitiques mafiques et ultramafiques et de leur encaissant à Essdalen, Vartdal, Sunnmöre (Norvège). Ph.D. Muséum-Univ. Paris VI, 319 pp.

Eskola, P. 1920. The mineral facies of rocks. *Norsk Geol. Tidsskrift, 6,* 43–194.

Eskola, P. 1921. On the eclogites of Norway. *Skrift. Videnskaps-selsk. Christiania Mat.-Naturv. Kl, 18,* 1–118.

Gebauer, D., Lappin, M. A., Grünenfelder, M., and Wyttenbach, A. 1985. The age and origin of some Norwegian eclogites. A U-Pb zircon and REE study. *Chemical Geology, 52,* 227–247.

Gee, D. G. 1966. A note on the occurrence of eclogites in Spitsbergen. *Aarbok Norsk Polarinstitut, 1964,* 240–241.

Gee, D. G., and Sturt, B. A. 1985. *The Caledonide Orogen–Scandinavia and Related Areas.* New York: John Wiley.

Gilotti, J. A. 1993. Discovery of a medium-temperature eclogite province in the Caledonides of North-East Greenland. *Geology, 21,* 523–526.

Gilotti, J. A., and Brueckner, H. K. 1993. A new eclogite province in the Caledonides of North-East Greenland: preliminary thermobarometry and geochronology. *Terra Abstracts suppl. 4 to Terra Nova, 5,* 11.

Gilotti, J. A., and Hull, J. M. 1991. A new tectonic model for the Central Scandinavian Caledonides. *Terra Nova Absracts Suppl. 4, 3,* 18.

Gilotti, J. A., and Hull, J. M. 1993. Kinematic stratification in the hinterland of the central Scandinavian Caledonides. *Journal Structural Geology, 15,* 629–646.

Gilotti, J. A., Hull, J. M., Higgins, A. K., Friderichsen, J. D., and Steenfelt, A. 1991. A new eclogite province in the Skaerfjorden Area, North-East Greenland, 77–78 North latitude. *Terra Abstracts suppl. 4 to Terra Nova, 3,* 18.

Gjelsvik, T. 1951. Oversikt over bergartene i Sunnmöre og tilgrensende deler av Nordfjord: *Norges Geol. Undersøkelse, 179,* 1–45.

Gjelsvik, T. 1952. Metamorphosed dolerites in the gneiss area of Sunnmore on the west coast of southern Norway. *Norsk Geologisk Tidsskrift, 30,* 33–134.

Gjelsvik, T. 1953. Det nordvestlige gneis-omraade i det sydlige Norge, alder, forhold og tektonisk-stratigrafisk stilling. *Norges Geol. Undersøkelse, 184,* 71–94.

Gjelsvik, T., and Gleditsch, C. 1951. Geologisk oversiktskart over Sunnmöre og til-grensende deler av Nordfjord. *Norges Geol. Undersøkelse, 179,* 1–45.

Green, D. H., 1969. Mineralogy of two Norwegian eclogites (in Russian). *Institute fiziki tverdogotela AN SSR,* 37–40.

Green, D. H., and Mysen, B. O. 1972. Genetic relationship between eclogite and hornblende + plagioclase pegmatite in Western Norway. *Lithos, 5,* 147–161.

Griffin, W. L. 1987. On the eclogites of Norway–65 years late. *Mineral. Mag., 51,* 333–343.

Griffin, W. L., Austrheim, H., Brastard, K., Bryhni, I., Krill, A. G., Krogh, E. J., Mørk, M.B.E., Qvale, H., and Törudbakken, B. 1981. High pressure metamor-phism in the Scandinavian Caledonides. *Terra Cognita, 1,* 48–49.

Griffin, W. L., Austrheim, H., Brastad, K., Bryhni, I., Krill, A. G., Krogh, E. J., Mørk, M.B.E., Qvale, H., and Törudbakken, B. 1981. High pressure metamor-phism in the Scandinavian Caledonides. *The Caledonide Orogen.* New York: John Wiley.

Griffin, W. L., and Brueckner, H. K. 1980. Caledonian Sm-Nd ages and a crustal ori-gin for Norwegian eclogites. *Nature, 285,* 319–321.

Griffin, W. L., and Brueckner, H. K. 1985. REE, Rb-Sr and Sm-Nd studies of Nor-wegian eclogites. *Chemical Geology, 52,* 249–271.

Griffin, W. L., and Carswell, D. A. 1985. In situ metamorphism of Norwegian eclog-ites: an example. *The Caledonide Orogen.* New York: J. Wiley, 813–822.

Griffin, W. L., and Qvale, H. 1985. Superferrian eclogites and the crustal origin of garnet peridotites, Almklovdalen, Norway. *The Caledonide Orogen.* New York: J. Wiley, 804–812.

Griffin, W. L., and Råheim, A. 1973. Convergent metamorphism of eclogites and dolerites, Kristansund area, Norway. *Lithos, 6,* 21–40.

Harris, A. L., and Fettes, D. J. 1988. *The Caledonian-Appalachian Orogen.* Oxford: Blackwell Scientific Publications.

Hernes, I. 1954. Eclogite-amphibolite on the Molde Peninsula, southern Norway. *Norsk Geologisk Tidsskrift, 33,* 163–184.

Hirajima, T., Zhang, R., Li, J., and Cong, B. 1992. Petrology of the nyböite-bearing eclogite in the Donghai area, Jiangsu Province, eastern China. *Mineral. Mag., 56,* 37–46.

Jacobsen, S. B., and Wasserburg, G. J. 1980. Nd and Sr isotopes of the Norwegian garnet peridotites and eclogites. *EOS Trans. Am. Geophys. Union, 61,* 389.

Jamtveit, B. 1986. Magmatic and metamorphic evolution of the Eiksunddal eclogite complex, Sunnmøre, Western Norway. Ph.D., Univ. Oslo.

Jamtveit, B. 1987. Metamorphic evolution of the Eiksunddal eclogite complex, West-ern Norway, and some tectonic implications. *Contrib. Mineral. Petrol., 95,* 82–99.

Jamtveit, B., Carswell, D. A., and Mearns, E. W. 1991. Chronology of the high-pressure metamorphism of Norwegian garnet peridotites/pyroxenites. *Journal of Metamorphic Petrology, 9,* 125–139.

Johansson, L., and Möller, C. 1986. Formation of sapphirine during retrogression of a basic high-pressure granulite, Roan, Western Gneiss Region. *Contrib. Min-eral. Petrol., 94,* 29–41.

Kechid, S.-A. 1984. Etude pétrologique et minéralogique des éclogites de Liset (Stad-landet, Norvège). Ph.D., Muséum–Univ. Paris VI, Paris.

Kechid, S.-A., and Smith, D. C. 1982. Nyböite-katophorite et taramite-pargasite dans la lentille d'éclogite de Liset, Région du Gneiss de l'Ouest, Norvège. *9th Réunion Annuelle Sciences Terre, Paris, Abstract,,* 333.

Kechid, S.-A., and Smith, D. C. 1985. The petrological evolution of the Liset eclogite pod, Norway. *Terra Cognita, 5*, 422.

Kolderup, N. H. 1960. Origin of Norwegian eclogites in gneisses. *Norsk Geologisk Tidsskrift, 40*, 73–76.

Krogh, E. J. 1977a. Evidence for a Precambrian continent-continent collision in western Norway. *Nature, 267*, 17–19.

Krogh, E. J. 1977b. Crustal and in situ origin of Norwegian eclogites. A reply. *Nature, 269*, 730.

Krogh, E. J., Andersen, A., Bryhni, I., and Kristensen, S. E. 1982. Tectonic setting, age and petrology of eclogites within the upppermost tectonic unit of the Scandinavian Caledonides, Tromsø area, Northern Norway. *Terra Cognita, 2*, 316.

Krogh, T. E., Mysen, B. O., and Davis, G. L. 1974. A Palaeozoic age for the primary minerals of a Norwegian eclogite. *Carnegie Institution Washington Yearbook, 73*, 575–576.

Kullerud, L., Tørudbakken, B. O., and Ilebekk, S. 1986. A compilation of radiometric age determinations from the Western Gneiss Region, South Norway. *Norges Geologiske Undersøkelse Bulletin, 406*, 17–42.

Lappin, M. A. 1962. The eclogites, dunites and anorthosites of the Selje and Almklovdalen districts, Nordfjord, S. Norway. Ph.D., Univ. Durham.

Lappin, M. A. 1966. The field relationships of basic and ultrabasic masses in the basal gneiss complex of Stadlandet and Almklovdalen, Nordfjord, southwestern Norway. *Norsk Geologisk Tidsskrift, 4*, 439–496.

Lappin, M. A. 1974. Eclogites from the Sunndal-Grubse ultramafic mass, Almklovdalen, Norway, and the T-P history of the Almklovdalen massif. *J. Petrol., 15*, 567–601.

Lappin, M. A. 1977. Crustal and in situ origin of Norwegian eclogites. *Nature, 269*, 730.

Lappin, M. A., Pidgeon, R. T., and Van Breemen, O. 1979. Geochronology of basal gneisses and mangerite syenites of Stadlandet, West Norway. *Norsk Geologisk Tidsskrift, 59*, 161–181.

Lappin, M. A., and Smith, D. C. 1978. Mantle-equilibrated orthopyroxene eclogite pods from the basal gneisses in the Selje district, western Norway. *J. Petrol., 19*, 530–584.

Lappin, M. A., and Smith, D. C. 1981. Carbonate, silicate and fluid relationships in eclogites, Selje district and environs, S. W. Norway. *Transactions Royal Society Edinburgh: Earth Sciences, 72*, 171–193.

Lavrova, L. D. 1991. New type of diamond deposits. *Prioda, 12*, 62–68 (in Russian).

Letnikov, F. A. 1983. Formation of diamonds within deep-seated tectonic zones. *Doklady Acad. Nauk SSSR, 371*, 433–435 (in Russian).

Lux, D. R. 1985. K/Ar ages from the Basal Gneiss Region, Stadlandet area, Western Norway. *Norsk Geologisk Tidsskrift, 65*, 277–286.

Mancktelow, N. 1993. Tectonic overpressure in competent mafic layers and the development of isolated eclogites. *J. Metamorphic Geol., 11*, 801–812.

McDougall, I., and Green, D. H. 1964. Excess radiogenic argon in pyroxenes and isotopic ages on minerals from Norwegian eclogites. *Norsk Geological Tidsskr., 44*, 183–196.

Mearns, E. W. 1984. Isotopic studies of crustal evolution in Western Norway. Ph.D. thesis, Univ. Aberdeen, 161 pp.

Mearns, E. W. 1986. Sm-Nd ages for Norwegian garnet peridotite. *Lithos*, 269–278.

Mearns, E. W., and Lappin, M. A. 1982. The origin and age of "external eclogites" and gneisses from the Selje District of the Western Gneiss Region, Norway. *Terra Cognita, First International Eclogite Conference (abstract)*, *2*, 324.

Mearns, E. W., and Lappin, M. A. 1982. A Sm-Nd isotopic study of "internal" and "external eclogites", garnet lherzolite and grey gneiss from Almklovdalen, Western Norway. *Terra Cognita, First International Eclogite Conference (abstract)*, *2*, 324–325.

Medaris, L. G. J. 1980a. Convergent metamorphism of eclogite and garnet-bearing ultramafic rocks in West Norway. *Nature, 283*, 470–472.

Medaris, L. M. J. 1980b. A tectonic melange of foreign eclogites and ultramafites in West Norway. *Nature, 287*, 366–368.

Medaris, L. G. J., and Wang, H. F. 1986. A thermal-tectonic model for high-pressure rocks in the Basal Gneiss Complex of Western Norway. *Lithos, Second International Eclogite Conference.* 323–332.

Messiga, B., Tribuzio, R., and Vannucci, R. 1990. Mafic and ultramafic pods with eclogite relics from the Proterozoic Nagssugtoqidian mobile belt of East Greenland. *Lithos, 25*, 101–118.

Milanovsky, E. E. 1983. Conference Report–Moscow Conference on the problems of the Earth's expansion and pulsation. *Terra Cognita, 3*, 22–24.

Mørk, M. B. E. 1985. A gabbro to eclogite transition on Flemsøy, Sunnmøre, Western Norway. *Chemistry and Petrology of Eclogites. Chemical Geology.* 283–310.

Mørk, M. B. E., and and Krogh, E. J. 1987. *Excursion guide for the eclogite field symposium in western Norway.* Spec. Pub. Univ. Tromsø, Norway, 115p.

Mørk, M. B. E., Kullerud, K., and Stabel, A. 1988. Sm-Nd dating of Seve eclogites. Norrbotten, Sweden–evidence for early Caledonian (505 Ma) subduction. *Contrib. Mineral. Petrol., 99*, 344–351.

Mysen, B. O., and Heier, K. S. 1971. A note on the field occurrence of a large eclogite on Hareid, Sunnmöre, Western Norway. *Norsk Geol. Tidsskrift, 51*, 93–96.

Mysen, B. O., and Heier, L. 1972. Petrogenesis of eclogites in high grade metamorphic gneisses, exemplified by the Hareidland eclogite, Western Norway. *Contrib. Mineral. Petrol., 36*, 73–94.

Nadezhdina, E. D., and Posukhova, T. V. 1990. The morphology of diamond crystals from metamorphic rock. *Mineralogitchesky Zhurnal, 12*, 3–15 (in Russian).

O'Hara, M. J., and Mercy, E. L. P. 1963. Petrology and petrogenesis of some garnetiferous peridotites. *Royal Society of Edinburgh Transactions, 65*, 251–314.

Oberti, R., Rossi, G. and Smith, D. C. 1985. X-ray crystal structure refinement studies of the TiO = Al(OH,F) exchange in high-aluminium-sphenes. *Terra Cognita, 5*, 428.

Oberti, R., Previde-Massara, E., Ungaretti, L., Kechid, S.-A., and Smith, D. C. 1989. Nyböite-taramite-sadanagaite trend in amphiboles from the Liset eclogite pod, Norway: crystal-chemical and petrogenetic implications. *Terra Abstracts, 1*, 17.

Oberti, R., Smith, D. C., Rossi, G., and Caucia, F. 1991. The crystal-chemistry of high-aluminium titanites. *Eur. Jour. Mineral., 3*, 777–792.

Oberti, R., Ungaretti, L., Tlili, A., Smith, D. C., and Robert, J.-L. 1993. The crystal structure of preiswerkite. *American Mineral., 78*, 1290–1298.

Ohta, Y., Hirajima, T., and Hiroi, Y. 1986. Caledonian high-pressure metamorphism in central western Spitsbergen. *Geol. Soc. Amer. Mem. 164*, 205–216.

Okay, A. 1993. Petrology of a diamond and coesite-bearing terrain: Dabie Shan, China. *Eur. Jour. Mineral., 5*, 659–675.

Perchuk, L. L., Podladchikov, Y. Y., and Polakov 1992. Hydrodynamic modelling of some metamorphic processes. *Jour. Metamorphic Geol., 10*, 311–319.

Pinet, M., and Smith, D. C. 1985. Petrochemistry of opaque minerals in eclogites from the Western Gneiss Region, Norway. *Chemical Geology (Chemistry and Petrology of Eclogites)*, 225–249.

Platt, J. P. 1986. Dynamics of orogenic wedges and the uplift of high-pressure metamorphic rocks. *Geol. Soc. Amer. Bull., 97*, 1037–1053.

Platt, J. P. 1993. Exhumation of high-pressure rocks: a review of concepts and processes. *Terra Nova, 5*, 119–133.

Rossi, G., Oberti, R., and Smith, D. C. 1986. Crystal structure of listite, $CaNa_2Al_4Si_4O_{16}$. *Am. Mineralogist, 71*, 1378–1383.

Rossi, G., Oberti, R., and Smith, D. C. 1989. The crystal structure of a K-poor Ca-rich silicate with the nepheline framework and crystal-chemical relationships in the compositional space $(K,Na,Ca)_8 (Al,Si)_{16}O_{32}$. *Euro. Jour. Mineral., 1*, 59–70.

Rossi, G., Smith, D. C., Ungaretti, L., and Domeneghetti, C. 1983a. Crystal-chemistry and cation ordering in the system diopside-jadeite: a detailed study by crystal structure refinement. *Contrib. Mineral. Petrol., 83*, 247–258.

Rossi, G., Smith, D. C., Ungaretti, L., and Domeneghetti, C. 1983b. Comparison of chemical analyses of sodic pyroxenes by X-ray structure refinement and electron microprobe techniques: In Mineralogical techniques applied to pyroxenes and amphiboles. *Peridico Mineralogia Roma, 52*, 342–354.

Rozen, O. M. 1973. About a diamonds presence in sedimentary metamorphic complexes. *Lithology and sedimentary geology of Precambrian*. Moscow: Acad. Nauk. SSSR Press, 382–384 (in Russian).

Rozen, O. M., Zayatchkovsky, A. A., Kljuev, Y. A., and Smirnov, V. L. 1979. Peculiarities of trace-minerals composition and conditions of the formation of the North Kazakhstan eclogites. *Problems of the sedimentary geology of the Precambrian*. Moscow: Nauka Press. 170–186 (in Russian).

Rozen, O. M., and Zorin, Y. M. 1972. Diamond occurrence related to eclogites in Precambrian of Kokchetav massif. *Doklady Acad. Nauk SSSR, 203*, 674–676 (in Russian).

Rutland, R.W.R. 1965. Tectonic overpressures. *Controls of Metamorphism*. London: Oliver and Boyd, 119–139.

Sanders, I. S. 1982. Eclogite and cofacial gneiss from Glenelg, N. W. Scotland. *Terra Cognita, 2*, 317.

Sanders, I. S., Daley, J. S., and Davies, G. R. 1987. Late Proterozoic high-pressure granulite facies metamorphism in the north-east Ox inlier, north-west Ireland. *Journal Metamorphic Geology, 5*, 69–85.

Schmitt, H. H. 1960. Geologic field investigations in the Eiksund-Jösak area, Sunnmøre, Norway. *Norges Geol. Undersøkelse Report*, 1–37.

Schmitt, H. H. 1964. Metamorphic eclogites of the Eiksund Area, Sunnmøre, Norway. *Trans. Amer. Geophys. Union, 45th annual meeting*.

Schrauder, M., and Navon, O. 1993. Solid carbon dioxide in a natural diamond. *Nature, 365*, 42–44.

Sharma, S. K., Mao, H. K., Bell, P. M. and Xu, J.A., 1985. Measurement of stress in diamond anvils with micro-Raman spectroscopy. *Jour. Raman Spectroscopy, 15*, 350–352.

Shatsky, V. S., Jagoutz, E., Kozmenko, O. A. and Sobolev, N. V., 1991. Geochemical characteristics of crustal rocks subducted into the upper mantle. *Terra Absracts (EUG 6), 3*, 83.

Smith, D. C. 1968. The petrology and chemistry of some eclogites in the gneiss of the Selje District, Nordfjord, S. W. Norway. Ph.D., Aberdeen Univ.

Smith, D. C. 1971. A tourmaline-bearing eclogite from Sunnmøre, Norway. *Norsk Geol. Tidsskrift, 51*, 141–147.

Smith, D. C. 1976a. The geology of the Vartdal area, Sunnmøre, Norway, and the petrochemistry of the Sunnmøre Eclogite Suite. Ph.D. thesis, Aberdeen, Univ.

Smith, D. C. 1976b. The recognition of two distinct pressure-temperature regimes of eclogite metamorphism in the Sunnmøre Eclogite Suite. *Fortschritte Mineralogie, 54*, 94.

Smith, D. C. 1977. Aluminiumholdig titanit i eklogitter fra Sunnmøre. *Geolognytt, 10*, 32–33.

Smith, D. C. 1980a. A tectonic melange of foreign eclogites and ultramafites in the Basal Gneiss Region, West Norway. *Nature, 287*, 366–368.

Smith, D. C. 1980b. Highly aluminous sphene (titanite) in natural high pressure hydrous-eclogite-facies rocks from Norway and Italy, and in experimental runs at high pressures. *26th Internat. Geol. Congr., Paris, Abstract, 02.31*, 45.

Smith, D. C. 1981a. The pressure and temperature dependence of Al-stability in sphene in the system Ti-Al-Ca-Si-O-F. *Fifth Progress Report on Experimental Petrology, Natural Environment Research Council, London*, 193–197.

Smith, D. C. 1981b. A reappraisal of factual and mythical evidence concerning the metamorphic and tectonic evolution of eclogite-bearing terrain in the Caledonides (abstract). *Terra Cognita, 1*, 73–74.

Smith, D. C. 1982a. Hjörundfjordite, un nouveau type de roche dans l'Ouest de la Norvège: quartz + microcline + almandine + hastingsite. *9th Réunion Annuelle Sciences Terre, Paris, Editions Soc. Géol. France*, 582.

Smith, D. C. 1982b. The essence of an eclogite: Al^{vi}, as exemplified by the crystal-chemistry and petrology of high-pressure minerals in Norwegian eclogites. *Terra Cognita (First International Eclogite Conference), 32*, 300.

Smith, D. C. 1982c. A review of the controversial eclogites in the Caledonides. First International Eclogite Conference (abstract). *Terra Cognita, 2*, 317.

Smith, D. C. 1982d. Petrochemical notes on the hydrous minerals in the Aarsheimne-set eclogite pod, Selje District, Norway. *Terra Cognita, 2*, 317.

Smith, D. C. 1982e. On the characterisation and credibility of supersilicic, stoichio-metric and subsilicic pyroxenes. *Terra Cognita, 2*, 223.

Smith, D. C. 1983. Nomenclature, fluids and deformation, geodynamics, and nodules. *Terra Cognita, 3*, 329–334.

Smith, D. C. 1984. Coesite in clinopyroxene in the Caledonides and its implications for geodynamics. *Nature, 310*, 641–644.

Smith, D. C. 1985a. Coesite in the Straumen eclogite pod, Norway. *Terra Cognita, 5*, 226–227.

Smith, D. C. 1985b. The coesite-eclogite-facies: from microtextures to megatectonics. *Terra Cognita, 5*, 418–419.

Smith, D. C. 1988. A review of the peculiar mineralogy of the "Norwegian-eclogite province," with crystal-chemical, petrological, geochemical and geodynamical notes and an extensive bibliography. *Eclogites and eclogite-facies rocks*. Amsterdam: Elsevier, 1–206.

Smith, D. C. 1989. A rhythmic/helicyclic approach to eclogite genesis. *Terra Abstracts, 1*, 22.

Smith, D. C. 1991. Book review of "Eclogite Facies Rocks" (ed. D. A. Carswell). *Mineral. Mag., 55*, 490–492.

Smith, D. C. 1992. *The P-T-t evolution of the Norwegian and Chinese coesite-eclogite provinces: comparisons and controversies: Lecture.* "Pressure and temperature evolution of orogenic belts," CNR Summer School, Univ. Siena, 261–299.

Smith, D. C. 1993. On "ultrametastability" and other problematical petrological/geo-dynamical models for the origin and evolution of UHPM in crust-derived ter-ranes. *Terra Abstracts suppl. 4 to Terra Nova, 5,* 24–25.

Smith, D. C., and Cheeney, R. F. 1980. Orientated needles of quartz in clinopyro-xene: evidence for exsolution of SiO_2 from a non-stoichiometric supersilicic "cli-nopyroxene." *26th Internat. Geol. Congress, Paris, Abstracts, 02.3.1,* 145.

Smith, D. C., and Cheeney, R. F. 1981. A new occurrence of garnet-ultrabasite in the Caledonides: a Cr-rich chromite-garnet-lherzolite from Tværdalen, Liv-erpool Land, East Greenland. *Terra Cognita, 1,* 74.

Smith, D. C., and Kechid, S.-A. 1983. Three rare Al- and Na-rich micas in the Liset eclogite pod, Norway: Mg-Fe-margarite, preiswerkite and Na-eastonite. *Terra Cognita, 3,* 191.

Smith, D. C., Kechid, S.-A., and Rossi, G. 1986. Occurrence and properties of liset-ite, $CaNa_2Al_4Si_4O_{16}$, a new tectosilicate in the system Ca-Na-Al-Si-O. *Am. Min-eralogist, 71,* 1372–1377.

Smith, D. C., Kienast, J.-R., Kornprobst, J., and Lasnier, B. 1982. Eclogites and their problems: an introduction to the First International Eclogite Conference (F.I.E.C.). *Terra Cognita, 2,* 283–295.

Smith, D. C., and Lappin, M. A. 1982. Aluminium- and fluorine-rich sphene and clino-amphibole in the Liset eclogite pod, Norway. *Terra Cognita, 2,* 317.

Smith, D. C., and Lappin, M. A. 1989. Coesite in the Straumen kyanite-eclogite pod, Norway. *Terra Research, 1,* 47–56.

Smith, D. C., Mottana, A., and Rossi, G. 1980. Crystal-chemistry of a unique jadeite-rich acmite-poor omphacite from the Nybö eclogite pod, Sörpollen, Nordfjord, Norway. *Lithos, 13,* 227–236.

Smith, D. C., Oberti, R., Ungaretti, L., and Cannillo, E. 1986. On petrogenetical classifications of clinoamphiboles using discriminant analysis (DA) of crystal-chemical data: the key roles of Na and Al. *Multivariate Statistical Workshop for Geologists and Geochemists.* Ulvik, Norway, Abstracts Book, Spec. Publ. Dept. Chemistry, Bergen Univ.

Smith, D. C., and Pinet, M. 1985. Petrochemistry of opaque minerals in eclogites from the Western Gneiss Region, Norway: II. Chemistry of the ilmenite min-eral group. *Chemical Geology, 50,* 251–256.

Smith, D. C., and Yang, J. 1991. Low-Al orthopyroxene in eclogite from gneiss and peridotite near to the "Su-Lu Coesite-Eclogite Province," China. *Terra Nova, 6,* 12.

Smith, D. C., Yang, J., Oberti, R., and Previde-Massara, E. 1990. A new locality of nyböite and taramite, the Jianchang eclogite pod in the "Chinese Su-Lu coesite-eclogite province," compared with the nyböite- and taramite-bearing Liset eclogite pod in the "Norwegian coesite-eclogite province." *15th IMA Assembly, Beijing, China, Abstracts,* 889–890.

Sobolev, N. V., and Shatsky, V. S. 1987. Carbon mineral inclusions in garnets of metamorphic rocks. *Geologiya i Geofizika, 7,* 77–80 (in Russian).

Sobolev, N. V., and Shatsky, V. S. 1990. Diamond inclusions in garnets from meta-morphic rocks. *Nature, 343,* 742–746.

Sobolev, N. V., Shatsky, V. S., Vavilov, M. A., and Goryainov, S. V. 1991. Coesite inclusion in zircon from diamondieferous gneiss of Kokchetav massif–first find of coesite in metamorphic rocks in the USSR territory. *Dokl. Akad. Nauk SSSR, 321,* 184–188 (in Russian).

Sobolev, N. V., Shatsky, V. S., Vavilov, S. V., and Goryainov, S. V. 1994. Zircon from ultra high pressure metamorphic rocks of folded regions as an unique container of inclusions of diamond, coesite and coexisting minerals. *Dokl. Akad. Nauk, 334,* 4 (in Russian).

Sobolev, V. S. 1960. Role of high pressure in metamorphism. *21st International Geological Congress Abst.,* 74.

Sörensen, H. 1955. A preliminary note on some peridotites from northern Norway. *Norsk Geol. Tidsskrift, 35,* 93–104.

Spear, F. S. 1993. Metamorphic phase equilibria and pressure-temperature-time paths. *Mineral. Soc. America Monograph, 1,* 1–799.

Spencer, D. A., and Spencer-Cervato, C. 1993. Himalayan eclogites II: one stratigraphic unit at five metamorphic grades? *Abstract Suppl. 4 to Terra Nova, 5,* 85.

Tlili, A., Smith, D. C., Bény, J.-M., and Boyer, H. 1989. A Raman microprobe study of natural micas. *Mineral. Magazine, 53,* 165–170.

Trommsdorff, V. 1991. *Guide Book for the excursion on high-pressure metamorphism in the Cima Lunga unit (Central Alps).* Summer School on "Pressure and Temperature Evolution of Orogenic Belts," Univ. Siena, Italy, 30 pp.

Trueb, L. F., and Butterman, W. C. 1969. Carbonado: a microstructural study. *Am. Mineralogist, 54,* 412–425.

Ungaretti, L., Oberti, R., Cannillo, E., and Smith, D. C. 1986. *Discriminant analysis (DA) of 400 X-ray refined clinoamphibole.* 14th Assembly, International Mineralogical Association, Stanford, USA.

Ungaretti, L., Oberti, R., and Smith, D. C. 1985. X-ray crystal structure refinements of ferro-alumino- and magnesio-alumino-taramites from the Liset eclogite pod, Norway. *Terra Cognita, 5,* 429–430.

Ungaretti, L., Smith, D. C., and Rossi, G.1980. *Crystal-chemistry by electron microprobe and X-ray structure refinement analysis of a series of calcic- to sodic-amphiboles from the Nybö eclogite pod, Norway.* 12th Assembly, International Mineralogical Association, Orléans, France, Abstracts, p. 193.

van Roermund, H. L. M. 1981. On the eclogites of the Seve Nappe, Central Scandinavian Caledonides. *Terra Cognita, 1,* 70.

Velain, C. 1891. Sur des sables diamantifères recuellis par M. Charles Rabot dans la Laponie russe (Vallée de Pasvig). *C. R. Acad. Sci., Paris, 112,* 112–115.

Wegmann, C. E. 1929. Zur kentins der tektonischen Beziehungen metallogen tisher Provinzen in der nordchsten Fennoskandia. *Zeitschr. für prakt. Geologie, Jahrgang, 37,* 193–202.

Wezel, F.-C. 1988. Earth structural patterns and rhythmic tectonism. *Tectonophysics, 146,* 1–45.

Xu, S., Okay, A. I., Ji, S. Y., Sengör, A.M.C., Su, W., Liu, Y., and Jiang, L. 1992. Diamond from the Dabie Shan metamorphic rocks and its implication for tectonic setting. *Science, 256,* 80–82.

Yang, J., and Smith, D. C. 1989. Evidence for a former sanidine-coesite-eclogite at Lanshantou, Eastern China, and the recognition of the Chinese 'Su-Lu coesite-eclogite province', East China [abs.]. *Terra Abst., 1,* 26.

10

UHPM Terrane in East Central China

XIAOMIN WANG, RUYUAN ZHANG, AND J. G. LIOU

Abstract

We provide up-to-date information on the study of coesite ± diamond-bearing ultrahigh pressure metamorphic (UHPM) rocks from east central China. Geologic, petrologic, mineralogic, geochemical and geochronologic data are systematically summarized and compared based on previous studies for the Jiangsu (Su)-Shandong (Lu), the Dabie, and the Hongan regions where high pressure (HP) and UHP metamorphic rocks have been described. We also attempt, for the first time, to correlate the regional and metamorphic geology of the Dabie Mountains with the Su-Lu region. We also propose further research work for this HP and UHP metamorphic terrane.

Introduction

The widespread occurrences of coesite- and/or diamond-bearing metamorphic rocks in the Dabie Mountains (Dabie), Jiangsu (Su), Shandong (Lu), and Henan regions of east central China have attracted a great deal of attention since the first discovery of coesite in the Dabie Mountains in 1989. The Dabie-Su-Lu terrane is one of a few localities (Coleman and Wang, Chapter 1, this volume), including the Western Alps in Italy (Chopin, 1984), Western Gneiss Region in Norway (Smith, 1984), and Kokchetav Massif in Russia (Sobolev and Shatsky, 1990), where UHPM continental rocks have been documented.

Eclogite from east central China was first described in the early 1960s (Wang, 1963). Since then, information about its distribution and general characteristics had been accumulated through provincial geologic surveys but had not been published until recently (e.g., Regional Geological Survey of Anhui, 1987, 1988; Regional Geological Survey of Henan, 1989; Regional Geological Survey of Hubei, 1990). Some mineralogic and petrologic studies were available in the late 1970s and early 1980s [e.g., Xu et al. (1979) and Ying et al.

356

(1981) on the Su-Lu region; Ye et al. (1980), Xie et al. (1983) and Xie et al. (1985) on the Dabie Mountains]. Compilation of the metamorphic map of China stimulated research on high-pressure metamorphism and provided a better understanding of the metamorphic geology of the region (Dong et al., 1986; Dong, 1993). Enami and Zang (1988) reported magnesium staurolite in eclogite and its associated rocks from Donghai of the Su-Lu region. Coesite-bearing eclogite was soon recognized in eclogite from the southeastern Dabie Mountains (Okay et al., 1989; Wang et al., 1989). Quartz pseudomorphs after coesite were reported (Yang and Smith, 1989; Enami and Zang, 1990; Zhang et al., 1990) prior to identification of relict coesite inclusions in eclogitic garnet in the Su-Lu region (Hirajima et al., 1990; Zhang and Cong, 1991; Wang et al., 1993). Microdiamond in the Dabie coesite-bearing eclogites was recognized as inclusions in garnet (Xu et al., 1992). Recently, coesite-bearing eclogite was reported in Xinxian of Henan, northwestern Dabie (Wang, 1993; Zhang et al., 1993a, 1994). Detailed petrologic studies of coesite-bearing eclogites and ultramafics have been conducted by Wang et al. (1990, 1992) and Okay (1993) for the Dabie Mountains and by Zhang (1992), Zhang et al. (1993b, 1994) and Yang (1991) for the Su-Lu region.

This chapter summarizes results of previous work on these regions and provides current understanding of this UHP metamorphic terrane. We also attempt, for the first time, to compare the regional and metamorphic geology of the Dabie Mountains and the Su-Lu region; an additional research focus is also suggested.

Regional and Metamorphic Geology

The coesite- and diamond-bearing metamorphic terrane (from east to west they are Su-Lu, Dabie, and Hongan blocks) in east central China is situated within the collision belt between the Sino-Korean and Yangtze cratons (Fig. 10.1). The collision belt extends in a WNW-ESE direction for more than 2000 km and contains abundant high-pressure metamorphic rocks including eclogite and blueschist (Liou et al., 1989). The eastern portion of the collision belt was displaced more than 500 km by the sinistral strike-slip Tanlu fault from the Dabie Mountains to the Su-Lu region to the northeast (Xu et al., 1987; Wang and Liou, 1989).

In the Su-Lu block, the UHP terrane is bounded on the northwest against the Sino-Korean craton by the Yiantai-Qingdao-Wulian fault (YQWF) and on the southeast against the Yangtze craton by the Xiangshui-Jiashan fault (XJF) (Zhang, 1992). The Dabie block is fault-bounded in all directions, in the east by the Tanlu fault (TLF), in the west by the Shanchen-Machen fault

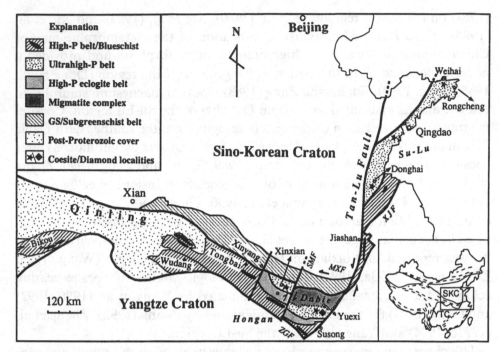

Figure 10.1. Tectonic setting of the Dabie Mountains (Dabie), Hongan block and Jiangsu-Shandong regions (Su-Lu) in east central China. The Dabie Mountains, the eastern portion of the collision zone between the Sino-Korean and Yangtze cratons, are offset by the Tanlu Fault (TLF) in the east. The eastern extension of the collisional zone is at the Su-Lu region. Star and diamond denote the areas where coesite and diamond, respectively, have been described. XJF = Xiangshui-Jiashan fault; SMF = Shanchen-Machen fault; MXF = Muzhitang-Xiaotien fault; ZGF = Ziangfan-Guangji fault.

(SMF), in the north by the Muzhitang-Xiaotien fault (MXF) against the Sino-Korean craton (SK), and in the south by the Ziangfan-Guangji fault (ZGF) against the Yangtze craton (YZ) (Xu et al., 1986; Xu et al., 1987; Liu and Hao, 1989) (Fig. 10.2). The Hongan block to the west of the Dabie block is bounded with the Tongbai block by faults. In these blocks, the coesite-bearing UHPM rocks occur to the north, whereas HPM rocks, including blueschist and/or eclogite, have been described in the southern tip of the Dabie Mountains and Tongbai-Wudang-Qinling areas (Zhou et al., 1989, 1993; Ernst et al., 1991; Eide, 1993).

Metamorphic rocks of the UHPM terrane include quartzofeldspathic gneiss, biotite-hornblende gneiss, pyroxene gneiss, and metapelite together with minor marble, eclogite, and ultramafic rocks; these rocks have been assigned to be Archean to Proterozoic in age with local stratigraphic names (Regional Geological Survey of Anhui, 1987; Regional Geological Survey of

Henan, 1989; Regional Geological Survey of Jiangsu, 1991). They have been considered to be metamorphosed under regional granulite facies to high-greenschist facies conditions. These Archean–Early Proterozoic units (the Dabie Group in the Dabie block, and the Donghai Group and Jiadong Group in the Su-Lu region) were reported to be nonconformably overlain by Middle–Later Proterozoic units (the Susong Group in the Dabie Mountains area, and the Haizhou Group and Fenzishan Group in the Su-Lu region). The Middle–Later Proterozoic units consist of bimodal volcanics, quartz schist, albite schist, phengite schist, and metabasite metamorphosed under greenschist-blueschist facies conditions (Dong et al., 1986; Eide, 1993).

Jurassic volcaniclastic rocks are confined to the area north of the Archean–Early Proterozoic unit in the Dabie Mountains, and lie only along faults in the Su-Lu region. Cretaceous–Quaternary sediments are ubiquitous as a cover sequence and occur mostly along the margins of metamorphic terrane. Postcollisional Mesozoic granites are abundant (Fig. 10.2).

The Dabie block has been considered to be a fragment of the Sino-Korean craton (Tectonic Compilation Group, Institute of Geology, Academia Sinica, 1974) or a microcontinent (Zhang et al., 1984). Inside the block, the Archean–Early Proterozoic unit can be subdivided into two subterranes (Wang and Liou, 1991; Wang et al., 1992; Okay, 1993) (Fig. 10.2). The north Dabie terrane (NDT) is composed of quartzofeldspathic rocks that display brittle deformation and contain an amphibolite-granulite facies assemblage of hornblende + epidote + biotite + muscovite + plagioclase + quartz ± garnet ± clinopyroxene ± orthopyroxene (Regional Geological Survey of Anhui, 1987). The NDT displays a SE-NW weak regional foliation in the Dabie Mountains area (Xu et al., 1986). Minor eclogites are associated with ultramafics whereas migmatites are abundant. The south Dabie terrane (SDT) is composed of biotite gneiss, metapelite, marble, and minor ultramafic rock. Contrary to the NDT, the SDT contains abundant eclogites occurring as blocks or boudins enclosed within gneiss or marble, or as layers associated with gneiss, metapelite, and marble. Gneisses constitute up to 80 vol% of the SDT; other lithologies are, in the order of abundance, metapelite, marble, eclogite, migmatite, and serpentinized ultramafics. These rocks exhibit well-defined E-W regional foliation which dips to the south and have experienced multiple ductile deformation. Post-collisional Mesozoic granites are widespread in both subterranes. In the Dabie Mountains, the boundary that separates these two terranes has been considered as either a nonconformity (Regional Geological Survey of Anhui, unpublished report) or a ductile tectonic melange belt in a regional detachment plane that formed at great depth (Xu et al., 1986; Dong et al., 1989). A shear zone has been located in the

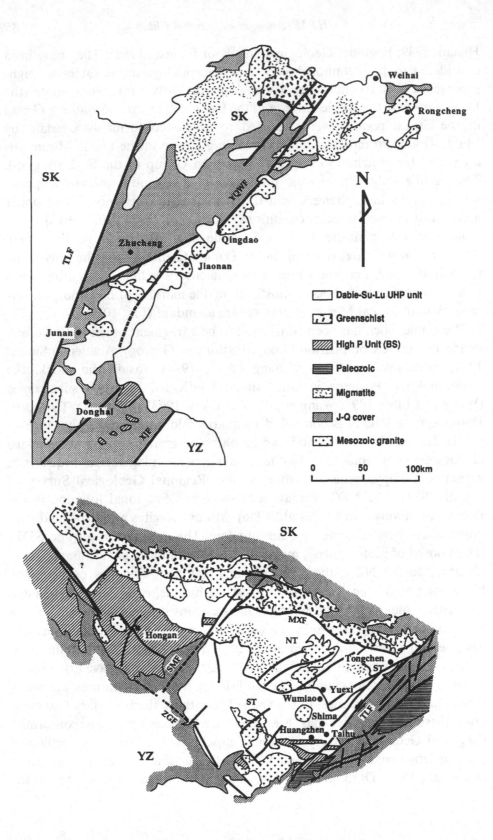

boundary between these two subterranes by Brad Hacker (personal communication, 1992).

Within the SDT subterrane, a northward increase in metamorphic grade has been suggested by Wang et al. (1992) based on *P-T* estimates of eclogites from three different areas. A similar prograde metamorphic temperature variation has also been described for Hongan high *P/T* blueschist and eclogite terrane (Zhou et al., 1993). However, Okay (1993) suggests tectonic breaks separating the epidote amphibolite, "cold" eclogite, and "hot" eclogite to the north.

Geologic correlation between the Dabie Mountains and the Su-Lu region suggests that the two terranes were continuous before they were offset by the Tanlu fault. They are comparable in many aspects, including tectonic settings, rock association, and metamorphic history (Table 10.1) (Wang et al., 1992). In the Su-Lu region, all the UHPM rocks occur in the area southeast to the Yantai-Qingdao-Wulian fault (YQWF) which separates them from the Sino-Korean craton (Fig. 10.2). The Su-Lu region is composed of a Later Proterozoic sequence of quartz-mica schist, chloritoid-kyanite-mica schist, greenschist, and marble, and an Early Proterozoic sequence of biotite-hornblende gneiss (most contain aegirine-augite), metapelite which includes mica schist and garnet-zoisite-kyanite quartz schist, and minor amounts of marble, amphibolite, eclogites, and ultramafic rocks (especially garnet-bearing ultramafic rocks) (Zhang, 1992; Zhang et al., 1994). Mesozoic granites are widespread. Volcanoclastic rocks from the Jurassic and Cretaceous ages constitute the cover sequence, and occur mainly along the major faults.

Recent geochronologic and petrologic studies have challenged the previous understanding of geology of central eastern China. Available Sm/Nd mineral-rock isochrons indicate that some eclogites and ultramafic rocks date to the early Mesozoic age (Li et al., 1989a; Li et al., 1989b; Li and Liu, 1990; Chen et al., 1992; Yin and Kroner, personal communication, 1991; Li et al., 1993). U-Pb dating of zircon and $^{40}Ar/^{39}Ar$ dating of phengite and hornblende of

Figure 10.2. Simplified tectonic map of the Dabie and Su-Lu regions in the same scale, showing the main lithologic units. The Dabie block is truncated by the Tanlu fault (TLF) in the east and by the Shanchen-Machen fault (SMF) in the west. It is bounded by the Muzhitang-Xiaotien fault (MXF) with the Sino-Korean craton (SK) to the north and by the Ziangfan-Guangji fault (ZGF) with the Yangtze craton (YZ) in the south (Xu et al., 1986, 1987; Liu and Hao, 1989). The dashed line in the middle of the Dabie Mountains is the boundary between the northern Dabie terrane (NT) and the southern Dabie terrane (ST). The UHP terrane in Su-Lu region is bounded by the Yiantai-Qingdo-Wulian Fault (YQWF) in the northwest. Xiangshui-Jiashan fault (XJF) separates the high-pressure unit (greenschist-blueschist) from the Yangtze craton. (Modified from Wang et al., 1992, and Zhang, 1992.)

Table 10.1. *Comparison between the Dabie Mountains and Su-Lu regions*

	Dabie Mountains	Su-Lu region
Field occurrence	Layers of eclogite (garnet-clinopyroxenite) interbedded with ultramafic rocks. Layers interbedded with gneiss and marble. Blocks within gneiss and marble.	Eclogite occurs as nodule or layer interbedded within ultramafic rocks; pod or block in gneiss and marble.
Mineral assemblage	High-temperature (>640°C): garnet, omphacite, phengite, kyanite, epidote, rutile, coesite/quartz, diamond. Low-temperature (580-640°C): garnet, omphacite, phengite, kyanite, paragonite, glaucophane, Mg-katophorite, epidote, rutile, quartz.	Except for a few, eclogites are "high-T" eclogite. Mineral assemblages of eclogite in gneiss are Grt+Omp+Rt+Cos / Qz+Ky+Zo+Ny+Phe+Ep+high-Al Ttn +Rt+Na-Ca Amp (Bar, Kato, taramite), +Mg-St +Pg; Grt+Omp+Rt+Qtz+Phn+Amp (colorless) in ultramafic rocks; Grt+Omp+Rt+Amp+Zo in marble.
Mineral chemistry	Garnet: varies greatly, $Gr_{23}Alm_{57}Prp_{18}Sp_1$ (W-3), $Gr_{37}Alm_{46}Prp_{15}Sp_2$ (MW-44), $Gr_{11-30}Alm_{45-55}Prp_{9-35}Sp_{1-16}$ (ST-15). Garnet has no consistent zonation in high-temperature eclogite and has a complicated prograde zonation. Omphacite: $Jd_{32}Di_{52}Aeg_{16}$ (W-3), $Jd_{53}Di_{42}Aeg_5$ (MW-44), $Jd_{44}Di_{45}Aeg_{11}$ (ST-15). Zonation of omphacite is not obvious. Primary (low-temperature eclogite facies) amphibole: Glaucophane: $Na^{M4}/(Na^{M4}+Ca) = 0.75$, $Mg/(Mg+Fe^{2+}) = 0.93$, $Fe^{3+}/(Fe^{3+}+Al^{VI}) = 0.23$. Mg-katophorite: $(Na+K)_A = 0.73$, $Na^{M4}/(Na^{M4}+Ca) = 0.45$, $Mg/(Mg+Fe^{2+}) = 0.67$. Phengite: $Si = 3.5 \sim 3.7$ pfu, $Mg^{2+} = 0.35 \sim 0.47$ pfu, $Mg/Fe = 0.72 \sim 0.84$.	Garnets show a wide compositional space: type I (in gneiss): $Alm_{20-61}Prp_{12-43}Gr_{14-40}$; type II (in ultramafic rocks): $Alm_{23-38}Prp_{40-46}Gr_{21-28}$; type III (in marble): $Alm_{20-42}Prp_{16-30}Gr_{45-55}$. Most garnets are homogeneous but some exhibit zonation. The Mg/Fe ratio decreases from core to rim. Omphacite: type I: $Jd_{37-71}Aeg_{0-20}Di_{11-51}$; type II: $Jd_{26-40}Aeg_{0-9}Di_{56-69}$; type III: $Jd_{22-31}Aeg_{0-5}Di_{65-74}$. Zonation of omphacite is not obvious. Amp: exhibits evolutionary trend from Na- through Na-Ca to Ca-Amp from early to late stage. Na-Amp, nyböite: Gln 1.48, Si=7.19; $Na^{M4}/(Na+Ca)^{M4}=0.76$; $Fe^{3+}/(Fe^{3+}+Al^{VI})=0.13$; $Mg/(Mg+Fe^{2+}) = 0.73$. Phe: $Si = 3.4-3.6$ & 3.2-3.4; $Mg=0.43-0.55$ and 0.26-0.41; $Mg/(Mg+Fe)=0.76-0.90$ and 0.64-0.84, for Jiangsu and Shandong, respectively.
P-T estimate	Peak eclogite facies metamorphism: High-temperature eclogite: 744 to 778°C (EG) and 697-813 °C (KR); 29-41 kb (Wumiao), 639 to 696°C (EG) and 630-710 °C (KR); 27-37 kb (Shima). Low-temperature eclogite: 582-611°C (EG) and 588 °C (KR); 21-25 kb (Huangzhen). Prograde blueschist facies P-T: ~400°C. Retrograde amphibolite facies P-T: 475-530°C; 6 kb.	Peak eclogite facies metamorphism: coesite eclogite in gneiss: 715-948°C at 30 kb except for one sample (633°C); eclogite in lherzolite at Chijiadian: 773-830°C at >35 kb Granulite facies retrograde metamorphism: 750°C at 8kb; amphibolite facies retrograde: 530-708°C (±75°C) at about 10 kb.
P-T evolution	Prograde through blueschist facies, peak metamorphism of eclogite facies followed by amphibolite and greenschist facies metamorphism. The prograde record is found only in the low-temperature eclogite.	Pre-eclogite event is "high-P" amphibolite facies and epidote amphibolite facies metamorphism; then through peak metamorphism of eclogite facies followed by granulite or amphibolite facies metamorphism in different localities.
Carbonate	Mineral assemblage: garnet ($Gr_{50-60}Alm_{19-28}Prp_{16-25}Sp_1$), clinopyroxene ($Jd_{22}Di_{74}Aeg_4$), calcite/aragonite, dolomite, phengite ($Si^{4+} = 3.6 \sim 3.7$ pfu), epidote, rutile, quartz/coesite. P-T conditions: peak metamorphism, 630-760°C (EG), 570-640°C (KR); >27 kb; retrograde metamorphism, 475-550°C, <7 kb.	Mineral assemblage: Olivine, calcite, diopside, muscovite.
Gneiss and metapelite mineral assemblage	Gneiss: K-feldspar, plagioclase, biotite, muscovite, epidote, garnet, rutile, quartz/quartz pseudomorph after coesite. Metapelite: garnet, phengite, epidote, rutile, quartz/quartz pseudomorph after coesite.	Gneiss: Kspar, Pl, Bt, Ms, Phe, Ep, Grt, Qz, Aeg, Hb, Ttn. Schist: Ky+Qz+minor Phe. Metapelite: Grt+Ky+Zo+Qz+Phe+Rt+Ttn.
Age of eclogite facies metamorphism	220-240 ± 10 Ma (Sm/Nd mineral and whole rock isochron on eclogite and ultramafic rocks). Similar ages were obtained on phengites based on $^{40}Ar/^{39}Ar$ method.	Sm-Nd: (individual mineral and whole rock or individual minerals) 210-230 Ma. $^{40}Ar/^{39}Ar$ (Phengite): 435 Ma for Mengzhong coesite eclogite.

eclogite and gneiss from the Dabie Mountains yield a similar age (Eide et al., 1992; Ames et al., 1993; see the next section). More and more data indicate that the UHP and HP metamorphic rocks in the collision belt are products of a Triassic collision between the Sino-Korean and Yangtze cratons.

Occurrence and Petrology of Eclogite

Classification, Field Occurrence, and Distribution of Eclogite

Various classification schemes on eclogite from east central China based on field occurrence and mineral assemblage have been proposed by Wang et al. (1990) and Okay (1993) for the Dabie eclogites and by Enami et al. (1992) and Zhang (1992) for the Su-Lu eclogites. A twofold division of eclogite is adopted here: (1) eclogite associated with ultramafic rocks, and (2) eclogite associated with gneiss, marble, or metapelite. Strictly speaking, most eclogites associated with ultramafic rocks in the Dabie Mountains are garnet clinopyroxenite with diopsidic pyroxene instead of omphacite. However, those eclogites associated with ultramafic rocks in the Su-Lu region contain omphacite. Since study on ultramafics and their associated eclogite from east central China is in an early stage (e.g., Zhang et al., 1994), in this chapter we place emphasis on eclogites associated with gneiss, marble, and metapelite. A short description of ultramafic rocks is given in the Discussion section.

Eclogites are widely distributed in the Dabie-Su-Lu region along a belt exceeding 1100 km from Weihai, to Donghai, to the Dabie Mountains and Hongan blocks (Figs. 10.1 and 10.2; Table 10.2). Wang et al. (1990) indicated that eclogite pods or blocks with sizes ranging from a few centimeters to 20 m are concentrated in certain horizon of the gneiss terrane, subparallel to regional foliation. These blocks vary in shape from lenses, typically 2 by 1 by 0.5 m in dimension, to rounded bodies of 1–2 m in diameter. A few hundred meters to 1-km-long blocks form an isolated hill. Some blocks were internally strongly foliated and display compositional bands with brown garnet-rich layers and dark green omphacite-rich layers. The altitudes of eclogite foliation or their compositional bands differ from one block to another (Wang et al., 1990). Blocks are wrapped by foliation of the country rocks. Ubiquitous post-eclogitic metamorphism results in amphibolitization around the rims and along certain layers of the eclogite blocks. A transition occurs from eclogite cores, through garnet amphibolite in the intermediate parts, to amphibolite on the margins of the eclogite blocks. Foliation due to the superimposed metamorphism is more intense rimwards. The same foliation is also well developed in gneissic country rocks and marble blocks, and is discordant with the original compositional bands of the eclogites.

Table 10.2. *Eclogite from different locations in east central China*

Locality	Occurrence	Mineral assemblage	Mineral chemistry	P-T conditions
Wumiao, Dabie	Blocks, layers enclosed in gneiss, marble, and metapelite	Garnet, omphacite, phengite, coesite/quartz, rutile, ilmenite, carbonate.	Garnet: $Gr_{23}Alm_{57}Prp_{18}Sp_1$ (W-3); omphacite: $Jd_{32}Aug_{52}Aeg_{16}$ (W-3); phengite: $Si = 3.5 \sim 3.7$.	T=744-778°C; P: 29-41 kbar.
Shima, Dabie	Blocks in gneiss	Garnet, omphacite, phengite, epidote, coesite/quartz, diamond, kyanite, rutile, carbonates.	Garnet: $Gr_{37}Alm_{46}Prp_{15}Sp_2$ (MW-44); omphacite: $Jd_{53}Aug_{42}Aeg_5$ (MW-44); phengite: $Si = 3.5 \sim 3.7$ pfu.	T=639-686°C; P >27-37 kbar
Huangzhen, Dabie	Layers interbedded with gneiss	Omphacite, garnet, glaucophane, kyanite, rutile, epidote, phengite, quartz.	Garnet: $Gr_{11-30}Alm_{45-55}Prp_{9-35}Sp_{1-16}$ (ST-15), garnet has a complicated prograde zonation; omphacite: $Jd_{44}Aug_{45}Aeg_{11}$ (ST-15); glaucophane: $Na^{M4}/(Na^{M4}+Ca)$=0.75, $Mg/(Mg+Fe^{2+})$=0.93, $Fe^{3+}/(Fe^{3+}+Al^{VI})$= 0.23; Mg-katophorite: $(Na+K)_A$=0.73, $Na^{M4}/(Na^{M4}+Ca)$=0.45, $Mg/(Mg+Fe^{2+})$= 0.67; phengite: Si =3.5 ~3.7 pfu.	T=582-611°C; P=21-25 kbar
Mengzhong, Su-Lu	Pod in gneiss	Garnet, omphacite, phengite, coesite/quartz, rutile, kyanite, zoisite.	Garnet: $Gr_{32-39}Alm_{20-42}Prp_{27-50}$; omphacite: $Jd_{30-58}Aug_{0-13}Aeg_{44-54}$; phengite: Si=3.5-3.6.	T=807-863°C; P>28 kbar
Jianchang, Su-Lu	Layers interbedded with gneiss and block in gneiss	Garnet, omphacite/jadeite, phengite, coesite/quartz, rutile, kyanite, zoisite, epidote, nyböite, high-Al titanite.	Garnet: $Gr_{15-38}Alm_{48-62}Prp_{12-26}$. Clinopyroxene: $Jd_{40-71}Aug_{13-47}Aeg_{3-20}$ nyböite: Na^{M4}=1.37-1.48, $(Na+K)_A$=0.86 -0.99, Si=7.12-7.18. High-Al titanite: Al_2O_3=7.4 wt%.	T=716-841°C; P>28 kbar
Datan, Su-Lu	Pod in gneiss	Garnet, omphacite, quartz, rutile, kyanite, zoisite.	Garnet: $Gr_{19-37}Alm_{23-38}Prp_{32-56}$; omphacite: $Jd_{26-36}Aug_{52-82}Aeg_{0-7}$; phengite: Si=3.5-3.6.	T=708-794°C; P>25 kbar

Layered eclogites from 5–100 cm thick are interlayered with gneiss, marble, and metapelite. Some layers extend for more than 1 km. Eclogites are massive compared to retrograded amphibolites, with foliation defined by the preferred orientation of amphibole and mica crystals. The foliation of amphibolite is concordant with layers of eclogite and amphibolite. Twenty layers of eclogite with consistent altitudes occur within a distance of 50 m at Huangzhen in the Dabie Mountains (Wang et al., 1990). Eclogite interlayered with marble in the northern part of the SDT of the Dabie Mountains have a greater thickness (~1 m) than those interlayered with gneiss and metapelite in the south. The contacts between eclogite and marble are usually sharp.

Between the two above-mentioned common occurrences, discontinuous boudins of eclogite often occur within the country rocks, especially marble. These eclogite boudins of dimensions 50 × 40 × 20 cm are spaced by a distance of 10–50 cm (Zhang, 1992; Okay, 1993; Wang and Liou, 1993). The spatial distribution indicates that these discontinuous boudins are derived from a single eclogite layer. The randomly distributed, isolated blocks are probably the result of scrambling of the discontinuous boudins during intensive deformation. Some smaller blocks were mixed into intensely deformed marble that fills the spaces between the eclogite blocks. The foliation of marble wraps around the eclogite blocks.

Mineral Assemblage and Petrography of Eclogite

Eclogite from east central China consists of minerals formed during at least two stages: the UHP (eclogite facies) metamorphic stage and the superimposed granulite-amphibolite-greenschist facies (post-eclogitic) metamorphic stage. Some eclogites from the southern part of the SDT have an additional relict prograde assemblage of blueschist facies (e.g., in Huangzhen of Dabie) and epidote amphibolite facies minerals (e.g., in Donghai of Su-Lu) formed in the pre-eclogite stage (Table 10.2).

UHP Assemblage

The UHP assemblage of eclogite includes garnet + omphacite + phengite + epidote/zoisite + rutile + apatite ± kyanite ± coesite/quartz ± diamond ± carbonate (Fig. 10.3a). Nyböite and high-Al titanite were reported in the UHP assemblage of the Su-Lu region (Fig. 10.3b) (Hirajima et al., 1992; Zhang, 1992). Microdiamond has been discovered as inclusion in eclogite from the Dabie Mountains (Xu et al., 1992).

Well-preserved eclogite is medium- to coarse-grained and granular in texture. Garnet (30–80%) and omphacite (15–50%) are the main phases; other

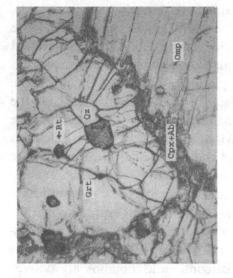

(a)

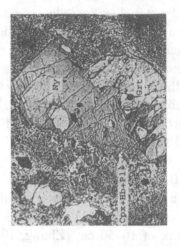
(b)

Figure 10.3. Photomicrographs of mineral assemblage of eclogite: (a) Common eclogite mineral assemblage from the Dabie Mountains with simple mineralogy of garnet (Grt), omphacite (Omp), rutile (Rt), quartz/coesite (Qz/Coe), ± phengite, ± kyanite, ± carbonate, and retrograde phases of clinopyroxene + albite (Cpx + Ab), and titanite. Field view is 1.5 mm. (b) Nyböite-bearing eclogite from the Su-Lu region with Nyböite (Ny), garnet (Grt), phengite, zoisite, and omphacite and retrograde assemblage of clinopyroxene (Cpx) + hornblende (Hb) + plagioclase (Pl). Field view is 6 mm. (c) Relict coesite (Coe) and low-relief quartz pseudomorph after coesite (Qz) as inclusions in garnet (Grt) from the Dabie Mountains. Radiating fractures are well developed around the inclusions. Retrograde assemblage after garnet is pargasitic hornblende and plagioclase (Hb + Pl). Field view is 1.2 mm. (d) Quartz pseudomorph after coesite as inclusions in epidote from the Su-Lu region. Field view is 1.2 mm. (e) Omphacite being replaced by clinopyroxene (diopside-rich) and albite (Cpx + Ab). Field view is 0.72 mm. (f) Garnet being replaced by granulite facies mineral assemblage of plagioclase, diopside, hypersthene and minor hornblende from Weihai in Su-Lu region. Field view is 0.8 mm.

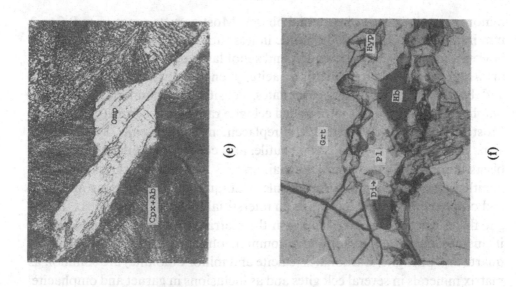

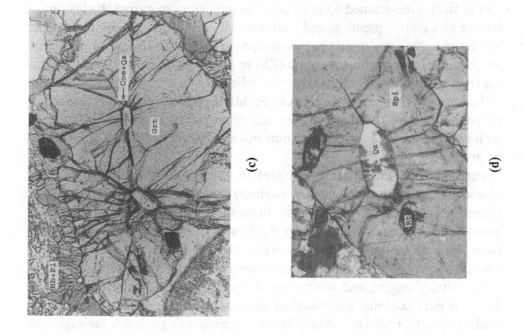

minor minerals occur as interstitial phases. Most garnets range from 0.5 to 2 mm in diameter. Garnet is idioblastic in less thoroughly retrograded eclogite; however, most garnets are irregular and xenoblastic due to retrograde replacement. Inclusions in garnet are omphacite, phengite, apatite, rutile, pargasitic and borroisitic amphiboles, carbonates, coesite, and diamond. Idioblastic omphacite crystals in less retrograded eclogite range in size from 0.1 to 1 mm. Most omphacites show symplectite replacement at their rims. Minor inclusions in omphacite include garnet, rutile, and coesite. Nyböite, as porphyroblasts in some eclogites from Donghai, up to 3 × 5 mm in size, is commonly mantled by a thin taramite rim. Rutile is ubiquitous as inclusions in garnet and omphacite, and also occurs as an interstitial mineral in the matrix. Phengite flakes are concentrated mainly in the matrix but a few are found as fine inclusions in garnet. Kyanite is a common phase in eclogite that contains quartz and phengite. Carbonates (calcite and minor dolomite) are common as matrix minerals in several eclogites and as inclusions in garnet and omphacite. Porphyroblastic epidotes are up to 5 mm in size; matrix epidotes are finer (<1 mm in size). Fine-grained apatites occur as inclusions in garnet. Inclusions of coesite and quartz pseudomorphs after coesite are rather common in garnet and omphacite; some quartz pseudomorphs after coesite also occur in epidote and kyanite from the Su-Lu region (Zhang et al., 1990; Zhang, 1992); radiating fractures are well developed around relict coesite or coesite pseudomorph inclusions in host minerals (Figs.10.3c,3d). Euhedral cubic or octahedral microdiamond inclusions have been described by Xu et al. (1992) in garnet with grain size from 10 to 60 μm; some microdiamond inclusions are replaced by graphite.

For some low-temperature eclogites in the Dabie Mountains, an early blueschist facies assemblage was partially preserved (Wang et al., 1992). Coarse-grained garnets (up to 5 mm in diameter) have at least two stages of growth with a distinct discontinuity visible at about two-thirds of the distance from the core to the margin; this break divides the garnet into two parts. The core part contains abundant inclusions of glaucophane, epidote, rutile, paragonite, phengite, and quartz, representing blueschist facies assemblage. The rim part contains inclusions of omphacite, phengite, Mg-katophorite, epidote, quartz, rutile, and minor ilmenite, representing eclogite facies assemblage. Glaucophane was also identified as the core of coarse-grained amphibole with barroisite in the intermediate part, and Mg-taramite or pargasite at the rim. Corresponding to the two-stage recrystallization, the eclogite shows two-stage foliation: The first foliation (F1) is defined by inclusions of Mg-katophorite, phengite, paragonite, and epidote in garnet; the second foliation

(F2) is defined by pargasitic amphibole, phengite, omphacite, and epidote in the matrix, and creates a large angle to the first foliation.

Post-eclogitic Assemblage

A spectrum of post-eclogitic metamorphic features was superimposed on the UHP assemblages of eclogite. This metamorphism occurred mainly under amphibolite and greenschist facies, and locally granulite facies conditions (e.g., in Weihai). Amphibole is the most common mineral formed during this stage. Symplectic amphiboles with epidote ± albite formed after garnet are calcic and range from silicic edenite, edenite, and edenitic hornblende, to ferroan pargasitic hornblende and ferropargasite, based on the classification of Leake (1978). Barroisite occurs in the cores of fine-matrix amphiboles rimmed by pargasite and actinolite. Fine-grained aggregates of pargasite and epidote ± albite occur around relict garnet (Fig.10.3c). The superimposed assemblage after omphacite consists of symplectic intergrowth of clinopyroxene ($\sim$Jd$_{20}$) + albite (An$_{2-8}$) (Fig. 10.3e). Other superimposed metamorphic features include (1) phengite being replaced by biotite, which in turn was replaced by chlorite; (2) muscovite and margarite + quartz forming after kyanite; (3) titanite and ilmenite forming after rutile; and (4) chlorite forming after garnet.

In the Weihai area of the Su-Lu region, post-eclogitic metamorphism occurred under granulite facies conditions (Zhang and Cong, 1991; Wang et al., 1993). All garnets are replaced by a thin corona of plagioclase (An$_{37-47}$) + hypersthene + diopside + minor hornblende (Fig. 10.3f), and omphacites by an intergrowth of diopside + plagioclase + quartz (Zhang and Cong, 1991). In the same thin section, a coarse-grained coesite of 0.3 mm is well preserved in garnet with distinct radiating fractures. The preservation of relict coesite and the occurrence of granulite facies assemblage suggest that granulite facies overprint must have been a very short-life feature since coesite would have converted to quartz readily under granulite facies.

Mineral Chemistry

UHP Minerals

Garnet covers wide compositional ranges as shown in Fig. 10.4. Such compositional variations resulted from differences in bulk rock composition and *P-T* conditions, and from the effect of superimposed metamorphism. Those garnets in eclogites associated with marble are distinctly higher in grossular component (44–60 mol%) than those in eclogites associated with gneiss. Most

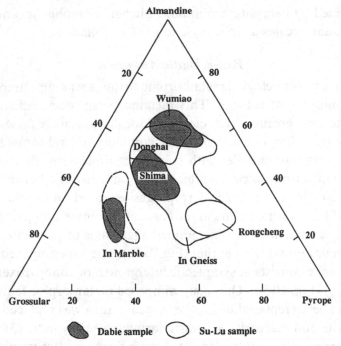

Figure 10.4. Garnet composition diagram. Solid lines are for eclogites associated with gneiss, whereas dashed lines are for eclogites associated with marble.

analyzed garnets in eclogites from gneiss fall in to the compositional range of $(Alm + Spe)_{20-70}Gr_{20-60}Prp_{10-55}$ and show distinct variation with respect to locality. For example, garnets from Rongcheng in the Su-Lu region contain 55 mol% pyrope component whereas those from Wumiao of the Dabie Mountains contain 70 mol% almandine component (Fig. 10.4).

Most garnets in eclogite from the Dabie Mountains show weak or no compositional zonation. However, low-temperature eclogites ($T<650°C$) from the southern tip of the Dabie Mountains contain coarse-grained garnets with a complicated prograde pattern with Mg/Fe ratio from 0.14 in the core to 0.7 in the margin (Wang et al., 1992) (Fig. 10.5). The almandine content increases from 43 mol% in the core to a maximum value of 58 mol% in the intermediate part and margin, whereas the pyrope content increases from 10 mol% in the core to 20 mol% in the intermediate part and 43 mol% at the margin. Grossular displays a typical bell-shape zoning pattern, with a high flat core (~25 mol%), decreasing rapidly to 12 mol% at the margin. Wang et al. (1992) noticed that most of these changes in the garnet occurred around the discontinuity in the intermediate part of the grain. Based on this observation and different mineral inclusions in the core and rim in garnets described in the

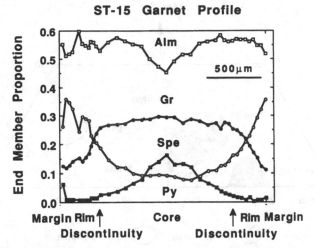

Figure 10.5. Garnet composition zonation from the Dabie Mountains. (After Wang et al., 1992.)

previous section, these garnets must have grown from blueschist facies to eclogite facies conditions. Prograde zonation patterns were also found in eclogites from Hongan block (Zhang and Liou, 1994) and the Su-Lu region (Enami et al., 1993). Zhang (1992) suggested that chemical zoning of most garnets in the Donghai eclogites revealed by x-ray mapping is caused primarily by Ca-Fe variation with nearly constant Mg content and a few by Fe-Mg variation. Enami et al. (1993) described a retrograde garnet zonation in eclogite from Junan in the Su-Lu region, which is believed to record later retrogression.

Omphacite shows similar compositional dependence as a function of bulk rock composition as well as *P-T* conditions, but has a much more restricted overall compositional variation compared with garnet (Fig. 10.6a). Eclogite associated with marble contains omphacite with high diopside component. Jadeite-rich omphacite from Donghai has been reported to contain 72 mol% of jadeite (Zhang, 1992). Such distinctly high jadeite content is apparently caused by the unique bulk chemistry; unfortunately, the original bulk chemistry cannot be determined because the rock is severely retrograded (Hirijima et al., 1992). Omphacites from other localities are compositionally within $Jd_{20-55}Di_{45-80}Aeg_{0-27}$. For example, eclogite from Shima in the Dabie Mountains contains omphacite with compositions ranging from about Jd_{29} to Jd_{53}. Compositional zonation in omphacite is absent or weak in almost all analyzed omphacites, even in those associated with garnets that have complicated zonation from the Dabie Mountains. Wang et al. (1992) examined a coarse-grained

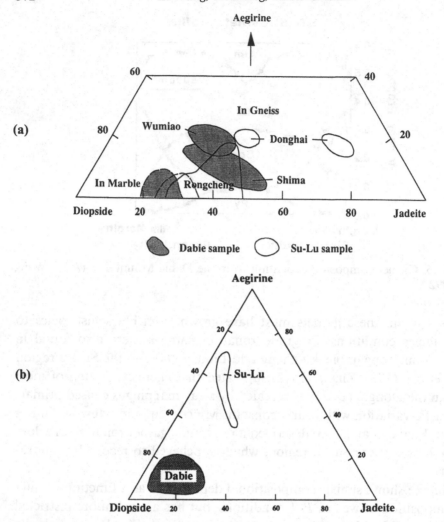

Figure 10.6. Clinopyroxene composition diagram. (a) Composition of clinopyroxene of UHP metamorphism (omphacite). Solid lines are for eclogites associated with gneiss, whereas dashed lines are for eclogites associated with marble. (b) Composition of clinopyroxene of retrograde metamorphism (diopside and aegirine-rich clinopyroxene).

omphacite with 0.7 mm in diameter and reported no apparent compositional zoning. A weak zonation was reported by Enami et al. (1993) with lower jadeite and higher aegirine contents in the core for eclogite from Donghai.

A close relationship between the composition of garnet and omphacite, and bulk rock chemistry in east central China was recognized. The most obvious feature of garnet and omphacite is the difference in composition between eclogites associated with gneiss and marble (Figs. 10.5 and 10.7). For example, eclogites associated with marble from the Su-Lu region are Ca-rich in bulk

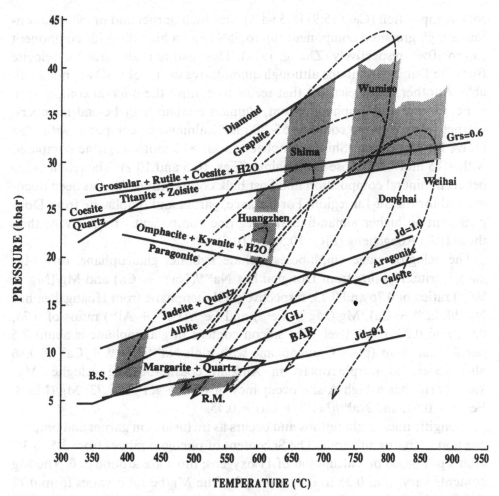

Figure 10.7. Pressure-temperature diagram for eclogite facies metamorphism and post-eclogite metamorphism for the Dabie and Su-Lu region and metamorphic *P-T* paths. One *P-T* path for eclogites from each area of Wumiao, Shima and Huangzhen of Dabie, and Donghai and Weihai of Su-Lu is defined. R.M. (area with shadow) = estimated *P-T* conditions for retrograde metamorphism (after Wang et al., 1990). B.S. (shadowed area) = estimated *P-T* conditions for prograde blueschist facies metamorphism. GL = glaucophane stability field. BAR = barroisite stability field. Data sources: Reaction jadeite + quartz = albite (Holland, 1980, 1983). Jd=1.0: Reaction jadeite + quartz = albite with jadeite content 10 mol% in clinopyroxene; paragonite = kyanite + omphacite (Newton, 1986); calcite = aragonite (Hacker and Bohlen, personal communication); coesite = quartz (Bohlen and Boettcher, 1982; Mirwald and Massonne, 1980); diamond = graphite (Kennedy and Kennedy, 1976); grossular + rutile + coesite + H_2O = titanite + zoisite (Berman, 1986); margarite + quartz stability field (Perkins et al., 1980); glaucophane stability field (Maresch, 1977).

rock composition (CaO 15.9–18.5 wt%), and their garnet and omphacite contain a high grossular component (up to 55%) and a high diopside component (up to 70%), respectively (Zhang, 1992). This feature is also true for eclogite from the Dabie Mountains although quantitative values of CaO are not available. Another major element that seems to control the mineral composition is Fe. Garnet and omphacite from Wumiao contain high Fe-end members. Garnet from Wumiao contains 55–70 mol% almandine compared with <55 mol% for those from Shima; omphacite has >12 mol% aegirine compared with <15 mol% for those from Shima (Figs. 10.5 and 10.7). The relationship between mineral composition and host bulk composition also has been documented for the Su-Lu region. For instance, garnet and omphacite from Donghai contain higher almandine and aegirine components, respectively, than those from Rongcheng (Figs. 10.5 and 10.7).

The eclogitic stage amphiboles include nyböite, glaucophane, and Mg-katophorite. Nyböite from Donghai has $Na^{M4}/(Na^{M4} + Ca)$ and $Mg/(Mg + Fe^{2+})$ ratios of 0.76 and 0.73, respectively. Glaucophane from Huangzhen has $Na^{M4}/(Na^{M4} + Ca)$, $Mg/(Mg + Fe^{2+})$ and $Fe^{3+}/(Fe^{3+} + Al^{VI})$ ratios of 0.75, 0.93, and 0.23, respectively; the silicon content in glaucophane is about 7.5 per formula unit (pfu). Glaucophane with high $Na^{M4}/(Na^{M4} + Ca)$ of 0.86 also occurs as porphyroblast in Xinxian coesite-bearing eclogite. Mg-katophorite has a high A-site occupancy of $(Na + K)^A = 0.73$, $Mg/(Mg + Fe^{2+}) = 0.67$, and $Na^{M4}/(Na^{M4} + Ca) = 0.45$.

Phengitic mica is ubiquitous and occurs as inclusions in garnet and omphacite and as matrix minerals. The Si content of phengite ranges from 3.5 to 3.7 atoms pfu based on calculation of 11 oxygens; most are around 3. 6. The Mg contents vary from 0.35 to 0.47 atom pfu. The Mg/Fe ratio varies from 0.72 to 0.84. No obvious difference in composition between phengite as inclusions versus those in the matrix has been observed.

For most eclogites from the Dabie Mountains, eclogitic epidote in the matrix has a Ps value $\{Ps = [Fe^{3+}/(Al + Fe^{3+})] \times 100\}$ of about 15 to 17%, which is similar to that of epidote inclusions in garnet. For low-temperature eclogite from the Dabie Mountains, epidote inclusions in garnet cores have a high pistacite content (Ps_{23}), which decreases continuously toward the rim of garnet (Ps_{15}). Higher Ps content (up to 25) in porphyroblastic epidote has been reported for eclogite from northern Jiangsu (Zhang, 1992). For example, epidote containing inclusion of quartz pseudomorph after coesite in Fig. 10.3d has Ps_{25}.

Coarse-grained calcite and dolomite in matrix of some eclogites have compositions of $Ca_{0.95}Mg_{0.02}Fe_{0.02}Mn_{0.01}CO_3$ and $Ca_{0.47}Mg_{0.42}Fe_{0.11}CO_3$, respectively. Rutile is very pure with >99 wt% of TiO_2. Minor impurities include SiO_2.

Post-eclogitic Metamorphic Minerals

Clinopyroxenes (superimposed) occur as symplectite around relict omphacite crystals. Compositions of these clinopyroxenes in the Dabie are very different from those from the Su-Lu region, especially in aegirine content. Those from the Dabie Mountains contain jadeite ranging from a few to 26 mol% with <20 mol% aegirine; average compositions are $Jd_{20}Di_{71}Aeg_9$. Those from the Su-Lu region contain distinctly high aegirine contents with a similarly high jadeite (Fig. 10.6b). For example, those from Donghai contain Jd_{6-26} $Di_{30-42}Aeg_{26-63}$. These clinopyroxenes have a distinctly lower jadeite content compared to primary omphacite. Enami et al. (1993), however, reported a high jadeite content of 40 mol% for post-eclogitic clinopyroxene from Donghai in the Su-Lu region.

Six late-stage amphibole species have been identified from the Dabie Mountains and Su-Lu region (Enami et al., 1992; Wang et al., 1992; Zhang, 1992). These amphiboles, based on classification of Leake (1978), include calcic amphiboles such as pargasite, edenite, and actinolite, and sodic-calcic amphiboles of Mg-taramite, winchite, and barroisite. Pargasite and edenite are most common, formed along with epidote after garnet. Pargasite has a $(Ca + Na^{M4})$ of around 2.0, a Na^{M4} from 0.1 to 0.5, and $Mg/(Mg + Fe^{2+})$ ratio of 0.3 to 0.7, and contains a very low Si content of about 6 pfu or less. Ferropargasite occurs instead of pargasite, as Fe content increases. Edenite or edenitic hornblende occurs as the Si content of amphibole increases to greater than 6.5 atoms pfu. Consistent variation in composition has not been found for amphiboles from different localities.

Mg-taramite is a sodic-calcic amphibole with $(Na + K)^A > 0.50$, whereas barroisite and winchite have $(Na + K)^A < 0.50$. Mg-taramite occurs in the rims around glaucophane and barroisite in the matrix. Barroisite occurs as rims on glaucophane in the matrix. Barroisite has values of $(Ca + Na^{M4}) = 1.90-1.94$, $Na^{M4} = 0.90-1.05$, $(Na + K)^A = 0.36-0.44$, and $Mg/(Mg + Fe^{2+}) = 0.78$. Winchite was identified in cores of coarse-grained matrix amphiboles in one eclogite sample.

Compositions of retrograde epidote are similar to those of the primary epidotes described in the previous sections. Sodic plagioclases range in composition from An_2 for those associated with symplectic pyroxene after omphacite, to An_{20} for those associated with pargasite and epidote after garnet. Chlorite after garnet has a $Mg/(Mg + Fe^{2+})$ ratio of 0.61. Biotite formed after phengite has a $Mg/(Mg + Fe^{2+})$ ratio of about 0.60. Most ilmenites have end-member compositions of $FeTiO_3$; however, in some samples, Mn-ilmenite with MnO content up to 7.9 wt% was identified as inclusions in garnet.

P-T Constraints for Eclogite Metamorphism

Various geothermometers and geobarometers have been applied to eclogites from east central China to estimate *P-T* conditions for eclogite formation (e.g., Xie, 1983, 1985; Enami et al., 1988, 1993; Enami and Zang, 1988; Okay et al., 1989; Wang et al., 1989, 1990, 1992; Yang, 1990; Zhang et al., 1990; Zhang, 1992). Prior to discovery of coesite, a minimum pressure of 11 kbar for eclogite from the Dabie Mountains was estimated mainly based on the jadeite content of omphacite and mineral assemblage in eclogite (Xie, 1983, 1985). Enami et al. (1988) estimated the formation pressure between 10 and 25 kbar for eclogite from the Su-Lu region. Recognition of coesite in eclogite allows use of the experimentally determined stability of coesite which yields minimum pressures of 28 kbar at 700°C and 30 kbar at 800°C (Mirwald and Massonne, 1980; Bohlen and Boettcher, 1982). The mineral assemblage kyanite + omphacite indicates a minimum pressure of 20–22 kbar at temperatures of 550–650°C (Enami et al., 1992; Wang et al., 1992). A recent report on the occurrence of microdiamond in eclogite from the Dabie Mountains yields a higher minimum pressure estimate of >40 kbar at 800°C (Xu et al., 1992) (Fig. 10.7).

The peak temperatures of UHP metamorphism have been well constrained. Wang et al. (1992) reported temperatures for eclogite from the Dabie Mountains to range from 580 to 780°C from the south to the north, based on garnet-clinopyroxene (Ellis and Green, 1979) and garnet-phengite (Krogh and Raheim, 1978) thermometers (see Tables 10.1 and 10.2). Enami et al. (1988) reported a temperature range of 700–850°C for eclogite from the Su-Lu region. Zhang (1992) pointed out that most eclogites from the Su-Lu region formed at higher temperatures ranging from 700 to 980°C with an average of 760°C based on 86 coexisting pairs of garnet-clinopyroxene, using the thermometer of Ellis and Green (1979). Based on temperature estimates for the Su-Lu region and the Dabie Mountains, Enami et al. (1993) and Zhang et al (1993a) speculated that there is an apparent decrease in UHP metamorphic temperatures from the Su-Lu region in the east to the Dabie Mountains and the Hongan Block in the west.

Most eclogite from east central China is devoid of a prograde mineral assemblage except for a few samples from the southern tip of the Dabie Mountains and some areas of the Su-Lu region; relict blueschist or amphibolite facies prograde assemblages occur as inclusions in garnet. For example, the assemblage glaucophane, paragonite, epidote, phengite, and quartz has been found in eclogite from the Dabie Mountains (Wang et al., 1992). *P-T* estimates for this assemblage was reported to be 7–10 kbar and 400°C based on the occurrence of paragonite and the garnet-phengite thermometer (Fig. 10.7). Zhang (1992) described some isolated inclusions of pargasitic hornblende, phengite, paragonite, and quartz; their occurrences suggest unusual amphibolite facies metamorphism prior to eclogite facies metamorphism. Also, an aggregate of Mg-chlorite + hornblende + zoisite + garnet and paragenesis of epidote, albite, barroisite/katophorite, and paragonite in eclogites from the Su-Lu region indicate epidote amphibolite facies with an estimated temperature of 477°C following the Grt-Amp geothermometer (Graham and Powell, 1984). Such a difference in prograde assemblage implies that eclogites from the Dabie Mountains went through different lower temperatures than those from the Su-Lu region.

Post-eclogitic Metamorphic P-T Conditions

The post-eclogitic metamorphism of these eclogites occurred mainly under amphibolite facies conditions (Wang et al., 1990, 1992; Zhang, 1992; Enami et al., 1993). An even later stage of greenschist facies metamorphism is also observed. In Weihai, the easternmost portion of the Su-Lu region, however, granulite facies metamorphism has been reported by Zhang and Cong (1991), Cong et al. (1992), and Wang et al. (1993). The *P-T* conditions of amphibolite facies retrogression are not well constrained owing to lack of index minerals and mineral pairs for thermobarometry. In spite of this, Zhang (1992), using the plagioclase-amphibole geothermometer of Blundy and Holland (1990), calculated the temperature of superimposed amphibolite facies metamorphism, deriving it to be from 540–710°C at 13 kbar for the Su-Lu eclogite. Wang et al. (1990) applied the garnet-biotite thermometer of Hoinkes (1986) to associated biotite gneiss and arrived at a temperature of 470–530°C. The occurrence of margarite + quartz as a superimposed assemblage in eclogite from the Dabie Mountains supports this estimate (Wang et al., 1992) (Fig. 10.7). The jadeite content of symplectic clinopyroxene after omphacite during the superimposed metamorphism constrains the minimum pressure of retrogression of less than 7 kbar (Enami et al., 1993) (Fig. 10.7). These *P-T* estimates are further constrained by *P-T* estimates for post-eclogitic metamorphism of associated marble based on the mineral assemblage (Wang and Liou, 1993).

P-T estimates for granulite facies superimposed metamorphism were reported by Zhang and Cong (1991) to be 8 kbar and 750°C based on the garnet-pyroxene-plagioclase-quartz geobarometer of Newton and Haselton (1981) and the two pyroxene geothermometers of Wood and Banno (1973), respectively. Wang et al. (1993) reported the granulite facies metamorphic conditions to be 10 kbar and 850°C. Similar *P-T* values were also reported by Enami et al. (1993) for the superimposed metamorphism in the Su-Lu region.

P-T Trajectory

Attempts were made by several authors to constrain the *P-T* path of metamorphic evolution of eclogite from the Dabie Mountains (Wang et al., 1992) and the Su-Lu region (Zhang, 1992; Enami et al., 1993). Their proposed *P-T* paths are summarized in Fig. 10.7. The records of prograde metamorphism have been mostly erased due to the high temperature of eclogite facies peak metamorphism. Only a few low-temperature eclogites have preserved the prograde blueschist, epidote-amphibolite, or amphibolite facies assemblage. The peak *P-T* conditions of UHP metamorphism of eclogite are well constrained as described in the previous sections. The superimposed metamorphic features are ubiquitous in eclogite with assemblages ranging from granulite, to amphibolite, to greenschist facies. Studies have concluded that eclogites from east central China have experienced a clockwise *P-T* path (Wang et al., 1992; Zhang, 1992; Enami et al., 1993) (Fig. 10.7).

It is also apparent that the peak metamorphic temperatures systematically decrease from east to west of the belt (Enami et al., 1993; Zhang et al., 1993b) and from the north to south in the Dabie Mountains (Wang et al., 1992), and that the superimposed metamorphism occurred at a higher temperature in the east (granulite facies) (Table 10.2). The prograde metamorphism for eclogite from the east may have experienced a higher temperature history (amphibolite to epidote-amphibolite facies in the Su-Lu region compared with blueschist facies in the Dabie Mountains). It may be inferred that eclogites from different localities have different *P-T* paths but possess similar trajectories, as shown in Fig. 10.7. This generalization assumes that eclogite from the east (eastern Shandong) has been subjected to higher *P-T* conditions for prograde, UHP, and superimposed metamorphism than eclogite from the Dabie Mountains. More detailed petrologic work is required to support this hypothesis.

Petrology of Country Rocks

Eclogite is associated with biotite gneiss, marble, and metapelite in this UHP metamorphic terrane in east central China. As evidence for UHP metamorphism becomes clear for eclogite, an immediate concern is whether the country rocks, gneiss, marble, and metapelite associated with eclogite have also been subjected to the same UHP metamorphism. Okay et al. (1989) reported a jadeite inclusion in garnet from gneiss from the Dabie Mountains. Jadeite gneissic rocks contain as high as 80 mol% jadeite component (Okay, 1993). Associated with this jadeite are quartz, garnet, and rutile. Phengite with Si of ~3.5 atom pfu was reported by Okay (1993) for the Dabie Mountains and Zhang (1992) for the Su-Lu region. Some phengites are zoned with higher Si content (3.45) in the core than in the rims (3.10). Wang and Liou (1991) reported the occurrence of coesite/quartz pseudomorphs after coesite, and calcite pseudomorphs after aragonite, as inclusions in garnet and/or clinopyroxene in gneiss, marble, and metapelite from the Dabie Mountains. Further study of marble revealed that marble from the Dabie Mountains contains UHP assemblage in addition to coesite and aragonite pseudomorphs; equilibrium *P-T* conditions and the *P-T* evolution of marble are compatible with those of eclogite from the same regions (Wang and Liou, 1993) (Table 10.1).

Eclogites are interlayered with their country rocks, gneiss, marble, or metapelite or occur as blocks enclosed in these lithologies. For example, in Huangzhen of the Dabie Mountains, layered eclogites of 5–50 cm thick are interbedded with biotite gneisses and amphibolites. These eclogite layers are coherent with their country rocks, are continuous on the kilometer scale, and contain the high-pressure assemblage of omphacite + kyanite (>22 kbar at 600°C) (Wang et al., 1992). In the Su-Lu region, three types of kyanite-bearing country rocks are recognized: kyanite-quartz schist, kyanite-garnet-zoisite-phengite-quartz schist, and kyanite-bearing gneiss consisting of garnet, zoisite, phengite, K-feldspar, biotite, epidote, amphibole, rutile, and quartz. Most eclogite layers have concordant contacts with kyanite-bearing rocks (Zhang, 1992). The concordant nature of this occurrence indicates that the country rocks must have been subjected to the same peak metamorphism.

Besides coesite/quartz pseudomorphs after coesite and calcite pseudomorphs after aragonite, these country rocks contain relict of high-pressure mineral assemblages (Wang and Liou, 1991, 1992). For example, marble contains calcite/aragonite (55 vol%), dolomite (10%), garnet ($Alm_{27}Gr_{54}Prp_{18}$-Spe_{01}) (10%), clinopyroxene ($Jd_{25}Di_{65}Hd_{10}$) (8%), phengite (3%), epidote (5%), quartz/coesite (5%), rutile (1%), and minor secondary minerals including pargasite and epidote (Fig. 10.8a). Phengite schist includes garnet, phengite, epidote, quartz/coesite, and rutile. Phengite contains as high as 3.56 Si atoms pfu in marble and up to 3.70 pfu in phengite schist. Biotite gneisses have assemblage of biotite + muscovite + garnet + epidote + K-feldspar + plagioclase + rutile + titanite + quartz/coesite. Associated gneiss and eclogite have simi-

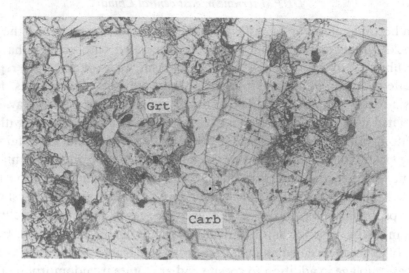

(a)

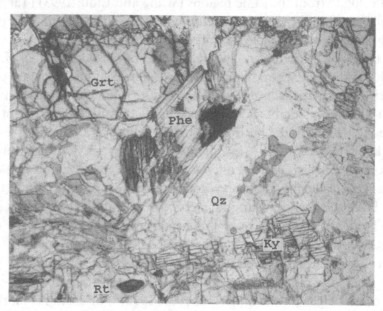

(b)

Figure 10.8. (a) UHP mineral assemblage of marble from the Dabie Mountains, with quartz pseudomorph after coesite inside garnet (Grt) and assemblage of grossular-rich garnet + omphacite (Omp) + rutile + quartz + calcite + dolomite (Carb) + epidote. Field view is 5 mm. (b) Kyanite and phengite gneiss from the Su-Lu region, with kyanite (Ky), phengite (Phe), garnet (Grt), rutile, quartz, and zoisite. Field view is 1.1 mm.

lar prograde garnet profiles although gneissic garnet shows weaker zonation (Wang and Liou, 1991).

Being consistent with eclogite paragenesis, marble and metapelite from the Dabie Mountains exhibit amphibolite and greenschist facies overprints. Garnet in marble has a corona of epidote and pargasite along the rim, and garnet in pelitic schist is partially replaced by biotite and chlorite (Fig. 10.8a). Phengites in both marble and schist are partially replaced by biotite. Garnet gneiss, however, displays only a greenschist facies superimposed assemblage. Chlorite occurs along fractures and rims of garnet.

The estimated peak *P-T* conditions for marble are comparable to those for eclogite; the minimum pressure for formation of marble, based on the jadeite content (25%) of omphacite (+quartz), is 12 kbar (Holland, 1980, 1983). Coexisting grossular-rich garnet, rutile, quartz (coesite), and epidote in the marble yield a minimum pressure of 28 kbar at 650°C (Wang and Liou, 1993), which is consistent with the occurrence of coesite. Garnet-clinopyroxene geothermometry yields temperatures of 630–770°C; the garnet-phengite geothermometer of Krogh and Raheim (1978) yields temperatures of 570–640°C for marble. The *P-T* conditions for superimposed metamorphism were estimated to be 475–550°C at ~7 kbar based on superimposed assemblage of marble (Fig. 10.7).

Discussion and Conclusions

The reported occurrences of coesite ± diamond in eclogites from many localities in east central China have confirmed the UHP metamorphism in the collision belt between the Sino-Korean and Yangtze cratons. The relics of coesite and/or aragonite or their pseudomorphs as inclusions in garnet and omphacite in gneiss, marble, and metapelite suggest the possibility of regional UHP metamorphism. Diamond may be found in the future in some Su-Lu eclogites, as their estimated *P-T* conditions for eclogite are higher than for Dabie eclogites.

The Geologic Relationship Between the Dabie Mountains and the Su-Lu Region

The geologic characteristics between the Dabie Mountains and the Su-Lu region have been correlated for at least 20 years; it has been suggested that the Su-Lu region is part of the collision belt and has been offset by the sinistral strike-slip Tanlu fault from the Dabie Mountains to its present position (e.g. Xu et al., 1987; Wang and Liou, 1990). The discovery of coesite and many petrologic characteristics in both regions further support the conclusion. As shown in Fig.10.1 and Table 10.1, these two terranes have similar geologic and lithologic units and metamorphic histories. The coesite-bearing UHP unit consists mainly of gneiss with marble, ultramafics, eclogite, and migmatite in both areas. To the south, the blueschist-bearing Susong Group in the Dabie Mountains is equivalent to the Haizhou Group in Jiangsu, to the Zhangbaling Group in Jiashan. Although blueschist has not been confirmed in the Haizhou Group, a report indicated that blueschist exists in a drill hole core sample in northern Jiangsu province (Zhang and Kang, 1989) (Fig. 10.2). The metamorphic ages and deformation features are also similar for both regions. Based on available geologic information, we conclude that the UHP terranes in the Dabie Mountains and the Su-Lu region can be matched, and they likely formed during the same collision event (see the following discussion).

Evidence for Regional UHPM

Studies have found that the country rocks of eclogite from east central China contain records of UHP metamorphism. Evidence includes field occurrence, mineral assemblage, mineral chemistry, and estimated P-T conditions and evolution (Okay et al., 1989; Wang and Liou, 1991, 1992; Wang et al., 1992; Zhang, 1992; Zhang et al., 1990, 1993a; Zhang and Liou, 1994). Wang and Liou (1991) proposed a regional UHP metamorphism for the south Dabie terrane based on available data and concluded that much of the country rocks have been subjected to the same P-T evolution as the eclogites they enclose. They reported that at least part of the Dabie Mountains has been metamorphosed through a clockwise P-T path as a coherent terrane in a subduction zone.

Immediately to the west of the Dabie Mountains, a coherent blueschist-bearing HP metamorphic terrane, characterized by progressive metamorphism from subgreenschist through blueschist/greenschist to epidote amphibolite and eclogite facies grade, occurs in the Hongan block (Zhou et al., 1993; Eide, 1993). A narrow coesite-bearing UHP belt has been documented recently (Zhang et al., 1993a; Zhang and Liou, 1994).

Zhang (1992) also described the occurrence of a high-pressure assemblage in some Su-Lu gneisses and concluded that these gneisses may share the identical P-T evolution as that of eclogite layers. However, she emphasized that some eclogite blocks might have a different origin. High-pressure evidence has been increasingly recognized in the country rocks in east central China and may indicate that at least some of these rocks have an identical P-T evolution path. The question is the extent of the occurrence of the regional UHP metamorphism. Further field, petrologic and geochemical studies should be carried out to address this question.

Age of UHPM

A long-standing idea has been held that the collision zone of east central China was subjected to an amphibolite-granulite facies regional metamorphism in the Late Archean to Early Proterozoic (Xu et al., 1987; Dong, 1988, 1993). This conclusion is based mainly on the U-Pb ages of 1952 to 2424 Ma from zircon of gneisses and ambiguous field relationships (Regional Geological Survey of Anhui, 1987; Regional Geological Survey of Anhui, 1982, unpublished report). An U-Th-Pb age of 2100 Ma from apatite was obtained for the Dabie gneisses (Regional Geological Survey of Hubei, 1990). In the Su-Lu region, U-Pb dating of zircon yields ages of 2233 Ma (Wang, 1986) and 2210 Ma (Lu et al., 1987). These ages have been interpreted to be the age of regional amphibolite facies metamorphism (e.g., Regional Geological Survey of Anhui, 1987; Xu et al., 1987) as the eclogite facies metamorphism event was not recognized then.

A debatable field relation that has been used to support the hypothesis of the Precambrian metamorphic age is that from an area southwest of Taihu (Fig. 10.2). The contact between the south Dabie terrane (SDT) lithology and the Middle to Late Proterozoic unit has been suggested as an unconformable contact (Fig. 10.2) (Sang et al., 1987; Ma and Zhang, 1988) or a fault (Y. Jing and W. Liang, personal communication). In the vicinity of the Dabie Mountains, most of the Dabie terrane rocks are in fault contact with the Middle to Late Proterozoic unit which was metamorphosed under transitional blueschist/greenschist facies conditions. In the northeastern corner of the Dabie Mountains, near Tongchen, the Proterozoic section and UHP in the SDT are unconformably overlain by unmetamorphosed Jurassic/Cretaceous clastic and volcaniclastic sequence. Thus, these Jurassic/Cretaceous volcanics place upper limits on the minimum age of UHP and HP metamorphism for the SDT.

Recent Sm/Nd isochron studies of eclogites from the Dabie Mountains provide new information on the age of eclogite formation. Li et al. (1989a, 1989b, 1990) reported several Nd-Sm mineral and whole-rock isochron ages ranging from 221 ± 20 to 224 ± 5 Ma for eclogites and from 231 ± 61 to 244 ± 11 Ma for garnet peridotites. Similar Sm/Nd ages (220–240 Ma) have been obtained for eclogites from Shandong (Cong et al., 1990; Li et al., 1993). If ages of 230 ± 10 Ma are assumed for the UHP metamorphism in the region, then amphibolite-greenschist facies regionally retrograde metamorphism must have occurred later. An U-Pb age of 240 Ma has been recently obtained for zircon from gneiss from the SDT of the Dabie Mountains by Ames et al. (1993). Recently, $^{40}Ar/^{39}Ar$ ages from the Dabie Mountains and Su-Lu region all indicate a young Mesozoic age for post-UHP metamorphism (Chen et al., 1992; Eide et al., in review).

The Early Proterozoic age, dated by various methods, therefore, probably represents the age of the protoliths of gneissic rocks in the SDT. The younger ages (230 ± 10Ma) may be the age of UHP metamorphism. Nevertheless, the ages of some old $^{40}Ar/^{39}Ar$ (>400 Ma) of phengite of coesite-bearing eclogite from the Donghai area of the Su-Lu region (Zhang, 1992) as well as from the Dabie Mountains (Brunel et al., 1991), remain to be evaluated.

Possible Tectonic Scenarios for Formation of UHPM in East Central China

The possible tectonic scenarios for formation of coesite-bearing eclogite in east central China have been discussed in several recent papers (Ernst et al., 1991; Maruyama et al., 1992; Okay and Sengor, 1992; Wang et al., 1992; Zhang, 1992; Zhang et al., 1993b; Enami et al., 1993). Most of these authors agree that the formation of eclogite in east central China is related to the Triassic collision between the Sino-Korean and Yangtze cratons.

Ernst et al. (1991) proposed a multiple subduction formation model and suggested that the UHP rocks of the Dabie Mountains formed during the Proterozoic Era whereas the high-pressure rocks (blueschist) in the south formed in the early Paleozoic. Wang et al. (1992) suggested a collision-subduction-thrust model (A-type subduction) for formation of eclogite of the Dabie Mountains. They believe that the Dabie terrane was a microcontinent with a Late Archean protolith; it was subducted northward beneath the Sino-Korean craton to subcrust-mantle depths during the collision about 230 ± 10 Ma (A-type subduction). Maruyama et al. (1992) proposed a wedge-extrusion model using the UHP rocks from the Dabie Mountains; they indicate that the UHP terrane between the relatively low-pressure units must have been extruded from the subcrust or mantle. Okay and Sengor (1992) attribute the uplift of the UHP unit of the Dabie Mountains to the movement of the Tanlu fault.

To the east of the collision zone in Su-Lu region, the UHP and superimposed metamorphism occurred at higher temperatures (probably higher pressure too) than in the Dabie Mountains. The different P-T paths for formation of eclogite may be related to the different depths to which the eclogite and its country rocks were subducted (Enami et al., 1992; Wang et al., 1992; Zhang, 1992), which in turn might have resulted from the difference in closure time of collision from the east to west. The collision between the Sino-Korean and Yangtze cratons began in the east (Shandong), and then continued to the west (Lin et al., 1985). Therefore, the UHP rocks from Shandong may have been subducted to a greater depth, hence have experienced higher *P-T* conditions. If this model is correct, a higher-pressure index mineral such as diamond might be found in Su-Lu region.

An alternative explanation for the apparent *P-T* variation over the region is differential uplift: the idea is that the Shandong area has been uplifted more than the areas west to it; therefore, rocks exposed in the Shandong area are from greater depth and, hence, higher *P-T* conditions (Wang et al., in preparation). More detailed work needs to be conducted to confirm the validity of these hypotheses.

Remaining Problems and Future Research

Although evidence for the regional UHP metamorphism has been recognized, the extent of regional UHP metamorphism in central China has yet to be documented. To answer this important question, the following studies must be carried out: (1) detailed field mapping program to document the distribution of the coesite- and diamond-bearing eclogite and its spatial relationship with its country rocks; (2) structural studies to provide data and information about the deformation history for both eclogite and its country rocks; (3) a mineral paragenetic study of country rocks, to document the UHP metamorphic history of country rocks; (4) systematic isotopic and geochemical work to determine the origin and protolith of eclogite. Information derived from these sources will provide important constraints for tectonic models.

A geochronologic study is currently being conducted in several independent institutes; some additional meaningful data are expected to be available soon, which will delimit further the timing of the UHP and superimposed metamorphic events and the timing of the collision between the Sino-Korean and Yangtze cratons. It might be useful to employ age dating of zoned zircons by the SHRIMP ion microprobe to clarify the polymetamorphic history of UHP metamorphic rocks, as did Claoue-Long et al. (1991) on diamond-bearing UHP rocks from the Kokchetav Massif in Russia.

Acknowledgments

The content of much of this chapter was based on Ph.D. theses by the first two authors (Wang at Stanford University; Zhang at Kyoto University). Wang thanks the Bureau of Geology and Mineral Resources of Anhui for the logistic arrangements; and S. Maruyama, G. Pan, Y. Jing, M. Xia, and W. Liang for field assistance in the Dabie Mountains. Wang appreciates the invaluable supervision and encouragement of Professor R. G. Coleman as a research committee member. Zhang would like to express her indebtedness to her advisor Professor S. Banno for many enlightening discussions as well as the supervision. She is extremely grateful to Professor Cong for both moral and practical support and also thanks Dr. Hirajima for helpful discussions and assistance during the Sino-Japanese cooperative research at Kyoto University. Financial support during the preparation of this manuscript was provided by the NSF (EAR92-04563 to Liou) and the Stanford-China Industrial Affiliate Program led by Professors S. A. Graham and R. G. Coleman. The manuscript has been greatly improved by constructive reviews by R. Compagnoni and E. Eide.

References

Ames, L., Tilton, G. R., and Zhou, G. 1993. Timing of collision of the Sino-Korean and Yangtse cratons: U-Pb zircon dating of coesite-bearing eclogites. *Geology, 21*, 339–342.

Blundy, J. D., and Holland, T.J.B. 1990. Calcic amphibole equilibria and new amphibole plagioclase geothermometer. *Contrib. Mineral. Petrol., 104*, 208–224.

Bohlen, S. R., and Boettcher, A. L. 1982. The quartz-coesite transformation: a pressure determination and the effects of other components. *J. Geophys. Res., 87*, 7073–7078.

Brunel, M., Maluski, H., Kienast, J. R., Matte, P., Smith, D. C., Xu, Z., and Mattauer, M. 1991. Proterozoic coesite-bearing eclogite from the Dabie Shan, east Qinling Rang, central China. *Terra Abstracts (EUG VI), 3*, 85.

Chen, W., Harrison, T. M., Heizerler, M. T., Liu, R., Ma, B., and Li, J. 1992. The cooling history of melange zone in north Jiangsu-south Shandong region: evidence from multiple diffusion domain $^{40}Ar/^{39}Ar$ thermal geochronology. *Acta Petrologica Sinica, 8*, 1–15.

Chopin, C. 1984. Coesite and pure pyrope in high-grade blueschists of the western Alps: A first record and some consequences. *Contributions to Mineralogy and Petrology, 86*, 107–118.

Claoue-Long, J. C., Sobolev, N. V., Shatsky, V. S., and Sobolev, A. V. 1991. Zircon response to diamond-pressure metamorphism in the Kokchetav massif, USSR. *Geology, 19*,

Cong, B. L., Wang, Q. C., Zhang, R. Y., Zhai, M. G., Zhao, Z. Y., and J. J., L. 1992. Discovery of the coesite-bearing granulite in the Weihai, Shandong Province, China. *Chinese Science Bulletin, 37*, 1494–1495.

Dong, S. 1993. Metamorphic and tectonic domains of China. *Jour. Metamorph. Geol., 11*, 465–481.

Dong, S., Shen, Q., Sun, D., and Lu, L. 1986. *Metamorphic map of China (1:4,000,000) with explanatory text*. Beijing: Geology Publishing House.

Eide, E. A. 1993. Petrology, geochronlogy, and structure of high-pressure metamorphic rocks in Hubei Province, east-central China, and their relationship to continental collision. Ph.D., Stanford University, USA.

Eide, E. A., Wang, X., Maruyama, S., Liou, J. G., and Zhou, G. 1989. Mineral paragenesis of eclogite blocks from a serpentinite melange belt, northern Hubei, China. *EOS, Transaction American Geophysical Union, 70*, 1379.

Ellis, D. J., and Green, D. H. 1979. An experimental study of the effect of Ca upon garnet. Clinopyroxene Fe-Mg exchange equilibria. *Contr. Min. Pet., 71*, 13–22.

Enami, M., and Zang, Q. 1988. Magnesium staurolite in garnet-corundum rocks and eclogite from the Donghai district, Jiangsu Province, east China. *Am. Mineralogist, 73*, 48–56.

Enami, M., and Zang, Q. 1990. Quartz pseudomorph after coesite in eclogites from Shandong Province, east China. *Am. Mineralogist, 75*, 381–386.

Enami, M., Zang, Q., and Yin, Y. 1993. High-pressure eclogites in northern Jiangsu-southern Shandong province, east China. *Jour. Metamorph. Geol., 11*, 589–603.

Ernst, W. G., Zhou, G., Liou, J. G., Eide, E., and Wang, X. 1991. High-pressure and superhigh-pressure metamorphic terranes in the Qinling-Dabie Mountain Belt, central China; Early- to Mid-Phanerozoic accretion of the western paleo-Pacific rim. *Pacific Science Association Information Bulletin, 43*, 6–14.

Graham, C. M., and Powell, R. 1984. A garnet-hornblende geothermometer: Calibration, testing, and application to the Pelona schist, southern California. *Jour. Metamorph. Geol, 2*, 13–31.

Heinrich, C. A. 1982. Kyanite-eclogite to amphibolite facies evolution of hydrous mafic and pelitic rocks, Adula Nappe, Central Alps. *Contrib. Mineral. Petrol., 81*, 30–38.

Hirajima, T., Ishiwatari, A., Cong, B., Zhang, R., Banno, S., and Nozaka, T. 1990. Coesite from Mengzhong eclogite at Donghai county, northeastern Jiangsu province, China. *Mineral. Mag., 54*, 579–583.

Hoinkes, G. 1986. Effect of grossular-content in garnet on partitioning of Fe and Mg between garnet and biotite. *Contrib. Mineral. Petrol., 92*, 393–399.

Holland, T. J. B. 1983. The experimental determination of activities in disordered and short-range ordered jadeitic pyroxenes. *Contrib. Mineral. Petrol., 82*, 214–220.

Holland, T. J. B. 1980. The reaction albite = jadeite + quartz determined experimentally in the range 600–1200 °C. *Am. Mineralogist, 65*, 125–134.

Krogh, E. J., and Råheim, A. 1978. Temperature and pressure dependence of Fe-Mg partitioning between garnet and phengite, with particular reference to eclogites. *Contrib. Mineral. Petrol., 66*, 75–80.

Leake, B. E. 1978. Nomenclature of amphiboles. *Am. Mineralogist, 63*, 1023–1052.

Li, S., Ge, N., Liu, D., Zhang, Z., Yie, X., Zhen, S., and Peng, C. 1989. Sm-Nd age of C-group eclogites in northern Dabie Mountains and its tectonic significance. *Bulletin of Sciences, 7*, 522–525 (in Chinese).

Li, S., Hart, S. R., Zhen, S., Guo, A., Liu, D., and Zhang, G. 1989. Sm-Nd age for the collision between the Sino-Korean and Yangtze cratons. *Academic Sinica, 2B*, 312–319 (in Chinese).

Li, S., and Liu, D. 1990. Isotopic chronological evidence for Indosinian orogeny in Dabie Mountain. *Geotectonica et Metallogenia, 14*, 159–163 (in Chinese with English abstract).

Li, S. G., Chen, Y. Z., Cong, B. L., Zhang, Z. Q., Zhang, R. Y., Liu, D. L., Hart, S. R., and Ge, N. J. 1993. Collision of the North China and Yangtze blocks and formation of coesite-bearing eclogite: timing and processes. *Chemical Geology, 109,* 70–89.

Lin, J. L., Fuller, M., and Zhang, W. 1985. Preliminary Phanerozoic polar wander paths for the North and South China blocks. *Nature, 313,* 444–449.

Liou, J. G., Wang, X., Coleman, R. G., Zhang, Z. M., and Maruyama, S. 1989. Blueschists in major suture zones of China Tectonics. *Tectonics, 8,* 609–619.

Liu, X., and Hao, J. 1989. Structure and tectonic evolution of the Tongbie-Dabie Range in the east Qinling collisional belt, China. *Tectonics, 8,* 637–645.

Ma, B., and Zhang, Z. 1988. The features of the paired metamorphic belts and evolution of paleotectonics in the east part of Dabie Mountains. *Seismology and Geology, 10,* 19–28 (in Chinese with English abstract).

Maruyama, S., Liou, J. G., Zhang, R., Wang, X., and Eide, E. 1994. Tectonic evolution of the ultrahigh-pressure (UHP) and high pressure (HP) metamorphic belts from central China. *Geology* (in press).

Mirwald, P. W., and Massonne, H.-J. 1980. The low-high quartz and quartz-coesite transition to 40 kbar between 600° and 1600°C and some reconnaissance data on the effect of $NaAlO_2$ component on the low quartz-coesite transition. *J. Geophys. Res., 85,* 6983–6990.

Okay, A. I. 1993. Petrology of a diamond and coesite-bearing terrain: Dabie Shan, China. *Eur. Jour. Mineral., 5,* 659–675.

Okay, A. I., Xu, S., and Sengor, A.M.C. 1989. Coesite from the Dabie Shan eclogites, central China. *Eur. Jour. Mineral., 1,* 595–598.

Regional Geological Survey of Anhui. 1987. *Regional geology of Anhui province.* Beijing: Geological Publishing House (in Chinese with brief English text).

Regional Geological Survey of Anhui. 1987. *Research on metamorphic geology of Anhui Province.* Hefei: Science and Technology Publishing House of Anhui (in Chinese).

Regional Geological Survey of Henan. 1989. *Regional geology of Henan province.* Beijing: Geological Publishing House (in Chinese with brief English text).

Regional Geological Survey of Hubei. 1990. *Regional geology of Hubei province.* Beijing: Geological Publishing House (in Chinese with brief English text).

Sang, B., Chen, Y., and Shao, G. 1987. The Rb-Sr ages of metamorphic series of the Susong group at the southeastern foot of the Dabie Mountains, Anhui province, and their tectonic significance. *Regional Geology of China, 4,* 364–370.

Smith, D. C. 1984. Coesite in clinopyroxene in the Caledonides and its implications for geodynamics. *Nature, 310,* 641–644.

Sobolev, N. V., and Shatsky, V. S. 1990. Diamond inclusions in garnets from metamorphic rocks. *Nature, 343,* 742–746.

Tectonic Compilation Group, T. C. 1974. A preliminary note on the basic tectonic features and their developments in China. *Geologica Sinica, 1,* 1–17 (in Chinese).

Wang, H. 1963. Characteristics and gneiss of eclogites in northern Jiangsu Province. *Journal of Nanjing University.*

Wang. Q., Ishiwatari, A., Zhao, Z., Hirajima, T., Enami, M., Zhai, M., Li, J. & Cong, B. L. 1993. Coesite-beaing granulite retrograded from eclogite in Weihai, eastern China: a preliminary study. *Euro. Jour. Mineral., 5,* 141–152.

Wang, X., Jing, Y., Liou, J. G., Pan, G., Liang, W., Xia, M., and Maruyama, S. 1990. Field occurrences and petrology of eclogites from the Dabie Mountains, Anhui, central China. *Lithos, 25,* 119–131.

Wang, X., and Liou, J. G. 1989. The large displacement of the Tanlu Fault: evidence from the distribution of coesite-bearing eclogite belt in eastern China. *Eos, Transaction American Geophysical Union, 70*, 1312–1313.

Wang, X., and Liou, J. G. 1991. Regional ultrahigh-pressure coesite-bearing eclogitic terrane in central China: evidence from country rocks, gneiss, marble, and metapelite. *Geology, 19*, 933–936.

Wang, X., and Liou, J. G. 1993. Ultrahigh-pressure metamorphism of carbonates in the Dabie Mountains, central China. *Jour. Metamorph. Geol., 11*, 575–588.

Wang, X., Liou, J. G., and Mao, H. K. 1989. Coesite-bearing eclogites from the Dabie Mountains in central China. *Geology, 17*, 1085–1088.

Wang, X., Liou, J. G., and Maruyama, S. 1992. Coesite-bearing eclogites from the Dabie Mountains, central China: Petrogenesis, P-T paths and implications to tectonics. *J. Geol., 100*, 231–250.

Wang, Z. 1986. *On division of the metamorphosed strata in the Jaionan uplift and its tectonic evolution*. In Proceedings of the International Symposium on Precambrian Crustal Evolution (Tectonics), Beijing, China, Geological Society of China.

Wood, B. J., and Banno, S. 1973. Garnet-orthopyroxene and orthopyroxene-clinopyroxene relationships in simple and complex systems. *Contrib. Mineral. Petrol., 42*, 109–124.

Xie, D., Mao, C., Shi, H., Jiang, Y., and Fong, J. 1983. Study on the mineralogy of eclogites in the Dabieshan. *Acta Petrologica Mineralogica et Analytica, 2*, 87–98.

Xie, D., Shi, H., Guo, K., and Fong, J. 1985. Study on the metamorphic minerals of the "the Dabieshan Complex" in the Dabieshan Mountain. *Geological Articles for International Exchange, 3*, 283–300.

Xu, H., Wang, W., and Go, J. 1979. Deep-seated eclogites in Junan and Rizhao areas, Shandong Province. *Seismology and Geology, 1*, 57–66 (in Chinese with English abstract).

Xu, J., Zhu, G., Tong, W., Cui, K., and Liu, Q. 1987. Formation and evolution of the Tancheng-Lujiang wrench fault system: a major shear system to the northwest of the Pacific Ocean. *Tectonophysics, 134*, 273–310.

Xu, S., Okay, A. I., Ji, S., Sengor, A.M.C., Su, W., Liu, Y., and Jiang, L. 1992. Diamond from the Dabie Shan metamorphic rocks and its implication for tectonic setting. *Science, 256*, 80–82.

Xu, S., Zhou, H., Dong, S., Chen, G., and Zhang, W. 1986. Deformation and evolution of the predominant structural elements in Anhui Province, China. *Scientia Geologica Sinica, 2*, 311–322 (in Chinese with English abstract).

Yang, J. 1991. *Eclogites, garnet pyroxenites and related ultrabasics in Shandong and north Jiangsu of east China*. Beijing: Geological Publishing House.

Ye, D., Li, D., Dong, G., and Qiu, X. 1980. 3T phengite and C-type eclogite in Xinyang metamorphic belt and tectonic significance. *Formation and development of the north China fault block region*. Beijing: Science Press, 122–132.

Ying, S., Yu, L., and Yang, Z. 1981. The metamorphism and genetic problems of eclogites from Hongtushan and Qinlongshan of the Donghai region, Jiangsu Province. *Seismology and Geology, 3*, 17–30.

Zhang, R. Y. 1992. Petrogenesis of high pressure metamorphic rocks in the Su-Lu and Dianxi regions, China. Ph.D., Kyoto University, Japan.

Zhang, R. Y., and Cong, B. 1991. UHP metamorphism and retrograde reaction of coesite-bearing quartz eclogite from Weihai, eastern China. *EOS Trans. Am. Geophys. Union, 72*, 559.

Zhang, R. Y., Takao, H., Banno, S., Ishiwatari, A., Li, J., Cong, B., and Nozaka, T.

1990. *Coesite-eclogite from Donghai area, Jiangsu Province in China. The 15th General Meeting of the International Mineralogical Association Abstract*s.

Zhang, R. Y., Cong, B. L., and Liou, J. G. 1993b. Su-Lu ultrahigh pressure metamorphism terrane and its tectonic implication. *Acta Petrologica Sinica, 9*, 1–13.

Zhang, R. Y., and Liou, J. G. 1994. Coesite-bearing eclogite in Henan Province, central China: detailed petrography, glaucophane stability and PT-path. *Eur. Jour. Mineral., 6*, 217–233.

Zhang, R. Y., and Kang, W. G. 1989. The character of the blueschist belt and discussion of the formation age in central China. *Journal of Changchun University of Earth Science, 1*, 9 (in Chineses with English abstract).

Zhang, Z. M., Liou, J. G., and Coleman, R. G. 1984. An outline of the plate tectonics of China. *Geol. Soc. Amer. Bull., 95*, 275–312.

Zhang, R. Y., Liou, J. G., Wang, X., Wang, Q. C. and Liu, X. H., 1993a. Discovery of coesite eclogites in Henan province, central China and its tectonic implication. *Acta Petrologica Sinica, 9*, 186–191 (in Chinese with English abstract).

Zhou, G., Kang, W., Gao, J., and Liu, X. 1989. The main features of blueschist belt in the northern Hubei province. *Journal of Changchun College of Geology*, 10–17.

Zhou, G., Liu, Y. J., Eide, E. A., Liou, J. G., and Ernst, W. G. 1993. High pressure-low temperature metamorphism in northern Hubei province, central China. *Jour. Metamorph. Geol., 11*, 561–574.

11

A Model for the Tectonic History of HP and UHPM Regions in East Central China

ELIZABETH A. EIDE

Abstract

The South Qinling-Dabie Mountains orogenic complex in east central China exposes deep sections of continental crust that were metamorphosed under high and ultrahigh pressure (HP-UHP) conditions during Triassic continental collision. This chapter presents a tectonic model for collision, metamorphism, and exhumation of the HP-UHP rocks in east central China that incorporates $^{40}Ar/^{39}Ar$ ages, geothermobarometric and structural data collected from this area, and published regional geophysical and geologic information.

Two new ideas proposed in this tectonic synthesis of east central China are, first, that exhumation of the HP-UHP rocks occurred in two episodes marked by different exhumation rates, and second, that these rates were probably associated with different tectonic processes. The first stage of exhumation was probably fairly rapid and was accomplished through continuous collision, stacking of continental crustal wedges via crustal-scale thrusts, and lateral translation in the upper crust along strike-slip faults. The second identifiable stage of exhumation, which finally brought the HP-UHP rocks to the surface, occurred at relatively slower rates than in the first stage, and was associated with processes of crustal extension, anatexis, and erosion.

Introduction

Discoveries in continental collision zones of metamorphic minerals generated under high and ultrahigh pressure (HP-UHP) conditions attest to former continental crustal thicknesses greater than any currently estimated on the Earth. In Tibet, for example, estimated crustal thicknesses range between 60 and 80 km (Dewey et al., 1986; Mattauer, 1986; Hirn, 1988), whereas crustal depths for coesite and/or microdiamond genesis are probably in excess of 85 km (Chopin, 1984, 1986; Okay et al., 1989; Wang et al., 1989; Xu et al., 1992).

The UHP minerals and their associated metamorphic products have raised fundamental questions regarding our knowledge of the physical and chemical properties of the continental crust and the mechanisms associated with deep subduction and exhumation of low density crustal materials. Unravelling the complex tectonometamorphic histories of these HP-UHP regions is fundamental toward understanding the processes of continental growth and destruction through time and providing constraints on current perceptions of the physicochemical properties of the Earth.

Collision between the Sino-Korean and Yangtze cratons of eastern China during the Triassic Period generated HP-UHP metamorphic rocks which are today exposed at the surface in the Dabie Mountains and Shandong Peninsula (Su-Lu region), the Hong'an and Tongbai regions, and the southern portion of the Qinling Mountains (Mattauer et al., 1985; Okay et al., 1989; Wang et al., 1989; Enami and Zang, 1990; Ernst et al., 1991; Hirajima et al., 1990; Eide, 1993; Zhou et al., 1993) (Fig. 11.1). These areas are collectively referred to as the South Qinling–Dabie Mountains HP-UHP complex and expose a fairly coherent and areally extensive sequence of subducted continental crust and cover.

This study assesses the evidence for timing of, and tectonic mechanisms associated with, continent–continent collision in east central China. This information is presented within the context of available field descriptions and geophysical, petrologic/chemical, and geochronologic data from the literature and structural, chemical, and geochronologic data collected by the author. This approach was adopted to address more thoroughly the complex interactions among mechanical, thermal, and chemical processes operative during continental collision and expressed in the wide variety of rock types exposed in this part of China. Other provocative tectonic models generated for this region of China (e.g., Liu and Hao, 1989; Hsü et al., 1990; Okay and Sengör, 1992) have addressed the general evolution of the region but have not interpreted the area specifically using an integrated geophysical, geochronologic, field, and petrochemical database.

The first section of the chapter is a compilation and review of regional geology and tectonostratigraphic units. This section provides the geologic justification for the tectonic synthesis presented in the latter half of the chapter. It is the hope of the author that this model will be quantified and revised further as more seismic, field-structural, and metamorphic data in east central China become available.

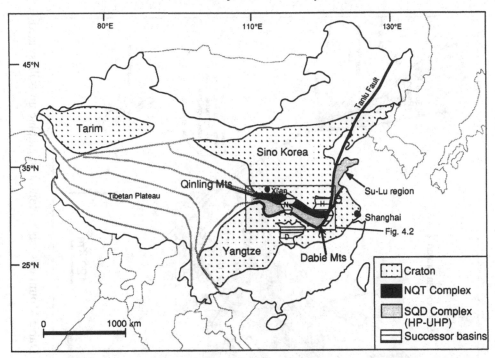

Figure 11.1. Map of China showing position of Qinling-Dabie Mountains orogenic zone between the Sino-Korean and Yangtze cratons. The Su-Lu region is displaced to the northeast of the Dabie Mountains by the Tanlu fault. H = Hefei Basin, D = Dongting Basin, N = Nanyang Basin. Major fault boundaries of the Qinling–Dabie area are black; those of the Tibetan Plateau are in gray. See text for NQT and SQD.

Principal Tectonostratigraphic Units

Six major tectonostratigraphic units will be discussed (Fig. 11.2). Although this division oversimplifies the evolutionary complexities of individual units, the divisions adopted here correspond to established temporal, structural, and tectonostratigraphic boundaries within the area and facilitate regional analysis and discussion. In addition, the attempt was made to correlate similar rock units which have been called by different names in different publications. A summary of the regional tectonostratigraphy has been given in Ernst et al. (1991).

Sino-Korean Craton

This craton is cored by Archean basement consisting of biotite-hornblende gneiss, tonalite, trondjhemite, and granodiorite gneiss, granulite, migmatite, marble, meta-peridotite, and banded iron formation (Figs. 11.1 and 11.2) (Zhang et al., 1984; Bureau of Geol. Anhui, 1987; Ma, 1989). The oldest

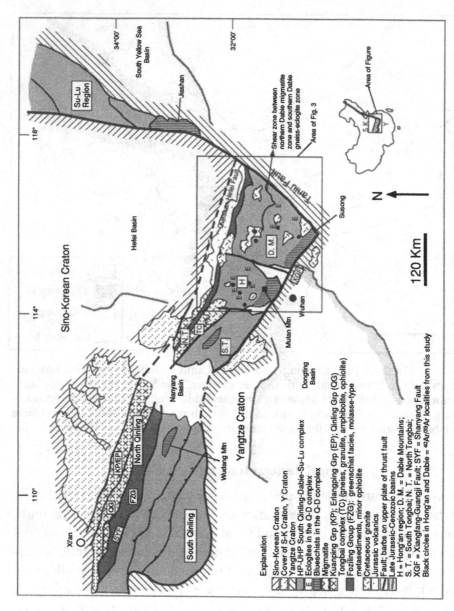

Figure 11.2. Geologic map of the Qinling–Dabie Mountains. Six tectonostratigraphic units illustrated are discussed in the text.

crustal components have been dated as greater than 3800 Ma (Liu et al., 1992), and amalgamation and stabilization of the craton was achieved by the Middle Proterozoic Era (Ernst, 1988a; Ren and Chen, 1989). Except for a hiatus in the Late Silurian and Devonian, the Precambrian basement is unconformably overlain by a continuous Cambrian through Carboniferous sedimentary section of marine carbonate, mudstone, and shale. The Permian Period marked a change to humid climatic conditions characterized by swamp facies with coal-bearing clastic sediments (Bureau of Geol. Anhui, 1987; Yang, 1989). The unconformity in the Siluro-Devonian Period may be correlated with orogenic processes that generated the North Qinling-Tongbai Complex (see below).

The Triassic Period began a Mesozoic through Cenozoic stage of terrestrial sedimentation on the craton including deposition of fluvio-alluvial clastic red beds, conglomerates, and sandstones. Late Jurassic deposition into fault-bounded basins surrounding the Dabie Mountains (Figs. 11.1 and 11.2) included conglomerate, coarse arkose, carbonaceous shale, and some intermediate volcanic tuff, and trachyandesitic, rhyolitic, and minor basaltic flows. During the Cretaceous Period, a thick sequence of continental clastics (conglomerate to mudstone) and minor alkaline volcanics was deposited largely within, and on the margins of, small successor basins surrounding the Dabie Mountains area (Fig. 11.2). The Tertiary through Neogene Periods were characterized by further continental-style deposition (Bureau of Geol. Anhui, 1987).

Yangtze Craton

The basement of the Yangtze craton is at least Early Proterozoic in age (Bureau of Geol. Hubei, 1990; Zhang et al., 1991) and is composed of biotite-hornblende gneiss, granulite, tonalite, trondjhemite, and granodiorite gneiss, and metamorphosed supracrustal rocks. The craton was probably consolidated by ~850 Ma (Lin et al., 1985; Ma and Ge, 1989). A volcano-sedimentary series of Late Proterozoic age was deposited on the basement and was composed of calc-alkaline basalt and rhyolitic eruptives and shallow marine clastics and carbonates (Zhang et al., 1991); these may have been products derived during the formation of a small rifted margin on the northern (present coordinates) side of the Yangtze craton, or of an island arc similarly associated with, and attached to, the northern edge of the Yangtze craton (Bureau of Geol. Hubei, 1990; Eide, 1993). The Cambrian through Permian Periods on the Yangtze craton were marked by extensive marine carbonate deposition, with occurrence of an unconformity in the Devonian Period (Bureau of Geol.

Anhui, 1987). Swamp and lacustrine facies characterize Late Permian sediments. The northern margin of the Yangtze craton is generally agreed to have been passive throughout the Paleozoic Era (Ma and Ge, 1989; Zhang et al., 1989), in contrast to the Sino-Korean craton which records Siluro-Devonian orogenesis on an active continental margin. Continental clastic deposition also dominated the Triassic through Neogene Periods on the Yangtze craton, with development of fault-bounded basins in the Jura-Cretaceous Period, characterized by fluvial and alluvial fan facies on basin margins and lacustrine facies toward the basin centers (Bureau of Geol. Anhui, 1987).

North Qinling–North Tongbai Complex (NQT Complex)

This complex forms the northern part of the Qinling–Dabie Mountains orogenic belt and is distinct from the metamorphic rocks of the South Qinling–Dabie Mountains HP-UHP complex both in rock type and age (Figs. 11.1 and 11.2). The NQT Complex represents the first period of post-cratonization tectonic activity in central China and is divided into six principal, fault-bounded zones from north to south (Fig. 11.2): (1) the Kuanping Group, alternatively called the Xionger Group (Zhang et al., 1991) or the Maoji Group (Ma, 1989); (2) the Erlangping Group; (3) the Qinling Complex; (4) the Danfeng ophiolite; (5) the Foziling Group (Ma, 1989) or, alternatively, the Xinyang Group (You et al., 1993), the Gumiao sequence (Kröner et al., 1993) or the North Huaiyang sequence (in the region north of the Dabie Mountains and Hong'an areas; Bureau of Geol. Anhui, 1987); (6) the Tongbai complex, which occurs in the North Tongbai area only (Kröner et al., 1993) (Fig. 11.2). In addition, two episodes of plutonic activity were recorded in the NQT Complex and were related to magmatic pulses during the Middle-Late Devonian and Mesozoic Periods, respectively (Kröner et al., 1993; You et al., 1993).

The Kuanping Group is characterized by greenschist facies metabasites and clastics with intercalated marbles (Ma, 1989; Zhang et al., 1991; Kröner et al., 1993). Ages between 975 and 920 Ma were derived from bulk-rock Sm-Nd isochrons on greenschists (Kröner et al., 1993 and references therein), and probably represent protolith ages. The Erlangping Group is a metapelitic-meta-arkosic sequence with associated chert, spilite-keratophyre, tholeiitic basalt, gabbro, and minor meta-carbonate; the sequence has a suggested Ordovician age based on fossil identification (You et al., 1993) or a more general, early Paleozoic affinity based on Rb-Sr and U-Pb isotopic ages which range from 357 to 500 Ma (Kröner et al., 1993 and references therein), or 505 to 438 Ma (You et al., 1993). This sequence has been interpreted by several authors to represent a back-arc basinal series containing ophiolitic rocks which were obducted onto the Sino-Korean margin during Siluro-Devonian time (Ma, 1989; Kröner et al., 1993).

The Qinling Complex in the NQT (Fig. 11.2) is a high-grade metamorphic complex composed of biotite gneiss, garnet-sillimanite gneiss, amphibolite, granulite, micaschist, marble, and metatuffaceous units (Mattauer et al., 1985; Ma, 1989; Kröner et al., 1993; You et al., 1993). $^{207}Pb/^{206}Pb$ ages from zircons in granulite and tonalitic gneiss exposed in the Tongbai area yielded ages of 470 and 435 Ma, respectively, and were interpreted to indicate the time of high-grade metamorphism (Kröner et al., 1993). $^{40}Ar/^{39}Ar$ ages on biotite and muscovite from mica schists yielded cooling ages of 328 and 348 Ma (Mattauer et al., 1985).

The Danfeng ophiolite sequence, which is exposed between the Qinling Complex and the Foziling Group in the North Qinling area, is composed of massive basalt, gabbro, dike, and turbidite rocks. Ages between 402 Ma (Sm-Nd on gabbro) and 447 Ma (Rb-Sr whole-rock on tholeiite) have been reported for rocks in this suite (Zhang et al., 1989). In the North Tongbai area, the Danfeng ophiolite occurs only as disaggregated blocks in a wide shear zone between the Qinling Complex and the Foziling Group (Kröner et al., 1993).

The Foziling Group has been described as a trough or flysch sequence dominated by conglomerates with high-grade metamorphic pebbles (presumably from the Qinling Complex), graywackes, and distal turbidites. The age of the sequence is reported to be Devonian (Mattauer et al., 1985; Kröner et al., 1993). In its exposure to the north of the Dabie Mountains and Hong'an areas, the Foziling has apparently been intruded and overlain by Jurassic volcanic rocks (see later sections) (Figs. 11.2 and 11.3).

Tongbai complex rocks lie to the south of the Foziling Group and occur in the Tongbai area only; the rocks are composed mainly of ortho- and paragneisses, amphibolite, and marble. $^{207}Pb/^{206}Pb$ ages from zircons in Tongbai gneisses are 746 and 776 Ma and were interpreted to represent protolith ages of these rocks. The ages and the rock types indicate a former Sino-Korean affinity (Kröner et al., 1993).

The sequences of the NQT Complex have been interpreted to represent various tectonostratigraphic fragments of a failed rifting event of Late Proterozoic or Early Paleozoic age on the southern (present coordinates) margin of the Sino-Korean craton or, alternatively, fragments of accreted continental crust and an island arc. The amalgamation of rifted materials, or collision of the island arc, was heralded by subduction and obduction of pieces of oceanic crust in the Silurian. Oceanic crustal fragments are suggested from the presence of both the Erlangping and Danfeng ophiolitic suites to the north and

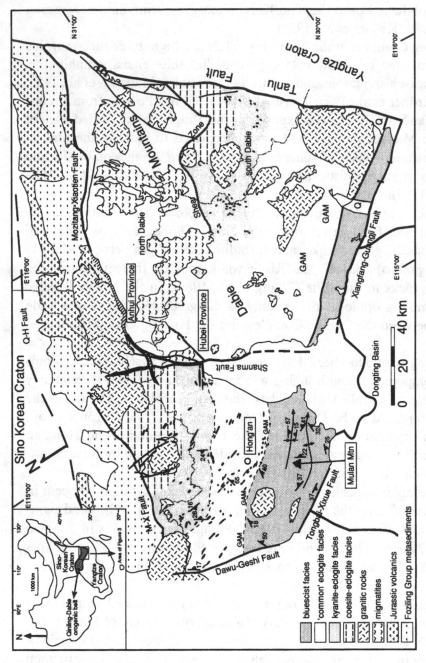

Figure 11.3. Geologic map of the HP-UHP zones of the Hong'an and Dabie Mountains. The Hong'an block is the semitriangular region with Hong'an city labelled. Both areas show a progressive increase in metamorphic grade from south to north. Eclogites are dark elliptical bodies. The Dabie Mountains has a wide northern tract of migmatite exposed north of the shear zone marked by a thick gray line. GAM = garnet-amphibolite regions; Q = Quaternary cover. See text for discussion.

south, respectively, of the Qinling Complex high-grade metamorphic rocks and by their associated sediments on the southern margin of the craton (Ma and Ge, 1989; Zhang et al., 1989; Kröner et al., 1993).

The literature indicates basic agreement on orientations of faults within the NQT. Most authors suggest the inter-unit NQT faults are north-dipping thrusts which intersect the surface at a steep angle and sole into the base of the lower crust at depths between 20 and 30 km (Fig. 11.2) (Gao et al., 1992; Huang and Wu, 1992). This information is based predominantly on interpretations of seismic refraction profiling through the Qinling area and will be described in a subsequent section. Cross sections through the NQT Complex nearly always include some south-dipping thrusts (back-thrusts?), and strike-slip and normal faults, interleaved with a predominance of southward-verging thrust sheets (Ma, 1989; Gao et al., 1992; Huang and Wu, 1992; Kröner et al. 1993).

In the Qinling region, however, the Shangdang fault zone, which is the main contact between high-grade Paleozoic metamorphic rocks of the NQT and the lower-grade rocks of the Foziling Group (Fig. 11.2), has been interpreted either as a steeply-dipping sinistral strike-slip fault (Mattauer et al., 1985) or as a northward-dipping listric thrust fault (Gao et al., 1992; Huang and Wu, 1992). In the Tongbai area the surface expression of the fault between the Paleozoic NQT and Mesozoic South Qinling-Dabie metamorphic units (Fig. 11.2) is not well-exposed, and the north-directed dip and southward vergence on schematic cross sections through the area have largely been inferred (Kröner, personal communication, 1993). The variable interpretations of the dip direction and sense-of-motion on some of the large faults in the NQT underscore the fact that more information regarding the nature and orientation of the faults in the orogenic zone has to be acquired. The orientation and sense of motion of faults within and surrounding the Qinling-Dabie orogenic belt are critical to the understanding of the tectonic history of the area, but their original orientations and degree of reactivation are the subjects of continued discussion.

South Qinling–Dabie Mountains Complex (SQD Complex)

This metamorphic complex comprises the southern and eastern portions of the Qinling-Dabie Mountains orogenic region (Fig. 11.2). The SQD contains rocks metamorphosed under HP-UHP and relatively low-temperature (LT) conditions; the rocks delineate a general increase in metamorphic grade from south to north with isofacial units occurring subparallel to the regional strike of the Qinling-Dabie Mountains (Figs. 11.2 and 11.3). The high-pressure nature of the metamorphism and polarity of the metamorphic gradient in the SQD support the idea that these rocks were generated in a northward-dipping subduction zone during collision of the Yangtze craton with the Sino-Korean craton (Ernst et al., 1991; Ames et al., 1993; Zhou et al. 1993). Details of the lithologies and pressure-temperature (*P-T*) histories of the rocks in the four fault-bounded SQD blocks, the South Qinling, South Tongbai, Hong'an, and Dabie regions, have been presented in numerous papers (Mattauer et al., 1985; Bureau of Geol. Anhui, 1987; Liu and Hao, 1989; Okay et al., 1989; Wang et al., 1989; Bureau of Geol. Hubei, 1990; Wang et al., 1990; Ernst et al., 1991; Sang, 1991; Wang and Liou, 1991; Wang et al., 1992; Xu et al., 1992; Ames et al., 1993; Eide, 1993; Zhou et al., 1993). Only a brief description of the metamorphic units within the blocks is presented here.

The South Qinling region (Fig. 11.2) is characterized by lower greenschist facies sequences in the south and west; the low-grade rocks are probably folded and slightly deformed parts of the volcano-sedimentary cover of the Yangtze craton. Higher pressure, blueschist facies rocks in the northern and eastern portion of the block are exposed at Wudang Mountain (Fig. 11.2). The protoliths of the blueschist facies rocks were metamorphosed argillaceous and arkosic sediments, minor carbonates, and volcano-sedimentary rocks; the latter contain bimodal volcanics (Bureau of Geol. Hubei, 1990; Ernst et al., 1991). The rocks of South Qinling are in fault contact with the NQT Complex (Fig. 11.2) and form E-W striking folds (F_1) with an overprinting F_2 fold series (Mattauer et al., 1985).

The South Tongbai region is characterized by a south-to-north progression of greenschist to blueschist facies metasedimentary and metavolcanic sequences and is separated from the South Qinling region by the Nanyang Basin (Fig. 11.2). A narrow zone of garnet amphibolite and eclogite facies rocks is apparently exposed to the north of the blueschists (Ernst et al., 1991) and is truncated by the fault contact with the NQT Tongbai Complex (see above). In the literature, the South Tongbai HP rocks have been referred to as the Suixian Group (Bureau of Geol. Hubei, 1990; Kröner et al., 1993).

The Hong'an block exposes even deeper portions of the HP subduction-collision complex and is separated from South Tongbai by a N-S striking fault (Fig. 11.2). Rocks distributed from south to north in Hong'an include transitional blueschist-greenschist and blueschist facies sequences (best exposed at Mulan Mountain), garnet-epidote amphibolites, eclogites, and kyanite-bearing, coesite-bearing, and microdiamond-bearing eclogites (Fig. 11.3). The amphibolites and eclogites occur as layers or lenses within regional phengite or biotite schist and gneiss. Protoliths of the blueschists are similar to those described for blueschists from South Tongbai and South Qinling; protoliths for the eclogites and country rocks are interpreted to be tholeiitic sills or dikes which intruded sialic, continental materials in an ensimatic island- or a back-arc basin setting (Eide, 1993). Protoliths for blueschists and eclogites may have the same Late Proterozoic origins along the northern margin of the Yangtze craton possibly during a rifting event. Confirmation of the protolith origins awaits interpretation of isotopic age determinations and bulk-rock chemical analyses of these rocks.

Contacts in the Hong'an block between isofacial metamorphic units may be tectonic, although poor exposure precludes conclusive field determination. The distance that corresponds to the metamorphic pressure gradient between coesite- and kyanite-bearing eclogite zones is greater than the presently exposed areal distance between these metamorphic units; therefore, some shortening and juxtaposition along faults within the Hong'an block has probably occurred during exhumation of the rocks.

The Dabie Mountains block contains a south-to-north progressive metamorphic sequence similar to the Hong'an block, but additionally exposes a wide northern zone of migmatite, granite, and gneiss and minor ultramafic rock, with few eclogites (Fig. 11.3). The migmatitic region has been described as the *North Dabie Terrane* (NDT), and the HP-UHP rocks to the south as the *South Dabie Terrane* (SDT) (Wang et al., 1992); the eastern end of the contact between the two "terranes" is apparently strongly mylonitized (Fig. 11.3). The northern Dabie Mountains are in fault contact with the Foziling Group and the Dabie Mountains are truncated on the east by the Tanlu fault.

The rocks of Shandong Peninsula (Figs. 11.1 and 11.2) contain a south-to-north succession of HP-UHP rocks similar in metamorphic *P-T* conditions and chemistry to the Dabie Mountains (Enami and Zang, 1990; Hirajima et al., 1990). This similarity has led investigators to conclude that the Tanlu fault is a sinistral strike-slip fault that has offset the Su-Lu region from the Dabie Mountains area (e.g., Okay and Sengör, 1992; Wang et al., 1992). Precise ages of inception and timing of motion on this fault are unknown; this chapter proposes that the Tanlu fault originated in Late Triassic time (see below).

Structural studies of the SQD Complex rocks are not abundant. A study of the blueschists at Mulan Mountain (Eide, 1993) indicates the blueschist facies rocks there underwent three deformational episodes: The first two were compressional, and the third was extensional. Some structural information from the region is plotted in Fig. 11.3. All the rocks in the SQD have experienced a retrograde, greenschist through amphibolite facies overprint, which indicates a clockwise *P-T* evolutionary path. Maximum *P-T* conditions calculated for the Hong'an area are (1) blueschists, 4–8 kbar and 300–475°C (Eide, 1993; Zhou et al., 1993); (2) eclogites, 9–12 kbar and 450–550°C (Eide, 1993; Zhou et al., 1993); (3) kyanite-bearing eclogites, 16.5 kbar and 550–640°C (Eide, 1993); (4) coesite-bearing eclogites, 29–31 kbar and 590–643°C (Zhang et al., 1993). Similar P-T conditions have been calculated for the southeastern Dabie Mountains (Wang et al., 1992).

The nature of faults surrounding the blocks of the SQD Complex and the geochronologic development of the metamorphic units are addressed in subsequent sections.

Granitic Rocks

The chemical features of granitic rocks in many regions have been used extensively to classify and interpret the tectonic setting in which granitic magmas may have been generated (e.g., Chappell and White, 1974; Loiselle and Wones, 1979; Pitcher, 1979; Chappell and White, 1984; Pearce et al., 1984; Chappell and Stephens, 1988). Granitic rocks of Late Jurassic–Late Cretaceous age (Li and Wang, 1991; Eide et al., 1994) in the Qinling-Dabie Mountains are exposed mainly in the Dabie and northern Hong'an blocks (Figs. 11.2 and 11.3) and have slightly unusual chemical characteristics. These plutons and their associated minor volcanics are significant with regard to the timing and mode of exhumation of the HP-UHP rocks in the SQD Complex. Most chemical information from the granites outlined in the following section was derived from a comprehensive study by Li and Wang (1991) on granites in the Tongbai-Hong'an-Dabie areas of Hubei Province.

The plutonic rocks are broadly granitic, but encompass quartz diorite, diorite, granodiorite, and monzonitic granite rock types (Shu, 1984; Li and Wang, 1991); for purposes of general description, all rocks will henceforth be called "granites." Except for minor calc-alkaline bodies to the southwest of the Dabie Mountains (Fig. 11.3) (Shu, 1984), the granites are almost exclusively peraluminous despite fairly high total alkali contents ($Na_2O + K_2O =$ 5.56–9.76 wt%) (Li and Wang, 1991). Other characteristics include variable SiO_2 contents (60.71–77.47 wt%), high, but variable Sr (15–1636 ppm, with most greater than 150 ppm), relatively low Rb (35.3–251.4 ppm, with most under 150 ppm), low Cr (<20 ppm), low Nb (3.6–20.2 ppm), and high Fe^{3+}/ ($Fe^{3+} + Fe^{2+}$) ratios (0.3–0.5). $^{87}Sr/^{86}Sr$ ratios are variable, but generally low, and range from 0.7029 to 0.7138. δO^{18} values range from 5.2 to 11.3‰, with just four out of 26 analyses exhibiting values >10‰. Mineralogically, the granites contain quartz, plagioclase, and pink K-feldspar, with additional Mg-biotite, magnetite, titanite, zircon, and apatite. Ilmenite and amphibole are rare (Li and Wang, 1991).

The Mesozoic granites of the Qinling-Dabie orogenic belt are I-types, after the classification scheme of Chappell and White (1974). Their variable SiO_2 values, high alkalis, high Fe^{3+} and magnetite (indicative of high fO_2), relatively low $^{87}Sr/^{86}Sr$, δO^{18} values generally <10‰, primary titanite, and lack of muscovite support this classification according to standard definitions (Chappell and White, 1984). However, the distribution and timing of emplacement of the Qinling-Dabie granites largely between 90 and 140 Ma, relative to the timing of collision and HP-UHP metamorphism (Mid–Late Triassic), may invite classification of these granites as A-types. It is emphasized, however, that on the basis of the chemistry and mineralogy of granites sampled by Li and Wang (1991) and those observed in Hubei Province in the course of this study, an A-type classification is not appropriate if the original definitions by Loiselle and Wones (1979) are used. A-type granites have some chemical characteristics similar to the Qinling-Dabie I-types described above which include high total alkalis, variable $^{87}Sr/^{86}Sr$ ratios (0.703–0.712), low values of trace elements (Co, Ni, Cr), low CaO, and relatively low SiO_2 (Loiselle and Wones, 1979). In contrast to the chemistry and mineralogy of the central China granites, however, A-types should also exhibit enrichment in incompatible elements (Zr, Nb, REE), relatively low fH_2O (and low to intermediate fO_2), high F^- and Cl^- values, and also contain Fe-rich biotites, Fe-rich olivine, hedenbergite, and hastingsitic amphibole. Because of the alkaline nature of the A-types, they may also contain Na-amphiboles and Na-pyroxenes. Generally, A-types lack magnetite or titanite, are mildly alkaline to subalkaline and are relatively low in Al_2O_3. These characteristics are not observed in Cretaceous granites from east central China.

The broadly peraluminous nature of the Qinling-Dabie granites is rather unusual for I-type granitic rocks, but similar, rare examples have been observed in some parts of the Lachlan Fold Belt in southeastern Australia (Chappell and White, 1984). The chemical variability exhibited by the Qinling-Dabie granites, especially regarding their alkaline nature, initial Sr ratios, SiO_2 values, and Rb contents, as well as their low CaO contents, has genetic significance because these do not conform strictly to "standard" I-type chemical signatures. Some Lachlan Fold Belt I-type granites exhibit variable $^{87}Sr/^{86}Sr$ ratios (0.704–0.712), the ranges for which are similar to the central China granites. The variability in $^{87}Sr/^{86}Sr$ in the Lachlan examples has been attributed to derivation of I-type source rocks over a protracted time period and may represent either prolonged assimilation of crustal materials or distinct periods of separation of the melt from the residual source material (Chappell and White, 1984; Chappell and Stephens, 1988). These processes may also account for some of the other chemical variations in the Hong'an-Dabie granitic rocks.

The classification schemes used for granites have become more detailed in recent years, so that straightforward tectonic and chemical divisions among I-, S-, and A-type granites are not necessarily distinct. In the case of the east central China granites, I-type chemical and mineralogic signatures overlap slightly with the tectonic scenario envisioned for a genetic situation that otherwise resembles an A-type origin. More detailed work on the nature of the east central China granites, and perhaps more refined classification of these rocks using the scheme for A-type granites proposed by Eby (1992) may be necessary. The tectonic significance of the genesis and interaction of these granitic melts with the Hong'an-Dabie HP-UHP rocks is clear, however, and will be addressed in the concluding section.

Post-Triassic Successor Basins

Three sedimentary basins occur immediately adjacent to the eastern Qinling-Dabie Mountains complex. The oldest sediments in the Hefei Basin (Figs. 11.1, 11.2, and 11.3) are apparently upper Jurassic (Watson et al., 1987; Xu et al., 1987) and are characterized by coarse red conglomerate, sandstone, minor shale, and siltstone which probably represent fluvio-alluvial and lacustrine facies; some volcanoclastic and extrusive volcanic flows also occur on the northern basin margin (Fig. 11.3) (Bureau of Geol. Anhui, 1987). The oldest sediments in the Dongting Basin, southwest of the Dabie-Hong'an region (Figs. 11.1, 11.2, and 11.3) are probably Jura-Cretaceous (Watson et al., 1987). Basin deposits include coal-bearing sediments and coarse conglomerate and sandstone of fluviolacustrine facies, in addition to minor volcanic extrusives (Bureau of Geol. Hubei, 1990). The Nanyang Basin to the west of the Tongbai area formed during the Cretaceous Period (Watson et al., 1987).

The major faults in the Hefei Basin are E-W trending, and the principal bounding fault is the Mozitang-Xiaotien fault which has an apparent 80°N dip immediately north of the Dabie Mountains (Figs. 11.2 and 11.3) (Xu et al., 1987). This fault has been interpreted as a northward-dipping thrust that was later reactivated with strike slip and normal components in Late Triassic through Early Jurassic time (Xu et al., 1987), but as with other faults in this orogen, its orientation and sense of motion are debated. The Queshan-Hefei fault, often interpreted as the main boundary between the Sino-Korean craton and the rocks of the NQT Complex, is speculated here also to have facilitated down-dropping of the Hefei Basin, probably aided by transtensional motion associated with the adjoining Tanlu fault. Extensional, Jurassic grabens mark the eastern side of the Tanlu fault (Figs. 11.1 and 11.3), with minor development of Jurassic through Tertiary terrestrial clastic deposits and eruption of minor extrusive flows during graben formation to the northeast of the Dabie Mountains (Figs. 11.2 and 11.3) (Xu et al., 1987).

The Dongting Basin trends predominantly east-west with the Xiangfang-Guangji fault at its northern boundary, although the basin margin is presently covered with alluvium (Watson et al., 1987; Bureau of Geol. Hubei, 1990). During the Cretaceous Period, the direction of principal extension in this region changed from N-S to predominantly E-W, which caused development of NNE-trending tensional faults in this basin and along all other E-W-trending Jurassic basins of the Yangtze craton. Development of the NNE-trending faults has been attributed to the effects of back-arc spreading during WNW subduction of the Kula Plate beneath the western Pacific continental margin beginning in Cretaceous time (Watson et al., 1987; Xu et al., 1989; Bureau of Geol. Hubei, 1990). The Nanyang Basin was apparently established during this period of E-W Cretaceous extension, and the configuration of the Hefei Basin was probably modified as the E-W extensional component was added (Watson et al., 1987; Bureau of Geol. Hubei, 1990).

Geophysical Constraints

Gravity and Magnetic Information

Gravity and aeromagnetic data have been collected within Hubei Province and across the Qinling-Dabie orogenic region (Yang et al., 1987; Han et al., 1989; Zhang, personal communication, 1990). Available information associated with the delineation and orientation of faults is limited, however, to the southern boundaries of the Qinling-Dabie complex in Hubei Province.

The Xiangfang-Guangji (X-G) fault is identified on magnetic anomaly maps by a high magnetic gradient; the interpretation of the data indicates that the fault is steeply north-dipping but shallows at approximately 10 km depth (Liu and Hao, 1989; Bureau of Geol. Hubei, 1990). The Tongbai-Xixue (T-X) fault (Fig. 11.3) also dips very steeply with an interpreted dip direction to the northeast (Zhang, personal communication, 1990). It is unclear at what depth, if any, the T-X and X-G faults coincide. Both gravity and aeromagnetic data imply depth to the Moho in the Dabie area is ~34 km; in Hong'an, this depth is estimated to be 28 km and in the Tongbai area, 31–34 km (Zhang, personal communication, 1990).

Seismic Information

Seismic refraction profiling has been conducted in a north to south transect across the Qinling Mountains area of the Qinling-Dabie orogenic complex (Gao et al., 1992 and references therein; Huang and Wu, 1992). Depth to the Moho has been described by a sharp P-wave velocity increase (6.5–6.8 to 7.8 km/s) in the vicinity of 31–34 km. Interpretive cross sections derived from refraction data display a three-layer structure between the ground surface and the Moho. The North and South Qinling Complexes are depicted with ramp-flat geometries characterized by north-dipping thrust faults (Gao et al., 1992; Huang and Wu, 1992). Gao et al. described the orogenic sequence between the cratons as representative of underthrusting of the Yangtze cratonal margin beneath the NQT Complex with resulting south-directed imbrication of sequences from both margins. This model adheres to the seismic interpretations which show a thin, imbricated set of upper crustal units (~12 km thick) thrust southward onto a combined, low-velocity (sheared?) layer of both NQT and South Qinling rocks (11 km thick), underlain by a lower crustal layer 8–9 km thick with velocities between 6.49 and 6.81 km/s (Gao et al., 1992). Both Huang and Wu (1992) and Gao et al. (1992) support their geophysical data with field interpretations, although it is not clear from their cross sections where they divide the South Qinling exposures and the NQT rocks on the ground surface. In this chapter, the adopted division coincides with the Shanyang fault, and the South Qinling rocks are interpreted to be the metamorphosed, fold-and-thrust cover of the Yangtze craton (Fig. 11.2). It is interesting to note that in map view (Figs. 11.1 and 11.2), the convex-to-the-south outline of the South Qinling area is reminiscent of the map expressions typically observed in thrust-dominated regions (in this case, with thrust vergence apparently toward the south).

Paleomagnetic Information

Paleomagnetic studies have been conducted on both the Yangtze and Sino-Korean cratons (McElhinny et al., 1981; Klimetz, 1983; Lin et al., 1985; Opdyke et al., 1986; Zhao and Coe, 1987; Metcalfe, 1988; Sharps et al., 1989; Li et al., 1990; Lin and Fuller, 1990; Zhu and Wan, 1991). All of the above studies concur that the Yangtze and Sino-Korean cratons did not have converging paleopoles until at least Late Triassic time. The timing of collision, or first statistically indistinguishable coincidence of paleopoles from the two cratons, varies slightly among investigators and is interpreted to have occurred either in Late Triassic or Late Triassic–Early Jurassic Periods. Cessation of suturing is interpreted to have occurred by Mid–Late Jurassic (Lin et al., 1985) or Late Jurassic–Early Cretaceous time (Klimetz, 1983; Lin and Fuller, 1990). Recently, precise ages have been placed on the timing of collision between the two cratons and exhumation of the metamorphic rocks using $^{40}Ar/^{39}Ar$, U-Pb, and Sm-Nd geochronologic techniques; these data are presented in the next section.

Lin et al. (1985), Opdyke et al. (1986), Zhao and Coe (1987), and Zhu and Wan (1991) discussed paleomagnetic evidence which indicates that at least through Middle Triassic time, the Yangtze craton was oriented between 60° and 90° counterclockwise from its present position relative to the Sino-Korean craton. These authors also proposed that the collision between the two cratons occurred at a point, possibly at the southeastern promontory of the Sino-Korean craton, and that the Yangtze craton subsequently rotated clockwise through 60° to 90° during collision to arrive in its current position (see Fig. 11.4a). In this way, collision propagated from east to west in a "scissors" fashion beginning in the Triassic Period.

Geochronologic Constraints

A brief review of the chronologic information available for the evolution of the Qinling-Dabie orogenic belt from isotopic data is presented here and summarized in Table 11.1. Further discussion of the various geochronologic data from the area is presented in Ames et al. (1993), Eide (1993), and Eide et al. (1994).

HP-UHP Rocks (SQD Complex)

Sm-Nd analyses on garnet and omphacite from eclogite, and one whole-rock eclogite from the Dabie Mountains section of the HP-UHP complex yielded ages of 209 ± 31 Ma, 224 ± 20 Ma, and 244 ± 11 Ma (Li et al., 1993). U-Pb analyses of zircons from two eclogites and zircon from a gneiss in the Dabie Mountains yielded ages of 208.6 ± 1.3, 210 ± 11, and 201 ± 10 Ma, respectively (Ames and Tilton, 1992; Ames et al., 1993). Both sets of data are interpreted by the authors to correspond with the timing of the HP-UHP metamorphic event in the Dabie Mountains.

$^{40}Ar/^{39}Ar$ cooling ages of 236 ± 5 and 217 ± 8 Ma were reported for phengite and riebeckite from rocks in the southern Qinling Mountains (possibly near Wudang Mountain, Fig. 11.2) by Mattauer et al. (1985). $^{40}Ar/^{39}Ar$ ages for phengites from schists and gneisses in the Hong'an block have yielded cooling ages between 230.1 and 195 Ma and are related to postmetamorphic

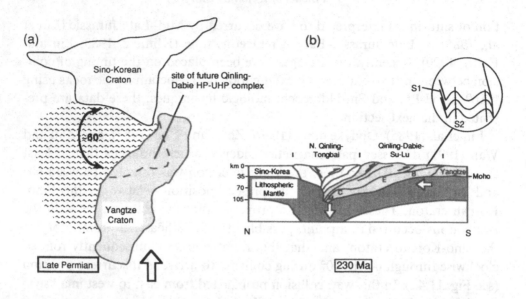

Figure 11.4 (a) Schematic diagram of the approximate position of the Yangtze craton relative to the stable Sino-Korean craton in the Late Permian; at this time the two were separated by a small ocean, and the Yangtze was disposed 60° counterclockwise its present relative position. (b) Schematic interpretation of the manner in which continental crust thickened during the Middle Triassic collision. Great thicknesses were achieved rapidly by stacking along crustal-scale thrust faults during shortening. The lithosphere is also proposed to have thickened beneath the crustal stack. M = proposed position of what is now the northern migmatite region in the Dabie Mountains; C = coesite eclogite; E = eclogite; B = blueschist. (c) Schematic diagram showing the clockwise rotation of the Yangtze craton during collision at ~200 Ma. Top: Development of strike-slip faults concurrent with shortening and thrusting. Bottom: Cross section showing possible positions of various metamorphic facies at this time (symbols as in Fig. 11.4b). Blueschists and kyanite-bearing eclogites of Hong'an were being exhumed and their temperatures were largely below the Ar-closure temperature for white mica at this time. Fold and thrust belt is forming on the southern edge of the crustal stack. The NQT Complex was experiencing deformation at this time also. Symbols M, C, E, and B as in Fig. 11.4b. (d) Cross section depicting the Dabie-Hong'an orogen in the Late Jurassic Period during exhumation and emplacement of granitic melts. Anatexis was accompanied by extension and development of sedimentary basins to the north and south of the orogen. Basin geometry may have been controlled by transtensional motion on strike-slip faults. Erosion became a factor during this late exhumation stage also. In contrast to rocks of the Dabie Mountains, Hong'an HP-UHP rocks were largely at shallow crustal levels and thereby were cool enough to escape the heating effects of rising magmas. Symbols M, C, E, and B as in Fig. 11.4b.

(c)

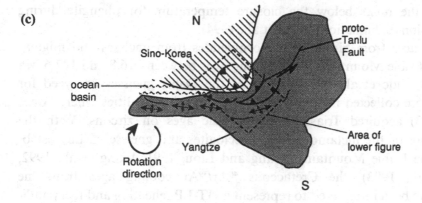

N

Sino-Korea

proto-
Tanlu
Fault

ocean
basin

Area of
lower figure

Yangtze

Rotation
direction

S

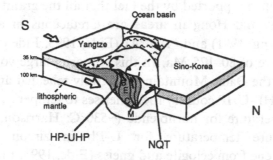

S

Ocean basin

Yangtze

Sino-Korea

35 km

100 km

lithospheric
mantle

HP-UHP

NQT

M

N

~200 Ma

(d)

S

Dongting
Basin

S. Qinling

NQT

Yangtze

Sino-Korea

35 km

100 km

lithospheric
mantle

Tanlu Fault

Hefei
Basin

M

Hong'an-Dabie

N

Late Jurassic (~140 Ma)

cooling of the rocks below the closure temperature for phengite during decompression (exhumation) (Eide et al., 1994).

$^{40}Ar/^{39}Ar$ ages from biotites and hornblendes from gneisses and migmatites in the Dabie Mountains yield cooling ages between 116.8 and 127.6 Ma (Eide, 1993; Eide et al., 1994). Some of the Dabie gneisses analyzed for $^{40}Ar/^{39}Ar$ were collected near the eclogite and gneiss localities where Ames et al. (1993) acquired Triassic metamorphic ages on zircons. With the *in situ* nature of the metamorphism of eclogites and gneisses firmly established in the Dabie Mountains (Wang and Liou, 1991; Wang et al., 1992; Ames et al., 1993), the Cretaceous $^{40}Ar/^{39}Ar$ cooling ages from the gneisses have been interpreted to represent a HT-LP reheating and recrystallization event in the Dabie Mountains region (Eide, 1993; Eide et al., 1994). This interpretation is supported by the fact that all the granitic plutons in the Dabie Mountains and Hong'an areas are Cretaceous in age, with U-Pb, K-Ar (Li and Wang, 1991) and $^{40}Ar/^{39}Ar$ (Eide, 1993; Eide et al., 1994) ages falling in the range of 80–175 Ma. In this scenario, large volumes of Cretaceous plutons in the Dabie Mountains effectively reheated and recrystallized the surrounding HP-UHP country rock gneisses to temperatures greater than the closure temperature for hornblende (~530°C; Harrison, 1981), but not above the closure temperature for U-Pb in zircon or Sm-Nd in garnet and pyroxene from eclogite and gneiss (Eide, 1993; Eide et al., 1994).

The time of crystallization of the mafic igneous protolith for one Dabie Mountain eclogite is suggested to be 685 ± 64 Ma (Ames et al., 1993). Protolith ages (U-Pb zircon) reported for the gneisses from the Dabie Mountains are reported between 1952 and 2424 Ma (Bureau of Geol. Anhui, 1987). U-Pb ages between 800 and 1500 Ma for rocks of the Wudang and southern Hong'an areas (blueschists) (Bureau of Geol. Hubei, 1990) may represent protolith ages for these rocks and correspond to Middle–Late Proterozoic deposition of bimodal volcanics, clastics, and minor carbonates on the north margin of the Yangtze craton, as discussed previously.

NQT Complex

The ages for most sequences in the NQT were presented in the tectonostratigraphic description of the rocks; these are reiterated in Table 11.1. Some disagreement exists in the literature with regard to tectonostratigraphic correlations along regional strike in the NQT, and also with regard to precise ages of these sequences. However, general consensus for the age of high-grade metamorphism and collision along the active southern Sino-Korean margin ranges from ~450 Ma (Kröner et al., 1993) to 328 Ma (cooling age of phengite

Table 11.1. *Time "line" (broken) for rocks of the South Qinling–Dabie (SQD) HP-UHP complex and North Qinling–Tongbai complex (NQT).*

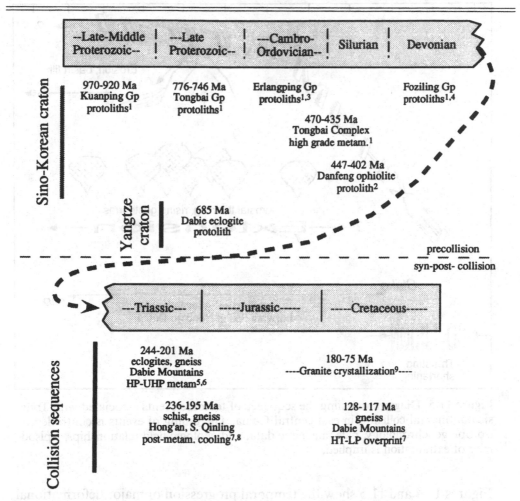

1= Kröner et al. (1993), 2 = Zhang et al. (1989), 3 = You et al. (1993), 4 = Ma (1989), 5 = Ames et al. (1993), 6 = Li et al. (1994), 7 = Eide et al. (1994); Eide (1993), 8 = Mattauer et al. (1985), 9 = Li and Wang (1991).

from micaschist; Mattauer et al., 1985); this range corresponds to the regional sedimentary hiatus observed on the Sino-Korean craton.

Paleotectonic Compilation

Boundary Conditions

The paleotectonic outline below focuses on the Triassic genesis of the HP-UHP SQD rocks and their exhumation; the history of the NQT Complex is not addressed, except where its interaction with the SQD Complex occurs.

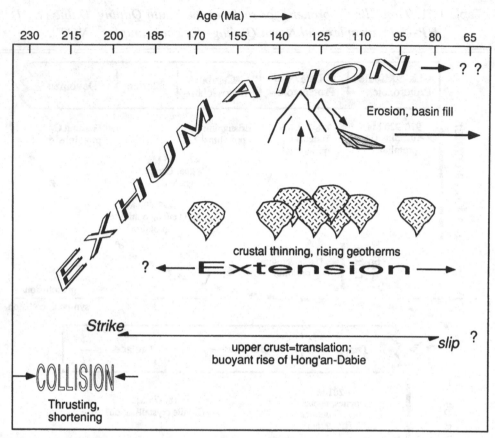

Figure 11.5. Diagram detailing the sequence of tectonic events associated with Triassic continental collision in east central China. The timing of events is controlled by isotope geochronology, paleomagnetic data, and geologic field relationships. Episodicity of exhumation is implied.

Figures 11.4 and 11.5 show the temporal progression of major deformational and tectonic events interpreted to have influenced the area. Several conditions have been assumed in the following discussion and related diagrams: (1) The deepest levels of subduction reached by the UHP Dabie and Hong'an rocks were between 85 and 100 km; (2) protoliths of the SQD Complex rocks were derived from the Yangtze craton; (3) protoliths for the NQT Complex rocks were derived from the Sino-Korean craton; (4) the age of metamorphism of the HP-UHP rocks was ~230 Ma; and (5) UHP metamorphism occurred as the continental crust was thickened owing to subhorizontal compression, shearing, and stacking of crustal wedges during craton–craton collision.

The first boundary condition places the UHP rocks within range of the stability pressures for coesite; crustal thicknesses greater than 100 km have been proposed for this part of east central China where microdiamonds have been discovered (Xu et al., 1992). Boundary conditions (2) and (3) have been justified in previous portions of this chapter. The age of metamorphism of 230 Ma for the HP-UHP sequences (condition 4) is an approximation based on the fact that Sm-Nd and U-Pb metamorphic ages from the rocks range between 201 and 244 Ma, and that $^{40}Ar/^{39}Ar$ cooling ages for phengites from blueschists are between 222 and 236 Ma (see section 'Geochronologic Constraints' above). In reality, it is probable that different minerals from rocks in different sections of the the HP-UHP belt passed through their metamorphic closure temperatures at various points during the Mid–Late Triassic collision. Boundary condition (5) is derived from an assessment of the two end-member mechanisms commonly proposed for massive thickening of continental crust: simple doubling of crust by emplacing one continental slab directly on another crustal slab, or intracrustal shortening via internal deformation (thrusting and stacking of crustal wedges) (England and Thompson, 1984; Dewey et al., 1986; Mattauer, 1986; Windley, 1988). To date, no geophysical evidence supports thickening by the first mechanism (Windley, 1988), so the second mechanism was adopted here.

Paleotectonic Compilation
Compression and Thrusting

By Middle Triassic time, the Yangtze craton had approached and begun to collide with the southeastern promontory of the Sino-Korean craton (Figs. 11.4a,b). The Yangtze was initially disposed counterclockwise 60–90° from its current position relative to Sino-Korea. Upon initial contact, the Yangtze began to rotate clockwise (Lin et al., 1985; Opdyke et al., 1986; Zhao and Coe, 1987; Zhu and Wan, 1991), and in this manner, collision effectively migrated from east to west (Fig. 11.4b). Collision focused at the Sino-Korean promontory facilitated massive, rapid crustal thickening by compression and stacking of crustal wedges (Fig. 11.4b). Crustal thicknesses reached ~85–100 km in the zone of impingement and coesite and/or microdiamonds in eclogites were generated. Coesite-free eclogites and blueschists were developed in successively higher levels of the collision zone (Fig. 11.4b). Rapid thickening accomplished with what was most likely a thick and cold, Proterozoic Yangtze continental crust would have maintained the lower portion of the doubled crust in an environment of relatively low heat flow (Molnar and Tapponnier, 1981). Mineral assemblages observed at the surface today necessitate low temperatures and high pressures for development and preservation. A rapid rate of thickening is also supported by theoretical modelling of heat transfer in continental crust, whereby only rapidly thickened crust is able to achieve high pressures at relatively low temperatures (England and Thompson, 1984).

As collision migrated obliquely westward due to the rotation of the Yangtze craton, the intensity of collision dissipated. Great crustal thicknesses were, therefore, not achieved outside the main locus of collision, and rocks progressively to the west of the initial contact zone were not as deeply subducted or as intensely metamorphosed (Fig. 11.4b,c). This suggestion is supported by the fact that UHP rocks are not found further west than northern Hong'an; HP blueschists and lower-pressure greenschists were the only collision-related sequences generated, and subsequently exposed, in the wide zone now comprising the South Qinling area (Figs. 11.2 and 11.4c).

Strike Slip Motion

In this model, it is assumed that deformation during initial collision was oriented perpendicular to the Sino-Korean margin and was characterized by shortening along thrust faults (Fig. 11.4a, b). Oblique convergence during cratonal rotation as proposed here would have generated a component of orogen-subparallel strike slip motion; strike-slip faults may have been introduced along previous thrust fault traces, and may have reached mantle lithospheric depths as suggested by Mattauer (1986). In this manner, shortening during collision in east central China could, therefore, have been accommodated by both thrust and strike slip deformation through at least ~201 Ma. The latter age represents the upper limit on collision-related metamorphic ages from Dabie Mountains eclogites (Ames and Tilton, 1992) (Figs. 11.4c and 11.5).

Strike-slip faults generated during continental convergence, superimposed upon former thrust faults, and operating simultaneously with thrust-related shortening processes, have been documented in the Himalayas (Mattauer, 1986; Tapponnier et al., 1986) and other convergent orogens (eastern Anatolia: Dewey et al., 1986; North Cascades: Brown and Talbot, 1989), to account for a change in orientation of a colliding plate, or as a means to absorb crustal shortening that cannot be accommodated through thrusting and stacking alone (Dewey et al., 1986; Mattauer, 1986; Tapponnier et al., 1986; Brown and Talbot, 1989). The main requirement for fault deformation character to change from one dominated by compression to one dominated by compression and/or translation as suggested here, is a slowing in the rate of convergence (Mattauer, 1986; Platt, 1986). In east central China, slowing in rate and change in style of convergence may be predicted by the continuous clockwise rotation of the Yangtze craton after its initial contact with Sino-Korea, documented by paleomagnetic data. Rotation resulted in migration of oblique convergence from east to west. Rapid propagation of strike slip motion on the heels of crustal thickening would have accommodated some of the crustal shortening. Initiation of exhumation processes was, therefore, a function of (1) slowing in the rate of crustal subduction; (2) lateral movement of the upper crust, and possibly the lithosphere; and (3) buoyant tendencies of the thickened proto-Dabie Mountains crustal package (Fig. 11.4c).

The overlap between HP-UHP metamorphic and cooling (exhumation-related) ages in central China implies that rotation, suturing, and initial exhumation of the rocks occurred in rapid succession: Rocks metamorphosed at temperatures >450°C in the Hong'an eclogite region (Eide, 1993; Zhou et al., 1993) were cooling below the assumed 350°C closure temperature for phengite between 5 and 35 million years (my) after UHP metamorphism was first recorded (Eide, 1993; see also Geochronologic Constraints above). A variety of mechanisms invoked to exhume HP and UHP rocks from continental collision zones include erosion, buoyant rise of low density, subducted crustal materials, lithospheric extension and lateral movement, and/or transport along lithospheric-scale thrust faults (England and Thompson, 1984; 1986; Burchfiel and Royden, 1985; Dewey et al., 1986; Platt, 1986; Ernst, 1988b; Blake and Jayko, 1990; Carmignani and Kligfield, 1990). Among these, strike slip motion coupled with buoyant rise of the low-density Dabie and northern Hong'an continental materials are the favored mechanisms to have controlled the earliest exhumation stages of these rocks, for the reasons outlined above (Figs. 11.4c and 11.5). In central China, erosion could not have been the primary mechanism to expedite initial rise of the HP-UHP rocks because basins surrounding the Dabie Mountains region do not record significant sedimentary deposition until the Late Jurassic Period (see Extension-Related Crustal Thinning below). Additionally, no older-on-younger or deeper-on-shallower tectonostratigraphic relationships have been documented in the area to suggest dominant movement along thrust faults.

Extension-Related Crustal Thinning

In many continental collision orogens, including east central China, extension-related fabrics and faults are apparently the youngest deformation features observed (Dewey et al., 1986; Mattauer, 1986; Huang and Wu, 1992). Many of the extension-related faults described are, at least initially, confined to upper crustal levels (Burchfiel and Royden, 1985). Development of Jurassic, possibly pull-apart, basins along, and associated with, preexisting strike-slip faults around the Dabie-Hong'an areas (Figs. 11.1 and 11.2) is in keeping with the timing of similar features developed in the Himalayas, eastern Anatolia, the northern Apennines, and the Western Alps following major continental collision and coincident with unroofing events (Burchfiel and Royden, 1985; Dewey et al., 1986; Mattauer, 1986; Carmignani and Kligfield, 1989; Blake and Jayko, 1990). Delays in onset of extension on the order of 30–40 my after collision have been estimated in the Himalayas (Burchfiel and Royden,1985); onset of extension in east central China in the Middle Jurassic would approximate the same time scale. Extension initiated at this time in east central China would just have preceded sedimentary basin deposition and would have been approximately synchronous with the timing of first granite formation (see below) (Fig. 11.4d).

Magma Genesis

Postmetamorphic exhumation of the HP-UHP rocks encompassed two different thermal regimes: (1) The HP-UHP rocks must have been maintained in a low heat flow environment in order to preserve HP mineral assemblages; and (2) some portion of the lower part of the thickened crust (now the northern Dabie migmatite region; see Fig. 11.4c,d) must have passed through the curve for vapor-absent melting (muscovite breakdown and dehydration) necessary to generate granitic magmas (England and Thompson, 1986) between 175 and 80 Ma (Li and Wang, 1991).

Maintenance of the low heat flow environment in the Hong'an-Dabie crust was achieved through rapid thickening of a cold, Proterozoic craton. Rapid initial rebound/exhumation of the HP-UHP rocks and their implied return to the surface along a cool thermal gradient are supported both geochronologically by cooling ages nearly overlapping the metamorphic ages, and petrographically by P-T paths exhibiting nearly isothermal, or cooling during, decompression (Wang et al., 1992; Eide, 1993). Geochronologic and petrologic constraints, therefore, necessitate a delay in, or slow initial rates of, relaxation of the geotherms in the lower crust as exhumation progressed.

Long residence time and delayed crustal melting are also indicated by the chemical signatures of the granitic rocks as outlined previously. Slow heating of crustal sources for periods on the order of 10^8–10^9 years before production of granitic melts has been postulated for some I-type granites in southeastern Australia (Chappell and Stephens, 1988); the Australian granites were proposed to have originated from delayed melting of older crustal rocks. A similar situation may be envisioned for Hong'an-Dabie granites which were apparently derived from a crustal and not a mantle source. The chemical information available from Li and Wang (1991) supports a delayed melting origin for the Cretaceous granites in central China, but thorough investigation of contact relationships, extent and type of thermal aureoles, and interpretation of granite chemistry in light of the information on the surrounding country rock gneisses must be conducted before this suggestion can be confirmed.

Later Features

The predominantly N-S orientation of Late Jurassic extension directions, most visibly expressed by the small sedimentary basins surrounding the Dabie Mountains (Bureau of Geol. Anhui, 1987; Bureau of Geol. Hubei, 1990), changed in Cretaceous–Cenozoic time as a result of the subduction of the Kula Plate beneath the west Pacific margin and the collision of India with Tibet (Tapponnier et al., 1986; Xu et al., 1989; Bureau of Geol. Hubei, 1990). These late tectonic activities were characterized by E-W back-arc extension due to subduction of the Kula Plate, and by lateral stresses that propagated eastward due to the India-Tibet collision (Sengör, 1985; Tapponnier et al., 1986; Fei, 1987; Xu et al., 1989). In this way, deformation unrelated to the collision between the Yangtze and Sino-Korean cratons may have remobilized the North Qinling–Tongbai Complex and the South Qinling-Dabie HP-UHP rocks along old fault zones, and yielded the present tectonostratigraphic configuration. Cenozoic strike slip displacement, for example, could account for some of the missing or uncorrelated sections between the North Qinling and North Tongbai areas of the NQT, and could have obscured the original faults along which the NQT and SQD were first brought into contact.

The Shanma fault between the Dabie Mountains and Hong'an areas is an important feature that deserves comment. The age, sense of motion, and dip of the fault are not well established. By virtue of the fact that the trend of the fault is subparallel to that of the Tanlu fault, the origin of the Shanma fault may be linked to the proposed origin for the Tanlu fault in this study (i.e., related initially to strike slip deformation). Alternatively, the Shanma fault may be a late, extension-related feature, linked to the differential exhumation these HP-UHP rocks have evidently experienced. The extent of the Shanma fault and its contact relationships with both the X-F and Q-H faults must be established before any reasonable genetic relationships can be proposed.

Discussion

The first-order applicability of the paleotectonic synthesis proposed in this study is demonstrated by investigation of three problems regarding fundamental spatial and temporal relationships within Hong'an and Dabie HP-UHP regions. Additional details from the following section may be found in Eide (1993).

1. Are faults between isofacial HP-UHP metamorphic sequences necessary to exhume the rocks? The field associations between the Hong'an and Dabie HP-UHP suites are very similar. In both of these areas, metamorphic gradient increases from south to north, and entails the same metamorphic progression from blueschist through to coesite-bearing eclogite facies rocks (Fig. 11.3). With possible exception of the northern Dabie Mountains migmatite region, the Hong'an and Dabie areas are clearly part of the same deeply metamorphosed complex. The Shanma fault presently separates these two blocks, but whether that separation was imposed early during the exhumation history of the HP-UHP units, or as a late extension feature, cannot be confirmed.

The relatively undisrupted, south-to-north exposures of progressively metamorphosed rocks in Hong'an and Dabie argue for their exhumation as a set of coherent units that lack major internal imbrication. Some juxtaposition along faults betwen isofacial units cannot be discounted, owing to the fact that the areal distance between the coesite- and kyanite-bearing regions in Hong'an, for instance, is not great enough to account for the distance corresponding to the pressures at which these rocks were metamorphosed (e.g., a 10-kbar-pressure difference corresponds to approximately 30–35 km in the subduction zone, and this distance is greater than the ground distance observed today, after correction for regional dip). This structural interpretation is unlike the pervasive south-vergent thrust and fold sequence observed in the Qinling area (Gao et al., 1992; Huang and Wu, 1992). The latter geometry may be a function of the Qinling area's greater distance from the proposed initial zone of collision and consequent decrease in metamorphic and deformational intensity experienced by the rocks there.

The shear zone contact (Figs. 11.2 and 11.3) between northern Dabie migmatites and southern Dabie gneisses and eclogites has previously been described as a possible "terrane" boundary between hotter, upper plate rocks (north Dabie) and colder, lower plate rocks (south Dabie) (e.g. Wang, 1991). However, similar to the above interpretation that the entire HP-UHP sequence was exposed fairly coherently and simultaneously, the interpretation favored here is that the Dabie Mountains block is an example of a relatively intact piece of continental material, with major internal faults and probably missing or telescoped sections, that now exposes progressively deeper levels of the continental collison zone (Fig. 11.4b-d).

Within the limitations of the current tectonic model, a "terrane" boundary is not necessary to account for the difference in P-T conditions exhibited by northern and southern Dabie rocks. The northern Dabie rocks were the most deeply buried portions in the subduction–collision zone and, therefore, experienced partial melting due to intersection with rising geotherms during exhumation. In this way, very few eclogites (HP or UHP) would be expected in this portion of the Dabie Mountains because any HP eclogites included in the northern zone would have been partly melted and/or recrystallized to higher temperature, lower pressure mafic mineral assemblages. The fault now apparently exposed between the northern and southern Dabie Mountains may be interpreted either as the trace of an early, collision-related thrust fault which aided in exhumation of the region (e.g., Fig. 11.4b), or as a later exhumation-related feature, generated along a zone of weakness between rocks with contrasting temperature, ductility, and rock strength characteristics.

Exhumation of the Dabie Mountains as a relatively composite block is the simplest, feasible explanation for the field relationships observed, although the possibility that the northern Dabie migmatite region was a different "terrane" from the southern Dabie rocks, cannot be completely excluded. Field descriptions and structural analysis of this fault zone, and isotopic characterization of the rock types in the northern and southern Dabie Mountains are necessary to address these issues.

2. Is rapid exhumation a necessary condition for the exposure and preservation of the HP-UHP continental collision-related rocks? Factors favoring rapid exhumation of the HP-UHP rocks in central China have been proposed and discussed previously. Rapid exhumation of previously thickened, cold continental crust offers the best means by which to preserve the Hong'an–Dabie HP-UHP sequences in the apparent absence of continuous subduction of cold oceanic crust. Thermal modelling of homogeneously thickened crust, such as that proposed for east central China, imposes a physical limit for crystallization of minerals of the HP facies. Thickened, old continental crust (>250 million years old) with an average assumed thermal conductivity and internal heat distribution generates and preserves HP assemblages only if the entire lithosphere has simultaneously been thickened (England and Thompson, 1986, figs. 2 and 3) (this study: Fig. 11.4b). England and Thompson (1986) further show that higher geotherms, with less preservation potential for HP-UHP assemblages, are necessarily developed when a long time interval elapses between crustal thickening and onset of crustal thinning/exhumation. These theoretical constraints lead to the suggestion that rapid exhumation of the central China rocks must have occurred in order to preserve the HP-UHP mineral assemblages, but further discussion will show that rapid exhumation and its associated tectonic style represented only the first episode in the unroofing of these deeply buried sequences (Fig. 11.5).

Constant, rapid exhumation operating as the sole unroofing mechanism would have exposed the HP-UHP rocks before the Cretaceous Period; it has been established that the Dabie and northern Hong'an rocks were still at depths of at least 10–12 km during Cretaceous granitic magma emplacement. Exhumation governed solely by erosion on the order of ≤1 mm/yr (e.g., England and Thompson, 1984; Hurford, 1986), alternatively, would have placed the UHP rocks in the appropriate time and depth window for invasion of Cretaceous plutons, but erosion-dominated exhumation is precluded for two reasons: (1) No evidence for erosion and its effects is present in the area until the Late Jurassic Period; (2) thermal modelling of continental collision zones indicates that long erosion duration (on the order of 100 my) does not allow thickened crust to remain ahead of relaxing geotherms (England and Thompson, 1984; 1986). This event would lead to characteristic features in the rocks including extensive anatexis and melting of large tracts of the crustal pile and petrographically discernible thermal overprint features and mineral assemblages. These are not observed in Hong'an-Dabie rocks, except for genesis of migmatites in northern Dabie, and *P-T* paths derived for Hong'an and Dabie HP-UHP rocks contrast with the *P-T* paths modelled in erosion-dominant exhumation scenarios by England and Thompson (1984, 1986).

The initiation of erosion in Late Jurassic time and nearly concomitant granitic magma genesis in central China place an upper limit on the cessation of rapid exhumation of the HP-UHP rocks (Fig. 11.5). Deposition of the basinal sediments also signals the onset of extensional and erosional processes during the last phase of exhumation of the metamorphic pile. According to simple theoretical constraints, rapid exhumation is required at least in the early stages of exhumation in a continental collision zone like this in order to preserve HP-UHP rocks. This area of central China demonstrates, however, that rapid exhumation is only one episode of the unroofing process and that slower exhumation rates and different tectonic processes are linked to the final exposure of the HP-UHP rocks.

Conclusions

The subduction–exhumation sequence for the South Qinling-Dabie HP-UHP rocks is characterized by processes of collision, compression, crustal thickening, and thrusting, later joined by strike slip motion and lateral translation. These processes are not unique to this collision zone. The lateral "extrusion" of metamorphic rocks in continental collision zones along strike-slip faults has been experimentally and geologically demonstrated by Tapponnier et al. (1986) for the collision between India and Tibet. Lateral translation proposed in central China is an extension of their model, and is founded on a similar premise that lateral motion may accommodate compression and shortening not achieved by crustal thickening. Lateral translation as presented in the central China case further suggests that strike slip motion may be an important part of the exhumation of deeply subducted crustal materials. The latter point was proposed first by Okay and Sengör (1992), where they attribute collision in central China to the formation of the Tanlu fault, and that strike slip motion on the fault instigated exhumation of the Dabie and Su-Lu UHP rocks. Except for the timing of the origin of the Tanlu fault, however, the Okay and Sengör (1992) model and the tectonic synthesis offered in this manuscript differ substantially.

In the model presented here, rotation relative to, and oblique collision of the Yangtze craton with the Sino-Korean craton, as suggested by paleomagnetic data, were critical factors in development of the strike-slip faults that aided in rapid exhumation of the HP-UHP rocks. Lateral translation is viewed as a means to create space in the upper crust into which deeply subducted, but buoyant, Hong'an-Dabie material was able to rise.

Extensional mechanisms have been proposed increasingly to explain the exposure of metamorphic rocks since the first descriptions of metamorphic core complexes and associated low-angle normal faults were presented by Coney (1979) and Davis and Coney (1979). Extension-related exposure of metamorphic rocks was extended directly to HP-LT metamorphic regions described by Lister et al. (1984), Wijbrans and McDougall (1988), Blake and Jayko (1990), Lee and Lister (1992), and Miller et al. (1992). Extension played a late role in exhumation in central China, most probably by providing the final impulse to raise the geotherms, attenuate the crust, and begin melting the lowermost, overly thickened portions of the collision zone. Other, younger continental collision zones have likewise begun to exhibit extension and active faulting related to crustal thinning relatively late in their tectonic histories (e.g. Tibet: Molnar and Tapponnier, 1981; England and Thompson, 1986). In central China, extension associated with, and related to transtension along, strike-slip faults may have thinned the upper crust sufficiently to facilitate final exposure of the HP-UHP rocks, but it is not determined to be the major factor in initial, rapid exhumation. Erosion as an exhumation mechanism became dominant only in the latter (Late Jurassic onward) stages of the history of this area. A tectonic model for central China that incorporates the chronologic and tectonostratigraphic associations, and the P-T constraints provided by the HP-UHP rocks, requires the incorporation of different tectonic mechanisms during distinct phases of the evolution of the orogenic belt.

Acknowledgments

The author would like to thank Bob Coleman, Gary Ernst, Rune Larsen, Juhn Liou, Mike McWilliams, and George Thompson for their comments on various drafts of this manuscript. Their frank discussions, suggestions, and encouragement with respect to many of the ideas presented here were also

extremely helpful. Yuan-Jun Liu, Gaozhi Zhou, Baocheng Xiong, and Yi
Zeng of the Hubei Bureau of Geology and Mineral Resources were instru-
mental in facilitating the author's three field seasons in central China, and
without whom the research could never have been accomplished. The international interest in this region of China, exemplified by the large number of both
Chinese and foreign researchers who have worked in the area, should provide
the impetus to coordinate an accessible database for further investigations of
this unique collision zone.

References

Ames, L., and Tilton, G. R. 1992. Timing of collision of the Sino-Korean and Yang-
    tse cratons. *EOS Trans. Am. Geophys. Union (abst), 73*, 652–653.
Ames, L., Tilton, G. R., and Zhou, G. 1993. Timing of collision of the Sino-Korean
    and Yangtse cratons: U-Pb zircon dating of coesite-bearing eclogites. *Geology,
    21*, 339–342.
Blake, M. C. J., and Jayko, A. S. 1990. Uplift of very high pressure rocks in the west-
    ern Alps: evidence for structural attenuation along low-angle faults. *Mem. Soc.
    geol. France, 156*, 237–246.
Brown, E. H., and Talbot, J. L. 1989. Orogen-parallel extension in the North Cas-
    cades crystalline core, Washington. *Tectonics, 8*, 1105–1114.
Burchfiel, B. C., and Royden, L. H. 1985. North-south extension within the conver-
    gent Himalayan region. *Geology, 13*, 679–682.
Bureau of Geology and Mineral Resources. 1987. Regional geology of Anhui prov-
    ince. *Geol. Mem., Geol. Pub. House, Beijing, 5*, 721pp.
Bureau of Geology and Mineral Resorces. 1990. Regional geology of Hubei prov-
    ince. *Geol. Mem., Geologic Publishing House, Beijing, 20*, 705 pp.
Carmignani, L., and Kligfield, R. 1990. Crustal extension in the northern Apen-
    nines: The transition from compression to extension in the Alpi Apuane core
    complex. *Tectonics, 9*, 1275–1303.
Chappell, B. W., and Stephens, W. E. 1988. Origin of infracrustal (I-type) granite
    magmas. *Trans. R. Soc. Edinburgh: Earth Sci., 79*, 71–86.
Chappell, B. W., and White, A. J. R. 1974. Two contrasting granite types. *Pacific
    Geol., 8*, 173–174.
Chappell, B. W., and White, A. J. R. 1984. I- and S-type granites in the Lachlan
    Fold Belt, southeastern Australia. In *Geology of granites and their metallogenic
    relations: Proceedings of the International Symposium, Nanjing University.*
    Beijing: Science Press, 87–101.
Chopin, C. 1984. Coesite and pure pyrope in high-grade blueschists of the western
    Alps: A first record and some consequences. *Contrib. Mineral. Petrol., 86*,
    107–118.
Chopin, C. 1986. Phase relationships of ellenbergerite, a new high-pressure Mg-Al-
    Ti-silicate in pyrope-coesite-quartzite from the Western Alps. *Geol. Soc. Amer.
    Mem., 164*, 31–42.
Coney, P. J. 1979. *Tertiary evolution of Cordilleran metamorphic core complexes.*
    Pacific Coast Paleogeog. Symp. 3; Cenozoic paleogeography of the western
    United States.

Davis, G. H., and Coney, P. J. 1979. Geologic development of the Cordilleran metamorphic core complexes. *Geology, 7*, 120–124.

Dewey, J. F., Hempton, M. R., Kidd, W. S. F., Saroglu, F., and Sengör, A. M. C. 1986. Shortening of continental lithosphere: The neotectonics of Eastern Anatolia–A young collision zone. *Collision Tectonics.* Oxford: Blackwell Scientific Publications, 3–36.

Eby, G. N. 1992. Characterization and petrogenetic subdivision of A-type granites (abs.). *Trans. R. Soc. Edinburgh: Earth Sci., 93*, 489.

Eide, E. A. 1993. Petrology, geochronology, and structure of high-pressure metamorphic rocks in Hubei Province, east-central China, and their relationship to continental collision. Ph.D., Stanford University, 235 pp.

Eide, E. A., McWilliams, M. O., and Liou, J. G. 1994. $^{40}Ar / ^{39}Ar$ geochronology and exhumation of high-pressure to ultra high-pressure metamorphic rocks in east-central China. *Geology, 22*, 601–604.

Enami, M., and Zang, Q. 1990. Quartz pseudomorph after coesite in eclogites from Shandong Province, east China. *Am. Mineralogist, 75*, 381–386.

England, P. C., and Thompson, A. B. 1984. Pressure-temperature-time paths of regional metamorphism. Heat transfer during the evolutions of regions of thickened continental crust. *J. Petrol., 25*, 894–928.

England, P. C., and Thompson, A. B. 1986. Some thermal and tectonic models for crustal melting in continental collision zones. *Collision tectonics.* Oxford: Blackwell Scientific Publications, 83–94.

Ernst, W. G. 1988a. Element partitioning and thermobarometry in polymetamorphic late Archean and Early-Mid Proterozoic rocks from eastern Liaoning and southern Jilin provinces, People's Republic of China. *Am. J. Sci., 288A*, 293–340.

Ernst, W. G. 1988b. Tectonic history of subduction zones inferred from retrograde blueschist P-T paths. *Geology, 16*, 1081–1084.

Ernst, W. G., Zhou, G., Liou, J. G., Eide, E., and Wang, X. 1991. High-pressure and superhigh-pressure metamorphic terranes in the Qinling-Dabie Mountain Belt, central China; Early- to Mid-Phanerozoic accretion of the western paleo-Pacific rim. *Pacific Science Association Information Bulletin, 43*, 6–14.

Fei, Q. 1987. A preliminary research into the plate collision, rotation, and divergence pattern of China and its periphery since the Mesozoic. *J. Wuhan Coll. Geol., 12*, 463–475.

Gao, S., Zhang, B.-R., Luo, T.-C., Li, Z.-J., Xie, Q.-L., Gu, X.-M., Zhang, H.-F., Ouyang, J.-P., Wang, D.-P., and Gao, C.-L. 1992. Chemical composition of the continental crust in the Qinling Orogenic Belt and its adjacent North China and Yangtze cratons. *Geochem. Cosmochim. Acta, 56*, 3933–3950.

Han, J., Zhu, S., and Xu, S. 1989. The generation and evolution of the Hehuai Basin. *Chinese Sedimentary Basins.* Amsterdam: Elsevier, 238 pp.

Harrison, T. M. 1981. Diffusion of ^{40}Ar in hornblende. *Contrib. Mineral. Petrol., 78*, 324–331.

Hirajima, T., Ishiwatari, A., Cong, B., Zhang, R., Banno, S., and Nozaka, T. 1990. Coesite from Mengzhong eclogite at Donghai county, northeastern Jiangsu province, China. *Mineral. Mag., 54*, 579–583.

Hirn, A. 1988. Features of the crust-mantle structure of the Himalayas, Tibet: A comparison with seismic traverses of Alpine, Pyrenean and Variscan orogenic belts. *Philos. Trans. R. Soc. Lond., A326*, 17–32.

Hsü, K. J., Li, J., Chen, H., Wang, Q., Sun, S., and Sengör, A.M.C. 1990. Tectonics of South China: Key to understanding West Pacific geology. *Tectopnophysics,* *183,* 9–39.

Huang, W., and Wu, Z. W. 1992. Evolution of the Qinling orogenic belt. *Tectonics,* *11,* 371–380.

Hurford, A. J. 1986. Cooling and uplift patterns in the Lepontine Alps south central Switzerland and an age of vertical movement on the Insubric fault line. *Contrib. Mineral. Petrol., 92,* 413–427.

Klimetz, M. P. 1983. Speculations on the Mesozoic plate tectonic evolution of eastern China. *Tectopnophysics, 2,* 139–166.

Kröner, A., Zhang, G. W., and Sun, Y. 1993. Granulites in the Tongbai area, Qinling belt, China: Geochemistry, petrology, single zircon geochronology, and implications for the tectonic evolution of eastern Asia. *Tectonics, 12,* 245–255.

Lee, J., and Lister, G. S. 1992. Late Miocene ductile extension and detachment faulting, Mykonos, Greece. *Geology, 20,* 121–124.

Li, S., and Wang, T. 1991. *Geochemistry of granitoids in Tongbaishan–Dabieshan, central China.* China University of Geosciences Press.

Li, S. G., Chen, Y. Z., Cong, B. L., Zhang, Z. Q., Zhang, R. Y., Liu, D. L., Hart, S. R., and Ge, N. J. 1993. Collision of the North China and Yangtze blocks and formation of coesite-bearing eclogite: timing and processes. *Chemical Geology, 109,* 70–89.

Li, Y., Sharps, R., McWilliams, M., Zhang, Z., Li, Y., Li, Q., Zhai, T., and Gao, Z. 1990. Apparent polar wander path from the Tarim Block in China. *Acta Geol. Sin., 3,* 1–13.

Lin, J.-L., and Fuller, M. 1990. Paleomagnetic constraints on relative motions of the Asian blocks and terranes. *Terrane analysis of China and the Pacific rim.* Houston: Circum-Pacific Council for Energy and Mineral Resources Earth Science Series, T. J. Wiley.

Lin, J. L., Fuller, M., and Zhang, W. 1985. Preliminary Phanerozoic polar wander paths for the North and South China blocks. *Nature, 313,* 444–449.

Lister, G. S., Banga, G., and Feenstra, A. 1984. Metamorphic core complexes of Cordilleran type in the Cyclades, Aegean Sea, Greece. *Geology,* 221–225.

Liu, D. Y., Nutman, A. P., Compston, W., Wu, J. S., and Shen, Q. H. 1992. Remnants of 3800 Ma crust in the Chinese part of the Sino-Korean craton. *Geology, 20,* 339–342.

Liu, X., and Hao, J. 1989.*The framework and tectonic evolution of east Qinling Belt. Advances in Geoscience, Contributions to 28th International Geological Congress, Washington, D.C.* China Ocean Press.

Loiselle, M. C., and Wones, D. R. 1979. Characteristics and origin of anorogenic granites. *Geol. Soc. Am. Abs. Prog., 11,* 468.

Ma, W. 1989. Tectonics of the Tonbai-Dabie fold belt. *J. SE Asian Earth Sci., 3,* 77–85.

Ma, X., and Ge, H. 1989. Precambrian crustal evolution of eastern Asia. *J. SE Asian Earth Sci., 3,* 9–15.

Mattauer, M. 1986. Intracontinental subduction, crust-mantle décollement and crustal-stacking wedge in the Himalayas and other collision belts. *Collision Tectonics.* Oxford: Blackwell Scientific Publications, 37–50.

Mattauer, M., Matte, P., Malavieille, J., Tapponnier, P., Maluski, H., Xu, Z. Q., Lu, Y. L., and Tang, Y. Q. 1985. Tectonics of the Qinling Belt: build-up and evolution of eastern Asia. *Nature, 317,* 496–500.

McElhinny, M. W., Embleton, B. J. J., Ma, X. H., and Zhang, Z. K. 1981. Fragmentation of Asia in the Permian. *Nature, 293,* 212–216.

Metcalfe, I. 1988. Origin and assembly of south-east Asian continental terranes. *Gondwana and Tethys–Geological Society Special Publication*. Oxford: Blackwell Scientific Publications, 101–118.

Miller, E. L., Calvert, A. T., and Little, T. A. 1992. Strain-collapsed metamorphic isograds in a sillimanite gneiss dome, Seward Peninsula, Alaska. *Geology, 20,* 487–490.

Molnar, P., and Tapponnier, P. 1981. A possible dependence of tectonic strength on the age of the crust in Asia. *Earth Planet. Sci. Lett., 52.,* 107–114.

Okay, A. I., and Sengör, A. M. C. 1992. Evidence for intracontinental thrust-related exhumation of the ultra-high-pressure rocks in China. *Geology, 20,* 411–414.

Okay, A. I., Xu, S., and Sengör, A. M. C. 1989. Coesite from the Dabie Shan eclogites, central China. *Eur. Jour. Mineral., 1,* 595–598.

Opdyke, N. D., Huang, K., Xu, G., Zhang, W. Y., and Kent, D. V. 1986. Paleomagnetic results from the Triassic of the Yangtze Platform. *J. Geophys. Res., 91,* 9553–9568.

Pearce, J. A., Harris, N. B. W., and Tindle, A. G. 1984. Trace element discrimination diagrams for the tectonic interpretation of granitic rocks. *J. Petrol., 25,* 956–983.

Pitcher, W. S. 1979. Comments on the geological environments of granites. *Origin of granite batholiths – geochemical evidence.* Kent: Shiva Publishing Ltd., 1–8.

Platt, J. P. 1986. Dynamics of orogenic wedges and the uplift of high-pressure metamorphic rocks. *Geol. Soc. Amer. Bull., 97,* 1037–1053.

Ren, J., and Chen, T. 1989. Tectonic evolution of the continental lithosphere in eastern China and adjacent areas. *J. SE Asian Earth Sci., 3,* 17–27.

Sang, L. 1991. The petrochemistry of the lower Proterozoic metamorphic rocks from the Dabieshan–Lianyungang area of the southeast margin of the North China platform. *J. China Univ. Geosci., 16,* 651–659.

Sengör, A.M.C. 1985. East Asian tectonic collage. *Nature, 318,* 16–17.

Sharps, R., McWilliams, M., Li, Y., Cox, A., Zhang, Z., Zhai, Y., Gao, Z., Li, Y., and Li, Q. 1989. Lower Permian paleomagnetism of the Tarim block, northwestern China. *Earth Planet. Sci. Lett, 92,* 275–291.

Shu, Q. 1984. Petrochemical characteristics and metallogenetic specialties of intermediate-acidic magmatic rocks in southeast Hubei. *Geology of granites and their metallogenic relations: Proceedings of the International Symposium.* Beijing.: Nanjing University, Science Press, Beijing.

Tapponnier, P., Peltzer, G., and Armijo, R. 1986. On the mechanics of the collision between India and Asia. *Collision Tectonics.* Oxford: Blackwell Scientific Publications, 115–157.

Wang, X. 1991. Petrology of coesite-bearing eclogites and ultramafic rocks from ultrahigh pressure metamorphic terrane of the the Dabie Mountains and implications to regional tectonics in central China. Ph.D., Stanford University.

Wang, X., and Liou, J. G. 1991. Regional ultrahigh-pressure coesite-bearing eclogitic terrane in central China: evidence from country rocks, gneiss, marble, and metapelite. *Geology, 19,* 933–936.

Wang, X., Jing, Y., Liou, J. G., Pan, G., Liang, W., Xia, M., and Maruyama, S. 1990. Field occurrences and petrology of eclogites from the Dabie Mountains, Anhui, central China. *Lithos, 25,* 119–131.

Wang, X., Liou, J. G., and Mao, H. K. 1989. Coesite-bearing eclogites from the Dabie Mountains in central China. *Geology, 17,* 1085–1088.

Wang, X., Liou, J. G., and Maruyama, S. 1992. Coesite-bearing eclogites from the Dabie Mountains, central China: Petrogenesis, P-T paths and implications to tectonics. *J. Geol., 100,* 231–250.

Watson, M. P., A. B., H., D. N., P., and Z. M., Z. 1987. Plate tectonic history, basin development and petroleum source rock deposition onshore China. *Marine and Petr. Geol., 4*, 205–225.

Wijbrans, J. R., and McDougall, I. 1988. Metamorphic evolution of the Attic Cycladic Metamorphic Belt on Naxos (Cyclades, Greece) utilizing $^{40}Ar/^{39}Ar$ age spectrum measurements. *Journal of Metamorphic Geology, 6*, 571–594.

Windley, B. F. 1988. Tectonic framework of the Himalaya and Tibet, and problems of their evolution. *Phil. Soc. Trans. Royal Soc. London: Series A, 326*, 3–16.

Xu, J., Tong, W., Zhu, G., Lin, S., and Ma, G. 1989. An outline of the pre-Jurassic tectonic framework in east Asia. *J. SE Asian Earth Sci., 3*, 29–45.

Xu, S., Okay, A. I., Ji, S., Sengor, A.M.C., Su, W., Liu, Y., and Jiang, L. 1992. Diamond from the Dabie Shan metamorphic rocks and its implication for tectonic setting. *Science, 256*, 80–82.

Xu, S., Zhou, H., Dong, S., Chen, G., and Zhang, W. 1987. *Deformation and tectonic evolution of the major structural elements in Anhui Province, China* (in Chinese with English summary). Beijing: China Ocean Press.

Yang, J. 1989. A survey of Cambrian palaeo-tectonogeography of East Qinling. *J. SE Asian Earth Sci., 3*, 203–210.

Yang, S., Chen, R., and Qian, X. 1987. Orogeny and orogen structure of Dabie Mountain in the Mesozoic. *J. Wuhan Coll. Geol., 12*, 495–502.

You, Z., Han, Y.]., Suo, S., Chen, N., and Z, Z. 1993. Metamorphic history and tectonic evolution of the Qinling complex, eastern Qinling Mountains, China. *Jour. Metamorph. Geol., 11*, 549–560.

Zhang, B., Ruyang, J., Lou, T., S., G., Li, Z., and Han, Y. 1991. Geotectonic subdivision and Paleozoic subduction and collision orogeny of East Qinling and its adjacent regions: Evidence from geochemical research. *J. China U. Geosci., 2*, 1–17.

Zhang, G., Yu, Z., Sun, Y., Cheng, S., Li, T., Xue, F., and Zhang, C. 1989. The major suture zone of the Qinling orogenic belt. *J. SE Asian Earth Sci., 3*, 63–76.

Zhang, R. Y., Liou, J. G., Wang, X., Wang, Q. C., and Liu, X. H. 1993. Discovery of coesite eclogites in Henan province, central China and its tectonic implication. *Acta Petrologica Sinica, 9*, 186–191 (in Chinese with English Abs.).

Zhang, Z. M., Liou, J. G., and Coleman, R. G. 1984. An outline of the plate tectonics of China. *Geol. Soc. Amer. Bull., 95*, 275–312.

Zhao, X., and Coe, R. 1987. Palaeomagnetic constraints on the collision and rotation of North and South China. *Nature, 327*, 141–144.

Zhou, G., Liu, Y. J., Eide, E. A., Liou, J. G., and Ernst, W. G. 1993. High pressure–low temperature metamorphic belt in northern Hubei Province, central China. *J. Meta. Geol., 11*, 561–574.

Zhu, H., and Wan, T. 1991. On kinematics of Late Proterozoic-Triassic paleoplates in eastern China. *J. China U. Geosci., 16*, 523–532.

12

Diamond-Bearing Metamorphic Rocks of the Kokchetav Massif (Northern Kazakhstan)

V. S. SHATSKY, N. V. SOBOLEV, AND M. A. VAVILOV

Abstract

Primary mineral assemblages preserved mostly as inclusions in garnets and zircons have been found as a result of detailed investigations of diamondiferous metamorphic rocks of the Kokchetav massif, Northern Kazakhstan. Garnet and clinopyroxene are the principal minerals of these assemblages along with mica, kyanite, rutile, sphene, coesite and diamond. Estimation of the equilibration conditions using the mineral compositions shows that diamondieferous rocks were metamorphosed at temperatures not lower than 900°C. In establishing independent indicators of very high pressures (K in cpx, Al_2O_3 in sphene, diopside-magnesite assemblage, coesite inclusion in zircon, presence of diamond itself), all testify to the crystallization conditions of diamond in its stability field.

Introduction

The Kokchetav massif is the first locality where microdiamonds were found within metamorphic rocks of the Earth's crust (Sobolev and Shatsky, 1987, 1990). Information on alluvial diamonds in Northern Kazakhstan was known long ago (Esenov et al., 1968). Two hundred fifty microdiamond grains were found during the investigation of the titanium-zircon placer samples from the northern part of the Kokchetav massif. Kashkharov and Polkanov (1972) assumed the relative closeness of the native sources, based on the preservation of very delicate, fine-grained aggregates. Rozen et al. (1972) published the data on the diamond found in the weathered crust of eclogites at the Kumdy-Kol site. Later, eclogites were considered to be the native source of alluvial diamonds for this area (Rozen et al., 1979).

During further investigations on this site, geologists of the Kokchetav expedition recognized several types of the rocks, including eclogites, which seemed to contain microdiamonds. However, the methods applied to determination of the diamond content, which were bound up with bulk rock decomposition, offered no opportunity for defining the actual process connected with diamond occurrence and their position within the metamorphic rocks. In this connection all new hypotheses were based only on the fact of general occurrence of diamond in metamorphic rocks (Sidorenko et al., 1978; Rozen et al., 1979; Letnikov, 1983).

Only by use of optical techniques was it possible to establish the diamond contents of rocks, to distinguish reliably the petrographic types of diamond-bearing rocks, and to obtain information on diamond-bearing associations (Sobolev and Shatsky, 1987; 1990; Shatsky et al., 1991; Sobolev et al., 1991). It permitted the study of microdiamond genesis in the metamorphic rocks of the Earth's crust and an oppurtunity to prove that the process of diamond formation was connected with high-pressure metamorphism of crustal rocks.

The results of these studies were published in a number of papers. The complex geologic structure of this region, the great variety of rocks, the processes of retrograde metamorphism, and the granitization of the Zerendin series rocks, however, require further study to model the ultrahigh pressure (UHP) metamorphism in the Caledonides of Northern Kazakhstan. In this chapter, we present the new data obtained during the study of the diamond-bearing rocks of the Kumdy-Kol site, and also of a new diamond locality, situated 17 km northwest of the Kumdy-Kol site. During the investigation of the primary associations of the high-pressure (HP) metamorphism, special attention was paid to the study of inclusions in garnets and zircons from diamondiferous rocks. A great number of silicate analyses of the Zerendin series rocks were carried out.

Distribution of rare and trace elements was studied to ascertain the diamond-bearing rock protolith and behavior of elements during the process of HP metamorphism and the later amphibolite facies metamorphism.

Geologic Structure

The Kokchetav Precambrian massif is situated within the Caledonides of Eastern Kazakhstan (Rozen, 1976, 1982), which are a part of the Central Asian belt (Fig. 12.1). According to modern concepts, the Caledonian structures of Kazakhstan formed during the collision of island arcs with Kazakhstan–Northern Tienshan Precambrian massif (Zonenshain et al., 1990). Today two separate geologic blocks with different histories are distinguished in the Kazakhstan plate (Rozen, 1982). In the western geoblock, the

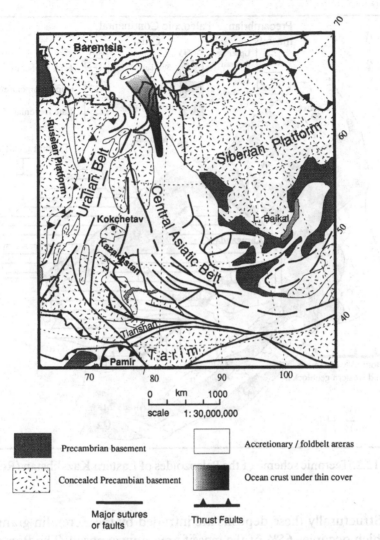

Figure 12.1. Tectonic scheme of the USSR (From Zonenshain et al., 1990)

continent was subjected to rifting during the Lower Paleozoic (Fig. 12.2); at
that time, an ocean was situated within the territory of the eastern block.

Formerly, the Kokchetav massif was considered as a part of the
Kazakhstan-Northern Tienshan massif. Bounded by the long lived faults in
the south and west, on the north it extends under the cover of the West-
Siberian lowland according to geophysical data (Rozen, 1979). The massif has
a complex heterogeneous structure (Fig.12.3). Its central part is composed of
the Zerendin series metamorphosed mainly under amphibolite facies condi-

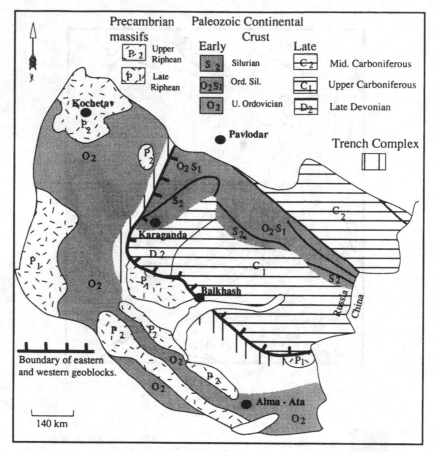

Figure 12.2. Tectonic scheme of the Paleozoides of Eastern Kazakhstan (Rozen, 1982)

tions. Structurally these deposits are intruded by the Zerendin granitoid pluton, which occupies 65% of the massif area now exposed. The Borovsk series is metamorphosed to green schist facies conditions, and the weakly metamorphosed Elektin series deposits are situated to the west of the ancient core.

Eclogites and diamond-bearing metamorphic rocks occur only within the deposits of the most ancient Zerendin series. This series is composed of kyanite-garnet-biotite gneiss, biotite gneiss, garnet-plagioclase gneiss, plagiogneiss, marble lenses, eclogite, and amphibolite. The rocks of this series border the Zerendin pluton, and they can be traced for a distance of about 100 km (Fig. 12.3).

A significant spread in P-T parameters of metamorphism was found in the eclogites from various sites (Sobolev et al., 1986). The highest-temperature eclogites are at the Kumdy-Kol site. Subsequently the high-temperature eclog-

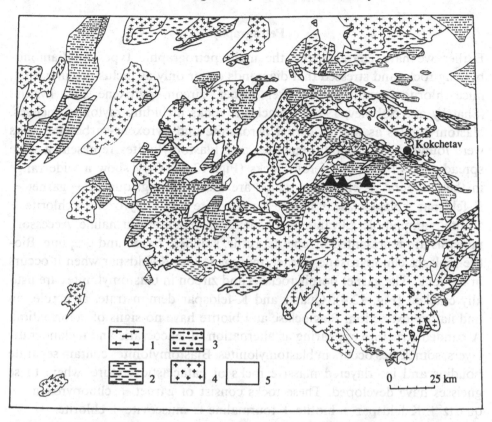

Figure 12.3. Simplified geologic map of Kokchetav massif. Filled triangles indicate diamond occurrence (Kumdy-Kol is the extreme right triangle). (1) Zerendin series (biotite gneies, garnet-biotite gneiss, garnet pyroxenite, marble, mica schists). (2) Borovsk series (epidote-actinolite schist, phorphrys, quartz-feldspar, and sericite-quartz schist, interlayered marble). (3) Elektin series (quartzite, quartz-sericitic schist, mica-quartz schist, marble). (4) Zerendin granitoid. (5) Undivided Phanerozoic sediments.

ites and diamond-bearing rocks were found 17 km northeast of the Kumdy-Kol site (Shatsky et al., 1991).

At the Kumdy-Kol site the diamondiferous rocks and eclogites occur among granite-gneiss and plagiogneiss. The eclogites occur as separated boudins from several centimeters to tens of meters. The site is poorly exposed and makes determination of the field relations between different rock types difficult to establish. However, it can be seen in the working adit and boreholes that certain types of rocks occur as relicts in granite-gneisses. At the site, situated 17 km to the northwest from Kumdy-Kol lake, only the amphibolized eclogite bodies are exposed. Amphibole-biotite and garnet-biotite plagiogneisses were uncovered in the boreholes.

Petrography

Earlier we have characterized the main petrographic types of diamond-bearing rocks and stressed that diamonds occur only as inclusions in garnet, mica-chlorite pseudomorphs after garnet, zircon, pyroxene, and kyanite (Shatsky and Sobolev, 1993). Cataclastic rocks including mylonite gneisses, blastomylonites as well as garnet pyroxenite and pyroxene-carbonate rocks were found there. Mylonite gneisses and blastomylonites represent a wide spread type of diamondiferous rocks (Fig. 12.4). They show a wide range in their mineral assemblages, which are the following: quartz + garnet + K-feldspar $\pm$plagioclase $\pm$ biotite $\pm$ muscovite $\pm$ clinopyroxene $\pm$chlorite + zoisite $\pm$amphibole $\pm$kyanite $\pm$sphene $\pm$carbonate $\pm$tourmaline. Accessory minerals are represented by apatite, rutile, sulfides, zircon and graphite. Biotite replaces clinopyroxene, garnet, muscovite and K-feldspar when it occurs in the rock. The garnet porphyroclasts and zircon in blastomylonites are usually cracked (Fig. 12.4). Quartz and K-feldspar demonstrate, as a rule, an undulatory extinction. Plagioclase and biotite have no signs of deformation. A banded structure appearing as alternation of leucocratic and melanocratic layers, sometimes occurs in blastomylonites. Blastomylonites contain separate boudins and interlayered massive rocks of cataclastic texture, where these gneisses have developed. These rocks consist of garnet + clinopyroxene + quartz + K-feldspar + biotite + tourmaline $\pm$ muscovite + chlorite.

The leucocratic gneisses alternate with mylonite gneisses, and a gradual transition between these lithologies has been observed. Within mylonite gneisses, the content of orthoclase and plagioclase increases as pyroxene and garnet disappear. The leucocratic gneisses are composed of quartz + ortho-clase + plagioclase + biotite + muscovite.

Primary fine-grained rocks with sutured structure make up another group. These rocks occur as interlayers 30 m thick among the diamondiferous gneissic rocks. They consist of a number of various minerals, but quartz, amphibole, chlorite, and sometimes tourmaline prevail with biotite, musco-vite, and zoisite also present. There are separate grains of garnet, which are replaced by amphibole, chlorite, and micas with chlorite developing over amphibole. Zoisite forms large grains with chlorite and amphibole inclusions. Individual sites may contain large grains of orthoclase. Pyroxene-carbonate and garnet-pyroxene rocks are present as interlayers up to 10 m thick and as lenses within granite and biotitic gneisses.

Pyroxene-carbonate rock consists of carbonates, pyroxene, garnet, and phlogopite. These rocks have variable structural-textural characteristics. The coarse-grained rocks are predominantly composed of pyroxene, carbonate,

Figure 12.4. Microstructures in blastomylonites. (a) Garnet porphyroclasts in strongly foliated biotite, quartz, and plagioclase (sample B93–155). Crossed nicols. (b) Garnet showing disruption (sample B93–160). Plane polarized light. (c) Mica-chlorite pseudomorph after garnet in deformed quartz. Pseudomorphs are formed after deformation (B93–38). Crossed nicols. (d) Garnet-quartz-chlorite-tourmaline blastomylonite. Tourmaline was probably formed during and after deformation (Sample 8–30). The field of view is about 5 mm for all.

Table 12. 1. *Bulk chemical composition of metamorphic rock from Kumdy-Kol locality.*

	Garnet-pyroxene rock		Pyroxene-carbonate rock		Garnet-pyroxene quartz rock		Granite-gneiss		Garnet-mica schist	
n	40		27		121		50		227	
	$\bar{x}$	σ	$\bar{x}$	σ	$\bar{x}$	σ	$\bar{x}$	σ	$\bar{x}$	σ
SiO_2	49.6	4.49	30.6	8.74	56.2	7.39	72.7	2.30	64.3	6.46
TiO_2	0.48	0.16	0.23	0.13	1.12	0.77	0.29	0.15	0.73	0.37
Al_2O_3	10.1	2.44	5.47	2.39	13.1	2.04	13.0	1.00	13.7	2.22
Fe_2O_3	10.7	4.67	3.41	1.45	10.2	3.97	3.86	1.62	7.72	2.64
MnO	0.31	0.17	0.17	0.06	0.21	0.~3	0.15	0.18	0.20	0.12
MgO	6.46	2.99	15.1	3.99	4.67	1.26	0.83	0.35	3.55	1.26
CaO	17.2	4.26	23.1	5.29	7.12	1.69	1.36	0.66	2.47	1.24
Na_2O	0.20	0.21	0.20	0.84	0.69	0.58	1.42	0.58	0.84	0.58
K_2O	1.77	1.96	1.02	0.64	2.28	1.34	4.99	0.99	3.23	1.12
P_2O_5	0.23	0.05	0.23	0.07	0.23	0.12	0.13	0.04	0.14	0.07

and phlogopite. Garnet occurs in some places, forming veinlike bodies or schlieren-like segregation. Within the medium-grained pyroxene-carbonate rocks, garnet content increases, and it is uniformly dispersed in the rock. Some samples show alteration of the pyroxene-garnet and carbonate layers.

Geochemistry

Silicate analyses of the rock varieties were carried out using the standard x-ray fluorescence (XRF) method. The average chemical compositions of different types of rocks are given in Table 12.1. As seen in ternary diagram of $(SiO_2/10)$-$(CaO + MgO)$-$(Na_2O + K_2O)$ (Fig.12.5), the metamorphic rocks of Kumdy-Kol form a continuous compositional sequence, extending from mafic to granite composition. This continuous sequence can be interpreted to indicate a mixture of mafic and leucocratic source rocks making up the protoliths of the Zerendin series rocks. However, our petrographic and field studies show that blastomylonite gneisses develop from the garnet-pyroxene-quartz rocks.

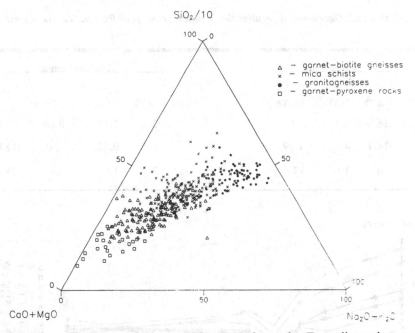

Figure 12.5. Plot of analyzed metamorphic rocks from the Zerendin series on the $(CaO + MgO)–(SiO_2/10)–(Na_2O + K_2O)$ ternary diagram. From Taylor and McLennan, 1985.

These rocks differ from other types of rocks in their higher CaO, FeO, and MgO contents and their lower SiO_2, K_2O, and Na_2O contents (Table 12.1) where they occupy the lower left part of the diagram.

The diamond-bearing rocks significantly differ from both the non-diamondiferous rocks of the Kumdy-Kol site and the Zerendin series rocks of the other site, when the ratios of La_N/Yb_N, $Sm_N/Yb/_N$, and Th/U are compared (Table 12.2). The slightly altered garnet-pyroxene-quartz and garnet-quartz rocks are greatly depleted in LREE (Fig. 12.6).

Mineralogy

Garnet

Garnets from metamorphic rocks of the diamondiferous block vary greatly in their composition. The Ca-Fe-Mg diagram shows three fields of the diamond-bearing mylonite gneisses and schists, garnet-pyroxene-quartz rocks, garnet-pyroxenites, and pyroxene-carbonate rocks of the non-diamondiferous granite gneisses (Fig.12.7). The fields of the non-diamondiferous and diamondiferous rocks partly overlap. Garnets from eclogites plot in a the field of diamondiferous rocks. Garnets from pyroxene-carbonate rocks differ significantly in their

Table 12.2. *Geochemical peculiarities of metamorphic rock from the Zerendin series.*

	Diamond free				Diamond bearing			
	La/Yb	Th/U	Sm/Yb		La/Yb	Th/U	Sm/Yb	
$\bar{x}$	16.9	6.2	2.5		2.5	1.32	0.88	0.97
s	14.1	4.5	1.09		1.1	0.87	1.0	0.84
n	16	19	19		7	13	18	18

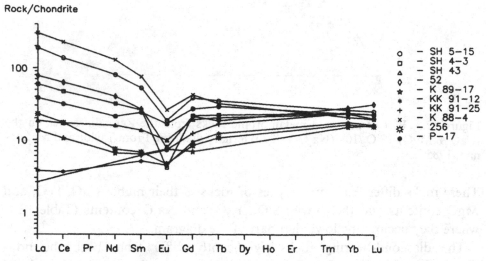

Figure 12.6. REE patterns of rocks of the Zerendin series (SH 5–15, 4–3, 52, K 88–4, 89–17: diamond-free; K-K 91–12, 91–25, 256, P–17: diamond-bearing).

composition from the garnets of other rock types. They are notable for their high calcium content and higher Mg/Mg + Fe ratio. The contents of Ca and Fe sometimes increase from the core to the rim when zoning is present, but in most cases garnets from diamondiferous rocks do not show zoning. Some samples of these rock types show garnets of various color. In sample 86-269 (Table 12.3) there are pinky-brown and pale-pink garnets. As seen from the given analyses, the pink garnet is weakly zoned. The pinky-brown garnet has a core that is similar in composition to the pink garnet. At the same time, it is characterized by complex zoning from the core to the rim. The opposite edges have different compositions. Mn and Fe content increases to one edge, and Mg and Fe content to the other. In sample 86-312, the intensely colored garnet, Fe and Mg content is greater compared to the central part of the

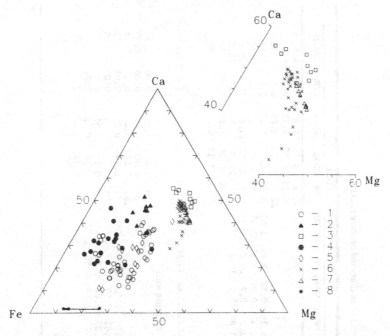

Figure 12.7. Compositional variations of garnet from the metamorphic rocks of Kumdy-Kol. (1) diamondiferous rock; (2) garnet pyroxenite; (3) pyroxene-carbonate rock; (4) diamond free gneiss; (5) eclogite: (6) inclusions in zircon from pyroxene-carbonate rock (K9–16); (7) garnet from rock (K9–16); (8) garnet-kyanite-muscovite schist (86–319).

colorless garnets and chemically close to the composition of colorless garnet rim. In the garnet-quartz-chlorite schist (sample 86-391), we can observe color-zonation garnet and evenly colored garnet. The color-zonation grains also show intense chemical zonation (Table 12.3, Fig. 12.8). The colored core is enriched in iron, titanium, and manganese. The calcium content is high in the pale pink rim. The same tendency is observed in the change of composition, but with smaller gradients in the colorless garnet. The colored garnet is weakly zoned, corresponding in composition to the core of the garnet which is zoned in its color. The reason for the distinction in composition between opposite edges of garnet from the blastomylonite (sample 86-269, Table 12.3) is that this is a fragment of a larger crystal. We speculate that garnets of diamondiferous rocks have been broken and dispersed during a later deformational stage.

The garnets of different color and chemical composition could be considered as two separate generations. A profile across the grain of the garnet having color zonation shows that the central part is not practically zoned, but a change in the chemical composition occurs in the rim 125 μm wide (Fig. 12.8).

Table 12.3. Representative analyses of minerals from diamond bearing rocks.

wt%

| | KK 91-16 | | | | | | | | | | | | | 81-21 | | | | |
	Ga1	Ga2	Ga1a	Ga2a	Ga3a	Cpx1	Cpx2	Cpx1a	Cpx2a	Maga	Cala	Dola	MgCala	Ga	Cpxc	Cpxr	Cpxcb	Cpxrb
SiO2	41.3	40.9	40.6	40.5	41.0	54.7	54.4	55.2	55.1					40.9	54.4	54.7	55.4	54.6
TiO2	0.13	0.01	0.16	0.24	1.66	0.00	0.00	0.01	0.02					0.25	0.07	0.02	0.05	0.07
Al2O3	22.4	22.2	22.1	22.1	22.2	1.18	1.37	1.97	1.75					22.6	2.69	1.85	2.56	3.26
FeO	8.68	8.29	14.3	10.7	7.74	2.26	2.12	1.61	2.08	6.35	0.38	2.83	3.00	6.03	1.37	1.53	1.27	1.20
MnO	0.84	0.83	2.03	1.35	0.90	0.17	0.14	0.04	0.10	0.43	0.08	0.27	0.38	0.54			0.04	0.04
MgO	11.6	10.2	10.3	11.1	9.25	16.9	16.7	16.5	16.5	41.0	0.76	21.0	10.9	10.2	16.5	16.8	15.8	15.6
CaO	15.5	17.5	10.2	13.4	17.9	23.8	23.6	23.7	23.5	1.88	62.5	31.2	43.7	19.3	22.8	23.6	23.4	23.8
Na2O	0.02	0.04	0.03	0.01	0.03	0.29	0.26	0.38	0.44					0.30	0.25	0.30	0.38	0.36
K2O	0.02	0.01	0.00	0.00	0.01	0.33	0.57	0.90	0.05					0.00	1.15	0.37	1.12	0.66
F																		
Cl																		
sum	100.49	100.03	99.67	99.37	100.71	99.65	99.18	99.61	99.46	49.64	63.68	55.26	57.93	100.12	99.23	99.17	100.02	99.59

Cations per formula unit

| | KK 91-16 | | | | | | | | | | | | | 81-21 | | | | |
	Ga1	Ga2	Ga1a	Ga2a	Ga3a	Cpx1	Cpx2	Cpx1a	Cpx2a	Maga	Cala	Dola	MgCala	Ga	Cpxc	Cpxr	Cpxcb	Cpxrb
Si	3.02	3.02	3.04	3.01	3.00	1.99	1.99	2.01	2.00					3.00	1.98	1.99	2.00	1.98
Ti	0.01	0.00	0.01	0.01	0.09	0.00	0.00	0.00	0.00					0.01	0.00	0.00	0.00	0.00
Al	1.93	1.93	1.95	1.93	1.91	0.05	0.06	0.08	0.07					1.95	0.12	0.08	0.11	0.14
Fe	0.53	0.51	0.89	0.67	0.47	0.07	0.06	0.05	0.06	0.08	0.00	0.04	0.04	0.37	0.04	0.05	0.04	0.04
Mn	0.05	0.05	0.13	0.09	0.06	0.01	0.00	0.00	0.00	0.01	0.00	0.00	0.00	0.03			0.00	0.00
Mg	1.27	1.12	1.15	1.22	1.01	0.92	0.91	0.85	0.89	0.89	0.02	0.46	0.25	1.12	0.89	0.91	0.85	0.84
Ca	1.21	1.38	0.82	1.07	1.41	0.93	0.93	0.92	0.91	0.03	0.98	0.50	0.71	1.52	0.89	0.92	0.91	0.92
Na	0.00	0.01	0.00	0.00	0.00	0.02	0.02	0.03	0.03					0.00	0.02	0.02	0.03	0.03
K	0.00	0.00	0.00	0.00	0.00	0.02	0.03	0.04	0.00					0.00	0.05	0.02	0.05	0.03
F																		
Cl																		

Table 12.3. (cont.)

Wt%	KK 91-25						86-319					86-269				
	Gac	Gar	Cpx	Bi	Phe	Fsp/am	Cpxb	Ga1c	Ga1r	Phe	Ky	Ga1c	Ga1r	Ga2c	Ga2r1	Ga2r2
SiO2	39.8	40.0	53.8	41.7	49.1	64.5	55.1	38.2	36.7	47.4	36.5	38.7	38.6	38.6	38.5	38.6
TiO2	0.11	0.13	0.00	1.69	2.05	1.00	0.10	0.01	0.02	1.59	0.00	0.08	0.00	0.14	0.07	0.02
Al2O3	22.1	22.1	2.35	15.7	27.6	17.7	4.06	21.1	20.3	32.4	62.4	21.4	21.5	21.0	21.2	21.3
FeO	14.1	14.4	4.55	9.23	0.88	0.17	3.42	32.8	36.4	1.62	0.23	22.7	22.2	20.9	23.4	23.4
MnO	0.70	0.67	0.13	0.10	0.00	0.00	0.03	0.28	1.83	0.01	0.00	1.42	1.69	1.62	5.39	1.48
MgO	8.39	8.46	14.8	17.8	3.53	0.14	13.6	6.57	2.89	1.33	0.00	5.89	6.30	5.22	3.71	6.49
CaO	15.1	15.2	23.1	0.19	0.18	0.53	21.8	0.75	0.59	0.02	0.00	9.42	8.84	11.5	8.11	7.91
Na2O	0.03	0.02	0.74	0.12	0.15	0.87	1.31	0.10	0.12	0.43	0.00	0.06	0.03	0.06	0.03	0.06
K2O	0.00	0.00	0.02	9.83	9.84	15.50	0.68	0.00	0.00	10.4	0.00					
F											0.07					
Cl																
sum	100.33	100.98	99.49	96.36	93.33	100.41	100.10	99.81	98.85	95.20	99.23	99.67	99.16	99.04	100.41	99.26
Cations per formula unit																
Si	2.99	2.99	1.98	2.97	3.32	2.97	1.99	3.01	3.01	3.16	0.99	3.01	3.01	3.02	3.02	3.01
Ti	0.01	0.01	0.00	0.09	0.10	0.03	0.00	0.00	0.00	0.08	0.00	0.00	0.00	0.01	0.00	0.00
Al	1.95	1.94	0.10	1.32	2.20	0.96	0.17	1.96	1.96	2.55	2.00	1.96	1.97	1.93	1.96	1.96
Fe	0.88	0.90	0.14	0.55	0.05	0.01	0.10	2.16	2.50	0.09	0.01	1.48	1.44	1.37	1.53	1.52
Mn	0.04	0.04	0.00	0.01	0.00	0.00	0.00	0.02	0.13	0.00	0.00	0.09	0.11	0.11	0.36	0.10
Mg	0.94	0.94	0.81	1.89	0.36	0.01	0.73	0.77	0.35	0.13	0.00	0.68	0.73	0.61	0.43	0.75
Ca	1.21	1.22	0.91	0.01	0.01	0.03	0.85	0.06	0.05	0.00	0.00	0.78	0.74	0.96	0.68	0.66
Na	0.00	0.00	0.05	0.02	0.02	0.08	0.09	0.02	0.02	0.06	0.00	0.01	0.01	0.01	0.01	0.01
K	0.00	0.00	0.00	0.89	0.85	0.91	0.03	0.00	0.00	0.89	0.00					
F																
Cl	0.00															

Table 12.3. (cont.)

	86-391									86-312							
Wt %	Ga1c	Ga1r	Ga2c	Ga2r	Ga3c	Ga3r	Carb	Ortb	Sphb	Ga1c	Ga1r	Ga2c	Ga2r	Pl	Bi	Bi1b	Bi2b
SiO_2	38.1	38.8	38.3	38.8	37.4	38.6	0.03	64.5	32.0	38.8	38.6	39.9	39.2	60.5	37.4	40.1	41.6
TiO_2	0.17	0.05	0.05	0.02	0.14	0.14	0.00	0.00	19.80	0.14	0.00	0.03	0.00	0.01	1.66	2.39	2.79
Al_2O_3	19.7	21.3	21.0	21.3	19.7	20.8	0.00	18.0	13.5	20.7	21.5	22.0	21.6	23.4	15.8	16.9	18.2
FeO	27.2	23.4	24.1	21.6	26.7	26.5	2.26	0.38	3.91	25.7	25.0	22.2	23.6	0.02	17.5	8.80	8.20
MnO	1.26	0.74	0.67	0.97	1.31	1.89	0.25	0.02	0.11	2.24	1.28	0.14	0.70	0.02	0.15	0.02	0.03
MgO	3.00	3.38	3.19	3.35	3.15	3.5	0.79	0.02	1.73	6.46	5.42	10.5	6.81	0.02	13.5	16.9	17.2
CaO	11.1	13.2	13.5	14.8	11.9	11.8	58.3	0.03	26.2	7.01	8.87	5.95	9.12	5.92	0.07	0.03	0.00
Na_2O	0.03	0.04	0.05	0.03	0.03	0.02	0.00	0.03	0.02	0.02	0.02	0.04	0.01	8.30	0.21	0.15	0.26
K_2O	0.00	0.00	0.00	0.00	0.00	0.00	0.00	16.3	0.00	0.00	0.00	0.00	0.00	0.35	9 13	9 78	9 35
F																	
Cl																	
Sum	100.56	100.91	100.86	100.87	100.33	103.25	61.63	99.28	97.27	101.07	100.69	100.76	101.04	98.54	95.42	95.07	97.63
Cations per formula unit																	
Si	3.02	3.01	2.99	3.00	2.98	2.97	0.00	3.01	1.05	3.00	2.99	3.00	3.00	2.73	2.81	2.89	2.89
Ti	0.01	0.00	0.00	0.00	0.01	0.01	0.00	0.00	0.49	0.01	0.00	0.00	0.00	0.00	0.09	0.13	0.15
Al	1.84	1.95	1.93	1.94	1.85	1.89	0.00	0.99	0.52	1.89	1.96	1.95	1.95	1.25	1.40	1.44	1.49
Fe	1.80	1.52	1.57	1.40	1.78	1.71	0.03	0.01	0.11	1.66	1.62	1.40	1.51	0.00	1.10	0.53	0.48
Mn	0.08	0.05	0.04	0.06	0.09	0.12	0.00	0.00	0.00	0.15	0.08	0.01	0.05	0.00	0.01	0.00	0.00
Mg	0.35	0.39	0.37	0.39	0.37	0.40	0.02	0.00	0.08	0.75	0.63	1.18	0.78	0.00	1.51	1.81	1.78
Ca	0.94	1.10	1.13	1.23	1.01	0.97	0.95	0.00	0.93	0.58	0.74	0.48	0.75	0.29	0.01	0.00	0.00
Na	0.00	0.01	0.01	0.00	0.00	0.00	0.00	0.00	0.00	0.00	0.00	0.01	0.00	0.73	0.03	0.02	0.04
K	0.00	0.00	0.00	0.00	0.00	0.00	0.00	0.97	0.00	0.00	0.00	0.00	0.00	0.02	0.88	0.90	0.83
F																	
Cl																	

Table 12.3. (cont.)

Wt.%	258							243					
	Ga	Cpx	Cpxb	Bi	Bi1b	Bi2b	Sphb	Ga	Pheb	Ga1a	Ga2a	Cpx1a	Cpx2a
SiO2	38.6	53.1	54.0	38.2	37.9	39.7	32.0	39.8	54.8	39.8	39.4	56.6	53.79
TiO2	0.11	0.02	0.04	2.38	2.52	1.85	22.08	0.25	0.24	0.15	0.09	0.18	0.15
Al2O3	20.6	1.60	3.39	14.9	14.6	15.0	11.7	21.8	25.9	21.4	21.4	14.7	12.2
FeO	25.1	8.75	7.11	17.1	11.3	10.7	0.88	16.6	1.90	17.6	20.4	3.31	3.57
MnO	2.16	0.30	0.25	0.21	0.21	0.07	0.05	0.57	0.02	0.46	0.71	0.11	0.11
MgO	6.07	11.8	12.3	14.9	16.2	17.5	0.05	6.33	4.00	5.35	5.66	6.61	8.62
CaO	7.37	22.3	20.4	0.23	0.37	0.1	28.5	14.6	0.18	14.3	11.8	11.0	15.6
Na2O	0.08	0.96	1.76	0.20	0.36	0.16	0.01	0.14	0.05	0.07	0.08	7.30	4.93
K2O	0.01	0.01	0.03	8.71	8.53	8.84	0.00	0.00	8.45	0.00	0.00	0.01	0.00
F	0.01	0.28	0.18	1.03	4.01	0.79	3.99	0.17	0.58				
Cl				0.44	0.39	0.00	0.00		0.00				
Sum	100.05	99.05	99.39	98.22	96.45	94.65	99.19	100.26	96.12	99.13	99.54	99.79	98.94

Cation per formula unit

	Ga	Cpx	Cpxb	Bi	Bi1b	Bi2b	Sphb	Ga	Pheb	Ga1a	Ga2a	Cpx1a	Cpx2a
Si	3.01	2.00	2.00	2.82	2.87	2.92	1.07	3.02	3.56	3.06	3.04	1.99	1.94
Ti	0.01	0.00	0.00	0.13	0.14	0.10	0.56	0.01	0.01	0.01	0.01	0.01	0.00
Al	1.90	0.07	0.15	1.30	1.30	1.30	0.46	1.96	1.98	1.94	1.95	0.61	0.52
Fe	1.64	0.28	0.22	1.06	0.72	0.66	0.02	1.06	0.10	1.13	1.31	0.10	0.11
Mn	0.14	0.01	0.01	0.01	0.01	0.00	0.00	0.04	0.00	0.03	0.05	0.00	0.00
Mg	0.71	0.66	0.68	1.64	1.82	1.91	0.00	0.72	0.39	0.61	0.65	0.35	0.46
Ca	0.62	0.90	0.81	0.02	0.03	0.01	1.02	1.19	0.01	1.18	0.97	0.41	0.60
Na	0.01	0.07	0.13	0.03	0.05	0.02	0.00	0.00	0.00	0.01	0.01	0.50	0.34
K	0.00	0.00	0.00	0.82	0.82	0.83	0.00	0.00	0.70	0.00	0.01	0.00	0.00
F	0.00	0.03	0.02	0.24	0.96	0.18	0.42	0.04	0.12				
Cl				0.06	0.05	0.00	0.00		0.00				

Note: Rock samples 91-16, and 81-21= pyroxene carbonate rock, 91-25 and 258 = garnet-pyroxene-quartz rocks; 86-319 = garnet-muscovite-kyanite schists; 86-312, 86-391, 86-269, and 243 = blastomylonite. Mineral samples 86-312 Ga1 = pink garnet, Ga2= pinky-brown garnet; 86-269 Ga1 = pale-pink, Ga2 = pinky brown,

Subscripts: c=core; r=rim. *a* - Mineral inclusions in zircons. *b*- Mineral inclusions in Ga,

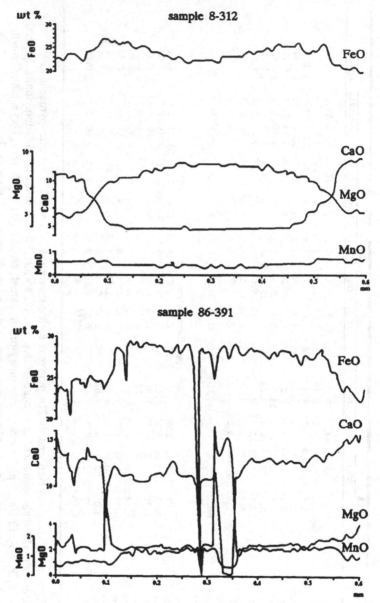

Figure 12.8. Compositional zoning profiles of individual garnets from metamorphic diamondiferous rocks: (a) 8–312; (b) 86–391.

The colored garnets are characterized by an increased content of titanium and ferric iron. It should be noted that when garnet zonation is absent, their compositions can vary within one thin section (Table 12.3).

The individual garnet grains in pyroxene-carbonate rocks from the matrix as well as those included in zircon tend to be homogeneous. Zircon mineral concentrates (sample 91–16) were prepared for the purpose of studying these inclusions. Garnets included in the zircons were analyzed from 34 separate zircon grains. These garnets have strong compositional variations as seen in Fig. 12.7 and Table 12.3. Calcium contents vary from 27% to 49%, and the Fe/Fe-Mg ratio ranges from 0.437 to 0.336. At the same time, in the garnet from the rock, compositional variations are insignificant.

Pyroxenes

Pyroxenes from various types of rocks were described earlier (Sobolev and Shatsky, 1990). It is observed that during processes of superimposed metamorphism, omphacite in eclogites is replaced by pyroxene-plagioclase symplectite (Shatsky et al., 1985a). It was suggested, on these grounds, that pyroxenes from gneisses with a low jadeite content had originally been omphacite.

The pyroxenes from garnet pyroxenites and pyroxene-carbonate rocks are different from other pyroxenes in their low jadeite content and high Mg content (Table 12.3). They are also distinguished by high K_2O. In most cases rock pyroxenes are characteristic, as a rule, of the lower K_2O and high Mg content when compared to the pyroxenes as inclusion in garnets and zircons. Orthoclase lamellae and silica needles are also found in the pyroxene grains in the matrix (Shatsky et al., 1985b). When exsolution structures are absent, the pyroxenes from the groundmass and those included in garnets, have the same composition.

The zoning patterns in pyroxene from pyroxene-carbonate rocks are commonly one of the following two types: (1) Al_2O_3 and K_2O decrease simultaneously from the core to the margin; (2) Al_2O_3 increases and K_2O decreases from core to rim. Type 1 zoning is observed in pyroxene from the matrix, whereas type 2 zoning is characteristic of pyroxene as inclusions in zircon; often pyroxenes included in zircon have no trace of zonation. At the same time, in one pyroxene grain from zircon, K_2O is significantly lower (0.08%) than that in the rock pyroxene (Table 12.3, K91-16)

The study of pyroxene inclusions in garnet and zircon from gneisses and pyroxene-quartz rocks provides support for the earlier assumption about the composition of primary pyroxene from rocks of this type. Though there is a tendency for an increase in jadeite content of the pyroxene inclusions in garnets, this Jd content does not exceed 14% (Table 12.3, sample 8-32, 258). Whereas we found that omphacite included in zircon from blastomylonite contains up to 50% of the Jd component (Table 12.3). The pyroxene from sample 91-25 has an exsolution texture similar to the carbonate rock pyroxenes.

The pyroxene from the matrix does not contain potassium admixture (Table 12.3). At the same time, in the pyroxene inclusions in garnet, the content of potassium admixture is up to 0.68%. As in other rocks, the potassium content in pyroxene decreases from the center to the margin of the grain. The jadeite content and Fe/Fe + Mg ratio remain the same.

Sample 260 seems to be rather interesting. We have described it earlier as eclogite, since the mineral assemblage includes garnet, pyroxene, quartz, and secondary biotite (Shatsky et al., 1993). The pyroxene from this sample contains 21% Jd and SiO_2 needles which is characteristic of the eclogite of the Kumdy-Kol site. However, the Nd-isotopes study in this rock indicates that the rocks have crustal characteristics (Et Nd–15, 5) and have acted as protolith of this rock (Shatsky et al., 1993). Potassium admixture is found only in pyroxene included in garnet. Not all the pyroxenes included in garnets, however, contain potassium admixture. Furthermore the pyroxenes without potassium admixture have the greater value of Fe/Fe + Mg ratio.

Micas

As noted earlier (Sobolev and Shatsky, 1990), diamond inclusions in garnets often form intergrowths with micas. Two generations of micas are found in the diamond-bearing rocks; they are the micas of the HP metamorphism, occurring as inclusions in garnets and forming intergrowth with diamonds and graphite, and the micas from the matrix that replace garnet and K-feldspar in the rock (Vavilov et al., 1991).

Mica inclusions in garnets and zircons occur as idiomorphic crystals with a size less than 40 μm. The representative analyses of the micas are given in Table 12.3. Both biotite and phengite were found. Si content in phengites varies from 3.22 to 3.56 pfu. Ti content decreases when Si content increases (Fig. 12.9). The phengites on the whole are characterized by a high Ti content (up to 0.28 pfu).

As seen in Fig. 12.10, Fe content in biotites for all types of rocks increases when Al^{VI} content decreases. The biotites included in garnets contain less Fe content than that in the matrix biotites. The biotites from the rocks that do not contain diamonds are characterized by the smaller Fe content and low Al^{VI} content.

Biotite inclusions in garnet show great variation within one sample. The changes in Fe content of the host garnets correlate with Fe variations in biotite. Thus in sample 8-47, Fe/Mg = 0.18, 0.20, 0.26, for the biotite, and Fe/Mg = 1.12 , 1.22, 1.54, for the garnet. Fe and Si contents also change in the phengites included within zircons.

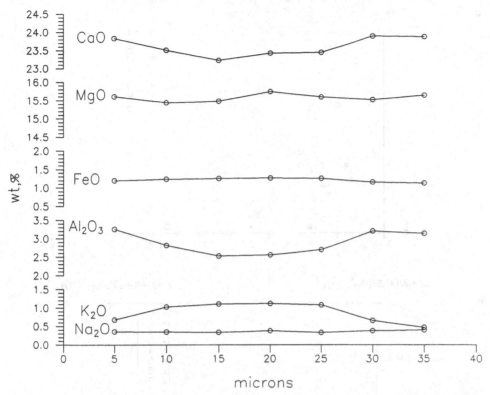

Figure 12.9. Compositional zoning profiles of clinopyroxene inclusions in garnet from pyroxene-carbonate rock (81–21).

Sphene

There are at least two generations of sphene in the rocks. The sphene included in garnet is characterized by high Al_2O_3 and F content (Sobolev and Shatsky, 1990). The maximum Al_2O_3 content is 13.5% (Table 12.3, 86-391). Matrix sphene contains less than 3% Al_2O_3.

Carbonates

There are dolomite, magnesite, and magnesian calcite inclusions in garnets (Table 12.3, K91-16). The $MgCO_3$ content in magnesian calcite reaches 23.5%. However, in most cases calcite contains minor Fe and Mg. Radial cracks are often observed around the calcite inclusions in garnet.

Coesite

Coesite is identified primarily as an individual inclusion in zircon, having a size of about 10 μm (Sobolev et al., 1991). Raman spectroscopy shows the

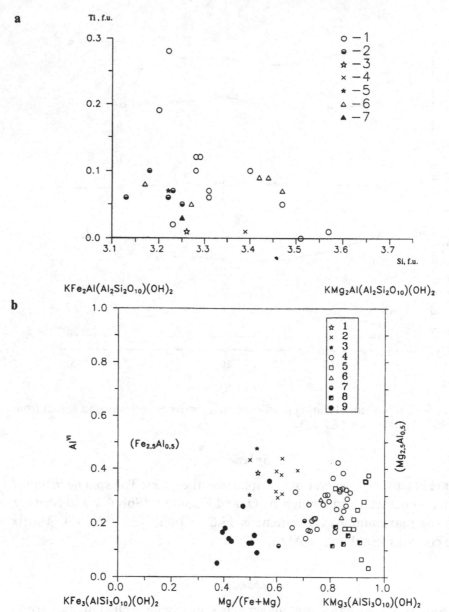

Figure 12.10. Plot of analyzed micas from metamorphic rocks of Kumdy-Kol. (a) 1, inclusions in garnets from diamond bearing gneiss; 2, from diamond free gneiss; 3, 4,5, diamond-bearing gneiss (303); 3, inclusions in zircon, 4, inclusions in garnet, 5, from rock; 6,7 diamond-bearing gneiss (8–27): 6, inclusions in zircon, 7, from rock; (b) 1–3 diamond-bearing gneiss (8–27): 1, inclusions in zircon; 2, inclusions in garnet; 3, from rock; 4–6 inclusions in garnet: 4, from gneisses; 5, from pyroxene-carbonate rock; 6, from garnet-pyroxene rock; 7,8 from diamond-bearing rocks: 7, gneiss; 8, pyroxene-carbonate rock; 9, diamond-free gneiss.

virtual absence of quartz in this grain which is quite different from relicts of coesite found as inclusions in pyropes (Chopin, 1984). The coesite inclusions found have been reproduced in a number of zircon grains including coesite-diamond intergrowth in a single zircon (Sobolev et al., 1994). Other associated minerals found are diamond, kyanite, zircon, K-feldspar, and rutile.

Metamorphic Conditions

We previously estimated P-T parameters of metamorphism of diamond-bearing rocks (Sobolev and Shatsky, 1990; Shatsky and Sobolev, 1993). The new data affirm these estimations. One must consider that diamondiferous rocks have minerals formed at least during three metamorphic stages (UHP, amphibolite, and greenschist facies). Here we are using only the minerals included within garnet porpyroblasts and zircon for the estimation of P-T parameters of the UHP stage related to diamond formation.

HP conditions evidenced within the rocks are discussed here as follows. If a diamond is not taken into account, one of the most significant indicators of HP is the presence of coesite inclusions in zircon and of potassium admixture in pyroxenes (Sobolev et al., 1991). The potassium content in pyroxene formed under HP conditions has been confirmed by experiments (Doroshev et al., 1992).

Al_2O_3 admixture in sphene (Sobolev and Shatsky, 1990) is considered to be another indicator of high pressure. Manning and Bohlen (1991), using the experimental data on reaction $Ca_3Al_2Si_3O_{12} + 3TiO_2 + SiO_2 = 3CaTiSiO_5 + Al_2SiO_5$, estimated the pressures for sample 258 (Table 12.3) of the diamondiferous rock. In this rock, inclusions of highly aluminous sphene and kyanite in garnet have been identified.

The estimated pressure values using the K-content of pyroxene and Al_2O_3 content of sphene are 45–46 kbar, well above the equilibrium line of diamond-graphite transition, indicating that diamond is crystallized within its stability field. Coesite inclusions found in zircon are also evidence of the HP metamorphism (Sobolev et al., 1991, 1994).

As mentioned, for pyroxene-garnet-carbonate rock (sample KK 91-16), pyroxenes, garnets, calcite, magnesite, magnesian calcite, and dolomite were found among the inclusions in zircons. $MgCO_3$ content in magnesisn calcite is about 23.5%. The Mg admixture is minor in matrix calcite and calcite inclusions in garnets. This fact can be accounted for by the increase in both temperature and pressure during carbonate crystallization.

At high pressure, when the temperature decreases, magnesian calcite will be broken down into aragonite and dolomite (Goldsmith, 1983). Thus, at a pressure of 30 kbar, the magnesian calcite disappears at a temperature below 850°C. A great variety of the garnet compositions in zircons can indicate a major variation in the temperature during their crystallization (Table 12.3). Considering the fact that pyroxenes are the predominant phase in the rock and their composition does not change noticeably, it is possible to estimate temperature variation during garnet growth. The values obtained are between 670 and 945°C. It should be noted that calcite inclusions in garnet, in equilibrium with dolomite, have no intergrowths.

The temperature at which magnesian calcite is in equilibrium with dolomite should not be higher than 900°C. Thus, it may be argued that garnet growth was completed below this temperature, and that during further increase of temperature, its composition became homogenized.

The presence of magnesite is additional evidence for high pressures at the formation of diamond-bearing rocks. Reaction Mg + Di = Dol + En has been investigated by experiments. The positions of univariant curves differ greatly. Nevertheless, if we take the curve of the reaction of Brey et al. (1983), then the pressure value should exceed 20 kbar.

The equilibrium temperatures, calculated according to several geothermometers, are given in Table 12.4. We have already mentioned that, for pyroxene-carbonate rocks, the temperature estimates obtained using the garnet-pyroxene geothermometer are obviously overestimated, which accounts for a high grossular content in garnet (Shatsky and Sobolev, 1993). The temperature estimations of Ellis and Green (1979) and Powell (1985) are practically identical. Pattison and Newton's geothermometer (1989) indicates, in a number of cases, considerably minor lower estimations. The biotite-garnet and phengite-garnet geothermometers show the values close to the values obtained by the first two geothermometers. The highest equilibrium temperatures, as seen from the given data, are obtained using garnet-pyroxene geothermometers. For the majority of samples the estimations exceed 950°C. Values from the biotite-garnet and phengite-garnet geothermometers are in the interval of 800–900 °C.

Our estimation of the P-T parameters during metamorphism of the diamondiferous rocks and eclogites at the Kumdy-Kol and Pakhar sites indicates that these rocks have been metamorphosed at a temperature not lower than 900°C and at a pressure exceeding 40 kbar.

Discussion of Results

The data presented in this paper together with the previous publications permit us to consider individual stages in the geological history of the Earth's crust rocks, which have undergone UHP metamorphism.

Table 12.4. Equilibrium temperature estimates (°C) for Zerendin series rocks (pressure = 40 Kbar)

SAMPLE	R.Powell (1985)	D.I.Ellis, D.H.Green (1979)	D.R.M. Pattison, R.C. Newton (1989)	G.Hopinkes. (1986)	T.H.Green, P.L.Hellmen (1982)
Pyroxene carbonate rock	990 ~ 1050	1000 - 1060	850 - 1130	960 - 950	
K91-16 a	990 - 1300	1000 - 1040	810 - 870		
K91-16 b	660 - 1020	680 - 1030	450 ~ 800		
Garnet-pyroxene quart rock	920 - 1020	680 - 1030	670 - 780		
Garnet-pyroxene rock	920 - 1040				
Gneisses	760 - 850			710 - 850	720 - 890
Eclogite	920 - 1000				

a Sample K91-16 pyroxene-carbonate rock
b Temperature estimates for mineral inclusions in zircon.

It should be noted, however, that some scientists do not support this hypothesis that the diamond in the Kokchetav massif metamorphic rocks was crystallized within the diamond stability field, but they did not take into account a number of mineralogic-geochemical features of the diamond-bearing rocks. According to our studies, the existing independent evidence of the UHPs in the studied rocks is a sufficiently convincing indicator that diamond has been crystallized within its stability field in crustal gneissic rock (Shatsky and Sobolev., 1993)

The isotopic-geochemical study of diamond-bearing rocks and the intercalated granite gneisses as well as the Zerendin rock series from different sites made it possible to suggest that diamond-bearing rocks had undergone partial melting stage (Shatsky et al., 1991, 1993). The data reported in this chapter provide some supplementary evidence in favor of this hypothesis. First of all, there is a depletion of incompatible elements in the diamond-bearing rocks. The mineral and chemical composition of these rocks as well as the estimated equilibration temperatures indicates their restite origin. According to experimental data during metapelite and tonalite gneisses melting at vapor-absent conditions at 900°C, the melt fraction is less than 30% (Veilzeuf and Holloway, 1988; Douce and Johnston, 1991; Skjerlie and Johnston, 1992). According to Douce and Johnston (1991), pressure produces a weak effect on the generation of melts during metapelite melting process. This phenomenon is explained by the fact that biotite dehydration has a very steep slope dP/dT. But for fluorine-bearing systems this slope may reach 140 bar/°C (Peterson et al., 1991). Biotites rich in fluorine (2.7%) may occur in the rock even at 1000°C (Skjerlie and Johnston, 1992). According to Peterson et al. (1991), experimental melting of phlogopite containing 4.57% fluorine begins at 15 kbar at 1020–1050°C, while at 8 kbar the melting process occurs at 950–1000 °C. Phlogopite free of fluorine in the presence of quartz melts at 850°C.

The study of micas included in garnets indicates wide variations in fluorine contents even within a single sample. For example, biotite from sample 258 contains from 0.8% to 4.0% of fluorine. According to Douce and Johnston (1991), biotites are enriched in Ti and F and have higher Mg content during melting at elevated temperatures. It may be concluded that micas were trapped by garnet at various stages of partial melting. But on the other hand, the difference in F content may be due to the gradient of F activity in the rock. To exclude this gradient the fluid-rock ratio would have to exceed 1000/1 (Wood and Walther, 1988). As mentioned above, among the inclusions of diamond-bearing rocks in garnets there are phengite inclusions. In some samples phengite occurs in association with biotite and K-feldspar, and the phlogopite barometer was used for this rock series (Massonne and Schreyer, 1987).

According to the barometer the pressure would not exceed 18 kbar even for phengite that contains a maximum of celadonite component (Si = 3.57 fe). In the presence of melt, according to Massonne's experimental data (1981), Si-isopleths in phengite acquire a steep slope, wheras phengites containing low Si may crystallize at pressure over 30 kbars.

As shown in the plots for $CaO + MgO-SiO_2/10-Na_2O + K_2O$ (Fig. 12.5), the rocks from the Kumdy Kol area make a continuous sequence from the field of rocks corresponding to mafic composition to that of leucocratic rocks. This trend may be the result of granitization of the diamond-bearing rocks, or the result of partial melting of metapelites. The latter seems to be more probable since it does not require abundant fluids, and besides it agrees with the fact that the diamond-bearing rocks and granite gneisses had the same Nd 143/144 ratio prior to the high pressure metamorphism (Shatsky et al., 1993).

Taking into account the high temperature of metamorphism and rock composition we may conclude that the rock melting has taken place in fluid absent conditions or in conditions when $P_{H_2O} < 1$. The previously cited experimental papers on rock melting under fluid absent conditions were carried out at pressures not higher than 20 kbar. In this case, garnet, alumosilicate, plagioclase, rutile, or orthopyroxene may be present as restite phases as a function of rock composition and P-T parameters (Vielzeuf and Holloway, 1988; Douce and Johnston, 1991; Skjerlie and Johnston, 1992).

Among the minerals of the Zerendin diamond-bearing rock series there are garnet, clinopyroxene, biotite, phengite, kyanite, rutile, sphene, and coesite.

The presence of melt would vary the fluid phase composition. With free carbon in the rocks, and at constant oxygen fugacity, the fluid phase composition would be a function of carbon and water reactions:

$$CO_2 + CH_4 = 2C + 2H_2O$$
$$CH_4 + 2CO = 3C + 2H_2O$$

When a melt is present, the reactions given above would be shifted toward the right side resulting in precipitation of C from the fluid phase. The reason is that the water dissolving is much higher in the melt compared with CO_2 and CH_4. Thus the appearance of such a melt could lead to a mass crystallization of diamonds or graphite. A great variety in morphology of diamond crystals even in polycrystalline agreggate (Sobolev and Shatsky, 1990) may be due to variation in this value of carbon supersaturation in the fluid phase as a consequence of contemporaneous of melt generation.

Abundant coexistence of diamond and graphite of the first generation could be explained in different ways. Graphite in the process of subduction of the Earth's crust rocks, if it does not move to the fluid phase as a whole, may exist in the diamond stability field. The graphite-diamond phase transition requires a much higher pressure when a catalyst is lacking. In this case diamond can be formed under conditions of metamorphic activity of diamond-bearing rocks of the Kokchetav massif only as a result of crystallization from the fluid phase. A diamond crystallized from the fluid phase would coexist with graphite that did not change to fluid phase. Another explanation for graphite's coexistence with diamond may be the increase in temperatures in subducted rocks at a constant pressure, resulting in a shifting of *P-T* conditions to the graphite stability field.

The contrasts in the diamond-grade of the diamondiferous rocks may also be explained by partial rock melting. The degree of partial melting would thus be a function of rock composition. Melt separation is possible only when certain partial melting value is reached and controlled by melt viscosity. In the number of subducted metapelites, there may be beds that have undergone various degrees of partial melting. In some cases, melt separation may be followed by its displacement and in some cases a mixture of restite and melt may occur. During partial melting a disequilibrium between the melt and the fluid phase may occur. Owing to this fact the diamond may interact with the melt, to be conserved only as inclusions in restite minerals. The rocks formed during crystallization of a separate melt would contain no diamonds. We should note that depending on the H_2O content of the melt and the rate of cooling, crystallization of these melts may take place at various levels of lithosphere. Because of this disparity we shall have a spatial coexistence of mineral assemblages that differ greatly in *P-T* parameters.

No doubt this model requires further proof and precise estimation of the *P-T* parameters at various stages in the history of high-pressure rocks of the Kokchetav massif. It is necessary to carry out isotopic study of both the Zerendin series rocks and overlying rocks of the Borovsk series. There is insufficient experimental study of melting of the rocks of pelitic composition at pressures over 20 kbar. The scales of UHPM activity require more detail. UHPM activity was found at the Kokchetav massif within an area of about 80 km^2. It is now clear that UHP metamorphism is one of the most important stages in the evolution of orogenic belts associated with the process of granitic magmatism and ore formation in this region.

Acknowledgments

We thank Robert G. Coleman for his assistance in editing this manuscript. This study was partly supported by grant 93-05-9613 from the Russian Science Foundation.

References

Brey, G., Brice, W. R., Ellis, D. J., Green, D. H., Harris, K. L., and Ryabchikov, I. D. 1983. Pyroxene-carbonate reactions in the upper mantle. *Earth Planet. Sci. Lett., 62*, 63–74.

Chopin, C. 1984. Coesite and pure pyrope in high-grade blueschists of the western Alps: A first record and some consequences. *Contrib. Mineral. Petrol., 86*, 107–118.

Doroshev, A. M., Sobolev, N. V., and Brey, G. 1992. Experimental evidence of high-pressure origin of the potassium-bearing pyroxenes. *29th Inter. Geol. Congress, 2*, 602.

Douce, A. E. P., and Johnston, A. D. 1991. Phase equilibria and melt productivity in pelitic system; implications for the origin of peraluminous granitoids and aluminous granulites. *Contrib. Mineral. Petrol., 107*, 202–218.

Ellis, D. J., and Green, D. H. 1979. An experimental study of the effect of Ca upon garnet-clinopyroxene Fe-Mg exchange equilibria. *Contrib. Mineral. Petrol., 71*, 13–22.

Essenov, S. E., Yefimov, I. A., Shlygin, E. D., Abdulkabirova, M. A., Vedernikov, N. N., and Nurlybaev, A. N. 1968. On the problem of diamond prospecting in Northern Kazakhstan. *Vestnik. Akad. Nauk Kazakh. SSR, 24*, 37–45 (in Russian).

Goldsmith, J. R. 1993. Phase relations of rhombohedral carbonates. In R. J. Reeder (ed.). Carbonates: mineralogy and geochemistry. *Reviews in Mineralogy, 11*, 49–76.

Kashkarov, I. F., and Polkanov, Y. A. 1972. Some specific features of diamonds from titaniferous placers of Northern Kazakhstan. *Novye dannye o mineralakh, 21*, 183–185 (in Russian).

Letnikov, F. A. 1983. Formation of diamonds within deep-seated tectonic zones. *Doklady Acad. Nauk SSSR, 371*, 433–435.

Manning, C. E., and Bohlen, S. B. 1991. The reaction titanite + kyanite = anorthite + rutile and titanite-rutile barometry in eclogites. *Contrib. Mineral. Petrol, 109*, 1–9.

Massone, H. J. 1982. Phengite: eine experimentelle unetzsuching ihres druck-temperatur-verhaltens in system K_2O-MgO-Al_2O_3-SiO_2-H_2O. Ph.D., Univ. Bocham, 75 pp.

Massonne, H. J., and Schreyer, W. 1987. Phengite geobarometry based on the limiting assemblage with K-feldspar, phlogopite and quartz. *Contrib. Mineral. Petrol., 96*, 212–224.

Newton, R. C. 1989. Metamorphic fluids in the deep crust. *Annual Review of Earth & Planetary Sciences, 17*, 385–412.

Pattison, D. R. M. and Newton, R. C. 1989. Reversed experimental calibration of the garnet-clinopyroxene Fe-Mg exchange thermometer. *Contrib. Mineral. Petrol., 101*, 87–103.

Peterson, J. W., Chacko, T., and Kuenher, S. M. 1991. The effects of fluorine on the vapor-absent melting of phlogopite + quartz: implications for deep-crustal processes. *Am. Mineralogist, 76*, 470–476.

Powell, R. 1985. Regression diagnostic and robust regression in geothermometer/geobarometer calibration: the garnet-clinopyroxene geothermometer revisited. *Jour. Metamorph. Geol. 3*, 231–243.

Rozen, O. M. 1982. Kokchetav massif. *Kazakhstan tectonics*. Moscow, Nauka, 9–12 (in Russian).

Rozen, O. M. 1976. The peculiarities of the inner structure and development of some Precambrian massifs of Paleozoides. *Tectonics of the median massifs*. Moscow, Nauka, 65–85 (in Russian).

Rozen, O. M., Zayatchkovsky, A. A., Klijuev, Y. A., & Smirnov, V. L. 1979. Peculiarities of trace-minerals composition and conditions of the formation of the North Kazakhstan eclogites. *Problems of sedimentary geology of Precambrian Sidorenko*. Moscow, Nauka Press, 170–186 (in Russian).

Rozen, O. M., and Zorin, Y. M. 1972. Diamond occurrence related to eclogites in Precambrian of Kokchetav massif. *Dokl. Akad. Nauk SSSR, 203*, 674–676 (in Russian).

Shatsky, V. S., Jagoutz, E., Kozmenko, O. A. and Sobolev, N. V., 1991. Geochemical characteristics of crustal rocks subducted into the upper mantle. 83. *Terra Absracts (EUG 6), 3*, 33.

Shatsky, V. S., Jagoutz, E., Kozmenko, O. A., and Sobolev, N. V. 1994. Sm-Nd isotopic systematics of diamondiferous metamorphic rocks in Kokchetav massif (in press).

Shatsky, V. S., and Sobolev, N. V. 1985. Pyroxene-plagioclase symplectites in eclogites of Kokchetav massif. *Geologiya i Geofizika, 9*, 83–89 (in Russian).

Shatsky, V. S., and Sobolev, N. V. 1993. Some aspects of origin of diamonds in metamorphic rocks. *Dokl. Akad. Nauk, 331*, 1217–219 (in Russian).

Shatsky, V. S., Sobolev, N. V., and Stenina, N. G. 1985. Structural pecularities of pyroxenes from eclogites. *Terra Cognita, 5*, 436–437.

Shatsky, V. S., Sobolev, N. V., Zayachkovsky, A. A., Zorin, Y. M., and Vavilov, M. A. 1991. New occurrence of microdiamonds in metamorphic rocks as a proof of regional character of ultra high pressure metamorphism in Kokchetav massif. *Dokl. Akad. Nauk, SSSR, 321*, 189–193 (in Russian).

Sidorenko, A. V. and Tenyakov, V. A. 1978. Exogenic, biogenic and metamorphic processes as factors in planetary evolution. *Dokl. Acad. of Sci. USSR, 241*, 131–133.

Skjerlie, K. P., and Johnston, A. D. 1992. Vapor-absent melting at 10 kbar of a biotite- and amphibole-bearing tonalite gneiss: Implications for the generation of A-type granites. *Geology, 20*, 263–266.

Sobolev, N. V., Bakumenko, I. T., Efimova, E. S., and Pokhilenko, N. P. 1991. Morphological features of microdiamonds sodium in garnets and potassium in clinopyroxenes contents, of two eclogite xenoliths of the Udachaja kimberlite pipe (Yakutia). *Dokl. Akad. Nauk SSSR, 31*, 585–591 (in Russian).

Sobolev, N. V., Dobretsov, N. L., Bakirov, A. B., and Skatsky, V. S. 1986. Eclogites from various types of metamorphic complexes in the USSR and the problems of their origin. *Geol. Soc. Amer. Mem., 164*, 349–360.

Sobolev, N. V., and Shatsky, V. S. 1990. Diamond inclusions in garnets from metamorphic rocks. *Nature, 343*, 742–746.

Sobolev, N. V., Shatsky, V. S., Vavilov, M. A., and Goryainov, S. V. 1991. Coesite inclusion in zircon from diamondieferous gneiss of Kokchetav massif–first find of coesite in metamorphic rocks in the USSR territory. *Dokl. Akad. Nauk SSSR, 321*, 184–188 (in Russian).

Sobolev, N. V., Shatsky, V. S., Vavilov, S. V., and Goryainov, S. V. 1994. Zircon from ultra high pressure metamorphic rocks of folded regions as an unique container of inclusions of diamond, coesite and coexisting minerals. *Dokl. Akad. Nauk, 334*, 488–492 (in Russian).

Sobolev, N. V., Shatsky, V. S., Zayachkovsky, A. A., Vavilov, M. A., and Sheshkel, G. G. 1989. Accessory diamonds in metamorphic rocks of North Kazakhstan. *Geology of metamorphic complexes*. Sverdlovsk (in Russian), 21–35.

628   Sobolev, N. V., and Shatsky, V. S. 1987. Inclusions of carbonmineral in the garnets
629   from metamorphic rocks. *Soviet Geol. Geophys., 27a*, 1–8.
630   Taylor, S. R., and McLennan, R. 1985. *The Continental Crust: Its Composition and*
631   *Evolution.* 312 pp. Oxford, Blackwell.
632   Vavilov, M. A., Sobolev, N. V., and Shatsky, V. S. 1991. Micas in diamondiferous
633   metamorphic rocks from Northern Kazakhstan. *Dokl. Akad. Nauk. SSSR, 319*,
          466–470 (in Russian).
      Vielzeuf, D., and Holloway, I. R. 1988. Experimental determination of the fluid-
          absent melting relations in the pelitic system. Consequence for crustal differen-
          tiation. *Contrib. Mineral. Petrol., 98*, 257–276.
      Wood, B. J., and Walther, J. V. 1986. Fluid flow during metamorphism and its impli-
          cations for fluid-rock ratios. *Fluid-rock interactions during metamorphism,*
          Advances in Physical Geochemistry. New York, Springer-Verlag, 87–108.
      Zonenshain, L. P., Kuzmin, M. I., and Natapov, L. M. 1990. *Geology of the USSR:*
          A plate-tectonic synthesis. Washington, D. C., American Geophysical Union.

13

Orogenic Ultramafic Rocks of UHP (Diamond Facies) Origin

D. G. PEARSON, G. R. DAVIES, AND P. H. NIXON

Abstract

Our physical understanding of tectonic processes is predominantly influenced by evidence from crustal rocks. There are few constraints on the coupling between crust and mantle and the degree to which mantle is physically involved in orogenic events. Occurrences of alluvial diamonds in mountain belts have caused episodic speculation that some tectonically emplaced ultramafic rocks may originate from within the diamond stability field. Since the mid–1980s substantive evidence has come to light to support this view. This chapter reviews the evidence for the tectonic emplacement of large fragments of mantle (up to 300 km²) from the diamond stability field into the crust. Orogenic peridotite massifs from the Betico-Rifean tectonic belt and ophiolitic peridotite bodies from the Indian–Tibetan suture zone are discussed in detail. Constraints are placed on the origin of these ultrahigh pressure (UHP) rocks, their *P-T* histories are constrained by geochemical and mineralogical data and their implications for orogenesis are discussed.

Pyroxenites within the Beni Bousera and Ronda orogenic peridotite bodies contain multicrystalline aggregates of graphite as octahedra and other forms of cubic symmetry that are interpreted as graphitized diamonds. Stable and radiogenic isotope data for the Beni Bousera pyroxenites indicate that some of them originated as high pressure (HP) cumulates from melts of subducted oceanic crust. Anomalously light carbon isotope values ($\delta13C = -16$ to $-27.6‰$) for the graphite suggest crystallisation of diamond from subducted kerogenous carbon.

Small, submillimeter diamonds have been recovered from two ophiolitic peridotite bodies in Tibet. Variation in the crystal morphology of the diamonds eliminate the possibility of a kimberlitic or synthetic origin. Moissanite occurs with the diamonds. The peridotite bodies are typical ophiolitic lherzolites/harzburgites. They appear to have been subducted into the diamond stability field and then obducted. The source of the carbon forming the diamonds is not yet clear.

Sub-solidus reequilibration has destroyed most evidence of HP mineral equilibria in these rocks. Equilibration temperatures calculated from the bulk compositions of exsolved pyroxenes in the Beni Bousera pyroxenites indicate derivation of the massif from the mantle at ~1400°C at 5 GPa. It is not clear what the exact mechanism of emplacement was for Beni Bousera and Ronda but the initiation of mantle up-welling may have been related to slab detachment. The associated up-welling could have then caused extension of the crust and mantle. Consideration of the physics of the graphitization of diamonds indicates that small diamonds such as those found in the Tibetan ophiolites can survive only if they formed, and were subsequently uplifted, in a cool environment such as a subduction zone. Their uplift path must have maintained low temperatures, 900°C, to prevent graphitization. Such uplift may have taken place along localized zones of shear reaching well into the mantle.

Evidence of the tectonic emplacement of peridotite bodies over 300 km^2 in outcrop into the crust from the diamond stability field (120–150 km deep) indicates the involvement of deep mantle in some orogenic processes. Further detailed study of ultramafic rocks in orogenic zones may produce more occurrences of diamond facies mantle. Current thermal/tectonic models require refinement to accommodate such possibilities.

Introduction

It has long been recognized from the occurrence of diamond in eclogite xenoliths and from mineral thermobarometry that volcanic eruption of kimberlite/lamproite magmatism can bring rocks to the Earth's surface from within the diamond stability field (Bonney, 1899; Boyd, 1973). More recent discoveries of pyroxene exsolved from garnet in both diamond inclusions and eclogite/lherzolite xenoliths from kimberlites indicate that the processes associated with volcanism may sample mantle from depths of over 400 km (Moore and Gurney, 1985; Wilding et al., 1989; Haggerty and Sautter, 1990; Moore et al., 1991; Sautter et al., 1991). In contrast, tectonic processes have traditionally been thought capable only of exposing rocks in the crust from ~90 km depth, either as metamorphosed fragments of subducted continental crust (Chopin et al., 1984) or as orogenic peridotite massifs (O'Hara, 1967).

The basal, peridotite-tectonite sections of ophiolites contain only plagioclase and spinel facies peridotite (Coleman, 1977) indicating maximum depths of origin of ~20 kbar (Fig. 13.1). The majority of orogenic peridotite massifs contain predominantly spinel facies peridotites but some contain garnet-

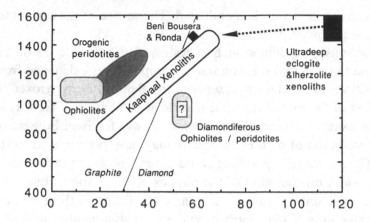

Figure 13.1. Estimates of the equilibration conditions of orogenic peridotites (Medaris and Carswell, 1990), ophiolites (Coleman, 1977), Kaapvaal peridotite xenoliths (Finnerty and Boyd, 1984; Brey et al., 1989a,b), ultradeep eclogite and lherzolite xenoliths, dashed line with arrow points from probable conditions of crystallization to later reequilibration (Haggerty and Sautter, 1990; Sautter et al., 1991). The points marked for Beni Bousera and Ronda and diamondiferous orogenic rocks are estimates of maximum pressures and temperatures of derivation based on the presence of diamonds or their graphite pseudomorphs (see text). (Diamond-graphite transition from Kennedy and Kennedy, 1976).

spinel or garnet-facies peridotites (Medaris and Carswell, 1990). Mineral thermobarometric calculations imply equilibration pressures in excess of 50 kbar for garnet peridotites from Alpe Arami (Ernst, 1978, 1981), various peridotites in the Bohemian massif (Medaris et al., 1990), and peridotites from the Western Gneiss Region of Norway (Medaris and Carswell, 1990). However, such equilibration pressures are thought to be spuriously high, and a result of incomplete mineral equilibration (Medaris and Carswell, 1990). Furthermore, it has been the view of Medaris and Carswell (1990), among others, that "it is difficult to envisage a tectonic process whereby upper mantle material from depths of 100–150 km could be injected into continental crust, even if the crust had attained a vertical dimension twice that of normal thickness due to continent–continent collision."

The discovery of graphitized diamonds in pyroxenites from the Beni Bousera peridotite massif, N. Morocco (Slodkevich, 1980; Pearson et al., 1989, 1991) and more recently from the Ronda massif, southern Spain (Davies et al., 1992a,b) together with reports of diamondiferous peridotites in ophiolites from Tibet (Fang and Bai, 1981; Bai et al., 1993) provides compelling evidence that tectonic forces *are* capable of emplacing large fragments of mantle (up to 300 km²) from within the diamond stability field into the crust. Additionally, crustally derived metamorphic rocks containing diamonds have now been described (Sobolev and Shatsky, 1991; Shatsky, Sobolev, and Vavilov, Chapter 12, this volume) and confirm that diamond-facies rocks may be created and exhumed by tectonic processes. These occurrences require re-evaluation of the details of collisional tectonics and the emplacement and exhumation of high density, high pressure (HP) rocks in the crust.

It is the purpose of this review to summarize the evidence for tectonically emplaced, diamond-facies ultramafic bodies of probable mantle origin, to place constraints on their origin, and pressure–temperature histories, and thus evaluate the implications for collisional tectonic models. The mineralogical characterization of ultrahigh pressure (UHP) metamorphism in ultramafic rocks is reviewed by Chopin and Sobolev (Chapter 3, this volume) while the mineral parageneses and phase equilibria of such compositions have recently been reviewed by Mottana et al. (1990) and Harley and Carswell (1990). In the light of such recent work these topics will not be dealt with here.

Orogenic or Alpine-Type Peridotites of UHP Origin: West Mediterranean Region

The Beni Bousera Peridotite Massif, Northern Morocco

Regional Geology and Petrology

The Beni Bousera peridotite massif is situated in the Rif Mountains of northern Morocco and forms part of the arcuate Betico-Rifean orogenic belt (Fig. 13.2). The peridotite body is a 70-km^2 SE-plunging antiformal dome, faulted along its northeastern margin, and largely overlain by migmatitic, graphite-garnet-sillimanite gneisses (kinzigites) belonging to the Sebtide sequence (Kornprobst, 1966, 1969, 1974; Reuber et al., 1982; Pearson, 1989). Evidence for high temperature emplacement of the massif into the crust is provided by the transformation of kyanite to sillimanite in the graphite-garnet gneisses close to the peridotite contact (Kornprobst, 1974). The surrounding gneisses are granulite facies that consist predominantly of garnet + K-feldspar + kyanite + plagioclase + quartz. $^{40}Ar/^{39}Ar$ analysis of a K-feldspar separate from the sillimanite grade contact aureole yielded a plateau age of 21.5 ± 1 Ma (Pearson et al., 1993). Garnet-clinopyroxene pairs within the garnet pyroxenites define isochronous relationships giving ages within error of the Ar-Ar cooling age (e.g., 20.6 ± 5 Ma; Pearson unpublished data). These ages are interpreted as cooling ages associated with emplacement of the massif into the crust and are identical to the 21.5 ± 1.8 Ma Sm/Nd isochron age obtained for a pyroxenite layer in the Ronda massif (Zindler et

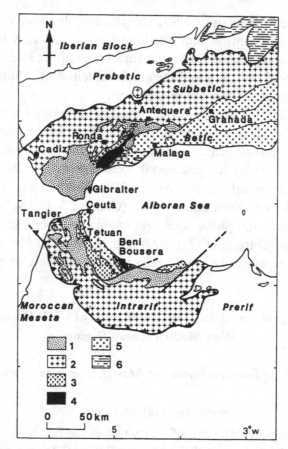

Figure 13.2. Regional geologic map of the western Mediterranean area, modified from Bonini et al. (1973). Key to legend: (1) Cretaceous/Tertiary flysch nappes. (2) Mesozoic sub-Betic deepwater carbonates and volcanics (Spain), IntraRif (Morocco). (3) Internal zone Dorsal calcaire (Morocco) Mesozoic and Tertiary epicontinental series. (4) Peridotite massifs. (5) Ghomaride (Morocco)/Malaguide (Spain) sequence of Late Paleozoic clastics, shaded area within this represents Sebtide complex (Morocco) and Alpujarride complex (Spain). (6) Late Cenozoic cover of the Pre-Rif (Morocco) and Pre-Betic (Spain).

al., 1983). The Beni Bousera and Ronda massifs lie along a continuous positive gravity high that underlies the western margin of the Alboran basin (Bonini et al., 1973). Stratigraphic equivalence of the crustal sequences comprising the Betics and Rif together with the similarity of emplacement ages suggests that the tectonic histories of the Ronda and Beni Bousera massifs are closely equivalent.

Detailed petrologic descriptions of the Beni Bousera pyroxenites and peridotites have been given by Kornprobst (1969, 1974), Pearson (1989), and

Pearson et al. (1993). The predominant lithology in the Beni Bousera massif is a variably altered spinel lherzolite. The spinel lherzolites contain between 2% and 15% clinopyroxene and show abundant evidence of subsolidus re-equilibration and high temperature plastic deformation. Patches of garnet-spinel lherzolites occur locally, often close to the margin of the massif, and plagioclase is a neoblast phase in many peridotites (J. Kornprobst, personal communication), but there are no peridotites containing plagioclase as the dominant aluminous phase. Peridotites become dunitic or harzburgitic close to the pyroxenite layer margins, probably as a result of reaction with the invading pyroxenite melt (Keleman et al., 1992). Pyroxenites bodies are predominantly sheet-like or lensoidal and are generally conformable to the local foliation in the peridotites (Kornprobst, 1974; Reuber, 1981). Some layers are isoclinally folded or boudinaged, the plunge of the pyroxenite fold hinges and boudin axes being approximately parallel to the plunge of folds in the surrounding crustal kinzigites, suggesting a late-stage, syn-emplacement origin for these structures rather than generation due to normal convective flow in the asthenospheric mantle as advocated by Allègre and Turcotte (1986). The pyroxenite bodies vary in thickness from sub-centimeters to 3 meters. The most common layers are websterite transitional to orthopyroxenite between 1 and 5 cm thick (Fig. 13.3). Garnet clinopyroxenites are the most abundant lithologies in layers over 50 cm thick. Measured sections indicate that pyroxenites may locally form over 90% of the outcrop and average over 10% in areas close to the southwestern contact of the massif (Pearson, 1989). However, the average pyroxenite abundance for the whole massif is between 2% and 5% (Allègre and Turcotte, 1986; Kornprobst, 1969; Pearson, 1989). All the pyroxenites display metamorphic textures, with granular to mylonitic textures (in areas of intense shearing). Pyroxene porphyroclasts exhibit varying degrees of exsolution due to subsolidus recrystallization. Some clinopyroxenes contain exsolved lamellae of orthopyroxene, together with blebs of exsolved garnet, locally breaking down to plagioclase. Plagioclase also occurs as a neoblast phase in many pyroxenites but never appears to have been part of the primary assemblage. A summary of the pyroxenite petrographic characteristics is provided in Table 13.1.

Many pyroxenite layers show pronounced mineralogic zonation, grading from orthopyroxenite or websterite at their margins to garnet websterite and finally garnet clinopyroxenite at the centers (Fig. 13.3). There is also significant mineral chemical zonation in some layers. Clinopyroxenes in the centers of layers are sodic Al-augites whereas those at the layer margins are more chrome-rich and tend towards chrome-diopside in composition. This varia-

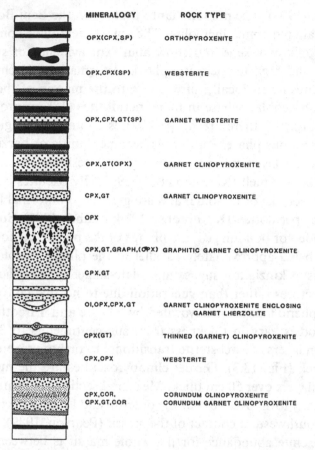

Figure 13.3. Idealized summary of pyroxenite lithologic variation at Beni Bousera. Not to scale. Note the bifurcating garnet clinopyroxenite layer surrounding a pod of garnet lherzolite, host is spinel lherzolite.

tion, and the accompanying trace element and isotopic variation appears to be the result of reaction and diffusive interchange with the host peridotites (see Pearson et al., 1993, for further details).

Diamond Facies Pyroxenites

The graphitic garnet pyroxenites consist of orange pyrope-almandine garnets and dark sodic augite porphyroclasts. Clinopyroxene usually forms >50% of the mode. Plagioclase neoblasts, sulfides (pyrite and pyrrhotite), and varying amounts of graphite are also present. Some clinopyroxene crystals contain inclusions of ilmenite. Graphite was first recorded in pyroxenites from Beni Bousera by Kornprobst (1974), but no detailed study was made of its form. Slodkevich (1980) recorded up to 15 vol% graphite in garnet clinopyroxenites and described multicrystalline aggregates of graphite occurring as sharp

Table 13.1. *Proxenite petrographic characteristics*

Group		Lithology	Porphyroclast Mineralogy Modal %	Texture	Thickness Range and average cm	Occurrence/Relations
Al	A u g i t e p y r o x e n i t e	Orthopyroxenite OPXITE	Opx 90-100 Cpx 0-10 Opx 10 30 ±Sp	Granular/ Porphyro.	<1 - 20 Av. 3	Individual Layers or at margins of zoned layers
		Clinopyroxenite CPXITE	Cpx 70-~0 Opx 10-30 ± Sp	Granular/ Porphyro.	<1 - 20 Av. 5	Inviddual Layers or parts of zoned layers
		Websterite WEB	Cpx 50-70 Opx 3 - 50 ±Sp	Granular/ porphyro	<1 - 80 Av. 5-20	Individual Layers or margtns of Gp and GGP
		Garnet Wbsterite GT WEB	Cpx 50 - 60 Opx 10 - 40 Gt 5 - 20 Fluidal ± Sp	Granular/ Porphyro./ Fluidal	1 - 20 Av. 5	Predominately parts of zoned Layers (Centers)
		Garnet Clinopyroxenite GP Graphite garnet Clinopyroxenite GGP	Cpx 55 - 65 Gt 35 - 45 Graph. 0-15	Porphyro./ Fluidal	5 - 260 Av. 20 200-260 Av. 230	Predominately individual layers or centers of zoned layers; GGP are slightly zoned.
		Corundum Garnet Pyroxenite CORP	Cpx GT ±Sp		Loose blocks	Relationships unknown
		Gametite GTITE	Gt 90 - 100 Cpx 0 - 10	Granular		Always part of a GP layer
Cr	P y r o x e n i t e	Cr-websterite WEB	Opx 60 - 70 ±Sp	Granular	< 1 - 15 Av. 3	Idividual layers, rarely zoned
		Wehrlite WER	Ol 60 -80 Cpx 20 - 40	Granular	1 - 20 Av..3	Individual layers rarely zoned

edged octahedra and other forms of cubic symmetry which he interpreted as pseudomorphs after diamond. The interpretation of the graphite aggregates as graphitized diamonds was subsequently confirmed by Pearson et al. (1989) using x-ray crystallography, Raman spectroscopy and scanning electron microscopy (SEM.). Graphite occurs in coarsely crystalline garnet clinopyroxenites, as aggregates in the form of octahedra (Fig. 13.4), modified octahedra possessing the forms {111} {100} and {011}, and contact twins (macles). X-ray crystallography and SEM show that {0001} of the graphite crystallites

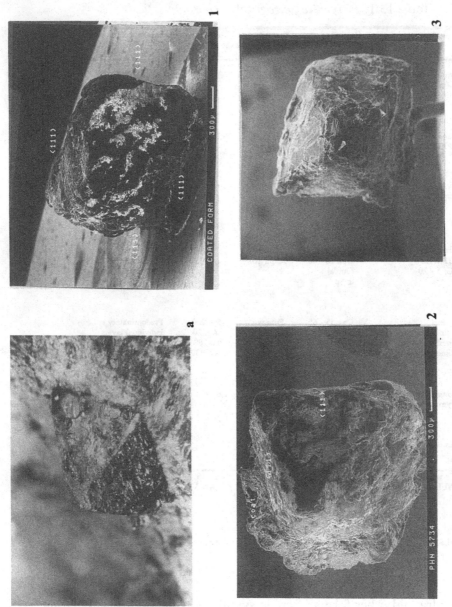

Figure 13.4. (a) Half of a sharp edged graphite octahedron protruding from the weathered surface of garnet clinopyroxenite GP147a. The edge length of the octahedron is 3 mm. (b) Secondary electron scanning electron microscope images of various graphite forms from the Beni Bousera pyroxenites. (1) Coated octahedral form showing rounded appearance, extracted by acid dissolution. (2) Octahedron showing partial retention of coat graphite in which {0001} of the graphite is perpendicular to {111} of the octahedral form. Sample extracted mechanically. (3) Uncoated octahedron showing preferred orientation of {0001} of graphite crystallites parallel to {111} of the octahedral form. Scale as in (2).

forming the aggregates is parallel to {111} of the octahedral forms. This is the preferred orientation expected for graphitized diamonds (Grenville-Wells, 1952). Cliftonite, a polycrystalline aggregate of graphite possessing cubic symmetry has been recorded from both terrestrial rocks (Disko Island, Goodrich and Bird, 1985) and meteorites (Haidinger and Partsch, 1846; Grenville-Wells, 1952; Brett and Higgins, 1969; Okada and Shima, 1972). However, the preferred orientation of the graphite crystallites in cliftonite (six fold axis of the graphite parallel to [001] and [113] axes) is markedly different from the Beni Bousera graphite aggregates and that expected for graphitized diamonds (Grenville-Wells, 1952). Experimental evidence indicates that cliftonite in meteorites results from the decomposition of cohenite, an orthorhombic, metallic carbide [$(Fe,Ni)_3C$], at low pressures (Brett and Higgins, 1969). Hence, natural occurrences of graphite aggregates with cubic symmetry found in low pressure rocks (basalts) and meteorites do not show the same preferred orientation as the Beni Bousera graphite aggregates, strengthening the conclusion that the Beni Bousera graphite aggregates represent graphitized diamonds. Furthermore, cliftonite often contains relict cohenite (Brett and Higgins, 1969); no evidence for any precursor mineral exists in the Beni Bousera graphite, consistent with the isochemical phase transition of diamond to graphite.

The Beni Bousera graphite aggregates contain inclusions of garnet, clinopyroxene, sulfides and ilmenite. Clinopyroxene and garnet grains possess cubo-octahedral faceting where {100} and {111} of the inclusion are parallel to {100} and {111} of the host octahedron (Slodkevich, 1980; Pearson et al., 1989). Such cubo-octahedral faceting is frequently observed on inclusions within natural diamond (Harris, 1968; Meyer and Tsai, 1976; Sobolev, 1979) and, together with the preferred orientation of the graphite is powerful evidence that the Beni Bousera graphite aggregates represent graphitized diamonds.

Since the transformation of diamond to graphite involves the breaking of C–C bonds in the diamond structure, the activation energy for graphitization of a {111} face is theoretically higher than for {110} faces, owing to the greater number of bonds to be broken on {111} faces (Evans, 1979). This difference in activation energy is thought to be the cause of the pronounced development of {110} surfaces on a diamond octahedron undergoing graphitization under experimental conditions, leading to the development of dodecahedra (Howes, 1962; Evans, 1979). Such experiments are frequently carried out on diamonds that have been pre-etched in acid and at significantly higher oxygen fugacity (fO_2) than prevalent in the mantle in order to promote graphitization on the time scale of the experiment. Graphitization experiments carried out under vacuum (Grenville-Wells, 1952) and at 10 GPa (Cooper, 1990) with no pre-etching and at low fO_2 do not result in pronounced changes in the shape of the diamonds, but show retention of sharp-edged octahedral forms, even on complete graphitization. Hence, the development of dodecahedra is not necessarily a natural consequence of graphitization and thus the presence of abundant sharp-edged octahedra in the Beni Bousera pyroxenites is consistent with their origin as graphitized diamonds.

Many of the graphite forms are surrounded by a rounded coat of fibrous graphite where {0001} of the coat graphite is perpendicular to {111} of the octahedral core (Fig. 13.5). Some suites of natural diamonds are dominated by forms consisting of well crystalline cores, often octahedra, surrounded by poorly crystalline, fibrous diamond coats which give the forms a rounded appearance (Kamiya and Lang, 1964). The coated graphite forms found at Beni Bousera are hence thought to represent graphitized coated diamonds (Pearson et al., 1989). Oxygen fugacity estimates based on peridotite spinel-orthopyroxene-olivine equilibria demonstrate that intrusion of magmas parental to the garnet pyroxenites caused reduction of the host peridotites upon intrusion (Woodland et al., 1992). This indicates that the pyroxenites are more reduced than the peridotite average of 1.5 ± 0.7 log units below the foresterite-magnetite-quartz (FMQ) buffer at 15 kbars. Such low fO_2 conditions are consistent with the original crystallization of diamond from the parental pyroxenite magmas as the CCO buffer is within 1 log unit fO_2 of the FMQ buffer at 50 kbars and 1100°C. At such pressures the average Beni Bousera peridotites and, by inference, the pyroxenites would be over 2 log units below the FMQ buffer. The low prevailing fO_2 conditions would also be conducive to preservation of graphite during graphitization and ascent of the massif into the crust.

It is interesting to note that if all the graphite in the graphitic pyroxenites represents graphitized diamonds, the resultant 15 vol% diamond would be 10^5 richer than the highest-grade kimberlites. The only other rocks with comparable diamond contents are some diamondiferous eclogites with up to 10% diamond (Gurney, 1991), suggesting that perhaps the graphitic pyroxenites have an origin similar to that of the carbonaceous eclogites (Pearson et al., 1991b). This hypothesis is strengthened by the occurrence of coated diamonds in an eclogite xenolith from a Siberian kimberlite (N. Pokhilenko, personal communication). No coated diamonds have ever been found in peridotites. The high carbon concentrations required for the growth of diamond in a fibrous form (Sunagawa, 1984) consistent with the high carbon concentrations evident in some diamondiferous eclogites and the Beni Bousera graphitic pyroxenites.

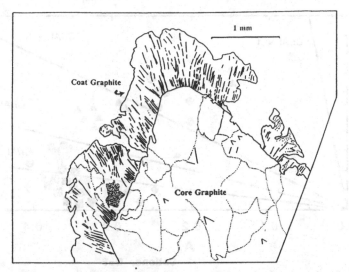

Figure 13.5. Line drawing from reflected light (crossed polars) photomicrograph of a coated graphite octahedron in sample PHN5734. Section cut almost parallel to {100} illustrates the sharp physical boundary between slightly deformed "fibrous" coat graphite and that of the octahedral core. Dotted lines define subgrain boundaries; the radiating structure of the coat graphite is defined by kink lamellae.

Mineral Chemistry

The pyroxenites at Beni Bousera and other orogenic peridotite massifs may be divided into two lithologic groups on the basis of their clinopyroxene chemistry (Wilshire and Shervais, 1975): (1) the Cr-diopside group, here called Cr-pyroxenites, and (2) the Al-augite pyroxenites. This mineralogic subdivision is clearly expressed on a plot of atomic Cr versus Al (Fig. 13.6), the division between the two groups occurring at a Cr/Al value of ~0.025 for Beni Bousera. Minerals in the pyroxenites are relatively unzoned, with slight Fe enrichments and Mg depletions at the rims of garnet and pyroxene porphyroclasts. Some domainal heterogeneity exists in the cores of sodic-augites in the graphitic pyroxenites. The Al-augite pyroxenites are characterized by sodic (up to ~3.5 wt% Na_2O) and highly aluminous (up to 11.5 wt% Al_2O_3) clinopyroxene with a considerable proportion of octahedrally coordinated Al (Fig. 13.7) and up to 30 mol% jadeite component (Pearson, 1989). Although the Na and Al contents of clinopyroxene and consequently their Jd and CaTs components generally increase with increasing pressure, the bulk composition of the parental melt is the dominant influence on clinopyroxene chemistry (Thompson, 1974). Clinopyroxenes from the corundum-bearing pyroxenites at Beni Bousera contain up to 19 wt% Al_2O_3 but have probably crystallized at only ~22 kbar (Kornprobst et al., 1990), supporting the conclusion that bulk composition is the major influence on clinopyroxene chemistry in these rocks.

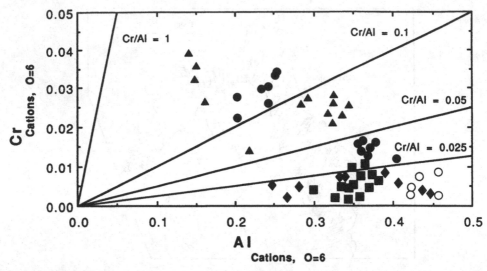

Figure 13.6. Cr versus Al cation proportion plot of clinopyroxenes from Beni Bousera pyroxenites and peridotites indicating the distinction between Cr-pyroxenites (Cr/Al > 0.025) and Al-augite pyroxenites (Cr/Al < 0.025). Solid triangles: lherzolites; solid circles: Cr-pyroxenites; open circles: non-garnetiferous Al-augite pyroxenites; solid squares: garnet pyroxenites; solid diamonds: graphitic garnet pyroxenites.

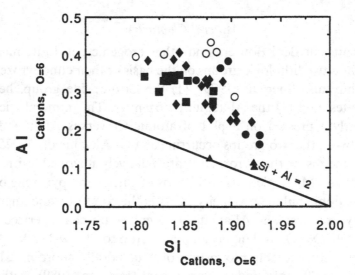

Figure 13.7. Al versus Si cation proportion plot of clinopyroxenes from Beni Bousera peridotites and pyroxenites. Symbols as in Figure 13.6.

Clinopyroxenes from the pyroxenites have relatively high Na_2O/Al_2O_3 ratios (some >0.25), and have generally higher Na_2O/Al_2O_3 than garnet pyroxenite xenoliths from alkali basalts (Fig. 13.8). Clinopyroxenes from the graphitic pyroxenites and clinopyroxenes are compositionally intermediate between clinopyroxenes in eclogite xenoliths from kimberlites and those in garnet pyroxenite xenoliths from alkali basalts. The Beni Bousera clinopyroxenes are anomalously rich in Mn for mantle derived rocks, up to 0.4 wt% MnO, comparable values being recorded only in eclogite suite diamond inclusions (Deines et al., 1987; Smith et al., 1989). This feature strengthens the similarity between the Beni Bousera garnet pyroxenites and eclogite xenoliths. High MnO contents may be due to crystallization from magma anomalously enriched in Mn. The generally lower Jd contents of many of the Beni Bousera clinopyroxenes compared to eclogite xenoliths may be ascribed to the more equilibrated nature of the former, as indicated by exsolution of orthopyroxene, plagioclase, and garnet. Mass balance calculations indicate that some clinopyroxenes contain over 30 wt% exsolved components (Pearson, 1989).

In terms of their mineral assemblage, the corundum-bearing pyroxenites described by Kornprobst et al. (1990) are broadly similar to grospydite xenoliths from kimberlite pipes. Grospydites are predominantly garnet-clinopyroxene assemblages of highly aluminous bulk composition, where the garnet contains a very high proportion of grossular component (Sobolev, 1979). However, clinopyroxenes from the corundum pyroxenites are markedly lower in Na (Fig. 13.8) and thus in Jd component and richer in CaTs than clinopyroxenes from grospydites. Together with the relatively low grossular contents of their garnets these characteristics discriminate the corundum pyroxenites from grospydite xenoliths from kimberlite pipes (Kornprobst et al., 1990). Pre-exsolution "bulk" clinopyroxene compositions calculated by mass balance using analyses of exsolved phases and their estimated volumes, and by defocused beam microprobe analyses, are very sub-calcic (<30 mol% Wo, Fig. 13.9) and aluminous (up to 10.3 % Al_2O_3; Fig. 13.8).

Garnets in the pyroxenite suite are pyrope-almandines (Fig. 13.9); those from the graphitic pyroxenites cover the range $Py_{57-35} Al_{49-31} Gr_{15-9}$ with minor (up to 4.3 mol%) spessartine components and are within the large compositional range of carbonaceous (graphite- and/or diamond-bearing) eclogite xenoliths from kimberlites. Some garnets from non-graphitic pyroxenites contain up to 59% pyrope, whereas those associated with the corundum pyroxenites are richer in pyrope (over 60 mol%) and have higher grossular content (Kornprobst et al., 1990). Garnets and clinopyroxenes from the margins of pyroxenite layers or from very thin, sub-centimeter, layers show enrichments in MgO and depletion in FeO, TiO_2, and Ca. These compositional changes are directly related to the thickness of the layer, or proximity to the layer margin and are thought to be due to high-temperature diffusional exchange between peridotite and pyroxenite in the mantle (Pearson, 1989; Kornprobst et al., 1990). This conclusion has been substantiated by radiogenic isotope

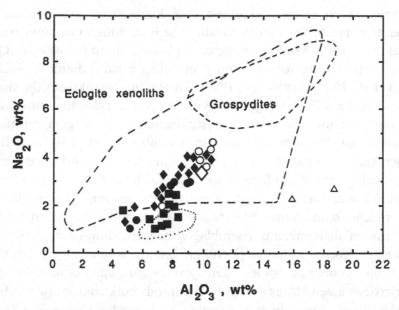

Figure 13.8. Na_2O versus Al_2O_3 wt% in clinopyroxenes from Beni Bousera pyroxe-nites. Symbols as in Figure 13.6 with additions of open triangles: corundum-bearing pyroxenites (Kornprobst et al, 1982); large open diamond: calculated pre-exsolution composition of clinopyroxenes from graphitic garnet pyroxenites. Dotted field defines the range of clinopyroxene compositions from garnet pyroxenite xenoliths in alkali basalts (from Wilkinson, 1976); eclogite and grospydite fields are from Sobolev (1979).

analyses which establish that there is elemental and isotopic exchange between the peridotite host rock and the margins of the pyroxenites layers.

Clinopyroxenes and garnet inclusions within the graphite octahedra show the same compositional variation as those of the host rocks. The slow emplacement of the peridotite massif has allowed re-equilibration of inclusion compositions (e.g., exsolution in the pyroxenes). Garnet inclusions analyzed by Slodkevich (1981) have compositions very similar to garnet core composi-tions in the host rocks. This compositional similarity is also evident in terms of the rare earth element content of the inclusion and host rock silicates (Pear-son, 1989). Mg-rich picroilmenite was identified by Slodkevich (1980) as an inclusion phase within the graphite octahedra. Ilmenite is a relatively rare inclusion in diamond (Meyer, 1987); however, those that have been analyzed are compositionally similar to the Beni Bousera picroilmenites. The similarity of both major and trace element compositions of garnet and clinopyroxene inclusions in the graphitized diamonds to those of the host pyroxenites, and the faceted nature of some of the inclusions indicate that the original dia-monds were not xenocrysts, but probably crystallized from the pyroxenite melt along with the silicates.

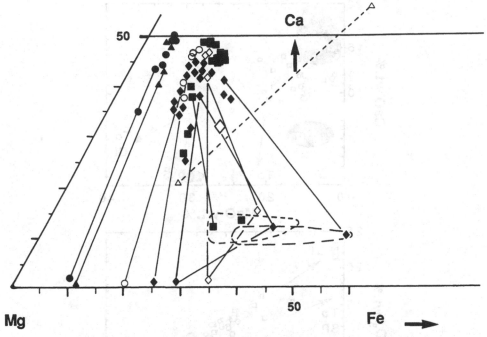

Figure 13.9. Ca-Mg-Fe diagram of clinopyroxene, orthopyroxene and garnet variation in Beni Bousera peridotites and pyroxenites (symbols as in Figures 13.6 and 13.8) and Ronda graphitic pyroxenites: open diamonds. Short-dashed field outlines garnet compositional variation of Beni Bousera nongraphitic pyroxenites, long-dashed field represents for garnet compositions in graphitic pyroxenites. Note the greater Mg and Ca content of the garnet from the corundum pyroxenite (from Kornprobst et al., 1982).

Petrogenesis

Large variations in compatible element concentrations, the correlation between elemental chemistry and modal mineralogy, together with symmetrical mineralogic zonation in some layers indicates that HP crystal–melt equilibrium was the dominant process causing their petrologic diversity (Kornprobst, 1969; Pearson et al., 1991a, 1993). Crystallization of the pyroxenites by segregation/plating onto magma conduit walls probably created the observed layers. Accumulation of the observed phases, i. e., orthopyroxene, clinopyroxene, and garnet from a picritic parental melt is capable of generating the major element variation displayed by the pyroxenite suite (Fig. 13.10). Fractional crystallization processes also best explain the factor of 10 variation in compatible element abundances within the pyroxenites, e.g., Ni contents of 50–3000 ppm (Pearson et al., 1993) and the variation in Cr/Ni (2–8). However, the low Sr contents of clinopyroxenes from the graphitic garnet pyroxenites imply crystallization from liquids with only ~10 ppm Sr. Such low incompat-

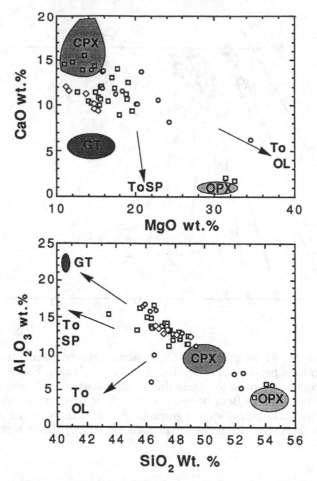

Figure 13.10. CaO versus MgO and Al_2O_3 versus SiO_2 wt% plots of Beni Bousera whole rock pyroxenite compositions. Shaded fields show the range of porphyroclast variation in constituent minerals. Open circles: websterites; open squares: garnet pyroxenites; open diamonds: graphitic garnet pyroxenites.

ble element abundances cannot be produced from melts of the peridotites by fractional crystallization of the observed phases. This evidence, together with the lack of correlation between highly incompatible elements (e.g., Zr) and fractionation indices (such as magnesium number, Mg#) within lithologic units or for the suite as a whole, indicates that the pyroxenite suite was not derived from genetically related magmas and that the parental magmas could not have been derived from the host peridotites as proposed by Kornprobst (1969). Rare earth element (REE) patterns of the bulk rock pyroxenites show variable enrichment in heavy REE (HREE), due to variable modal abundance of garnet. All samples are light REE (LREE)-depleted relative to chondrites: $(Ce/Sm)_n$ 0.91–0.016. The graphitic pyroxenites are extremely LREE-depleted,

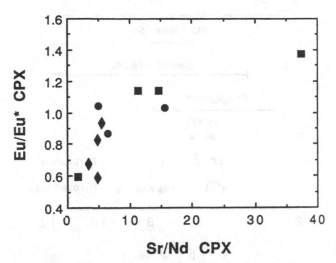

Figure 13.11. Eu/Eu* versus Sr ppm for clinopyroxenes from Beni Bousera pyroxenites. Symbols as in Figure 13.10.

$(Ce/Sm)_n$ 0.374–0.016. REE patterns for whole rocks and minerals show positive and negative Eu anomalies (Eu/Eu* 0.54–3.54), and there is a positive correlation between Eu/Eu* and Sr content for wholerocks or clinopyroxenes (Fig. 13.11). The correlation is best within lithologic groups and indicates the involvement of plagioclase at some stage in the pyroxenite genesis. However, the presence of graphitized diamonds in the graphitic pyroxenites and the absence of primary plagioclase in the pyroxenites demonstrate crystallization at pressures within the diamond stability field, i. e., outside the range of plagioclase stability. This dichotomy may be resolved if the pyroxenites crystallized at high pressure from melts of subducted oceanic crustal precursors that experienced plagioclase fractionation prior to subduction (Pearson et al., 1991a; Pearson et al., 1993). Derivation of the pyroxenites from crustal precursors is supported by their oxygen isotope compositions. $\delta^{18}O$ values of hand-picked, acid washed clinopyroxenes vary from + 4.9 to + 9.3‰ vs. SMOW. This contrasts with the much more restricted range of the host peridotites (+ 5.3 to + 6.1‰, Fig. 13.12) which are within the range typical of mantle (Kyser, 1990). Pearson et al. (1991a) showed that mineral pairs in the pyroxenites were in high temperature oxygen isotope equilibrium. Intermineral oxygen isotopic equilibrium and the lack of correlation between $\delta^{18}O$ and fractionation indices such as Mg#, rule out generation of the pyroxenite oxygen isotope variation solely within the mantle as a result crystal fractionation. Later metasomatism/alteration in the crust is also unlikely to preserve isotopic equilibrium. The range of $\delta^{18}O$ values in the pyroxenites is within that found in sections of altered oceanic lithosphere now exposed as ophiolite

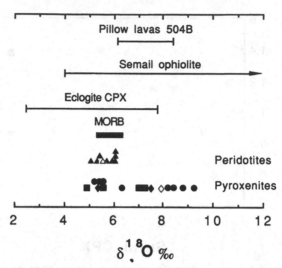

Figure 13.12. Oxygen isotope variation (in units per mil vs. SMOW) of clinopyroxenes from Beni Bousera and Ronda pyroxenites and peridotites. Open diamond: clinopyroxene from Ronda graphitic pyroxenites; open triangle: clinopyroxene from Beni Bousera garnet lherzolite; closed triangles: clinopyroxenes from Beni Bousera and Ronda spinel lherzolites; closed circles: clinopyroxenes from Beni Bousera websterites; closed squares: clinopyroxenes from Beni Bousera garnet pyroxenites; closed diamonds: clinopyroxenes from Beni Bousera graphitic pyroxenites. Data from Pearson et al. (1991a) and Javoy (1980). The MORB range is from Ito et al (1987); the eclogite clinopyroxene range from summary by Kyser (1990); Semail ophiolite range from Gregory and Taylor (1981); the pillow lava range, from Hole 504B, from Alt et al. (1987).

massifs (Gregory and Taylor, 1980; Schiffmann et al., 1991). The $\delta^{18}O$ enriched values of some of the pyroxenite minerals are characteristic of the upper sections of oceanic lithosphere that have undergone low-temperature seawater alteration. Mafic portions of the such crust altered at epidote-actinolite facies have $\delta^{18}O$ values between 6.0–10.6‰.

Sulfides included in clinopyroxene from two pyroxenites exhibit considerable sulphur isotope heterogeneity ($\delta^{34}S_{PDB}$ = 4.1–13.1‰). The latter value is considerably heavier than the range for MORB or OIB (0–5 ‰) and consistent with interaction with sulfur derived from either seawater ($\delta^{34}S$ ~20‰) or sediment-derived sulfate in equilibrium with seawater (Pearson et al., 1993). The upper portions of oceanic crust are hydrous as a result of seawater alteration and, hence, have the lowest solidi; consequently, they are the most likely to melt during subduction. The oxygen and sulfur isotope data are thus consistent with the hypothesis that the pyroxenite suite crystallize from melts of different portions of hydrothermally altered, subducted oceanic crust. Garnet, orthopyroxene and clinopyroxene crystallizing from melts of the oceanic crust would produce the trace element variation and REE patterns observed in the pyroxenites.

Most of the pyroxenites from Beni Bousera show major and trace element geochemistry consistent with their formation via HP "cumulate" processes involving garnet fractionation, probably via crystal plating onto the walls of magma conduits in the mantle. However, Kornprobst et al. (1990) identified a group of garnet pyroxenites, including corundum-bearing pyroxenites that had markedly higher Mg/Fe than the main suite and REE patterns with positive Eu anomalies but no HREE enrichments. This group of pyroxenites (group II of Kornprobst et al., 1990) were interpreted as oceanic gabbros directly subducted back into the mantle and metamorphosed to a HP assemblage. An alternative interpretation of these rocks is that they originally crystallized as highly aluminous clinopyroxenites or corundum clinopyroxenites at high pressure (> 30 kbar) and subsequently exsolved most of their garnet upon cooling, producing relatively HREE-depleted garnet pyroxenites. Such an origin has been proposed for some eclogite xenoliths from kimberlites (Smyth et al., 1990). In order to explain their positive Eu anomalies, the parental melts to the Group II pyroxenites may also have been derived from subducted oceanic lithosphere.

Detailed radiogenic isotope studies of the pyroxenites have been carried out, concentrating exclusively on hand-picked mineral separates that were acid leached to remove the effects of secondary alteration (Pearson et al., 1993). Clinopyroxenes from the pyroxenites show extreme Nd-Sr (ε Nd = + 26 to –9, $^{87}Sr/^{86}Sr$ = 0.7025 – 0.7110, Fig. 13.13) and Pb ($^{206}Pb/^{204}Pb$ = 18.21– 19.90) isotope variability (Fig. 13.14) compared to the mid-oceanic ridge basalt (MORB) source mantle. Most of the Beni Bousera pyroxenites have $^{207}Pb/^{204}Pb$ and $^{208}Pb/^{204}Pb$ ratios considerably elevated above the Northern Hemisphere Reference line (NHRL) defined by Hart (1984) (Fig. 13.14), and hence have high $\Delta7/4$ and $\Delta8/4$ values (a measure of displacement above or below the NHRL; Pearson et al., 1993), indicating the presence of a component with high time-integrated U/Pb and Th/Pb in their source.

Sm/Nd, Rb/Sr, and U/Pb are uncorrelated with isotopic compositions, whereas the low Rb/Sr ratios of most of the pyroxenites do not permit their derivation from the host peridotites, even over the age of the Earth. Nd isotope model ages for the pyroxenites indicate that the most recent Sm/Nd fractionation probably took place less than 150 Ma and may have occurred during emplacement of the massif from the mantle into the crust. Incorporation of local continental crust during emplacement of the massif cannot explain the radiogenic $^{87}Sr/^{86}Sr$ yet low Rb/Sr of most samples or the very LREE-depleted

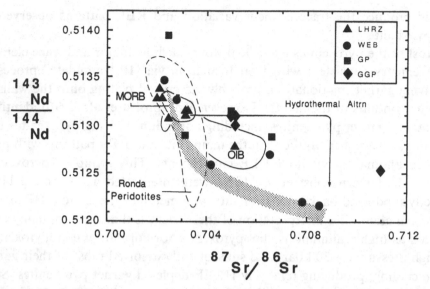

Figure 13.13. Sr-Nd isotope compositions of hand-picked, acid-washed clinopyroxenes from Beni Bousera pyroxenites and lherzolites. The MORB-OIB range from
Zindler and Hart (1986), the Ronda peridotite range is from Reisberg and Zindler
(1986), and Reisberg et al. (1989). The shaded field represents a family of mixing
curves between pyroxenite and surrounding kinzigite crust (see Pearson et al., 1993).

nature of the graphitic pyroxenites. Hydrothermal alteration of the crustal
precursors may be invoked to explain some of the Sr and ^{206}Pb/^{204}Pb isotope
variation. This process could also produce the elevated ^{207}Pb/^{204}Pb of some
samples if the alteration took place over 1.7 Ga. However, this process cannot
explain the unradiogenic Nd isotope compositions and high Δ8/4 values of
most of the pyroxenites. Hydrothermal alteration of oceanic crust results in
low Th/U which, with time, produces unradiogenic ^{208}Pb/^{204}Pb at a given ^{206}Pb/
^{204}Pb. Oceanic sediments have high Δ7/4 and Δ8/4 and low Sm/Nd and hence
^{143}Nd/^{144}Nd compared to mantle such that incorporation of subducted marine
sediment into a pyroxenite source consisting of variably altered oceanic crust
during melting would produce the observed radiogenic and stable isotope
characteristics of many of the pyroxenites (Pearson et al., 1993).

Incorporation of sediment rich in kerogenous organics into the source of
the pyroxenites may also explain the anomalously light carbon isotope systematics of the graphitized diamonds (Pearson et al., 1991b). The carbon isotope composition of this graphite is depleted in ^{13}C (δ^{13}C = −16 to −27.6‰,
Fig. 13.15) compared to "typical mantle" carbon exemplified by MORBs
(δ^{13}C = −4 to −8‰) and graphites from peridotite xenoliths (Pearson et al.,
1990). Disseminated graphite in the garnet sillimanite gneisses surrounding
the peridotite body and from graphite-mineralized veins that cross-cut the

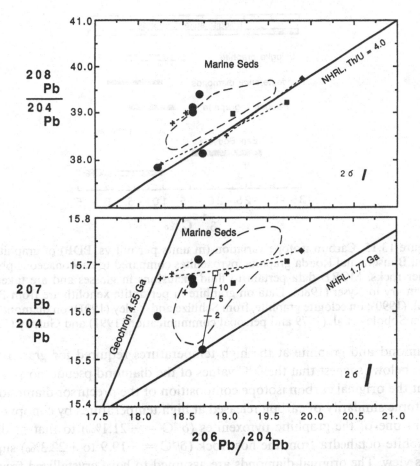

Figure 13.14. Pb isotope composition of hand-picked, acid-washed clinopyroxenes from Beni Bousera pyroxenites, dashed lines join pyroxene residues to leachate compositions (crosses). Mixing line shows effect of adding marine sediment to a pyroxenite composition lying on the NHRL reference line. Marine sediment data from Ben Othman et al. (1988) and White and Dupre (1986).

peridotites have ^{13}C-depleted isotope compositions which overlap the range for the graphitic pyroxenites (δ^{13}C $= -13.5$ to $-22.5‰$). The graphite veins in the peridotites appear to have precipitated from a CO_2-rich fluid phase originating from the graphitic gneisses during devolatization and migmatization associated with peridotite emplacement (Pearson et al., 1991b). Disseminated graphite is absent from the surrounding peridotites and graphite is not observed to cross-cut the pyroxenite/peridotite contacts. Thus, petrographic and field evidence argue against isotopic equilibration of the graphitized diamonds with a pervasive, late stage carbonaceous fluid phase such as that which deposited the graphite veins. The small fractionation factor between

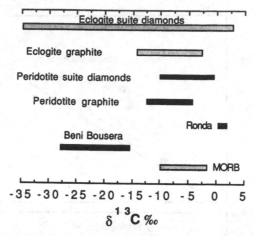

Figure 13.15. Carbon isotope variation (in units per mil vs. PDB) of graphite from Beni Bousera and Ronda graphitic pyroxenites compared to carbonaceous phases in other rocks. MORB data pertain to fluid inclusions in glasses and are taken from summary in Kyser (1986). Data on graphite in peridotite xenolith are from Pearson et al. (1990); on eclogite graphite, from Schulze and Valley (1991); on diamond range, from Sobolev et al. (1979 and personal communication, 1991) and Gurney (1991).

diamond and graphite at the high temperatures required for graphitization (see below) suggest that the $\delta^{13}C$ values of the diamond pseudomorphs represent the original carbon isotope composition of the precursor diamonds. The isotopic similarity of carbon released at high temperatures by clinopyroxenes from one of the graphitic pyroxenites ($\delta^{13}C = -21.1‰$) to that of discrete graphite octahedra from the host rock ($\delta^{13}C = -19.9$ to $-22.3‰$) supports this view. The original diamonds are assumed to have crystallized from subducted, ^{13}C-depleted, crustally derived carbon, in the form of kerogenous carbon-bearing sediments (Pearson et al., 1991b). Mantle-derived eclogite xenoliths and eclogite suite inclusions in diamond have a radiogenic and oxygen-carbon-sulphur isotopic diversity comparable to that of the Beni Bousera pyroxenites, with values for all systems indicative of interaction with crustal reservoirs. A subducted oceanic lithosphere source plus or minus organic-rich sediment has also been proposed as the origin of eclogite xenoliths and eclogite suite diamonds (e.g., Ringwood and Green, 1966; Sobolev et al., 1979; Jagoutz et al., 1984; MacGregor and Manton, 1986; Shervais et al., 1988; Neal et al., 1990; Smith et al., 1991). However, some authors have also stressed the evidence for igneous fractionation in many eclogite suites (e.g., Smyth et al., 1991). The current debate as to whether some eclogite xenolith suites represent subducted, metamorphosed oceanic crust or HP igneous products from mantle melts may be largely resolved if those eclogites with crustal isotopic signatures crystallized as crystal segregates from *melts* of subducted oceanic crust/lithosphere (Pearson et al., 1991b).

Having invoked a model involving melting of oceanic lithosphere, we must examine what physical parameters are required to do this. Melting of a subducted oceanic slab is controlled by the temperature increase of the descending slab which is primarily a function of age of the slab and speed of subduction whereas the temperature at which the oceanic crust begins to melt is dominantly a function of its water content. Thermal modeling of slab subduction indicates that melting of the upper portion of the descending oceanic slab may take place only in a scenario where a young slab is being slowly subducted, at depths of ~150 km, i. e., into the diamond stability field if most of its water content is removed by dehydration at shallow levels (Sacks and Kincaid, 1990). This is consistent with the presence of graphitized diamonds in the Beni Bousera pyroxenites, which, from geochemical constraints, appear to have crystallized from melts of oceanic crust (Pearson et al., 1991, 1993). Experiments indicate that melting of quartz eclogite at high pressures (>30 kbar) is capable of producing melts saturated in garnet, clinopyroxene, and orthopyroxene (Green, 1982). Accumulation of such phases, perhaps by crystal plating on conduit walls, is capable of producing the lithologic and geochemical variation observed in the Beni Bousera pyroxenites.

The Ronda Peridotite, Southern Spain

Tectonic Setting and General Geology

The Ronda peridotite massif is the largest and possibly the best studied peridotite massif in the world (Dickey, 1970; Obata, 1980; Frey at al., 1985; Reisberg and Zindler, 1987; Suen and Frey, 1987). Garnet-spinel-, spinel-, and plagioclase facies peridotites are all represented within the 300 km² of outcrop (Obata, 1980). As noted earlier, the Ronda and Beni Bousera massifs are tectonically related, and there is stratigraphic correspondence between the internal zones of the Betics and the Rif orogens. Thus, the Alpujarride sequence into which the Ronda peridotite was emplaced is stratigraphically equivalent to the Sebtide sequence surrounding the Beni Bousera massif (e.g., Chalouan and Michard, 1990; Weijermars, 1991): see Fig 13.2. The peridotites, predominantly lherzolites, have been shown to represent a series of residues from partial melting of a fertile garnet lherzolite (Frey et al., 1985). The peridotites consist of olivine, enstatite, diopside and an aluminous phase. Cr-Al spinel occurs throughout the massif, in the east and southeastern portions calcic

(An_{60-80}) plagioclase predominates, whereas pyrope-rich garnet is present in the northwest (Obata, 1980). Some peridotites have transitional features such as spinel surrounded by plagioclase. Much of the peridotite has a porphyroclastic texture. The degree of peridotite deformation (decreasing grain size, foliation, stretched porphyroclasts, etc.) increases toward the northwest contact. Mylonitic lenses are locally developed, especially at the peridotite–pyroxenite boundary. The peridotites show a west to east decrease in CaO, Al_2O_3, and REE content across the massif, which was interpreted by Frey et al. (1985) to be a gradient in degree of melting.

The pyroxenites may be divided into garnet-bearing (clinopyroxene + garnet ± orthopyroxene ± plagioclase ± quartz ± spinel) and spinel-bearing (clinopyroxene + olivine + plagioclase + orthopyroxene + spinel) varieties. The petrology of the Ronda massif has been described in detail by Dickey (1970), Obata (1980), and Suen and Frey (1987); hence the reader is referred to these publications for more information. The primary purpose of this section is to document the evidence for an UHP origin for the massif in the form of the newly discovered graphitic pyroxenites (Davies et al., 1992) and to compare them to the Beni Bousera occurrence.

Graphite-Bearing Pyroxenites

Diamond was reported from the Ronda massif in the last century (Calderon, 1890) although the validity of the find is doubtful. Obata (1980) first noted the occurrence of graphite in a loose block of pyroxenite from the Ronda massif. Following the discovery of graphitized diamonds at Beni Bousera (Slodkevich, 1980; Pearson et al., 1989), a renewed search was made at Ronda, leading to the discovery of graphite in two pyroxenite layers (Davies et al., 1992, in press). At one locality graphite occurs predominantly as coated octahedra in a quartz-bearing garnet clinopyroxenite within the ariegite subfacies (spinel + orthopyroxene + clinopyroxene + garnet) of the peridotites. Upon breaking the rocks, the coat often separates from the octahedral core to yield the upper half of a sharp-edged octahedron which protrudes (see Fig. 13.16). The layer is approximately 2 m thick but varies along strike. The graphite aggregates range in size from ~1 mm to over 20 mm, many occurring as distorted octahedra but also as irregular forms with no obvious crystal symmetry (Davies et al., 1992). SEM observations show that graphite {0001} is parallel to {111} of the octahedra, as in the Beni Bousera aggregates, indicating an origin similar to that of graphitized diamonds. Graphite concentrations are also comparable to those recorded in the Beni Bousera pyroxenites, locally up to 15 vol%.

In contrast to the graphitic pyroxenites from Beni Bousera, those at Ronda

Figure 13.16. Upper half of a graphite octahedron protruding from the surface of a Ronda pyroxenite.

are extensively recrystallized, with ubiquitous developments of symplectitic intergrowths of subsolidus reaction products. The most common reaction appears to have been:

garnet + clinopyroxene + quartz = orthopyroxene + plagioclase + spinel.

Secondary olivine appears to have been generated by a reaction of the type:

clinopyroxene + orthopyroxene + spinel = olivine + plagioclase.

Davies et al. (1992) note that the occurrence of graphitized diamonds in the pyroxenites implies that the free silica phase may have originally been coesite; however, the extensive textural reequilibration at high temperatures has obscured any unambiguous evidence of this. Quartz is documented as an accessory phase in other Ronda pyroxenites (Suen and Frey, 1987). Free silica may be dissolved in an eskolite component ($Ca_{0.5}AlSi_2O_6$) within clinopyroxene at pressures around 50 kbar at temperatures above 1400°C (Irifune and Ringwood, 1987). Lower pressure/temperature equilibration would cause exsolution of silica, as coesite, in a garnet pyroxenite. However, the large volume (>10%) of quartz in the Ronda rocks implies crystallization of coesite as a discrete phase. Coarse, relict clinopyroxene is the major porphyroclast phase in the pyroxenites and clinopyroxene is the dominant inclusion phase within the graphite. Mineral compositions in the graphitic pyroxenites fall within the range shown by Beni Bousera pyroxenites (Fig.13.9). Garnets are pyrope-almandines (Py_{47}-Al_{36}-Gr_{17}) and are markedly enriched in Fe at their margins. Clinopyroxenes are aluminous (up to 9.3 wt% Al_2O_3), high in CaTs but relatively low in Jd components (<20%), broadly resembling the Beni Bousera clinopyroxenes in composition (Davies et al., 1992). Vermiform spinel, 2–4 mm wide, commonly occurs as fine intergrowths with orthopyroxene-plagioclase symplectites but also forms Cr-rich inclusions in the symplectitic olivines. Rods of spinel up to 50 mm long also occur in clinopyroxenes which are hercynitic, Cr# ~9 (Cr# = [100Cr/(Cr + Al)].

Petrogenesis

The Ronda pyroxenite suite shows an equally diverse range of compositions as the Beni Bousera suite but with the addition of lower pressure, olivine gabbros which are thought to be retrogressed from HP pyroxenite parents (Suen and Frey, 1987). As at Beni Bousera, major and trace element data indicate that the pyroxenites do not have the geochemical characteristics of primary mantle melts. Suen and Frey (1987) showed that the majority of the pyroxenite suite can be interpreted as crystal segregates from HP (>19 kbar) melts migrating from the hot interior of a rising peridotite diapir to the cooler exterior. Although the low incompatible element concentrations of the Ronda pyroxenites are consistent with this hypothesis there is no correlation between different incompatible-elements or between incompatible elements and fractionation indices (Suen and Frey, 1987). The complex elemental variation observed could result from multistage crystal-melt fractionation, pyroxenite interaction with the peridotites, and subsolidus recrystallization (McDonough and Frey, 1990). Preliminary isotopic data indicate that much of the variability may also be due to crystallization from non-cogenetic melts as at Beni Bousera. The plagioclase-garnet-pyroxenite analyzed by Zindler et al. (1983) had Sr-Nd isotope systematics similar to the MORB reservoir and within the large isotopic range of the peridotites (Reisberg and Zindler, 1987; Reisberg et al., 1989). Although the radiogenic $^{187}Os/^{186}Os$ ratios of some of the pyroxenites (1.7–47.9) do not rule out the possibilty that pyroxenites are derived from melts of the host peridotite (Reisberg et al., 1991), the pyroxenites cannot be generated from the host peridotites during emplacement of the massif into the crust (within the last 100 Ma) as required by Suen and Frey (1987). The graphitic pyroxenites are more LREE-depleted than the peridotites and have negative Eu anomalies of a magnitude similar to that of the anomalies of the Beni Bousera graphitic pyroxenites.

Preliminary oxygen isotope data (Davies et al., in press) indicate that the graphitic pyroxenites are enriched in ^{18}O ($\delta^{18}O$ of clinopyroxene and garnet = 8.0). As at Beni Bousera, such oxygen isotope compositions preclude derivation of the graphitic pyroxenites from the host peridotites and may indicate derivation from melts of subducted oceanic lithosphere. The isotopically heavy C isotope compositions of the graphite from the Ronda graphitized diamonds ($\delta^{13}C$ = + 1‰) is outside the range for "typical mantle" carbon exemplified by MORB and OIB (Fig. 13.15) and may be a product of subducting sea floor carbonate into the mantle, possibly in the form of hydrothermal calcite veins in the oceanic crust. Additionally, the presence of graphitized diamonds in a pyroxenite layer requires that some of the pyroxenites crystallized in the diamond stability field (~50 kbar).

Occurrences of Diamondiferous Ophiolites and Orogenic Peridotites

Introduction

There are many occurrences of alluvial diamonds in orogenic belts worldwide where no known volcanic sources exist (e.g., western United States; Alaska; eastern United States; New South Wales, Tasmania and Victoria, eastern Australia; southeast Kalimantan and the Urals of Russia). Mantle-derived peridotites occur in many of these orogenic belts in the form of ophiolites and orogenic peridotites massifs, providing a speculative link between the diamonds and the orogenic ultramafic rocks (e.g., Pavlenko et al., 1974; Dawson, 1983; Pearson et al., 1989; Dawson and Carswell, 1990; Nixon et al., 1991; Janse, 1991; Nixon et al., 1986). This association has been frequently confirmed by prospecting. For example, using mineral chemical data, Kaminskii and Vaganov (1977) calculated equilibration conditions of 1170–1300°C and 30–65 kbars for ophiolitic peridotites from the Koryak Mountains in Kamchatka, Russia. From this they predicted the occurrence of diamonds which were subsequently found in a lherzolite and a harzburgite during prospecting activities initiated on this basis (Shilo et al., 1978).

Well-documented occurrences of diamond in orogenic ultramafic rocks are those from Alpine-type lherzolite-harzburgite massifs in the Tulameen district, British Columbia (Camsell, 1911); the Little Caucasus, Russia (Pavlenko et al., 1974); the Koryak Mountains, Russia (Shilo et al., 1978); and from ophiolites in Tibet (Fang and Bai, 1981; Bai et al., 1993). Diamonds have also been found in serpentinized dunite/peridotite "pods" of uncertain genesis, from Burkina Faso (Haut et al, 1984), on the West African craton. Of these occurrences the Tibetan ophiolites are the best described. In most of the cases cited above the diamonds were recovered from heavy mineral concentrates after processing, rather than being observed in situ within the host rock. It is possible that the very low diamond concentrations reported in some cases were the result of contamination during processing, but most studies undertook stringent precautions to prevent this. The increasing number of such occurrences and their profound implications for the physical processes operating during collisional orogenesis warrant discussion in a review of this scope.

Diamondiferous Tibetan Ophiolites

Exploitation of podiform chromites in ophiolitic peridotites has yielded over 100 diamonds from the Loubusa massif, of the Yarlungzangbo ophiolite belt in southern Tibet, and the Donqiao massif, of the Bongong-Nujiang ophiolite belt in northern Tibet (Fang and Bai, 1981; Bai et al., 1993). The two massifs form parts of ophiolite belts which occur along suture zones that separate the Indian subcontinent from the Lhasa block, in the south, and the Lhasa block from the Tanggula block, in the north (Girardeau and Mercier, 1988).

The Luobusa massif is probably of Cretaceous age and consists of tectonized spinel-bearing peridotites, mainly dunite and harzburgite, with minor lherzolites (Bai et al., 1993). The peridotites are separated tectonically from cumulate dunites and pyroxenites of probable crustal origin. Podiform chromite deposits are associated with the harzburgite/dunite sections. Dunitic lenses within chromite-bearing harzburgites commonly enclose stratiform chromitite bodies.

The Donqiao massif, probably of Upper Jurassic age, consists predominantly of tectonite spinel-harzburgite with small dunite pods associated with chromite deposits (Bai et al, 1993). The chromitite bodies are 200 m long, 30 m wide and 50 m thick. Chromites from this massif and the Luobusa massif are mainly of the high-Cr magnesiochromite variety (Bai and Zhou, 1988) with Cr# >60 and Cr_2O_3 contents reaching up to 60 wt%. Diamonds were first identified in the heavy mineral fraction of chromites and then confirmed to be from the chromite-bearing peridotite massifs by Fang and Bai (1981).

Diamonds recovered from heavy mineral concentrates of the chromitites and chromite-bearing harzburgites of both massifs, identified by x-ray analysis, are commonly 0.1–0.2 mm in diameter, but some grains reach 0.5 mm (Fang and Bai, 1981). Six diamonds recovered from the chromitites and harzburgites by crushing and processing samples over 1000 kg are illustrated in Fig 13.17 (from Bai et al., 1993). Although no diamonds have been found in situ yet, it should be noted that no diamonds were found in meticulous searches of equivalently sized control samples. Additionally, the presence of

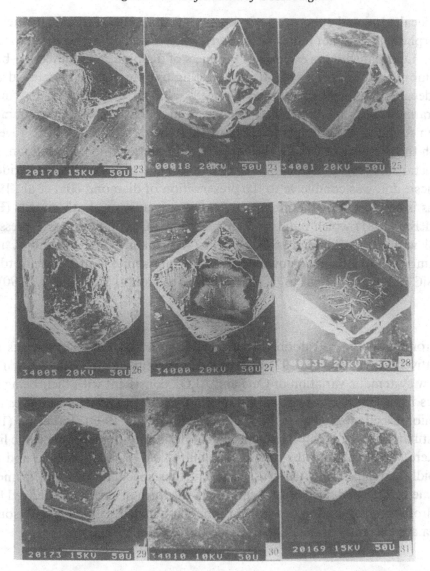

Figure 13.17. Secondary electron images of microdiamonds from heavy mineral concentrates of peridotites from the Luobusa and Donqiao massifs. (Reproduced by permission of Dr Mei-Fou Zhou; see Bai et al., 1993)

diamonds in alluvials on the southern side of the Donqiao massif argues against the view that contamination is the source of the diamonds found by crushing the peridotites and chromitites. Diamonds from the Luobusa massif are transparent, whereas those from the Donqiao massif are yellow-green. The diamonds show little signs of chemical or physical corrosion, in contrast to those obtained from kimberlites (e.g., Robinson, 1978). The morphology of the diamonds range from sharp-edged octahedra to what appear to be dodecahedral and cubo-octahedral growth forms (Fig. 13.17). The latter two morphologies are not represented as growth forms in diamonds from kimberlites. Dodecahedra are exclusively products of dissolution in kimberlite-borne diamonds (Robinson, 1978). It is interesting to note that cubo-octahedral to dodecahedral forms are common in the alluvial diamonds found in the ultramafic belts of Armenia (Pavlenko et al., 1974). Native chromium has been found in the ultramafic rocks of the Donqiao massif, which, together with the presence of Cr^{2+} in chromites (Cr^{2+} in tetrahedral coordination, Fe^{2+} in octahedral coordination), indicates that low fO_2 prevailed during peridotite genesis and was conducive to the preservation of diamond (Bai et al., 1993). This is supported by the presence of SiC (moissanite) in both massifs (Fang and Bai, 1981) which indicates extremely low fO_2 conditions and pressures well within the diamond stability field (Woermann and Rosenhuaer, 1984; Leung, 1990). It is not known whether the moissanite is of the well-ordered β-SiC polytype thought to be characteristic of natural SiC (Leung, 1990).

Petrogenesis

Chromite-bearing peridotites from the Luobusa and Donqiao massifs have relatively high Mg# (90–92) and low Al_2O_3 contents (0.25–0.97 wt%) and show systematic variation of CaO and Al_2O_3, consistent with their being residues from large degrees of partial melting, similar to some peridotites from ophiolite belts (e.g., Papua New Guinea; Bai et al., 1993). Bai et al. (1993) postulate that the highly depleted Tibetan peridotites represent oceanic lithosphere subducted into the diamond stability field, partially melted and then rapidly emplaced into the crust. They further speculate that the diamonds formed from subducted crustal carbon, in the form of carbonate derived from hydrothermal alteration or from carbon in subducted sediments. No isotopic data are currently available to examine this possibility.

Thermobarometry

The application of mineral geothermobarometers to ultramafic rocks in orogenic peridotites is not straightforward. All the lithologies contain polymetamorphic mineral assemblages that have suffered multistage equilibration during, and in some cases after, emplacement into the crust (Medaris and Carswell, 1990). The constituent minerals are frequently compositionally zoned and exsolved on various scales such that identification of equilibrium mineral assemblages is difficult. Additionally, the interdependence of mineral thermometers and barometers combined with differences in diffusion rates for elements employed in the various formulations may yield spurious results. A good example is provided by the application of Fe-Mg exchange thermometers together with a barometer based on the alumina content of orthopyroxene to garnet peridotites of the Western Gneiss Region of Norway. Temperature and pressure estimates for these rocks vary from 800°C, 30 kbars to 1000°C, 55 kbar. Medaris and Carswell (1990) proposed that the equilibration pressures of up to 55 kbar are artificially high owing to an isobaric heating event following initial equilibration at 600–700°C and 15–25 kbars which caused reequilibration and homogenization of stage II olivine, orthopyroxene, clinopyroxene and garnet in terms of Mg-Fe exchange but left the more slowly diffusing Al species in the cores of orthopyroxene porphyroclasts unequilibrated. This suggestion is supported by the much slower diffusion of Al than Mg and Fe in pyroxenes (e.g., Smith and Barron, 1991)

All the Beni Bousera and Ronda lithologies are extensively reequilibrated and exsolved such that it is not possible to obtain a direct estimate of the original crystallization conditions via mineral thermobarometry. Pressure estimation in spinel facies lherzolites is very unreliable; Al_2O_3 isopleths for enstatite in the CMAS system have gentle, negative dT/dP slopes which become steeper only in the garnet stability field. Furthermore, the essentially biminerallic nature of most garnet clinopyroxenites prevents any estimation of equilibration pressure using mineral barometry. Probably the best indication of the crystallization pressures is the presence of graphitized diamonds in some pyroxenites. The fact that the graphite aggregates contain inclusions of clinopyroxene and garnet with cubo-octahedral facets indicates that diamond crystallization was synchronous with the host pyroxenite. This implication will be used later, in conjuction with equilibration temperatures to estimate likely temperature–pressure conditions for the crystallization of the graphitic pyroxenites.

It is theoretically possible to use both the two-pyroxene solvus and garnet-clinopyroxene Fe^{2+}-Mg exchange thermometers to obtain estimates of equilibration temperatures of olivine-free pyroxenite assemblages. The two-pyroxene thermometer has been used extensively for thermometry of high-temperature garnet lherzolite xenoliths from kimberlites (Finnerty and Boyd, 1984; Carswell and Gibb, 1987); however, below 900°C the diopside limb of the solvus becomes insensitive to temperature, limiting its use at low temperatures. The lack of primary orthopyroxene coexisting with garnet in many pyroxenite assemblages prevents application of this barometer here. Additionally, although recent, high-quality experimentally reversed data are available for the two-pyroxene solvus in the CMAS system (Bertrand and Mercier, 1985) and for natural peridotite compositions (Brey et al., 1989), it is doubtful whether such formulations can be applied to the olivine-free pyroxenite assemblages which have much more aluminous pyroxenes with lower Cr contents (see Carswell and Harley, 1990). Because of these reservations we have chosen to apply the garnet-clinopyroxene Fe-Mg exchange formulation of Powell (1985) to obtain temperature estimates from the pyroxenites. This thermometer appears to yield the most consistent results for eclogite facies assemblages between 1000° and 1200°C compared to either the Ellis and Green (1979) or Krogh (1988) formulations (Carswell and Harley, 1990). Below 1000°C the formulation of Krogh (1988) is probably superior, whereas above 1200°C the Ellis and Green (1979) formulation appears to give more accurate results. The differences between these formulations are slight and probably within the errors induced from lack of knowledge of Fe^{2+}/Fe^{3+} in the minerals (Luth et al., 1989). All temperatures calculated in this review assume $Fe^{2+} =$ Fe total, and as such should be viewed as maxima.

Equilibration Pressures

The presence of graphitized diamonds at both Beni Bousera and Ronda indicate that the graphitic pyroxenites crystallized in their peridotite hosts at pressures of ~50 kbar. This is supported by the detection of very small but significant amounts of Na in garnets (0.05 wt%) from the graphitic pyroxenites (Pearson et al., 1989). Comparable or much higher Na values have been recorded in eclogitic garnet inclusions in diamond (Sobolev and Lavrentev, 1971; Moore et al., 1991) and are thought to be due to pyroxene solid solution in the garnet structure which becomes significant at ~50 kbar (Chopin and Sobolev, Chapter 3, this volume). The pyroxenites and peridotites in both massifs have clearly undergone extensive equilibration during their emplacement into the crust such that it is not possible to estimate their original depth of derivation from mineral thermobarometry.

The presence of both spinel- and garnet-facies lherzolites at Beni Bousera and Ronda implies re-equilibration of both massifs close to the garnet-spinel transition. The transition between spinel and garnet lherzolite as represented by the univariant reaction in the CMAS system that is,

$$2Mg_2Si_2O_6 + MgAl_2O_4 = Mg_2SiO_4 + Mg_3Al_2Si_3O_{12} \qquad (R1)$$
$$\text{Opx} \qquad\quad \text{Spinel} \qquad\quad \text{Olivine} \qquad \text{Garnet}$$

is a potentially useful geobarometer (O'Neill, 1981) and may be used to constrain equilibration pressures. The garnet-bearing peridotites at both Ronda and Beni Bousera are five-phase garnet-spinel lherzolites that occur in highly strained areas of both massifs where abundant evidence exists for mixing

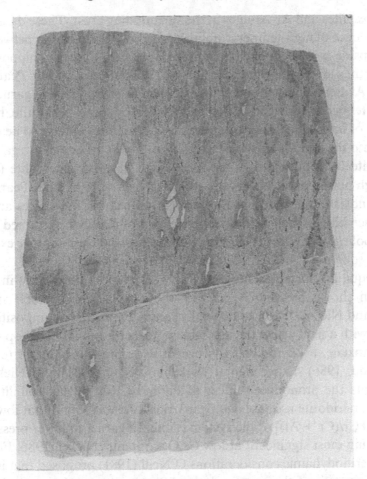

Figure 13.18. Photograph of a thin section (7.5 by 5 cm) showing highly sheared pyroxenite layers with remnant garnet porphyroclasts in a matrix of spinel lherzolite. Garnet porphyroclasts are surrounded by rims of kelyphite.

between the host peridotites and tectonically disrupted pyroxenite layers (Fig. 13.18). This mode of occurrence has led to conflicting opinions on the origin of the garnet-bearing peridotites. Kornprobst (1966) and Schubert (1982) have proposed that garnet in the Beni Bousera and Ronda peridotites is derived directly from the pyroxenites by tectonic disruption of pyroxenite layers and as such the rocks represent disequilibrium assemblages. Obata (1980, 1982) maintains that the garnet-spinel lherzolites at Ronda represent equilibrium assemblages. Thus, the garnet-spinel peridotites are largely due to preferential cooling of the exterior of a diapir during emplacement of the massif into the crust, allowing preservation of garnet-facies peridotites. Isochronous relationships for Nd isotopes defined by diopside and garnet in the garnet lherzolites from Ronda give the same cooling ages as mineral isochrons for the pyroxenites and Ar-Ar ages in the surrounding crust. These data, together with evidence of high-temperature oxygen isotope equilibrium between garnet and diopside in garnet-spinel lherzolites from Beni Bousera, support the theory that they are equilibrium assemblages (Reisberg et al., 1989; Pearson et al., 1991a). At Beni Bousera and to a lesser extent at Ronda, the garnet-bearing peridotites occur in patches, often surrounded by spinel peridotite, but always in areas of intense shearing where garnet pyroxenites appear to be physically mixed into the peridotites on all scales (Fig. 13.18). In one case a garnet clinopyroxenite layer bifurcates to enclose a pod of garnet lherzolite (Fig.13.3). Although petrographic observations and oxygen isotope data (Pearson et al., 1991a) indicate that the Beni Bousera garnet-bearing lherzolites are equilibrium assemblages, their field occurrence cannot readily be ascribed to preferential cooling of the garnet-bearing peridotites, and thus another explanation must be sought.

The equilibrium represented by R1, which marks the garnet-in reaction, has been shown by MacGregor (1970), O'Neill (1981), Webb and Wood (1986), and Nickel (1986), to depend strongly on the bulk composition of the system, with a field where the assemblage garnet + spinel + clinopyroxene + orthopyroxene + olivine coexists over a wide pressure interval (e.g., Webb and Wood, 1986). Experimental work by O'Neill (1981) and Nickel (1986) documents the pronounced effect of Fe and Cr_2O_3 on the stability field of garnet in peridotite assemblages. In particular it was found that lowering the Cr# [100 Cr/(Cr + Al)] of the system stabilized garnet to lower pressures, this effect being most significant at low Cr_2O_3 contents (Nickel, 1986; Fig. 13.19). From thermodynamic considerations O'Neill (1981) proposed that increasing the Fe^{2+} content of the system also stabilized garnet to lower pressures, but this has not been calibrated experimentally. The presence of Fe^{3+} will expand the spinel stability field (MacGregor, 1970); however, the relatively low levels of Fe^{3+} in mantle assemblages should minimize this effect such that increasing the Fe content of the system should depress the garnet-in reaction by up to 2 kbar in natural assemblages (O'Neill, 1981; Harley and Carswell, 1990).

The field relations of garnet-bearing peridotites at Beni Bousera suggest that they formed by mixing peridotites with pyroxenite layers, followed by high temperature equilibration to give an equilibrium five-phase spinel-garnet lherzolite assemblage. Introduction of a typical pyroxenite into a spinel lherzolite by tectonic disruption of pyroxenite layers would decrease Cr_2O_3, and increase Al_2O_3 and FeO in the bulk system, thus decreasing the Cr# and Mg# (Fig.13.20). Both of these chemical changes favor garnet stability. The relatively low Cr_2O_3 contents and hence low Cr# of the Beni Bousera peridotites act to enhance the effect of changing the bulk composition of the system on

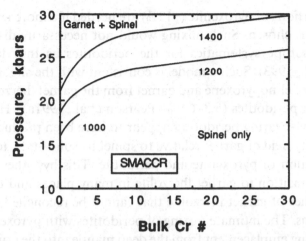

Figure 13.19. Pressure versus bulk rock Cr# = (100Cr/Cr + Al) showing the "garnet-in" reaction boundary as a function of temperature in the system SiO_2-MgO-Al_2O_3-CaO-Cr_2O_3 (SMACCR). (Modified after Nickel, 1986)

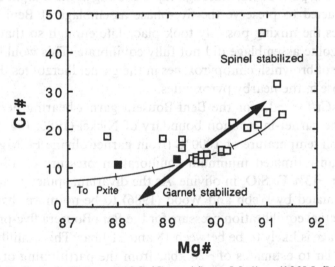

Figure 13.20. Bulk rock Cr# = (100Cr/Cr + Al) vs. Mg# = (100Mg/Mg + Fe) for Beni Bousera peridotites. Mixing line shows the effect of isothermal, isobaric mixing between a typical garnet pyroxenite and spinel lherzolite; tick marks are 5% increments. Garnet is stabilized at low Cr# and low Mg#. Open squares: spinel peridotites; closed squares: garnet-spinel lherzolites.

garnet stability. The bulk compositions of two spinel-garnet lherzolites from Beni Bousera (Pearson, 1989) plot in the most "fertile" region of Fig. 13.20 and thus most strongly favor garnet stabilization. Such compositions may be produced by ~25–30% addition of a typical garnet pyroxenite to a residual spinel lherzolite with Cr# ~24 and Mg# ~91 (Pearson, 1989; Fig. 13.20).

Lower proportions of pyroxenite (~12%) are needed if a more fertile peridotite end member is chosen. Such mixing would not necessarily disrupt the trace element or isotope systematics for the peridotites as they largely overlap (Pearson et al., 1993). Such a model is consistent with the slightly higher $\delta^{18}O$ value (6.1‰) of clinopyroxene and garnet from the garnet lherzolite compared with the other peridotites (5.3–6.0‰; Pearson et al., 1991a) . Hence, the Beni Bousera garnet-bearing peridotites appear to have been produced by increasing the stability field of garnet relative to spinel by solid-state, tectonic mixing and equilibration of pyroxenite and peridotite. This hypothesis explains the very sharp transition to garnet lherzolite in many places and the occurrence of small patches of garnet lherzolite that cannot be reconciled with changing *P-T* conditions. The intimate mixing of peridotites with pyroxenites probably occurred during emplacement from the deep mantle into the crust. This chemically enhanced garnet peridotite stability contrasts with the dynamic cooling model proposed by Obata (1980, 1982) for "preservation" of the Ronda garnet-bearing peridotites; although rapid cooling at the margin of the massif may have acted to preserve the five-phase assemblage at Beni Bousera. In some places the mixing possibly took place late enough so that the garnet-spinel lherzolite assemblage did not fully equilibrate. This would explain the occurrence of brownish clinopyroxenes in the garnet lherzolites that are identical to those in the nearby pyroxenites.

Using a Cr# of ~12 for the Beni Bousera garnet-bearing peridotites and applying the garnet-in reaction boundary of Nickel (1986; Fig. 13.19) at an equilibration temperature of 1000°C (from garnet-olivine Fe-Mg thermometry) gives an estimated minimum equilibration pressure of 18 kbar, if we account for ~13% Fe_2SiO_4 in olivine. As the divariant spinel-garnet lherzolite field is estimated by Webb and Wood (1986) to be no more than 2–3 kbars wide, the actual equilibration pressure for the Beni Bousera five-phase lherzolite assemblage is likely to be between 18 and 21 kbar. This equilibration pressure is similar to estimates of ~20 kbar from the partitioning of Cr between garnet and spinel in pyroxenites, and 20–22 kbar estimated on the basis of the coexistence of garnet, clinopyroxene and corundum in some pyroxenites (Kornprobst et al., 1990). Analogous equilibration conditions (~1000°C, 21.7 kbars) have been estimated for the Ronda garnet-bearing peridotites by Medaris and Carswell (1990) using Fe-Mg garnet-olivine exchange thermometry and the Al-in orthopyroxene barometer of Nickel and Green (1985).

The equilibration pressures estimated above obviously represent subsolidus equilibria recorded by the response of mineral equilibria to polybaric cooling during the emplacement of these massifs into the continental crust. As such they indicate conditions that the peridotite bodies must have experienced at some time and thus place constraints on the *P-T-t* evolution of the massifs. Much higher pressures of origin are indicated by the presence of graphitized diamonds in both massifs which indicate their ultimate derivation from mantle at least 150 km deep.

Equilibration Temperatures

Compositions of garnets included in graphite, which have been isolated from other silicates, are similar to core compositions of garnets in the host rocks implying that the garnet cores may preserve compositions imposed during crystallization at 50 kbar. It is therefore possible that the combination of garnet core compositions with calculated pre-exsolution clinopyroxene compositions are indicative of the temperature conditions during initial crystallization, or at least close to solidus conditions in the graphitic pyroxenites. This is supported by observation of faceted garnet inclusions within the graphitized diamonds at Beni Bousera (Slodkevich, 1980) which suggests that garnet was a HP liquidus phase during crystallization of the graphitic pyroxenites. However, it should be noted that petrographical evidence and REE partitioning data between garnet and clinopyroxene (Pearson et al., 1993) indicate that *some* garnet in the pyroxenites probably exsolved from very aluminous liquidus clinopyroxene at subsolidus temperatures and pressures below 30 kbars. Given the above caveats it remains an interesting exercise to try to estimate the original crystallization conditions of the graphitic pyroxenites.

As precise pressures of crystallization cannot be estimated for the biminerallic graphitic pyroxenites from mineral equilibria, 50 kbar will be assumed to be a crystallization pressure at which diamond was a stable phase. Temperatures estimated using the garnet-clinopyroxene Fe-Mg exchange formulation of Powell (1985) change by approximately 50°C for every 10 kbars variation in the assumed pressure. Hence, 50°C may be taken as a minimum error estimate. Temperatures of 1450–1500°C are calculated if the pre-exsolution clinopyroxene compositions and garnet cores were in equilibrium at the time of diamond formation (Fig. 13.21). This temperature range is anomalously high in that it plots well above present day continental or oceanic geotherms at 50 kbar, but it is below the anhydrous peridotite solvus at 50 kbar (Fig. 13.22). No estimates of bulk clinopyroxene compositions for the Ronda graphitic pyroxenites are yet available, hindering estimation of pre-exsolution temperatures; however, the overall similarity of bulk compositions of the graphitic pyroxenites at Beni Bousera and Ronda (Davies et al., 1992) and the presence of graphitized diamonds in both lithologies supports a similar petrogenic history. The high temperatures apparently prevailing during diamond crystallization in the pyroxenites are consistent with the observation that octahedra are

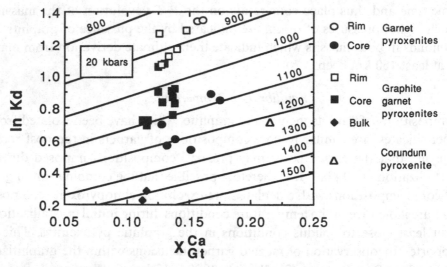

Figure 13.21. ln Kd versus X_{Ca} in garnet for garnet-clinopyroxene pairs in the Beni Bousera pyroxenites at 20 kbar. Isotherms were calculated using the formulation of Powell (1986).

the most common morphologic forms displayed by the graphitized diamonds at Beni Bousera (Pearson et al., 1989) and Ronda (Davies et al., 1992). Octahedra are the high temperature growth form of synthetically grown diamonds (Robinson, 1978). Such high, close-to-solidus temperature estimates for the graphitic pyroxenites are comparable to crystallization temperatures of ~1450°C deduced experimentally (at ~28 kbar) from non-graphitic Beni Bousera pyroxenites (Kornprobst, 1969, 1970). These temperature conditions are also similar to temperatures of 1550°C at 30 kbar (or 1450°C at 20 kbar) estimated by Morten and Obata (1983) using the pyroxene solvus thermometer on pre-exsolution compositions of clinopyroxene "megacrysts" from pyroxenites within a garnet peridotite body from northern Italy. These authors note that this may be a minimum temperature estimate owing to the effect of Al_2O_3 on the pyroxene solvus. Such high estimated temperatures indicate that some orogenic peridotite bodies may have been derived from anomalously hot mantle.

Coincidence of subsolidus equilibration pressures established from phase equilibria constraints on the corundum pyroxenites and garnet-spinel lherzolites leads to the conclusion that the Beni Bousera massif experienced significant re-equilibration at ~20 kbar during its pressure–temperature evolution. If we now assume that the garnet core compositions were reequilibrated with the core compositions of exsolved clinopyroxenes at 20 kbars, equilibration temperatures range from 938°–1205°C (Pearson, 1989; Fig. 13.21). This range

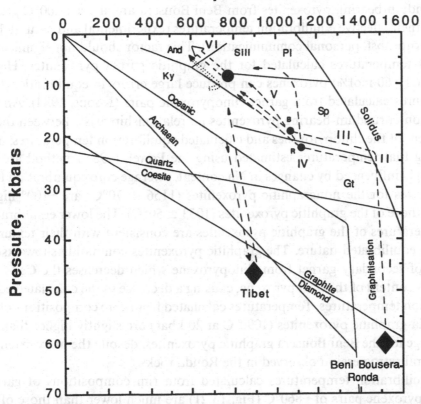

Figure 13.22. Pressure–temperature evolution (dashed lines) of the Beni Bousera (B) and Ronda (R) massifs as constrained by thermobarometry and a speculative trajectory for the Tibetan diamondiferous ophiolites that may apply to any orogenic diamondiferous peridotite body. Peridotite solidus constrained from experiments on natural compositions (McKenzie and Bickle, 1988), oceanic and Archaean geotherms after McKenzie and Bickle (1988). Quartz-coesite transition after Bohlen and Boettcher (1982). Dotted arrow indicates trajectory followed by surrounding crustal gneisses entrained during faulting (Kornprobst and Vielzuf, 1984). See text for detailed discussion.

is lower than that estimated by Kornprobst et al. (1990) for garnet-clinopyroxene pairs from the corundum-bearing pyroxenites at 20 kbar (1200–1350°C). The high apparent temperatures given by the corundum pyroxenites may be a function of their highly aluminous clinopyroxene compositions, which are significantly different from the compositions of the pyroxenes produced in the Ellis and Green (1979) experiments on which the Powell (1985) formulation is based. The effect of anomalously high alumina on equilibration temperatures calculated using formulations based on the experimental data of Ellis and Green (1979) is to produce considerable overestimates. The calculated equilibration temperatures for the aluminous, corundum-bearing pyroxenites from Beni Bousera are at least 200°C greater than their effective equilibration temperatures (experimental and natural data; J. Kornprobst, personal communication). This factor should not significantly affect temperatures calculated for the graphitic garnet pyroxenites. Highly jaditic (>60 mol%) pyroxenes can produce large errors in equilibration temperatures calculated from garnet-clinopyroxene pairs (Koons, 1984). Among the non-corundum-bearing pyroxenites no relationship exists between the Jd content of the clinopyroxenes and calculated equilibration temperatures, indicating that temperatures estimated using the Powell (1985) method are not strongly influenced by changes in Na content. Average core equilibration temperatures for the nongraphitic pyroxenites (1156 ± 70°C) are ~100° higher than those of the graphitic pyroxenites (1053 ± 50°C). The lower equilibration temperatures of the graphitic pyroxenites are consistent with their texturally more equilibrated nature. The graphitic pyroxenites commonly show exsolution of secondary garnet from clinopyroxene which decreases the CaTs and MgTs content of the host pyroxene, causing a decrease in the calculated equilibration temperatures. Temperatures calculated from core compositions of the Ronda graphitic pyroxenites (1090°C at 20 kbar) are slightly higher than the average for the Beni Bousera graphitic pyroxenites, despite the more extensive textural retrogression observed in the Ronda rocks.

Equilibration temperatures calculated from rim compositions of garnet-clinopyroxene pairs of ~860°C (Fig. 13.21) are much lower than those of the core compositions but comparable to temperatures obtained from exsolved garnet in equilibrium with immediately adjacent host clinopyroxene (850°C) at 20 kbar. This indicates significant re-equilibration of grain margins, probably in response to progressive cooling during uplift and emplacement. Garnet-clinopyroxene pairs from the margins of pyroxenite layers yield lower calculated temperatures than those from pairs in the center of layers (Pearson, 1989; Kornprobst et al., 1990). These differences are interpreted to result from the significantly higher stress levels experienced by layer margins during layer parallel shearing at the pyroxenite-peridotite interfaces. The accompanying grain size reduction acts to enhance diffusive re-equilibration.

Pressure – Temperature – Time Evolution and Emplacement Mechanisms

Indications from garnet-clinopyroxene thermometry using calculated, pre-exsolution clinopyroxene compositions and experimental petrology lead to the conclusion that the ultramafic rocks of the Beni Bousera massif were derived from at least 150 km deep, from an anomalously hot area of mantle

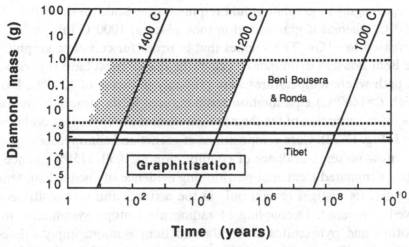

Figure 13.23. Time required to graphitize half the mass of a diamond octahedron by conversion on {110}, theoretically the fastest mechanism, at 20 kbar (modified after Eggler, 1989). Range for Beni Bousera and Ronda graphite aggregates is indicated between dots, range of diamond size reported from Tibetan ophiolites (Fang and Bai, 1981; Zhou et al, 1993) indicated between lines. Shading indicates range of equilibration temperatures demostrated from thermobarometry for Beni Bousera and Ronda.

with a high geothermal gradient (Pearson, 1989; Kornprobst et al., 1990). There is no geologic/tectonic evidence for the presence of a mantle plume in this region that may have been capable of supplying hot jets of rising mantle. Furthermore, if the pyroxenites are generated in a subduction zone environment (Pearson et al., 1991, 1993) then the cause of such high temperature conditions is unclear at the moment. Graphite is the stable, high temperature allotrope of carbon, and as such the transformation of diamond to graphite may be used to place constraints on the temperature conditions and possibly ascent rates experienced by the graphitic pyroxenites and hence the Beni Bousera massif. It can be shown (Davies and Evans, 1972) that the rate of graphitization per unit time can be expressed as:

$$dx/dt = C_{exp}^{-(\Delta E + P\Delta V)/RT}$$

where C is a constant, ΔV is the activation volume and ΔE is the activation energy. The low activation volume for graphitization ($\sim 10 cm^3 \cdot mol^{-1}$) means that the influence of pressure on the reaction is small. The rate of graphitization is thus largely dependent on temperature. If one uses the experimentally determined activation energy for graphitization of {110} faces (760 kJmol^{-1}; Davies and Evans, 1972) as a maximum indication of graphitization rate (the activation energy for {111} faces in 1060 kJ.mol^{-1}), it is apparent from Fig. 13.23 that conversion of the largest Beni Bousera diamond octahedra (~ 10

mm edge length) to graphite would require a time scale on the order of 1 Ma at 1200°C, whereas if graphitization took place at 1000°C this period would be increased to ~1Ga. This implies that in order for complete graphitization of the Beni Bousera diamonds to occur, the massif must have experienced an uplift path where temperatures were probably in excess of 1200°C. Evidence for high (>1400°C) equilibration temperatures in the graphitic pyroxenites comes from estimation of the their pre-exsolution clinopyroxene compositions (Stage I, Fig. 13.22). Core compositions of garnet and clinopyroxene porphyroclasts now preserve evidence of equilibration at 1050–1150°C; hence diffusion has eliminated elemental partitioning evidence of such a high temperature origin. Its vestiges remain only in the textures and compositions of the exsolved pyroxenes. Decoupling of radiogenic isotope systematics in both peridotites and pyroxenites at Ronda and Beni Bousera imply a late-stage melting event (Pearson et al., 1993; Reisberg et al., 1991), probably during adiabatic upwelling associated with emplacement. This means that both massifs crossed the (anhydrous) peridotite solidus at some time during their ascent into the crust. If upwelling occurred approximately parallel to calculated mantle adiabats (Fig. 13.22), and the pyroxenite bodies originated from >50 kbar at ~1400°C, the upwelling mass cannot have overstepped the solidus for very long as such high temperatures would have caused extensive melting (McKenzie and Bickle, 1988). The geochemistry of the peridotite at both Ronda and Beni Bousera indicate that they are predominantly residual after <20% melting (Frey et al., 1985; Pearson, 1989). Re-Os isotope systematics for the Ronda peridotites indicate that the melting event segregated them from the asthenosphere ~1.3 Ga and caused their main geochemical diversity (Reisberg et al., 1991). Therefore, melting associated with emplacement was probably minor.

At Beni Bousera the generation of five-phase garnet-spinel lherzolites occurred owing to tectonic mixing of peridotites and pyroxenites during uplift. Phase equilibria indicate that this assemblage equilibrated at ~18–20 kbar and 1000°C. Crystallization of the corundum-bearing pyroxenites at ~22 kbar and 1200°C suggests a sharp change in the *P-T* path following melting, with a period of almost isobaric cooling (Fig. 13.22, Stage III). This cooling could have taken place as result of incorporation of the peridotite body into a thickened continental lithosphere during the collision of Africa and Europe (e.g., Platt and Vissers, 1989). Whatever the uplift mechanism after this point, the Ronda massif appears to have undergone the dynamic cooling proposed by Obata to produce the garnet-, spinel-, and plagioclase facies peridotites observed (Stage V). Although small plagioclase neoblasts are observed in some Beni Bousera peridotites (J. Kornprobst, personal communication), there is no area where plagioclase forms the major aluminous phase in the peridotite. The less extensive development of plagioclase at Beni Bousera, in contrast to Ronda, is possibly due to the smaller size of the Beni Bousera massif which facilitated more rapid cooling. Finally, both massifs appear to have entrained 200–300 m of surrounding granulite facies crustal rocks (kinzigites) from about 10 kbars (Kornprobst and Vielzeuf, 1984) by movement along shear zones/faults at high crustal levels. Kyanite in some of these rocks was converted to sillimanite whereas cordierite was also produced by a variety of reactions and the rocks probably underwent the clockwise *P-T* path (Stage VI, Fig. 13.22) suggested by Kornprobst and Veilzeuf (1984). In the Ronda area the kinzigitic assemblages are less well developed, and there is more extensive low-pressure recrystallization to gneisses and schists.

Loomis (1972) and Obata (1980) proposed that the crustal metamorphism in the Ronda region was due to peridotite emplacement in the crust. However, Kornprobst and Vielzeuf (1984) conclude that the HP, kyanite-bearing metamorphic basement existed prior to peridotite emplacement and equilibrated under granulite facies conditions of ~10 kbar and 750°C during pre-Triassic continental collision.

The *P-T* path outlined for the ultramafic rocks above requires a tenable tectonic model that is able to emplace dense, HP rocks into the crust from 150 km deep. In the Beni Bousera case, original, pre-exsolution temperatures of 1400°C in the pyroxenites and evidence for late-stage melting associated with decompression indicates derivation from the asthenosphere. Mechanisms of emplacement involving lithospheric (crust + mantle) extension followed by collision, such as those proposed for the emplacement of deeply derived ultramafic massifs in the Alps, Pyrenees, Betico-Rif, and Galicia margin (Kornprobst and Vielzeuf, 1984; Nicolas, 1984; Boillot et al., 1988), appear to favor upwelling of the immediately underlying asthenosphere and its incorporation into the crust during collision. Unless the lithospheric mantle has been previously greatly thickened to ~150 km, it is unlikely that the mantle emplaced into the crust via these models will be from the diamond stability field. The same argument can be made against the model of Doblas and Oyarzun (1989).

Other models involving delamination of a tectonically thickened lithospheric mantle (Platt and Vissers, 1988) appear to shed the deepest portions of the lithospheric mantle, precluding their incorporation into the crust, but may induce counter flow of hot, convecting mantle which may get tectonically entrained and exhumed along shear zones reaching into the mantle.

An alternative mechanism of inducing upwelling of hot asthenospheric material is by detachment of a subducted oceanic slab such as that thought to have been present under the western Mediterranean during the Oligiocene

(Blanco and Spackman, 1993). This upwelling could have emplaced the peridotites directly from the asthenosphere (Davies et al., 1993). A more extensive review of possible tectonic scenarios for the emplacement of the Ronda and Beni Bousera massifs is given by van der Wal (1993).

Tibetan Diamondiferous Ophiolites

There is insufficient published mineralogic data for the Tibetan diamondiferous ophiolites to make useful thermobarometric calculations. However, several characteristics enable constraints to be placed on the likely *P-T* evolution of the diamondiferous ultramafics. The most pertinent observation is that small (<0.5 mm) diamonds in the Tibetan case and other occurrences such as Tulameen (British Columbia) and the Koryak mountains appear to survive tectonic emplacement into the crust apparently without significant graphitization. None of the diamonds from the Tibetan peridotites appear partially graphitized although this may be an artifact of the separation procedures employed during mineral processing. Some of the diverse shapes of the Tibetan diamonds (Fig. 13.17) may be due to partial graphitization. Fang and Bai (1981) and Bai et al. (1993) report the occurrence of graphite in concentrates from the peridotites, but it is not known whether this is a result of graphitization of diamond or graphite precipitation during serpentinization (e.g., Pasteris, 1988). Assuming that the present size of the Tibetan diamonds is their original size, their ungraphitized nature suggests that the host peridotites did not reside outside the diamond stability field at temperatures near 1200°C for an extended period. Fig. 13.23 indicates that diamonds of the size found in the Tibetan peridotites (and at other localities) would graphitize on the order of 10,000 years at 1200°C. Only at temperatures of ~1000°C could such small diamonds remain ungraphitized for the several million of years potentially required to emplace them into the crust from the deep mantle.

As the peridotite sections of ophiolites are generally derived from pressures <20 kbar a mechanism is required to produce pressure–temperature conditions conducive to the formation of diamonds. The surrounding crustal units show no sign of prograde ultrahigh pressure metamorphism (Bai et al., 1993), hence it seems likely that during plate collision, the oceanic lithosphere was temporarily subducted into the diamond stability fields where diamonds formed from subducted carbon (Bai et al., 1993). Physical modeling of the thermal evolution of a cool, old oceanic slab being rapidly subducted beneath Asia, prior to collision with India, is such that the basal portion of the subducted oceanic lithosphere may attain temperatures only of 850–900°C when well into the diamond stability field (Sacks and Kincaid, 1990). These temperatures are sufficient for diamond formation on geological time scales but insufficient to induce graphitization during uplift if emplacement back into the crust is achieved by a relatively fast mechanism such as movement along deep mantle shear zones related to the closed subduction zone. A suggested *P-T* trajectory consistent with such a model is depicted in Fig. 13.22.

Economic Implications

The occurrence of graphitized diamonds in the Beni Bousera and Ronda peridotite massifs and the presence of diamonds in the Tibetan ophiolites provide strong support for previous suggestions that the source of alluvial diamonds in many orogenic belts may be tectonically emplaced orogenic ultramafic rocks of ultradeep origin (Pavlenko et al., 1974; Shilo et al., 1981; Dawson, 1983; Nixon et al., 1986; Pearson et al., 1989). Although the original diamond concentrations in the Beni Bousera and Ronda graphitic pyroxenites were on the order of 10^5 greater than the richest kimberlites, the limited extent of the pyroxenite bodies studied so far does not indicate significant economic potential unless the layers can be traced for several kilometers. In contrast, the diamond concentration reported from the Tibetan ophiolites is very low, but the massifs are large in size such that alluvial concentrations may yield economic concentrations. However, the diamonds reported from the Tibetan ophiolites (Bai et al., 1993) are only in the micro-diamond size range and, hence, may not be attractive economically. Despite these reservations, economically exploitable concentrations of diamonds do occur at Copeton Bingara, eastern Australia, in alluvials on the western slopes of the Urals and at SE Kalimantan. Several large (>100 carat) stones have been recovered from the latter deposit (A. Janse, personal communication). All these deposits are distant from any proven volcanic source but are in relatively close proximity to ophiolite belts or orogenic peridotite massifs, e.g., the orogenic ultramafics of the Urals, the Bobaris ophiolite in SE Kalimantan, and the Great Serpentinite Belt of New South Wales. Such deposits indicate that in some instances processes of secondary concentration may be effective in creating economically exploitable diamond deposits.

Exploration for "anomalous" diamond deposits requires a completely different approach in terms of the "indicator" minerals commonly utilized in the search for kimberlite and lamproitic diamond primary sources. Most of the indicator minerals that characterize lamproite (Ti-rich phlogopite, K-Ti-richterite, low Al-diopside, priderite, and wadeite) and kimberlite (low-Ca high-Cr pyrope garnets, magnesian ilmenite, the megacryst suite, etc.) occurrences are scarce or absent in confirmed or suspected orogenic diamond occurrences. The only "typical" indicator mineral present in many cases may be diamond itself. Anomalous diamond occurrences may otherwise be frequently associated with the presence of pyrope-almandine garnets, platinoid mineralization (in particular, osmiridiums), rutile, corundum, and various metamorphic minerals including staurolite and florencite (Kaminskii, 1984; Nixon et al., 1991). Moissanite is associated with the Tibetan ophiolitic diamonds and has also been reliably identified as an inclusion in some diamonds (Moore et al., 1986; Jaques et al., 1989). One mineral that classical kimberlite-related and some anomalous diamond occurrences, such as Tibet, may have in common is high-Cr chromite, particularly if some of the Cr is present as Cr^{2+}. Chromites containing tetrahedrally coordinated Cr^{2+} are found as diamond inclusions (Haggerty, 1979) and in chromite associated with diamond in the heavy mineral concentrates from the Tibetan ophiolite massifs (Bai et al., 1993).

Conclusions

Our physical understanding of tectonic processes is influenced by evidence we see in the rocks exposed at the Earth's surface. The role of the continental crust during orogenesis is thus relatively well understood compared to that of the mantle, but new discoveries indicate ever deeper levels of exhumation of crust (Chopin, 1984; Sobolev and Shatsky, 1990). The speculative association of diamonds with orogenic ultramafic rocks has long hinted at the involvement of deep mantle during collisional events. Evidence discussed above indicates that tectonic processes involved in the genesis of orogenic belts are capable of bringing large volumes of mantle to the surface of the Earth from depths of ~150 km, i.e., within the diamond stability field. In the face of such evidence, we should now develop tectonic models to account for these occurrences. Two distinct scenarios are involved, both associated with collisional orogenesis. In one situation, anomalously hot mantle appears to have risen, possibly diapirically from ~150 km, causing graphitization of original diamonds present in pyroxenites. In the other situation oceanic lithosphere may have been subducted into the diamond stability field, and then, while still cool, thrust back up rapidly, perhaps along deep mantle shear zones and with their diamonds preserved. The exact mechanisms for emplacement of these deeply derived, large fragments of high density rocks into the crust is probably complex, but their elucidation should lead to a much better understanding of the processes involved in orogenic events. Developing this understanding should be a major goal of tectonophysics.

Acknowledgments

We would like to express our gratitude to Dr. Mei-Fou Zhou and Dr. Wen-Ji Bai for allowing us to use Fig. 13.16 and for supplying details of unpublished work. The manuscript was greatly improved by reviews and suggestions from Drs. H. Wilshire, J. Kornprobst, and A. J. Janse. The work on Beni Bousera was carried out while D.G.P. was under tenure of an NERC research studentship. We thank Rod Green, Eric Condliffe, and Tom Oddy for skilled technical assistance.

References

Allegre, C. J., and Turcotte, D. L. 1986. Implications of a two component marble-cake mantle. *Nature, 323*, 123–126.

Alt, J. C., Muehlenbachs, K., and Honnorez, J. 1986. An oxygen profile through the upper kilometer of the oceanic crust, DSDP Hole 504B. *Earth Planet. Sci. Lett., 80*, 217–229.

Ater, P. C., Eggler, D. H., and McCallum, M. E. 1984. Petrology of mantle eclogite xenoliths from Colorado-Wyoming kimberlites recycled oceanic crust? *Kimberlites II : The mantle and crust-mantle relationships.* Amsterdam: Elsevier, 309.

Bai, W. J., and Zhou, M.-F. Chemical compositions of chrome spinels from Hongguleleng ophiolite, Xinjiang, China, and their significance. *Acta Mineral. Sinica, 8*, 313–323 (in Chinese).

Bai, W.-J., Zhou, M. F., and Robinson, P. T. 1993. Diamond-bearing mantle peridotites and podiform chromitites in the Luobusa and Donqiao ophiolite massifs, Tibet. *Contrib. Canadian Jour. Earth Sci, 30*, 1650–1659.

Ben Othman, D., White, W. M., and Patchett, J. 1989. The geochemistry of marine sediments, island arc magma genesis and crust – mantle recycling. *Earth Planet. Sci. Lett., 94*, 1–21.

Bertrand, P., and Mercier, J.-C. C. 1986. The mutual solubility of coexisting ortho and clinopyroxene: Towards an absolute geothermometer for the natural system? *Earth Planet. Sci. Lett., 76*, 109–122.

Blanco, M. J., and Spackman, W. 1991. The P-wave velocity stucture of the mantle below the Iberian Peninsula: Evidence for subducted lithosphere below southern Spain. *Tectopnophysics, 221*, 13–34

Bohlen, S. R., and Boettcher, A. L. 1982. The quartz-coesite transformation: A pressure determination and the effects of other components. *J. Geophys. Res., 87*, 7073–7078.

Boillot, G., Girardeau, J., and Kornprobst, J. 1988. Rifting of the Galicia margin: Crustal thinning and emplacement of mantle rocks on the seafloor. *Proc. Ocean Dril. Prog., Sci. Results, 103*, 741–756.

Bonini, W. E., Loomis, T. P., and Robertson, J. D. 1973. Gravity anomalies, ultramafic intrusions and the tectonics of the region around the Strait of Gibralter. *J. Geophys. Res., 78*, 1372–1382.

Bonney, T. G. 1899. The parent-rock of the diamond in South Africa. *Geol. Mag., 6*, 309–321.

Boyd, F. R. 1973. A pyroxene geotherm. *Geochem. Cosmochim. Acta, 37*, 2533–2546.

Brett, R., and Higgins, G. T. 1969. Cliftonite: A proposed origin and its bearing on the origin of diamonds in meteorites. *Geochem. Cosmochim. Acta, 33*, 1473–1484.

Brey, G. P., and Kohler, T. 1990. Geothermobarometry in four-phase lherzolites II. New thermobarometers, and practical assessment of existing thermobarometers. *J. Petrol., 31*, 1353–1378.

Brey, G. P., Köhler, T., and Nickel, K. G. 1990. Geothermobarometry in four-phase lherzolites I. Experimental results from 10 to 60 kb. *J. Petrol., 31*, 1313–1352.

Calderon, D. S. 1890. La region epijenica de Andalucia y el origen de su ofitas. *Bol. Bom. Mapa geol. Espana, 13*, 499–526.

Camsell, C. A. 1911. A new diamond locality in the Tulameen District, British Columbia. *Econ. Geol., 6*, 604–611.

Carswell, D. A. 1991. The garnet-orthopyroxene A1 barometer: problematic application to natural garnet lherzolite assemblages. *Mineral. Mag., 55*, 19–31.

Carswell, D. A., and Gibb, F.G.F. 1986. Evaluation of mineral thermometers and barometers applicable to garnet lherzolite assemblages. *Contrib. Mineral. Petrol., 95*, 499–511.

Carswell, D. A., and Harley, S. H. 1990. Mineral barometry and thermometry. In *Eclogite Facies Rocks*. Glasgow and London: Blackie, 83–109.

Chalouan, A., and Michard, A. 1990. The Ghomarides nappes, Rif coastal range, Morocco: A Variscan chip in the Alpine belt. *Tectonophysics, 9*, 1565–1583.

Chopin, C. 1984. Coesite and pure pyrope in high-grade blueschists of the Western Alps: A first record and some consequences. *Contributions to Mineralogy and Petrology, 86*, 107–118.

Coleman, R. G. 1977. *Ophiolites*. Berlin: Springer Verlag.

Cooper, G. 1990. Infrared microspectroscopy of diamond in relation to mantle processes. Ph.D., University of London.

Davies, G., and Evans, T. 1972. Graphitization of diamond at zero pressure and high temperature. *Proceedings Royal Society, 328*, 413–427.

Davies, G. R., Nixon, P. H., Pearson, D. G., and Obata, M. 1992a. *Graphitised diamonds from the Ronda peridotite massif, S. Spain*. Proc. 5th Int. Kimberlite Conf., Brazil.

Davies, G. R., Nixon, P. H., Pearson, D. G., and Obata, M. 1992b . The tectonic implications of graphitised diamonds from the Ronda peridotite massif, S. Spain. *Geology, 21*, 471–474.

Dawson, J. B. 1983. New developments in diamond geology. *Naturwissenschaften, 70*, 586–593.

Dawson, J. B., and Carswell, D. A. 1990. High temperature and ultra-high pressure eclogites. In *Eclogite Facies Rocks*. London: Blackie, 315–349.

Deines, P., Gurney, J. J., and Harris, J. W. 1987. Carbon isotopic composition, nitrogen content and inclusion composition of diamonds from Roberts Victor Kimberlite, South Africa. Evidence for 13C depletion in the mantle. *Geochem. Cosmochim. Acta, 51*, 1227–1243.

Dickey, J. S. 1970. Partial fusion products in Alpine peridotites: Serrania De La Ronda and other examples. *Min. Soc. Am. Pap., 3*, 33–49.

Doblas, M., and Oyarzun, R. 1989. "Mantle core complexes" and neogene extentional detachment tectonics in the western Betic Cordilleras, Spain: an alternative model for the emplacement of the Ronda peridotite. *Eart Planet. Sci. Lett., 93*, 76–84.

Eggler, D. H. 1986. Kimberlites: How do they form? *Geol. Soc. Amer. Spec. Pap., 14*, 489–504.

Ellis, D. J., and Green, D. H. 1979. An experimental study of the effect of Ca upon garnet-clinopyroxene Fe-Mg exchange equilibria. *Contrib. Mineral. Petrol., 71,* 13–22.

Ernst, W. G. 1978. Petrochemical study of lherzolitic rocks from the W. Alps. *J. Petrol., 19,* 341–353.

Ernst, W. G. 1981. Petrogenesis of eclogites and peridotites from the Western and Liguria Alps. *Am. Mineralogist, 66,* 443–472.

Evans, T. 1979. Changes produced by high temperature treatment of diamond. In *Properties of Diamond.* NewYork: Academic Press, 403–424.

Fang, C., and Wenji, B. 1981. The discovery of Alpine-Type diamond bearing ultrabasic intrusion in Xizang (Tibet). *Int. Geol. Rev., 27,* 455–457.

Finnerty, A. A., and Boyd, F. R. 1984. Evaluation of thermobarometers for garnet peridotites. *Geochem. Cosmochim. Acta, 48,* 15–27.

Frey, F. A., Suen, C. J., and Stockman, H. W. 1985. The Ronda high temperature peridotite: Geochemistry and petrogenesis. *Geochem. Cosmochim. Acta, 49,* 2469–2491.

Girardeau, J., and Mercier, J.-C. C. 1988. Petrology and texture of the ultramafic rocks of the Xigaze ophiolite (Tibet): Constraints for mantle structure beneath slow-spreading ridges. *Tectopnophysics, 147,* 33–58.

Goodrich, C. A., and Bird, J. M. 1985. Formation of iron-carbon alloys in basaltic magma at Uivfaq, Disko Island: The role of carbon in mafic magmas. *J. Geol., 93,* 475–492.

Green, D. H. 1967. High temperature peridotite intrusions. *Ultramafic and Related Rocks.* New York: Wiley, 212–222.

Green, T. H. 1982. Anatexis of mafic crust and high pressure crystallisation of andesite. *Andesites: Orogenic Andesites and Related Rocks.* Chichester: Wiley. 465–487.

Gregory, R. T., and Taylor, H. P. 1981. An oxygen isotope profile in a section of Cretaceous oceanic crust, Samail ophiolite, Oman: Evidence for $\delta^{18}O$-buffering of the oceans by deep (< 5 km) seawater-hydrothermal circulation at mid-ocean ridge. *J. Geophys. Res., 86,* 2737–2755.

Grenville-Wells, H. J. 1952. The graphitization of diamond and the nature of cliftonite. *Mineral. Mag., 24,* 803–817.

Gurney, J. J. 1991. The diamondiferous roots of our wandering continents. *S. Afr. J. Geol., 93,* 423–437.

Haggerty, S. E., and Sautter, V. 1990. Ultradeep (greater than 300 kilometers) ultramafic upper mantle xenoliths. *Sciences, 248,* 993–996.

Haidinger, W.K.V., and Partsch, P. 1846. Pseumomorph nach schwefelkies. *Ann. Phys., 67,* 437–439.

Harley, S. L., and Carswell, D. A. 1990. Experimental studies on the stability of eclogite facies mineral parageneses. In *Ecologite Facies Rocks.* Glasgow: Blackie. 53–82.

Harris, J. W. 1968. The recognition of diamond inclusions P.1 Syngenetic mineral inclusions. *Industr. Diam. Rev., 334,* 402–410.

Hart, S. R. 1984. A large-scale isotope anomaly in the Southern Hemisphere mantle. *Nature, 309,* 753–7.

Haut, F. R., Levin, P. and Eisenburger, D. 1984. Diamantfuhrende in Obervolta. *Geol. Jb., A75,* 361–378.

Howes, V. 1962. The graphitisation of diamond. *Phys. Soc., 80,* 648–662.

Irifune, T and Ringwood, A. E. 1987. Phase transformations in primitive MORB and pyrolite compositions to 25 GPa and some geophysical implications. In: *High Pressure Research in Geophysics,* M. Manghnani and Y. Syono, ed., pp. 231–242. Washington, D.C.: Am. Geophys. Union.

Jagoutz, E., Dawson, J. B., Hornes, S., Spettel, B., and Wanke, H. 1984. Anorthositic oceanic crust in the Archean. *Lunar. Planet. Sci., 15*, 395–396.

Janse, A. J. A. 1991. *Non-kimberlitic diamond source rocks.* Extended Abstracts 5th International Kimberlite Conference, Araxa, Brazil, 199–201.

Jaques, A. L., Hall, A. E., and Sheraton, J. W. 1989. *Composition of crystalline inclusions and C-isotopic composition of Argyle and Ellendale diamonds.* Fourth International Kimberlite Conference, Perth, Australia, 966–989.

Javoy, M. 1980. $^{18}O/^{16}O$ and D/H ratios in high temperature peridotites. *Colloq. Int. C.N.R.S., 272*, 279–287.

Kaminskii, F. V. 1984. *The diamond content of non-kimberlitic eruptive rocks.* Moscow: Izd-vo Nedra (in Russian).

Kaminskii, F. V. and Vagnov, V. E. 1976. Petrology precondition carrier of diamond type of Alpine ultrabasic rock. *Izv. Akad. Nauk. ser geol., 6*, 35–47 (in Russian).

Kamiya, Y., and Lang, A. R. 1964. On the structure of coated diamonds. *Phil. Mag., 11*, 347–356.

Keleman, P. B., Dick, H. J. B. and Quick, J. E. 1992. Formation of harzburgite by pervasive melt / rock reaction in the upper mantle. *Nature, 358*, 635–641.

Kennedy, C. S., and Kennedy, G. C. 1976. The equilibrium boundary between graphite and diamond. *J. Geophys. Res., 81*, 2467–2470.

Koons, P. O. 1984. Implications to garnet clinopyroxene geothermometry of non-ideal solid solution in jadeitic pyroxenes. *Contrib. Mineral. Petrol., 88*, 340–347.

Kornprobst, J. 1966. A propos des péridotites du massif des Beni Bouchera (Rif septentrional, Maroc). *Bull. Soc. fr. Mineral. Cristallogr., 89*, 399–404.

Kornprobst, J. 1969. Le massif ultrabasique des Beni Bouchera (Rif Interne, Maroc). *Contrib. Mineral. Petrol., 23*, 283–322.

Kornprobst, J. 1970. Les Peridotites et les pyroxenolites du massif ultrabasique des Beni Bouchera: Une étude experimentale entre 1100 et 1550 C sous 15 à 30kb de pression seche. *Contrib. Mineral. Petrol., 29*, 290–309.

Kornprobst, J. 1974. Contribution à l étude petrographique et structurale de la zone interne du Rif (Maroc Septentrional). *Notes. Serv. Geol. Maroc., T251*, 2556 pp.

Kornprobst, J., Piboule, M., Roden, M., and Tabit, A. 1990. Corundum-bearing garnet clinopyroxenites at Beni Bousera (Morocco): Original plagioclase-rich gabbros recrystallized at depth within the mantle? *J. Petrol., 31*, 717–745.

Kornprobst, J., Piboule, M., and Roux, L. 1982. Corundum bearing garnet pyroxenites at Beni Bousera (Morocco): An exceptionally Al-rich clinopyroxene from grospydites associated with ultramafic rocks. *Terra Cognita, 2*, 257–259.

Kornprobst, J., and Vielzeuf, D. 1984. Transcurrent crustal thinning: A mechanism for the upper lift of deep continental crust/ upper mantle associations. *Kimberlites II: The mantle and crust-mantle relations.* Amsterdam: Elsevier, 347–355.

Krogh, E. J. 1988. The garnet-clinopyroxene Fe-Mg geothermometer – a reinterpretation of existing experimental data. *Contrib. Mineral. Petrol., 75*, 387–393.

Kyser, T. K. 1986. Stable isotope variations in the mantle. *Reviews in Mineralogy, Miner. Soc. Amer., 16*, 141–164.

Kyser, T. K. 1990. Stable isotopes in the continental lithospheric mantle. In *Continental Mantle.* Oxford: Clarendon Press, 127–156.

Leung, I. S. 1990. Silicon carbide cluster entrapped in a diamond from Fuxian. *Am. Mineralogist, 75*, 110–119.

Loomis, T. P. 1972. Diapiric emplacement of the Ronda high temperature intrusion, southern Spain. *Geol. Soc. Amer. Bull., 83*, 24–75.

Loomis, T. P. 1975. Teritary mantle diapirism, orogeny and plate tectonics east of the Strait of Gibralter. *Am. J. Sci., 275*, 1–30.

Lundeen, M. T. 1978. Emplacement of the Ronda peridotite, Sierra Bermeja, Spain. *Geol. Soc. Amer. Bull., 89*, 172–180.

Luth, R. W., Virgo, D., Boyd, F. R., and Wood, B. J. 1989. Iron in mantle derived garnets. *Terra Absracts*, 136.

MacGregor, I. D., and Manton, W. I. 1986. Roberts Victor eclogites: Ancient oceanic crust. *J. Geophys. Res., 91*, 14063–14079.

McDonough, W. F., and Frey, F. A. 1990. Rare earth elements in upper mantle rocks. Geochemistry and mineralogy of rare earth elements. *Min. Soc. Amer. Rev. Mineral., 21*, 99–145.

McKenzie, D. P. and Bickle, M. J. 1988. The volume and composition of melt generated by extension of the lithosphere. *J. Petrology, 29*, 625–679.

Medaris, L. G., Jr., and Carswell, D. A. 1990. The petrogenesis of Mg-Cr garnet peridotites in European metamorphic belts.In *Eclogite Facies Rocks*. Glasgow and London: Blackie, 260–290.

Medaris, G.L.J., Wang, H. F., Misar, Z., and Jelinek, E. 1990. Thermobarometry, difussion modelling and cooling rates of crustal garnet peridotites: Two examples from the Moldanubian Zone of the Bohemian Massif. *Lithos, 25*, 189–202.

Meyer, H. O. A., and Tsai, H. M. 1976. Mineral inclusions in diamond: temperature and pressure of equilibration. *Science, 191*, 849–851.

Moore, R. O., and Gurney, J. J. 1985. Pyroxene solid solution in garnets included in diamond. *Nature, 318*, 553–555.

Moore, R. O., Gurney, J. J., Griffin, W. L., and Shimizu, N. 1991. Ultra-high pressure garnet inclusions in Monastery diamonds: Trace element abundance patterns and conditions of origin. *Eur. Jour. Mineral., 3*, 213–230.

Moore, R. O., Otter, M. L., Rickard, R. S., Harris, J. W., and Gurney, J. J. 1986. *The occurrence of moissanite and ferropericlase as inclusions in diamond*. The International Kimberlite Conference, Extended Abstracts, *Abstr. Geol. Soc. Aust., 16*, 300–302, Perth, Aust.

Morten, L., and Obata, M. 1983. Possible high-temperature origin of pyroxenite lenses within garnet peridotite, northern Italy. *Bull. Minéral., 106*, 775–780.

Mottana, A., Carswell, D. A., Chopin, C., and Oberhänsli, R. 1990. Eclogite facies mineral parageneses.In *Eclogite Facies Rocks*. London: Blackie, 14–52.

Neal, C. R., Taylor, L. A., Davidson, J. P., Holden, P., Halliday, A. N., Nixon, P. H., Paces, J. B., Clayton, R. N., and Mayeda, T. K. 1990. Eclogites with oceanic crustal and mantle signatures from the Bellsbank kimberlite, South Africa, Part 2: Sr, Nd and O isotope geochemistry. *Earth Planet. Sci. Lett., 99*, 362–379.

Nickel, K. G. 1986. Phase equilibria in the system SiO_2-MgO-Al_2O_3-CaO-Cr_2O_3 (SMACCR) and their bearing on spinel/garnet lherzolite relationships. *N. Jb. Mineral. Abh, 155*, 259–287.

Nickel, K. G. and Green, D. H. 1985. Empirical geothermobarometry for garnet peridotites and implications for the nature of the lithosphere, kimberlites and diamond. *Earth Planet. Sci. Lett., 73*, 158–170.

Nicolas, A. 1984. Lherzolites in the western Alps: a structural review. *Kimberlites. Developments in Petrology, Series IIB*. Amsterdam: Elsevier, 333–345.

Nixon, P. H., Davies, G. R., Slodkevich, V. V., and Bergman, S. C. 1986. *Graphite pseudomorphs after diamond in the eclogite peridotite massif of Beni Bousera, Morocco and a review of anomalous diamond occurrences*. Fourth Int. Kimberlite Conf. Perth, Extended Abst. Geol. Soc. Aust. , No. 16 , 412–414.

Nixon, P. H., Pearson, D. G., and Davies, G. R. 1991. Diamonds: the oceanic lithosphere connection with special reference to Beni Bousera, North Morocco.In

Ophiolite Genesis and Evolution of the Oceanic Lithosphere. Dordrecht: Kluwer Academic Publishers, 275–289.

O'Hara, M. J. 1967. Mineral facies in ultrabasic rocks. In *Ultramafic and Related Rocks.* New York: John Wiley, 7–18.

O'Neill, H. St. C. 1981. The transition between spinel lherzolite and garnet lherzolite, and its use as a geobarometer. *Contrib. Mineral. Petrol., 77*, 185–194.

Obata, M. 1980. The Ronda peridotite: garnet, spinel and plagioclase lherzolite facies and the P-T trajectories of a high temperature mantle intrusion. *J. Petrol., 25*, 533.

Obata, M. 1982. Reply to W. Schubert's comments on "The Ronda peridotite: Garnet-, spinel-, and plagioclase-lherzolite facies and the P/T trajectories of a high temperature mantle intrusion" by M. Obata (*J. Petrology, 21*, 533–72, 1980). *J. Petrol., 23*, 296–298.

Okada, A., and Shima, M. 1972. Crystallographic study of cliftonite: A new internal structure found in the inclusion of the Campo Del Cielo meteorite. *J. Japan Assoc. Min. Petr. Econ. Geol., 67*, 45–49.

Pasteris, J. D. 1988. Secondary graphitisation in mantle derived rocks. *Geology, 16*, 804–807.

Pavlenko, A. S., Gevorkin, R. G., Aslanyan, A. T., Gulyan, E. K., and Yegorov, O. S. 1974. On the diamonds in the ultramafic belts of Armenia. *Geokhimiya, 3*, 366–379.

Pearson, D. G. 1989. The petrogenesis of pyroxenites containing octahedral graphite and associated mafic and ultramafic rocks of the Beni Bousera peridotite massif, N. Morocco, Ph.D., Univ. of Leeds.

Pearson, D. G., Boyd, F. R., and P.H., N. 1990. Graphite-bearing mantle xenoliths from the Kaapvaal craton: Implications for graphite and diamond genesis. *Carnegie Inst. Washington Yearbook*, 11–19.

Pearson, D. G., Davies, G. R., and Nixon, P. H. 1993. Geochemical constraints on the petrogenesis of diamond facies pyroxenites from the Beni Bousera peridotite massif, N. Morocco. *Journal of Petrology, 34*, 125–172.

Pearson, D. G., Davies, G. R., Nixon, P. H., Greenwood, P. B., and Mattey, D. P. 1991. Oxygen isotope evidence for the origin of pyroxenites in the Beni Bousera peridotite massif, North Morocco: derivation from subducted oceanic lithosphere. *Earth Planet. Sci. Lett., 102*, 289–301.

Pearson, D. G., Davies, G. R., Nixon, P. H., and Mattey, D. P. 1991. A carbon isotope study of diamond facies pyroxenites and associated rocks from the Beni Bousera peridotite massif, North Morocco. *J. Petrol.*, 175–189.

Pearson, D. G., Davies, G. R., Nixon, P. H., and Milledge, H. J. 1989. Graphitized diamonds from a peridotite massif in Morocco and implications for anomalous diamond occurrences. *Nature, 338*, 60–62.

Platt, J. P., and Vissers, R. M. 1989. Extensional collapse of thickened continental lithosphere: A working hypothesis for the Alboran Sea and Gibralter arc. *Geology, 17*, 540–543.

Powell, R. 1985. Regression diagnostic and robust regression in geothermometer/geobarometer calibration: the garnet-clinopyroxene geothermometer revisited. *Jour. Metamorph. Geol., 3*, 231–243.

Reisberg, L., and Zindler, A. 1987. Extreme isotopic variations in the upper mantle: evidence from Ronda. *Earth Planet. Sci. Lett., 81*, 29, 45.

Reisberg, L. C., Allegre, C. J., and Luck, J. M. 1991. The Re-Os systematics of the Ronda Ultramafic Complex of southern Spain. *Earth Planet. Sci. Lett., 105*, 196–213.

Reisberg, L. C., Zindler, A., and Jagoutz, E. 1989. Sr and Nd isotopic compositions of garnet and spinel bearing peridotites in the Ronda Ultramafic Complex. *Earth Planet. Sci. Lett., 96*, 161–180.

Reuber, I., Michard, A., Chalcouan, A., Juteau, T., and Jermoumi, B. 1982. Structure and emplacement of the Alpine type peridotites from Beni Bousera, Rif Morocco: A polyphase tectonic interpretation. *Tectopnophysics, 82*, 231–251.

Ringwood, A. E. 1990. Slab–mantle interactions 3. Petrogenesis of intraplate magmas and structure of the upper mantle. *Chemical Geology, 82*, 187–207.

Ringwood, A. E., and Green, D. H. 1966. An experimental investigation of gabbro-eclogite transformation and some geophysical consequences. *Tectopnophysics, 3*, 383–427.

Robinson, D. N. 1978. The characteristics of natural diamond and their interpretation. *Miner. Sci. Engineer., 10*, 55–72.

Sacks, I. S., and Kincaid, C. 1990. On slab melting in subduction zones. *EOS Trans. Am. Geophys. Union, 71*, 1714.

Sautter, V., Haggerty, S. E., and Field, S. 1991. Ultradeep (>300 kilometers) ultramafic xenoliths: petrological evidence from the transition zone. *Science, 252*, 827–830.

Schiffmann, P., Evarts, R. C., Williams, A. E., and Pickthorn, W. J. 1991. Hydrothermal metamorphism in oceanic crust from the Coast Range Ophiolite of California: fluid-rock interaction in a rifted island arc. In: Peters, T. J., Nicolas, A. and Coleman, R. G. (eds.), *Ophiolite Genesis and Evolution of the Oceanic Lithosphere, 399–426.* Dordrecht: Kluwer.

Schubert, W. 1982. Comments on "The Ronda peridotite: Garnet-, spinel-, and plagioclase-lherzolite facies and the *P/T* trajectories of a high temperature mantle intrusion" by M. Obata (*J. Petrol., 21*, 533–72, 1980). *J. Petrol., 23*, 293–295.

Schulze, D. J., and Valley, J. W. 1991. *Carbon isotope composition of graphite in mantle eclogites.* 5th Int. Kimberlite Conference, Araxa, Brazil.

Shervais, J. W., Taylor, L. A., Lugmair, G. W., Clayton, R. N., Mayeda, T. K., and Kordiev, R. L. 1988. Early Proterozoic oceanic crust and the evolution of subcontinental mantle: Eclogites and related rocks from Southern Africa. *Geol. Soc. Amer. Bull., 100*, 411–423.

Shilo, N. A., Kaminskiy, F. V., Palandzhyan, S. A., Til'man, S. M., Tkachenko, L. A., Lavrova, L. D., and Shepeleva, P. 1978. First diamond finds in Alpine-type ultramafic rocks of the northeastern USSR. *Doklady Akad. Nauk SSSR., 241*, 933–936.

Slodkevich, V. V. 1980. Polycrystalline aggregates of octahedral graphite. *Doklady. Akad. Nauk SSSR, 253*, 697–700.

Slodkevich, V. V. 1983. Graphite paramorphs after diamond. *Internat. Geol. Rev., 25*, 497–514.

Smith, C. B., Gurney, J. J., Harris, J. W., Otter, M. L., Kirkley, M. B., and Jagoutz, E. 1991. Neodymium and strontium isotope systematics of eclogite and websterite paragenesis inclusions from single diamonds, Finsch and Kimbereley Pool, RSA. *Geochem. Cosmochim. Acta, 55*, 2579–2590.

Smith, D., and Barron, B. R. 1991. Pyroxene-garnet equilibration during cooling in the mantle. *Am. Mineralogist, 76*, 1950–1963.

Smyth, J. R., Caporuscio, F. A., and McCormick, T. 1989. Mantle eclogites: evidence of igneous fractionation in the mantle. *Earth Planet. Sci. Lett., 93*, 133–141.

Sobolev, N. V. 1979. *The Problem of the Constitution of the Earth's Upper Mantle.* Washington, AGU Press (translated from Russian).

Sobolev, N. V., Galimov, E. M., Ivanovskaya, I. N., and Yefimova, E. S. 1979. Isotopic composition of carbon of diamonds containing crystalline inclusions. *Doklady. Akad. Nauk SSSR, 249*, 1217–1220.

Sobolev, N. V., and Lavrent'ev, J. G. 1971. Isomorphic sodium admixture in garnets formed at high pressures. *Contrib. Mineral. Petrol., 31*, 1–12.

Sobolev, N. V., and Shatsky, V. S. 1990. Diamond inclusions in garnets from metamorphic rocks. *Nature, 343*, 742–746.

Suen, C. J., and Frey, F. A. 1987. Origins of the mafic and ultramafic rocks in the Ronda peridotite. *Earth Planet. Sci. Lett., 85*, 183–202.

Sunagawa, I. 1984. Morphology of natural and synthetic diamond crystals. In: *Materials Science of the Earth's Interior*, Sunagawa, I. (ed), 303–330. Tokyo: Terra Scientific Publishing Co.

Thompson, R. N. 1974. Some high pressure pyroxenes. *Mineral. Mag., 39*, 768–787.

van der Wal, D. 1993. Deformation processes in mantle peridotites. University of Utrecht.

Vielzeuf, D., and Kornprobst, J. 1984. Crustal splitting and the emplacement of Pyrenean lherzolites and granulites. *Earth Planet. Sci. Lett., 67*, 87–96.

Webb, S.A.C., and Wood, B. J. 1986. Spinel-pyroxene-garnet relationships and their dependence on Cr/Al ratio. *Contrib. Mineral. Petrol., 92*, 471–480.

Weijermars, R. 1985. Uplift and subsidence history of the Alboran Basin and a profile of the Alboran diapir (W. Meditteranean). *Geol. Mijnbouw, 64*, 349–356.

Weijermars, R. 1991. Geology and tectonics of the Betic Zone, S. E. Spain. *Earth Planet. Sci. Lett., 31*, 153–236.

White, W. M., and Dupre, B. 1986. Sediment subduction and magma genesis in the Lesser Antilles: isotopic and trace element constraint. *J. Geophys. Res., 91*, 5927–5941.

Wilding, M. C., Harte, B., and Harris, J. W. 1989. Evidence of asthenospheric source for diamonds from Brazil. *Abstracts Vol 3, 28th Int. Geol. Congress Washington D. C., 3*, 359.

Wilkinson, J.F.G. 1976. Some subcallic clinopyroxenes from Salt Lake Crater, Oahu, and their petrological significance. *Contrib. Mineral. Petrol., 58*, 181–202.

Wilshire, H. G., and Shervais, J. W. 1975. Al-augite and Cr-diopside ultramafic xenoliths in basalt host rocks from the western United States. *Phys. Chem. Earth, 9*, 257–272.

Woermann, E., and Rosenhauer, M. 1985. Fluid phases and the redox state of the Earth's mantle. *Fortsch. Mineral., 83*, 263–349.

Woodland, A. B., Kornprobst, J., and Wood, B. J. 1992. Oxygen thermobarometry of orogenic lherzolite massifs. *J. Petrol., 33*, 203–230.

Zindler, A., and Hart, S. 1986. Chemical geodynamics. *Ann. Rev. Earth Planet. Sci. Lett., 14*, 493–571.

Zindler, A., Staudigel, H., Hart, S. R., Endres, R., and Golstein, S. 1983. Nd and Sr isotopic study of a mafic layer from Ronda ultramafic complex. *Nature, 304*, 226–230.

Author Index

Ackermand, D., 42
Adam, J., 195, 193
Adams, G.E., 48
Agoshkov, V.M., 37
Agrinier, P. 318, 324, 329
Akaogi, M., 37, 49
Akella, J., 47
Akimoto, S., 36, 37, 39, 49, 124
Allegre, C.J., 461
Althaus, E., 198
Ames, L., 13, 164, 363, 383, 400, 407, 410, 414
Ampferer, O., 23
Anders, E., 335
Andersen, D.J., 69
Andersen, T. B., 25, 82, 146, 148, 258, 262, 263,
    264, 265, 272, 283, 285, 312, 320, 321, 343
Andreasson, P.G., 245, 252, 253, 266, 267, 309
Andresen, E., 265, 269, 270, 271
Apted, M.J., 198
Araki, T., 120
Argand, E., 137
Austreheim, H.A., 73, 245, 262, 263, 264, 265,
    283, 285, 309, 312, 320, 321, 322, 340
Avigad, D., 25, 133, 148, 149, 151, 152, 172,
    182,183

Backlund, H.G., 245, 247, 248
Bai, W., 458, 483, 484, 486, 500
Bakirov, A., 207
Bakker, E., 266, 267,
Bakun-Czubarow, N., 18
Baller, T., 38
Ballévre, M., 15, 25, 68, 119, 149, 150, 151, 152,
    172, 177, 183
Baltatizis, E., 68
Barnicoat, A.C., 135, 150, 151, 164
Barron, B.J., 186
Barron, B.R., 487
Bauer, J.F., 37
Becker, H., 17, 103, 163, 175
Bell, P.M., 44
Beller, U., 60
Ben Othman, D., 477
Berhmann, J.H., 148, 149

Berman, R.G., 33, 35, 36, 52, 55, 63, 66, 224
Bernoulli, D., 208
Berry, H.N., 259
Berthé, D., 147
Bertrand, P., 487
Bickle, M.J., 495
Bigi, G., 135
Biino, G., 144, 145, 213, 214, 215, 222
Binns, R.A., 254, 307
Birch, F., 44
Bird, J.M., 465
Birkeland, T., 265
Biryukov, V.M., 111
Bishop, F.C., 48, 49, 63
Blake, M.C.Jr, 25, 148, 172, 182,183, 415, 416, 421
Blacic, J.D., 169
Blanco, M.J., 500
Blundy, J.D., 377
Bocquet, J., 14, 17, 236
Boettcher, A., 2, 35, 44, 45, 51, 52, 167, 279, 376,
    495
Bohlen, S., 2, 35, 63, 66, 167, 279, 376, 447, 494
Boillot, G., 499
Bollingberg, H.J., 256
Bone, R., 175
Bonini, G., 460
Bonney, T.G., 457
Borghi, A., 139, 235, 236
Boundy, T.M., 263, 321
Bovenkerk, H.P., 334
Boving, R., 68
Boyd, F.R., 43, 110, 457, 487
Boyer, H., 312, 324
Bozo, E., 135
Brace, W.F., 338
Brastad, K., 245, 249
Brett, R., 465
Brey, G.P., 47, 48, 49, 99, 276, 448, 487
Broks, T.M., 269
Brown, E.H., 414
Brown, M., 97
Brown, T.H., 224
Brueckner, H.K., 249, 258, 259, 270, 272, 275, 304,
    305, 322

511

Brunel, M., 384
Brunet, F., 117
Brunfelt, A.O., 249
Bryhni, I., 245, 248, 256, 258, 300, 303, 307, 318, 319, 324
Bulanova, G.P., 106
Bundy, F.P., 35
Burchfiel, B.C., 415, 416
Butler, R.W.H., 149, 150, 152
Butterman, W.C., 107, 335

Calderone, D.S., 480
Camsell, C.A., 483
Capdoppi, P., 139
Carman, J.H., 47, 224
Carmignani, L., 415, 416
Carlson, W.D., 167
Caron, J.M., 208
Carpenter, M., 310, 325
Carswell, D.A., 7, 18, 245, 248, 249, 250, 256, 260, 272- 276, 283, 287, 300, 306, 310, 312, 317, 318, 319, 320, 341, 458, 459, 483, 487, 488, 490, 492
Chalouan, A., 479
Chapman, D.S., 24, 169, 173
Chapman, N.A., 47
Chappell, B.W., 402, 403, 416
Chatterjee, N.D., 35, 36,.39, 44, 52, 61, 67, 198
Chauvet, A., 249, 255, 285
Cheeney, R.F., 314, 315, 322
Chen, T., 395
Chen, W., 361, 384
Cheng, W., 63
Chesnokov, B.V., 2, 310
Cho, M., 68
Chopin, C., 2, 15, 17, 33, 39, 40, 41, 57, 97, 105, 112, 113, 115, 117, 118, 119, 121, 122, 123, 135, 137, 138, 141, 143, 151, 159, 161, 162, 163, 165, 174, 175, 183, 186,199, 207, 209, 210, 212, 214, 216–221, 227–229, 231, 233, 234–236, 310, 312, 323, 324, 341, 356, 391, 457, 459, 488, 502
Church, W.R., 323
Claue-Long, J.C., 10, 104, 385
Cloos, M., 150
Coe, R., 406, 407, 413
Coes, L., 38
Cohen, A.S., 264, 322
Coleman, R.G., 73, 337, 356
Comodi, P., 112, 121
Compagnoni, R., 14, 113, 118, 122, 123, 138, 144, 145, 213, 214, 215, 219, 220, 221, 222, 224, 230, 235, 236
Coney, P.J., 421
Cong, B., 320, 357, 369, 377, 378, 383
Cooper, G., 466
Copeland, C., 25
Cotkin, S.J., 257, 260
Cowan, D.S., 150
Coward, M., 133
Cuthbert, S.J., 5, 8, 25, 245, 248, 249, 254, 260, 272, 283, 287, 300, 317, 318, 319, 321, 341

Dachille, F., 167
Dal Piaz, G., 14, 135, 207, 235
Dallmeyer, R.D., 252, 253, 267, 270, 271, 322
Daly, J.S., 253
Danckwerth, P.A., 42, 44, 69
Davidson, P.M., 63, 66, 69, 99
Davies, G.R., 458, 480, 481, 482, 494, 497, 500
Davis, G.A., 176
Davis, G.H., 421
Davy, P., 174
Dawson, J.B., 107, 483, 501
De Roever, W.P., 337
Debelmas, J., 207
Deines, P., 110, 469
Delany, J.M., 52
Dercourt, J., 152
Dergachev, D.V., 111
Desmon, J., 208
Deville, E., 149, 150, 151, 152
Dewey, J., 149, 152, 175, 391, 413, 414, 415, 416
De Wys, E.C., 107, 108
Dickey, J.S., 479
Dietrich, D., 133
Dietrich, V., 14
Doblas, M., 499
Dobrestov, N.L., 10, 20, 101, 183, 186, 337
Dobrizhinetskaya, L.F., 8, 111, 330, 333–335, 341
Dodson, M.H., 260
Dong, S.B., 357, 359, 383
Doroshev, A.M., 103
Douce, A.E.P., 450, 451
Draper, G., 175
Dunning, G.R., 250
Dupre, B., 477
Durr, S., 77
Dutrow, B.L., 118

Easterling, K.E., 168
Eby, G.N., 404
Eggler, D.H., 46, 52
Eide, E.A., 358, 359, 363, 382, 395, 400, 401, 407, 410, 415, 416, 417
Ekimova, T.E., 333, 334
Ellis, D.J., 48, 67, 193, 376, 448, 488, 495
Elvevold, S., 253
Enami, M., 118, 119, 164, 165, 207, 357, 363, 371, 372, 375, 376, 377, 378, 383, 401
Endell, K., 269
Engi, M., 124
England, P.C., 24, 43, 150, 169, 171, 173, 413, 414, 415, 416, 419, 420, 421
Erlank, A.J., 101
Ernst, W.G., 11, 14, 68, 148, 150, 164, 174, 175, 358, 383, 458, 392, 395, 400, 415
Eskola, P., 245, 247, 271, 300, 310, 313, 314, 318, 324
Esnov, S.E., 427
Essene, E.J., 39, 50, 198
Essenov, S.E., 107
Evans, B.W., 47
Evans, T., 465, 497

Fang, C., 22, 458, 483, 484, 486, 500
Færseth, R., 265
Fei, Q., 417
Ferraris, G., 218
Ferry, J.M., 69
Fettes, D.J., 303, 312
Fiala, J., 99
Finnerty, A.A., 487
Flux, S., 67
Fockenberg, T., 42, 120
Fossen, H., 172
Franke, W., 17, 19
Fry, N., 149, 150, 151, 164, 479, 481, 498
Fujisawa, H., 39
Fuller, M., 406, 407

Gaal, 107
Galimov, E.M., 107, 108
Ganeev, I.I., 103
Ganguly, J., 43, 62, 63, 66
Gao, S., 399, 406, 417
Garanin, V.K., 111
Garcia-Celma, A., 148
Garfunkel, Z., 148, 152
Gasparik, T., 37, 38, 42, 44, 45, 48, 61, 66, 67, 69
Ge, H., 395, 396, 399
Gebauer, D., 163, 175, 275, 258, 212, 305, 309
Gee, D.G., 250, 252, 266, 267, 303, 312, 322
Ghiorso, M.S., 69
Gibb, F.G.F., 272, 276, 487
Gil Ibarguchi, J.I., 118
Gilbert, M.C., 47, 224
Gillet, P., 2, 106, 165, 168, 174, 218
Gillotti, J.A., 322
Girardeau, J., 484
Gjelsvik, T., 245, 248, 257, 309
Godard, G., 73
Goffe, B., 15, 135
Goldsmith, J.R., 44, 447
Goodell, H.G., 167
Goodge, J.W., 118
Goodrich, C.A., 465
Graham, C.M., 61, 377
Green, D.H., 47, 48, 51, 67, 73, 258, 305, 307, 318, 376, 448, 478, 488, 492, 495
Green, T.H., 49, 118, 195, 193, 479
Gregory, R.T., 474
Grenville-Wells, H.J., 465, 466
Grew, E.S., 119
Griffin, W.L., 5, 7, 73, 249, 254–260, 262, 263, 265, 270, 274, 275, 282, 300, 305, 306, 309, 312, 318, 321, 329
Gruney, J.J., 457, 466, 478
Guiraud, M., 33, 58, 62
Gustavson, M., 260

Hacker, B.R., 40, 163, 167, 176
Hackler, R.T., 69
Haggerty, S.E., 457, 502
Haidinger, W.K.V., 465
Halbach, H., 52

Hamilton, W., 73, 146
Han, J., 405
Handy, M.R., 143
Hao, J., 358, 392, 400, 405
Hariya, Y., 49
Harley, S.L., 48, 272, 276, 459, 488, 490
Harlov, D.E., 50
Harlow, G.E., 102, 103
Harms, T.A., 186
Harris, A.L., 303, 312
Harris, J.W., 104, 107, 108, 465
Harrison, C.G.A., 23, 260
Harrison, T.M., 410
Hart, S., 475, 476
Haselton, H.T., 63, 378
Haut, F.R., 483
Hawthorne, F.C., 118
Hays, J.F., 44
Heier, L., 245, 249, 257, 258, 260, 263, 307, 328, 329
Heinrich, W., 198
Helgeson, H.C., 52
Hellman, P.L., 49, 118
Henry, C., 123, 136, 137, 138, 139, 144, 147, 150, 172
Hermans, G.A.E.M., 265
Hernes, I., 245, 247, 248, 309
Herzberg, C.T., 37, 66
Higgins, G.T., 465
Hill, R.E.T., 51
Hirajima, T., 13, 72, 73, 144, 122, 123, 164, 207, 209, 214, 219, 220, 221, 224, 236, 316, 357, 401
Hirn, A., 391
Hodges, K.V., 149
Hoinkes, G., 377
Holdaway, M.J., 55, 118
Holland, T.J.B., 33, 44, 45, 47, 48, 52, 58, 59, 101, 135, 150, 221, 224, 234, 377, 381
Holloway, I.R., 450, 451
Hölscher, A., 50
Howes, V., 465
Hsu, K., 25, 175, 392
Hu, K., 73
Huang, W., 399, 406, 415, 417
Huang, W.L., 39, 46, 47, 51, 224
Hull, J.M., 322, 343
Hunziker, J.C., 17, 133, 135, 151, 163, 212, 225
Hurford, A.J., 419
Huth, J., 47
Hy, C., 212

Indares, A., 69
Ingrin, J., 165
Irifune, T., 50, 99, 481
Irving, A.J., 40

Jackson, S.L., 62
Jacobsen, S.B., 260, 306
Jagoutz, E., 10, 478

Jamveit, B., 25, 245, 249, 257, 262, 273, 274, 275, 276, 279, 283, 306, 312, 319, 329
Janse, A.J.A., 483, 501
Jaques, A.L., 101, 104, 502
Jayko, A.S., 25, 148, 172, 182, 183, 186, 415, 416, 421
Jeanloz, R., 35
Jenkins, D.M., 42, 49, 61
Johannes, W., 44
Johansson, L., 257, 322
Johnston, A.D., 450, 451

Kaminsky, F.V., 111, 483, 502
Kamiya, Y., 466
Kanzaki, M., 37
Kashkharov, I.F., 106, 107, 108, 427
Katagas, C., 68
Kechid, S.-A., 307, 310, 311, 316, 326, 327
Keleman, P., 461
Kennedy, C.S., 35, 38, 224
Kennedy, G.C., 35, 44, 45, 224
Khodyrev, O.Y., 37
Kienst, J.R., 15, 71, 123, 138, 187,189, 195, 214, 220, 221, 231, 233, 237
Kincaid, C., 479, 501
Kinny, 104
Kitahara, S., 42
Klaper, E.M., 263
Klaska, R., 218
Kleppa, O.J., 63
Kligfield, R., 415, 416
Klimetz, M.P., 406, 407
Koch-Muller, M., 69
Köhler, T.P., 49, 276
Kohn, M.J., 257
Kolderup, N.H., 245, 247, 262, 309
Kolle, J.J., 169
Koons, P.O., 496
Korikovskiy, S.P., 119
Kornprobst, J., 460, 461, 469, 471, 484, 492, 494–499
Kosygin, Y.A., 111
Koziol, A.M., 44, 63
Kretz, R., 220, 224
Krill, A.G., 266
Krogh, E.J., 8, 73, 104, 195, 245, 247, 249, 253–258, 260, 269, 270, 275, 282, 283, 300, 305, 318, 319, 322, 376, 381, 408
Kroner, A., 396, 397, 399, 400, 410
Krossw, S., 51
Kullerud, L., 245, 259, 262, 266, 267, 269, 303, 304, 305
Kunzler, R.H., 167
Kushiro, I., 48, 51
Kvashitsa, V.N., 106
Kyser, T.K., 473, 479

Lambert, I.B., 49, 51
Landmark, K., 245, 269
Lane, D.L., 43   103
Lang, A.R., 466   104

Langer, K., 113, 218
Lappin, M.A., 7, 8, 207, 245, 248, 254, 256, 257, 272, 282, 283, 301, 303, 304, 305, 307–310, 313, 315–318, 324–331, 333, 338, 341
Lardeaux, J.M., 146
Lattard, D., 50
Laubscher, H.P., 23, 208
Lavrova, L.D., 333, 334
Le Pichon, X., 152
Leake, B.E., 189, 220, 369, 375
LeComte, P., 44
Lee, J., 421
Lemoine, M., 133, 207
Letnikov, F.A., 333, 334, 428
Leung, I.S., 486
Li, S., 402, 403, 416
Li, S.G., 13, 164, 361, 407
Li, Y., 406
Lin, J., 385, 395, 406, 407, 413
Lindsley, D.H., 63, 66, 69, 124
Liou, J.G., 3, 12, 77, 164, 165, 166, 167, 168, 198, 207, 357, 359, 365, 371, 377, 379, 381, 382, 400, 410
Lister, G.S., 147, 148, 421
Liu, D., 361
Liu, J., 66
Liu, L., 38
Liu, X., 11, 12, 73, 357, 383, 392, 400, 405
Loiselle, M.C., 402, 403
Lombardo, B., 135,187,189, 207
Loomis, T.P., 499
Luth, R.W., 103, 488
Lux, D.R., 259, 304
Lyon-Caen, H., 172

Ma, B., 11, 383, 392
Ma, W., 396, 397
Ma, X., 395, 396, 399
MacGregor, I.D., 42, 43, 44, 274, 478, 490
MacKenzie, W.S., 52
Mahwin, B., 139
Maijer, C., 266
Major, A., 37, 39
Malm, O.A., 260
Maluski, M., 210
Manckletov, N., 338, 343
Manning, C.E., 447
Manton, W.I., 478
Marthaler, M., 133, 137
Martignole, J., 69
Martinotti, G., 17
Maruyama, S., 25, 77, 198, 383
Massone, H. -J., 2, 33, 35, 36, 38, 40, 42, 44, 45, 46, 55, 57, 59, 61,63, 67, 68, 71, 73, 77, 138, 162, 214, 224, 234, 376, 450, 451
Mattauer, M., 391, 397, 399, 400, 407, 411, 413, 414, 416
Matte, P., 17, 18, 19, 20
Mattioli, G.S., 63
McDonough, W.F., 482
McDougall, I., 258, 305, 421

McElhinny, M.W., 406
McKenzie, D.P. 495
McLenman, R., 435
Mearns, E.W., 258, 259, 273, 304, 305, 306
Medaris, L.G.J., 8, 245, 272, 275, 318, 319, 458, 487
Mercier, J.-C.C., 484, 487
Mercy, E.L.P., 245, 272, 309, 310, 318, 338
Merle, O., 140, 150, 151, 152
Merrill, R.B., 47, 49, 51
Messiga, B., 322
Metcalfe, I., 406
Meyer, H.O.A., 97, 98, 101, 104, 465. 470
Michard, A., 15, 137, 148, 149, 150–152, 209, 218, 227, 236, 479
Milanovsky, E.E., 338
Miller, E.L., 421
Millhollen, G.L., 51
Mirwald, P.W., 2, 35, 36, 44, 224
Molchanova, T.V., 333
Möller, C., 257, 259, 323
Molnar, P., 169, 171, 172, 413, 421
Monié, P., 17, 151, 163, 174, 186, 212, 235
Moore, P.B., 117, 120
Moore, R.O., 457, 488, 502
Mori, T., 47, 48
Mørk, M.B.E., 7, 245, 249, 252, 253, 258, 259, 262, 273, 300, 309, 318, 322
Morten, L., 494
Mottana, A., 459
Mposkos, E., 68
Mukhopadhyay, D.K., 69
Murr, L.E., 168
Mysen, B.O., 51, 73, 245, 249, 257, 307, 318, 328, 329

Nadezhdina, E.D., 333, 334
Navon, O., 337
Navrotsky, A., 71
Neal, C.R., 478
Nell, J., 69
Newton, R.C., 42, 44, 45, 47, 48, 50, 51, 61, 66, 67, 69, 448
Nicholson, R., 266
Nickel, K.G., 47, 49, 490, 492
Nicolas, A., 25, 133, 499
Nicollet, C., 118
Nisio, P., 189, 192, 195, 200
Nitsch, K . -H., 198
Nixon, P.H., 483, 501, 502

O'Hanley, D.S., 42
O'Hara, M.J., 245, 272, 309, 310, 318, 338, 457
O'Neill, H.S.C., 48, 50, 274, 488, 490
Obata, M., 479, 480, 489, 492, 494, 498, 499
Oberhansli, R., 71, 212
Oberti, R., 326, 327
Oh, C.W., 200, 250, 254
Ohta, Y., 322
Okay, A.I., 3, 13, 25, 73, 105, 110, 111, 124, 148,

207, 331, 341, 357, 359, 361, 376, 379, 382, 383, 391, 392, 400, 401, 420, 421
Okrusch, M., 19, 78
Ormaasen, D.E., 260
Opdyke, N.D., 406, 407, 413
Oyarzun, R., 499

Pacalo, R.E.G., 37, 38
Padera, K., 99
Paquette, J.R., 17, 163, 212
Parise, J.B., 38
Partsch, P., 465
Pasquare, G., 207
Pasteris, J.D., 500
Patrick, B., 139
Pattison, D.R.M., 48, 66, 67, 448
Pavlenko, A.S., 483, 486, 501
Pawley, A.R., 49, 123
Peacock, S.M., 169, 170
Pearce, J.A., 402
Pearson, D.G., 22, 97, 98, 111, 458, 460–467, 469–471, 473–480, 483, 488, 490–494, 496–498, 501
Pedersen, R.B., 250
Perchuk, L., 10, 48, 66, 336, 343
Perdikatzis, V., 68
Perkins, D.I., 42, 43, 47, 224
Petersen, J.W., 450
Pettersen , K., 245, 269
Peucat, J.-J., 304
Philipot, P., 17, 133, 135, 139, 147, 149, 151,182, 183, 186
Pinardon, J.L., 119
Pinet, M., 314, 327, 328
Pitcher, W.S., 402
Platt, J. P., 148, 151, 185, 172, 207, 319, 336, 414, 498, 415
Pognante, U,. 15, 17
Pokhilenko, N., 466
Polino, R., 208
Polkanov, Y.A., 106, 107, 108, 427
Pollack, H.N., 169, 173
Popov, V.A., 310
Porter, D.A., 168
Posukhova, T.V., 333, 334
Powell, R., 33, 52, 58, 59, 276, 377, 448, 488, 492, 495, 496
Pownceby, M.I., 63

Qvale, H., 245, 249, 274, 312

Råheim, A., 49, 245, 249, 257, 258, 264, 309, 376, 381
Ratschbacher, L., 148, 149
Regional Geological Survey, Anhui, 356, 358, 359
Regional Geological Survey, Henan, 356, 358
Regional Geological Survey, Hubei, 356
Regional Geological Survey, Jiangsu, 358
Reinecke, T., 2, 72, 15, 21, 105, 135, 150, 161, 164, 165, 186, 207
Reisberg, L., 476, 479, 482, 490, 498

Ren, J., 395
Reuber, I., 461
Reynard, B., 68, 176
Ridley, J., 200
Ringwood, A.E., 37, 39, 99, 478, 481
Roberts, D., 250, 252
Robertson, E.C., 45
Robinson, D.N., 485
Robinson, P., 266
Romey, W.D., 262
Rosenfeld, J.L., 167
Rosenhuaer, M., 52, 486
Rossi, G., 121, 122, 310, 326
Rossman, G.R., 162
Royden, L.H.R., 415, 416
Rozen, O.M., 333, 427, 428, 429
Rubie, D.C., 165, 167, 174, 175
Rutland, R.W.R., 338
Ryabchikov, I.D., 52, 103
Rykkelid, E., 172

Sacks, I.S., 469, 79, 501
Sanders, I.S., 323
Sandiford, M., 119
Sandrone, R., 15, 235, 236
Sang, B., 13, 383, 400
Santallier, D.S., 266, 267,
Sautter, V., 34, 457
Saxena, S.K., 62, 63, 66
Scaillet, S., 151, 163
Schertl, H.-P., 2, 41, 113, 132, 143, 162, 176, 214,
    217–219, 220, 221, 227, 233, 234, 237
Schiffman, P., 474
Schliestedt, M., 77, 78, 80
Schmadicke, E., 20, 207
Schmid, S.M., 143, 147
Schmitt, H.H., 329
Schrauder, M., 337
Schreyer, W., 2, 24, 36, 38, 39, 41, 42, 45, 46, 50,
    55, 57, 58, 67, 118, 119, 120, 124, 138, 162,
    163, 214, 217, 219, 234, 235, 237, 450
Schubert, W., 489
Schulien, S., 69
Schulze, D.J., 478
Sclar, C.B., 35, 37
Seifert, F., 119
Seki, Y., 38
Sekine, T., 42, 51
Sengor, A.M.C., 25, 148, 383, 392, 401, 417, 420
Senior, A., 121
Sharp, Z., 51, 162, 234
Sharps, R., 406
Shatsky, V.S., 3, 8, 11, 72, 97, 101–105, 107, 108,
    110, 111, 207, 311, 333, 334, 341, 356, 427,
    428, 431, 432, 443, 444, 445, 447, 448, 450,
    451, 459, 502
Shau, Y.-H., 68
Shervais, J.W., 467, 478
Shilo, N.A., 483, 501
Shimizu, N., 101
Shui, T., 11

Sidorenko, A.V., 428
Silling, R.M., 150
Simpson, C., 147
Sjostrom, H., 252
Skjerlie, K.P., 249, 450, 451
Slodkevich, V.V., 458, 462, 465, 470, 480, 493
Slutskiy, A.B., 37
Smith, C.B., 469
Smith, D.C., 2, 7, 8, 73, 98, 99, 106, 112, 118, 121,
    122, 123, 161, 165, 207, 215, 218, 220, 245,
    248, 256, 257, 282, 283, 303–338, 340, 341,
    356, 357, 478, 487
Smith, J.V., 107
Smyth, J.R., 106, 478
Snoke, A.W., 147
Sobolev, N.V., 3, 9, 72, 97–106, 108, 110, 111, 207,
    311, 333, 334, 341, 356, 427, 428, 430, 432,
    443, 444, 445, 447, 448, 450, 451, 459, 467,
    469, 470, 478, 488, 502
Sobolev, V.S., 98, 99, 108, 111
Spackman, W., 500
Spear, F., 69, 257, 338
Spencer, D.A., 176, 323
Spencer-Cervato, C., 323
Stabel, A., 253
Stampfli, G.M., 133
Staugigel, H., 42, 57
Stephens, M.B., 250, 252, 260, 266, 267, 269
Stephens, W.E., 402, 403, 416
Stern, C.R., 51
Stiberg, J.-P., 259
Stockhert, B., 17
Stolen, L.K., 245, 266
Sturt, B.A., 252, 303, 312
Su, Q., 402
Sudo, A., 50
Suen, C.J., 479, 481, 482
Sunagawa, I., 107, 466
Svisero, D.C., 104

Tagiri, M., 207
Takahashi, E., 47
Takasu, A., 250
Talbot, J.L., 414
Tapponnier, P., 413, 414, 417, 420, 421
Tatsumi, Y., 50
Taylor, H.P., 474
Taylor, S.R., 435
Terada, S., 49
Terhart, L., 61
Theye, T., 50
Thompson, A.B., 24, 35, 61, 167, 173, 413, 414,
    415, 416, 419, 420, 421
Thompson, R.N., 467
Thon, A., 250
Tilton, G.R., 17, 135, 151, 162, 163, 175, 176, 212,
    407, 414
Tlili, A., 327
Tørudbakken, B., 245, 258,
Tricart, P., 149
Trommsdorf, V., 341

Trueb, L.F., 107, 108, 335
Trumpy, R., 207
Trouw, R.A.J., 266
Tsai, H.M., 465
Tucker, R.D., 259
Tull, J.F., 260
Turcotte, D.L., 461

Ungaretti, L., 122, 310, 314, 316

Vagonov, V. E. 483
Valley, J.W., 478
Van der Molen, I., 2, 166, 168
Van der Wal, D., 500
Van Roermund, H.L.M., 2, 73, 168, 245, 252, 266, 267, 268, 280, 281, 322
Van Wyck, N., 263
Varshavsky, A.V., 106
Vavilov, M.A., 444, 459
Veblen, D.R., 102, 103
Velain, C., 335
Vialon, P., 137, 210, 217
Vidal, O., 50
Vielzeuf, D., 450, 451, 495, 499
Visser, D., 121
Vissers, R.M., 498
Vuichard, J.P., 15

Wade, S.J.R., 262
Wakabayashi, J., 77
Walcott, R.I., 172
Walker, J.D., 149
Walther, I.V., 450
Wan, T., 406, 407, 413
Wang, H.F., 8, 319
Wang, Q., 106, 163, 357, 369
Wang, T., 164, 402, 403, 416
Wang, X., 3, 11, 12, 13, 161, 164–168, 207, 356, 357, 359, 361, 363, 365, 368, 370, 371, 375, 376, 377, 378, 379, 381, 382, 383, 384, 391, 400, 401, 410, 416
Ward, C.M., 118
Wasserberg, G.J., 260, 306
Wastrum, E.F., 63
Waters, D., 328
Watson, M.P., 404, 405
Webb, S.A.C., 274, 490, 492
Wegmann, C.E., 335
Weijermars, R., 479
Wendlandt, R.F., 46
Wendt, A.S., 105, 106, 152, 166
Wenji, B., 22

Wezel, F.-C., 336
Wheeler, J., 139, 148, 151, 172
White, A.J.R., 402, 403
White, W.M., 477
Wijbrans, J.R., 421
Wilding, M.C., 457
Wilkinson, J.F.G., 470
Williams, P.F., 148, 266
Wilshire, H.G., 467
Wilson, M.R., 252
Windley, B.F., 413
Woermann, E., 486
Wones, D.R., 402, 403
Wood, B.I., 48, 63, 69, 274, 450, 490, 492
Woodland, A.B., 50, 466
Wu, Z.W., 399, 406, 415, 417
Wunder, B., 36, 38, 124
Wyllie, P.J., 39, 40, 42, 45, 46, 47, 49, 51, 224

Xie, D., 13, 357, 376
Xu, J., 357, 358, 417
Xu, S., 12, 13, 111, 164, 207, 331, 356, 358, 359, 368, 381, 383, 391, 400, 404, 413
Xue, H., 11

Yagi, K., 49
Yamada, H., 47
Yamamoto, K., 36, 37, 124
Yang, J., 3, 207, 316, 320, 357, 376, 395
Yang, S., 11, 405
Ye, D., 357
Ying, S., 356
Yoder, H.S.Jr., 41
You, Z., 396, 397

Zachrisson, E., 266
Zanazzi, P.F., 112, 121
Zang, Q.,118, 119, 164, 165, 357, 401
Zhang, B., 395, 396
Zhang, G., 397, 399
Zhang, R., 13, 357, 361, 363, 365, 368, 369, 371, 374–379, 382, 383, 384
Zhang, S., 382
Zhang, Z.M., 11, 359
Zhao, X., 406, 407, 413
Zhou, G., 358, 361, 382, 400, 401, 415
Zhou, M.F., 458, 483, 484, 486, 500, 501, 502
Zhu, H., 406, 407, 413
Zindler, A., 459, 476, 479, 482
Zinner, E., 335
Zonenshain, L.P., 8, 9, 428, 429
Zwart, H.J., 7, 266

Subject Index

Bold numbers refer to figures. Italics refer to tables.

Aarsheimneset, headland, 304, 329
  eclogite pod, 327
activity
  of H₂O, 162
Adria plate, 208
Adula nappe, Switzerland, 341
African margin, 23
Alboran basin, 460
Almklovdalen, Norway, 306, 307, 338
alluvial diamonds, 501
  Alaska, 483
  Dongiao, massif, 485
  Kalimantan, 483
  Kokchetav, 427
  New South Wales, 483
  Tasmania, 483
  Urals, 483
  USA, 483
  Victoria, 483
amphibole
  thermodynamic mixing property, 68
amphibolite
  eclogite, Munchberg klippe, 19
  facies, 21
  garnet bearing, Kokchetav, 10
  grade, Dabie Shan
  P-T stability, 49–50
anatexis, 21, 22, 23, 25
Anhui province, China, 10
  geologic map, 11
Apulian plate (Africa), 13, 175
  continental margin, 14
aragonite
  Dabie Mountains, 164, 379
  transition to calcite, 40, 167
Armorican massif, 73
Austroalpine nappes, 133
authochthon, Norway, 250, 253

Baltic plate, 5

Baltoscandian continental margin, 5
Bamble area, Norway, 121
Barradian terrane, 19
bearthite, 96
  Monte Rosa massif, Western Alps, 116
  occurrence, 115–116
  properties, 116–117
  stability, 117
  Vestana deposit, Waermland, Sweden, 116
Beni Bousera peridotite, massif, 22
  age, 459
  diamond facies pyroxenite, 462
  emplacement mechanism, 496–500
  graphite aggregates, 480
  graphite octahedra, 470
  petrology, 459–462
  picromelanite, 470
  pyroxenite lithology, **462**
  pyroxenite composition, 467
  pyroxenite-peridotite, 466, 479, 489, 498
  radiogenic isotopes, 475–476
  regional geology, 459–462
  tectonic model, 499
  thermobarometry, 487–488
Bergen, Norway, 73, 301, 305
Betico-Rifean tectonic belt, 22, 459, 479
Bitlis massif, 73
blastomylonite
  Kokchetav, **433**, 437
blueschist facies, 15, 175
  Dora-Maira, 15, 139, 146, 186, 191, 199
  overprint, 199
Bohemian massif
  age, 19
  description of, 17–20
  volume and thickness, 21
Borovsk series, Kokchetav, 430
  geologic map, **431**
Brossasco-Isasca unit (BIU), 210, 207
  Alpine history, 232

519

Brossasco-Isasca unit (*cont.*)
  climax conditions, 233
  decompression path, 234
  ellenbergerite, 113
  geologic map, **211**
  metamorphic evolution, 232–235, **209**, **211**
  pre-Alpine history, 232
  radiometric age, 212
Brossasco-Isasca petrography and petrology
  clinopyroxene composition, **214**
  deformed paraschists, 227–228
  eclogite, 230
  garnet composition, **215**
  kyanite paraschists, 227
  marble, 228
  marginal igneous facies, 216
  metagranitoids, 213–216
  monometamorphic unit, 213–222, 235
  photomicrograph, relics and metagranitoid, **225**
  polymetamorphic complex, 222, 235
  pyrope-bearing white schist, 217
  quartz-coesite-jadeite-phengite, 221
  relict Variscan contact metamorphic schist, 223
  relict Variscan paraschist, 222
  sodic whiteschist, Case Ramello, 221
  whiteschist, Case Ramello, 219
Briançonnais-Grand Saint Bernard nappe, 137, 149
  Sub-Briançonnais, 149
Burkino Faso, West Africa, 483

Cadominian basement, 19
calc-alkaline rock, 22
calderite, 50
Caledonian Orogeny
  collision, 8
  FIF model, 343
  high-pressure history, 7, 306
  Kazakhstan, 428
  Norway, 5, 207, 301, 305, 312, 323
CaO-Al$_2$O$_3$-SiO$_2$-H$_2$O (CASH) system, 44–45
CaO-FeO-MgO-Al$_2$O$_3$-SiO$_2$ (CFMAS) system, 47–58
CaO-MgO-SiO$_2$-H$_2$O (CMSH) system, 488
CaO-Na$_2$O-MgO-Al$_2$O$_3$-SiO$_2$ (CNMAS) system, 47–58
carbon isotopes
  Beni Bousera, 476–478
  Brazilian carbonados, 108
  for diamond, 108–110, **109**
  Grytting, 324
  isotope variation, 478
  mantle vs crust, 476
  Ronda, 483
  Ulstein-Dimnöy, 328
carbonados, 107
Case Ramello, near Parigi, 113
Central Africa, 107
Central Alps, 17
Central Asian fold belt, 8, 428
  tectonic scheme, **429**

Chaos Theory, 338
chloritoid, 60, 71
chromium number [100Cr/(Cr+Al)], 490, **491**
Cima Lunga nappe, Switzerland, 341
cliftonite, 465
  Beni Bousera, 465
  Disko Island, 465
  meteorites, 465
clinopyroxene
  Al content in, 47–48, 49
  Cr solubility, 103
  potassium in, 96, 101–103, **102**
  thermodynamic mixing property, 63–67, *64–65*
  in UHPM rocks, 101–104
clockwise *P-T* path
  Dora-Maira, 15
coesite, 1–3, 21, 25, 162, 481
  Alps, 3, 14, 17, 207,
  Aosta valley, 186
  Bohemia, 3
  Bohemian massif, 17
  Brossaco-Isasca unit, 215, 218, 227
  Dabie Shan, 3, 13, 161, 362, 365, **367**, 402, **408**
  Dora-Maira, 138, 141, 142, 145, 148, 161
  Grytting, Slje, 7, 324
  identification of, 105–106
  Kokchetav, 3, 495
  Makabal, 3
  Maksutov, 3
  Mali, 3
  Norway, 3, 161
  stability boundary, 35
  Straumen eclogite, 325
  Su-Lu region, 3
coesite eclogite province, Norway, 299, 305–312, 330, 331, 334, 341
continental collision, 21, 23
  consumption, 23
  Europe-Africa (Apulia), 133
  intracontinental thrusting and wedging, 25
continental crust
  old, cold, dry, 24
  partitioned flow, 146
crustal shortening, 287
crustal thickening, 173

Dabie group, 359
Dabie Mountains, eastern China (also see Dabie Shan)
  age of UHPM, 383–384, 407, 410–411
  amphibole, 375
  aragonite, 379
  blueschist facies, 368
  coesite in marble, 379
  cold eclogite, 361
  collision, 384
  comparison with Su-Lu, *362,* 381–382
  country rock, 378
  description of, 164, 357, 392, 399–402
  eclogite, 356, 363, *364*
  exhumation, 176, 419

foliation, 363
geochronology, 361
geophysical constraint, 405–407
glaucophane, 368
hot eclogite, 361
in situ model, 382–383, 410
I-type granitic rocks, 402–404
jadeite, 379
mafic rocks, 410
map, **358, 360**
microdiamond, 368
mineral assemblage, 365
mineral chemistry, 369–376
paleomagnetic, 406–407
petrology, 365, 378
pressure-temperature path (trajectory), **373,**
376–378
quartz pseudomorph after coesite, 368
regional UHPM, 382
retrograde metamorphism, 375–376, 377
tectonic history, 391
tectonic scenarios, 384
terrane (north and south), 359, 418
Dabie Shan, eastern China (also see Dabie Mountains), 10, 207, 320, 333, 41
age of metamorphism, 13
basement, 11, 13
description of, 10–13
diamond verification, 110–111
low angle thrusting, 12
map, 11
metamorphic mineral assemblages, 13
orogen, 13
peak metamorphism, 13
regional scale metamorphism, 164–165
southern margin, 12
decompression, 173
deerite, 50
deformation
buoyancy, 150
coaxial, 143
collapse, Dora-Maira, 149
crustal extension, Dora-Maira, 149
Doria-Maira, late, 146–148
heterogeneous, Brossaco, granite, 145
Dent Blanche nappe, 212
detachment zones, 25
diabase
Fjordane complex, 5
diamond, 2, 148, 201
alteration to graphite, 465
bearing UHP rocks, 9, 207, 458, 498
Burkma Faso, 483
crystallization of, 452
economics, 501–502
experimental *P-T* conditions, 35
exploration, 98, 501
facies in pyroxenite, 461
facies in ultramafics, 459
graphitized, 458, 473, 479, 483, 488, 493, 502
mica in, 444

Ronda, 480, 483
temperature of crystallization, 493
Tibetan ophiolite, 484
Donghai County, Shandong, eastern China
description of, 72
Mg-staurolite, 357
microdiamond, 73
Dora-Maira massif, western Alps, 2,13, 14, 132,
182, 207, 310, 312, 340
age, 15, 17, 186, 212
case study, conclusion, 153
coesite-bearing eclogite, 71
decompression and cooling, 149
description, 13–17, 161–164,
exhumation, 148–152, 172, 175, 176
experimental study, 41
geologic map, **136, 184, 209**
geologic setting, 133–137, **134,**
Mg-staurolite, 117, 118, 119
monometamorphic complex, 210
nappes, 14, 183
overburden, 149, 183
P/T determination, 193, 200–201
P/T of mylonite, 139
peak metamorphism, 15, 138
polymetamorphic complex, 210
structure, **136**
talc-chloritoid-kyanite schist, 138
tectonic scenario, **151**
Dronero-Sampeyre unit (DS), 137
Drosendorf terrane, 18

East Sudeten, 18
Eastern Alps, 13, 135
Austro-alpine nappe, 13
eclogite facies, 15, 138, 191, 209
assemblage temperature, 488
eclogites, 20, 301
Aarsheimneset headland (microcoesite),
327
Brossasco-Isasca unit, 230
coesite in, 3
"cold", Dora-Maira, 199
Dabie Mountains, 356, 363, *364*
Dabie Shan, 12, 13
Dora-Maira, 135, 138, 149, 186, 230
Drage (microcoesite?), 326
Erbendorf-Vohenstraus zone (ZEV), 19
exotic origin, 21
Fjordane gneiss, 5, 7
Flöha syncline, 20
Gföhl terrane, 18
Grytting (microcoesite), 323–325
Hessdalen (microcoesite?), 329
Jiangsu and Shandong (Su-Lu), 3, 356, 363
Kokchetav complex, 10, 430
Liset (microcoesite?), 326
minerals, associated with, 316–317
Munchberg klippe, 19
Norwegian, 245, 250–271, 301, 304, 305, 309,
323–326

Norwegian controversy, 245–249, 317–318
  origin of, 248–249
  orthopyroxene-bearing, 7
  quartz, high pressures, 479
  Sesia, 230
  spatial distribution, 250–253
  Straumen (microcoesite), 325–326
  structures in, 143–144
  Ulstein-Dimnöy (microcoesite?), 328
  Western Alps, 14, 128
  Winklaren, 19
ECORS-CROP profile, Europe, **134**
Eiksundalen, Norway, 312, 306
Elektin series, Kokchetav, 430
  geologic map, **431**
ellenbergerite, 39–40, **36**, 96
ellenbergerite group
  chemistry, 114–115, **114**
  occurrence, 113–114
  stability, 115, **116**
  structure, 112–113
Enbek-Berlyk, Kokchetav
  metamorphic assemblages, 10
enstatite
  Al_2O_3 content, 43–44
enstatite eclogite, 271
Erbendorf-Vohenstraus zone (ZEV), 19
erosion
  rates of, 23
Erzgebirge, 19
  high-grade metamorphics, 20
European plate, 13, 183
  Europa, 152
  lithosphere, 133
  margin, 23
  Tethyan hiatus, 152
exhumation
  Bohemian massif, 20
  cooling during, 172
  Dabie Mountains, China, 159, 176
  decompression, 173
  granulite, 25
  mechanism, 171–172
  rates of, 5, 22–23, 175–176, 177
  tectonics, Dora-Maira, 147, 148–152, 159
  Western Gneiss Belt, Norway, 285, 343

fabrics
  discussion of, Dora-Maira, 146
  foliation and lineation, Dora-Maira, 146
  L-S tectonite, 143, 237
  stretching lineation, Dora-Maira, **147**
  subsequent, Dora-Maira, 144, **144**
Fe-Mg distribution
  garnet-clinopyroxene, 48
  garnet-phengite, 49
Fe_2SiO_4 spinel, 39
Fenzishan group, eastern China, 59
Fjordane complex, Norway, 5, 7
  gneissic rocks, 5
  supracrustal rocks, 5

Fjörtoft, Norway, 8, 306, 329, 333
  deduced microcoesite in gneiss, 330
Flemsøy, Norway, 309
Flöha syncline, 20
fluid inclusion, 258, 263
"foreign model", 248–249
Fosen peninsula, Trondheim, 322
Franciscan complex, California, 34
  eclogite, 74–75
  P-T conditions and P-T path, 75, **76**, 77
  thermodynamic calculation, 71, 73–77

gabbro
  in Fjordane complex, Norway, 5
  in Kokchetav complex, Kazakhstan, 10
garnet
  CaO versus Cr_2O_3 in, 98–100, **100**
  as container of mineral inclusions, 97
  in Dabie Mountains, 369, **370**
  sodium in, 96, 100–101
  thermodynamic mixing property, 63
  in UHPM rocks, 98–101
  zonation, **371**
Garnet peridotite, 18, 22
  Flöha syncline, 20
  Gföhl terrane, 18
  Winklaren, 19
geochemistry
  Kokchetav, Zerendin series, 434- 435, 450
geochronology, 5, 21
  Brossaco-Isasca Unit, 213
  Brossaco meta-granite, 144
  Dora-Maira, 5, 163, 186, 201
  retrograde growth, 25
  Western Gneiss unit, 5, 303–307
geodynamic models
  creation and exhumation, 336
  Dora-Maira, **151**
  Norwegian eclogite, 317
  ultramafic rocks, 496–500
  Western Gneiss region, 343
geologic setting
  Brossasco-Isasca unit, 207
  Dora-Maira, 207
  UHPM localities, 5–20
  Western Alps, 183
Gföhl terrane, 18
  estimated P-T conditions, 18
Glarus Alps, 17
glaucophane, **36**, 40
  in Dabie Mountains, 368
  in Su-Lu, 374
  stability field, 47
Glomfjorden, Norway, 323
Gondwana, 17
Gran Paradiso, 207, 212, 235
  amphibole, 123
  transect, 149
granites
  Brossaco metagranite, 144–146, *216*
  melting P-T conditions, 51

Mesozoic granite, Dabie Shan, 12, 402
wet solidus, Tonalite, **24**
granulite facies, 22
Granulitegebirge, 19
P-T estimates, 20
structure, 20
graphite, 462
aggregates, 465, 480
bearing pyroxenite, 480, 498
forms, 466
graphitized diamonds, 462, 498, 500
octahedron, **464, 467,** 480, **481**
from peridotite xenoliths, 476
rate of graphitization, 497
shapes, **464**
time for graphitization, **497**
Greek Cyclades, 34, 71, 77–81
Greenland plate, 5
greenschist facies
Dora-Maira, 138, 139, 146, 149, 162, 192
Grytting, Norway, 304, 305, 312
field photograph, **313**
Gunskebotn, Norway, 306

Haizhou group, eastern China, 359
Hardangervidda-Ryfylke Nappe Complex, Norway, 265
Hefei Basin, Anhui, 404
Hefei-Jiaoli basin, China, 11
Helvetic nappes, 175
Henan province, China, 10
geologic map, 11
Hercyian basement, 14
Holsnöy, Norway, 309, 312, 320
Hongan block, Hubei, China, 358, 392, 400
structure of blueschist, 401
Huangzhen, Dabie, China, 365, 379
Hubei province, China, 10
geologic map, 11
hydrothermal alteration of radiogenic isotopes, Beni Bousera, 476
hydrous Mg-silicates, 35–38

"in situ model," 248–249
Isasca formation, 14, 140, 210
geologic map, 211
Ivera zone, 14
gravity anomaly, **134**
magnetic anomaly, **134**

jadeite
Brossco-Isasca unit, Western Alps, 213, 219, 227
Dabie mountains, China, 3, 379
Dora-Maira, Western Alps, 138, 149
Kokchetav, Kazakhstan, 443
P-T stability field, 44–45
Jæmtland, Norway, 267–268
Jiadong group, eastern China, 359
Jiangsu-Shandong area, China, 13
Junction School eclogite, California, 73–77, **74, 75**

Kazakhstan block, 9, 331, 333, 335, 428
Kazakhstan-Northern Tianshan belt, 429
tectonic scheme of Paleozoids, **430**
K-cymrite, 38, 45
KD plots
garnet-clinopyroxene, Beni Bousera, 494
kimberlite, 457, 466, 469
Siberian, 466
South Africa, 99
Kokchetav massif, Kazakhstan, 8, 333, 334, 341, 427
alluvial diamonds, 427
description of, 8, 9, 427, 428
diamondiferous metamorphic rocks, 72, 427
expedition for diamonds, 427
geologic map, **9, 431**
geologic structure, 428
metamorphic conditions, 447, 450
Pakhar site, 448
partial melting, 452
sodium in garnet, 100–101
tectonic reconstruction, 23
K2O-MgO-Al₂O₃-SiO₂-H₂O (KMASH) system, 45–47
kinetic studies, 167, 168
Kristiansund area, 73
Kumdy-Kol, Kokchetav, 334, 427, 428
diamondiferous rocks, 431
eclogites, 430, 431
lake, 431
mineral assemblage, 10, 432
mineral composition, *438–441*
P-T parameters, 448
partial melting, 451
pyroxene, 103, 444
rock composition, *434*
kyanite gneiss
Flöha syncline, 20

Lago Superiore, 191
metamorphic temperatures, 199
lamporite, 457, 501
Laurasia, 17
Leknes group, Norway, 262
Lepontine event, 17
lherzolite
garnet, 479, 490, 498
sheared pyroxenite, **489**
spinel, 461, 490, 498
spinel facies, Beni Bousera and Ronda, 488
subsolidus equilibration, 494
Lingorka massif, 99
Liset eclogite, 121
lisetite, 121–122
Liverpool Land, East Greenland, 314, 322
Lofoten-Vesteralen area, 253
Rendeelvan area, 314
Tværdalen area, 314

magmatic rocks
  mafic underplating, 21
  magma underplating, 24
magnesiodumortierite, 120–121
magnesiostaurolite, see Mg-staurolite
magnesium number (100Mg/Mg+Fe), **491**
mangerite, 5, 301
mantle, upper, 1, 494
marble (UHP), 379, 381
Mediterranean geologic map western part, **460**
melange
  antigorite serpentine, 186
  serpentinite, 182
  serpentinite matrix, 186
melting relations, 51
Mg-carpholite, 40, 41, 58
Mg-chloritoid, 39, 41, 57
MgMgAl-pumpellyite, 38, 57
MgO-Al$_2$O$_3$-SiO$_2$-H$_2$O (MASH) system
  mineral assemblage, 41–44,
  petrogenetic grids, **53**, **54**, 55
Mg-staurolite, 41, 57–58
  composition, 118
  occurrence, 117–118
  properties, 118–119
micaschist, 256
microcoesite
  deduced Norwegian localities, 326–331
  Norwegian localities, 323–326
microdiamond
  coesite-eclogite province, Norway, 111, 341
  Dabie Shan, China, 13, 110–111, 357
  description of, 106–112
  Donghai county, Shandong, China, 73
  Fjörloft Island, Norway 8, 331–336
  Kokchetav, Kazakhstan, 10, 106–110, 428
mineral assemblage
  Kokchetav, 432–433, **433**
  marble (BIU), 228
  Mont Viso, lower oceanic unit, 187, **188**, 200
  Norwegian eclogites, 30
  peak metamorphism, Case Ramello, 219
  polymetamorphic complex (BIU), 222–223
  UHPM, 310, 311, 315
  white schist, Case Ramello, 217, 219
mineral composition
  Beni Bousera, 467; clinopyroxene, 468, **468**, 469,
  470, garnet, 469; pyroxene in Ca-Mg-Fe
  space, **471**; Pb-isotopes for clinopyroxene, **477**
  Brossasco-Isasca: garnet, **215**, 218; clinopyroxene, **214**; amphibole, **220**; phengite, **222**;
  omphacite, 231; garnet, 231; phengite, 231
  Kokchetav: garnet, **437**, *438*; phengite, *438*; clinopyroxenite, *438*, **445**; biotite, *438*; mica; **446**
  Mont Viso: eclogite minerals, *195*; lower oceanic
  unit, 187, **189**, **190**, *190, 191, 192, 195, 196;*
  retrograde minerals, *196*
  Ronda: graphitic pyroxene, 482; garnet, 482
minerals from Kokchetav
  biotite, 444
  carbonates, 445

  coesite, 445
  garnet, 435, 436, 437, *438*
  garnet zoning, **442**
  mica, 444
  mica inclusions, 444
  phengite, 444
  pyroxene, 443–444
  pyroxene zoning, **445**
  sphene, 445
Modum deposit, Norway, 115
moissanite
  Tibetan ophiolite, 502
Moldanubian zone, 18, 17
  tectonics, 19
Moldefjorden district, Norway, 272, 331
Mont Rosa, Central Alps, 116, 207, 212, 235
Mont Viso ophiolite, 15
  geologic map, **184**
  higher sub units, 191
  Lower oceanic unit, 183, 200
  Median oceanic unit, 183
  mineral assemblage, lower unit, 187
  steep normal fault, 183
  temperature of metamorphism, 199
  transect, 183
Motalafjella HP complex, 73
Munchberg gneiss, 19
Munchberg klippe, 19, 310
  equilibrium assemblage, 19
  radiometric age, 19
mylonite
  blastomylonites, Kokchetav, 433
  Dora-Maira, 148, 139
  Kokchetav, 432

Nanyang Basin, eastern China, 404, 405
Naqssuqtoqidian mobile belt, Greenland, 322
Na$_2$O-FeO-MASH (NFMASH) system, 58
Na$_2$O-MgO-Al$_2$O$_3$-SiO$_2$-H$_2$O (NMASH) system,
  47
Na$_2$O-K$_2$O-Al$_2$O$_3$-SiO$_2$-H$_2$O (NKASH) system, 47
neodymium isotopes
  Beni Bousera, 473
  Kumdy-Kol, 444
new or future research, 26
  field studies, 345
  high pressure granite systems, 26, 256
  large scale mapping, 344
  petrologic study of gneiss, 344
  stable and radiogenic isotopes, 26, 345
  tomography, 26
Nordfjord area, 73, 272
North Dabie Terrane (NDT), 359, 401, 418
Northern Territory, Australia, 108–109
North Huaiyang belt, China, 11, 396
Norway
  age constraint, 258–260, 264–265, 267, 269,
  270–271
  allochthon, 252, 262, 266
  Bergen arcs, 247, 258, 262, 320
  coesite-eclogite province (CEP), 282, 300

diamond gneiss province (DGP), 300
eclogite controversy, 245–249, 341
facies grid, **251**
fluid inclusion, 258
garnet-ultrabasic province (GUP)
microcoesite, 300
microdiamonds, 300, 304
mineral assemblage, *251,* 254
pressure-temperature (-time) evolution, **255,**
    260, **261, 265,, 268,** 270, **271**
protolith, 254, 257
tectonic model, 283–287
tectonostratigraphy, 253
Ulstein-Dimnöy, 330–331
Vestranden gneiss (VGC), 322
nyböite, 122–123, 310, 315, 325, 365, 368, 374

Oberplatz klippen, 19
OH-Topaz, 38
olivine, 68–69
olivine eclogite, 273
Oman mountains, 135, 152
omphacite
    in Dabie Mountains, 371–374,
    Na content in, 48
ophiolite, 21, 137, 323, 457
    Bongong-Nujiang belt, 484
    diamondiferous: Tibet, 458, 483, 484, 500; Kor-
        yak Mts., 483, 500; Bobaris belt, 501
    Great Serpentine Belt, 501
    Mont Viso, 139
    ocean crust, 22
    on top of Dora-Maira, 183
    Piemonte zone, 207
    Tulameen, British Columbia, 500, 501
    UHP-HP, 150
    Yarlungzangbo belt, 484
    Zermatt, 135
Omsk, Russia, 8
orthopyroxene
    Al content in, 47–48, 49
    thermodynamic mixing property, 68–69
overpressures, 337, 338, 340, 343
    regional tectonic, 339
    in situ, 339
    taboo, 340
oxygen fugacity, 466
    CCO buffer, 466
    FMQ buffer, 466
    Tibetan ophiolite has a low, 486
oxygen isotopes
    Beni Bousera, 473, **474**
    Grytting, 324
    Ronda, **474,** 482–483
    Ulstein-Dimnöy, 328

palisade texture, 141
Pan-African basement, 17, 18
    Apulia, 152
Pangea, 9
parauthochthon, 250, 253

paragonite, **36,** 40, 44
Parigi gneiss, 137, 142
Passo Gallarino, 191
    metamorphic temperature, 199
    minerals, **194**
Pennine nappes, 13, 14, 15, 21, 133, 207, 212
    Hohe Tauren window, 135
peridotite
    age, 273–275
    Alpine-type (orogenic), 272, 459
    garnet, 458, 271–281
    garnet-spinel facies, 458
    Mg-Cr type, 273, 274, **275**
    orogenic with diamond, 483;
    orogenic, *P/T* conditions, 458, 494
    pressure-temperature (-time), 276–280
    solidus, 51
    spatial distribution, Scandinavian Caledonides,
        271–281, **273**
    spinel facies, 457
    thermobarometry, 486–488
    Urals, Tulameen, Little Caucasus, Koryyak
        Mts., 501
peridotite massif, 457
    Alpe Arami, garnet-bearing, 458
    basal tectonite, 457
    Beni Bousera, 458
    Ronda, 458
petrogenesis
    Beni Bousera 471–479
    Luobusa and Donqiao massifs, 486
    Ronda, 482–483
    Tibetan diamondiferous ophiolite, 486
petrogenetic grid
    Brossasco-Isasca unit *P-T-t* paths, 224
    calculation of, 51
petrography-petrology
    Beni Bousera ultramafics, 460–462, *463*
    Brossasco-Isasca unit, 213–232
    Kokchetav diamond-bearing rocks, 432–433
    Mont Viso-Dora Maira, 201–203
    Ronda, 480–481
petrologic models
    foreign in-situ (FIF), 320, **332,** 342
    Norwegian ophiolites, 317
    regional in-situ, 318
    ultrametastability, 320
    ultrareactivity, 319
phase A, 35–38
phengite, 45–46
    Si content in, 46
    thermodynamic mixing property, 67–68, *65*
phlogopite, 45–46, 51–52
phyllosilicates, 42
Piemonte zone, 207
piezotite, 38
Pinerolo unit, 210
plate tectonics
    Africa-convergence, 152, 323, 488
    Asia-India collision, 500
    Baltica-Laurentia convergence, 319, 343

plate tectonics (*cont.*)
  Baltica-Laurentia vs. microcontinent, 322
  collision of island arcs, Kokchetav, 428
  paradigm, 23
  UHPM convergence rate, 23
Po Basin (Plain), Italy, 13, 207
Polish Czechslovakia, 18
pressure
  discontinuties, 15
  equilibrium, Beni Bousera and Ronda, 488
  estimated, Kokchetav, 447, 452
  overpressures, 23
pressure-temperature paths
  Beni Bousera, **495**
  clockwise, 173
  continent-continent collision, **24**
  Dora-Maira, 162, 193, 196, *197*, **198**
  estimates, 310
  Kokchetav, 448, *449*
  prograde, 169, **170**
  retrograde, 174
  Ronda, **495**
  subduction, 169
  thermal modeling, 169–176
  Tibetan, **495**
  with time, Beni Bousera, 496
  ultramafic rock, 486
  Western Gneiss Belt, **311**
  Zerendin series, Kokchetav, 448, *449*
protolith
  igneous, 138
  Kumdy-Kol, Kazakhstan, **444**
  UHPM rocks, 21–23
  Western Gneiss Region, Norway, 307, 309
pyrope garnet, 2
  megablast, 143
  pseudomorph, Dora-Maira, 142
  P-T stability, 41, **43**
  in quartzite, 41, 137
  in Venesca unit, Dora-Maira, 14, 17
pyroxenite
  Beni Bousera, 466, 488
  garnet with graphite, 462, 463
  graphitic, 466, 473
  Ronda, 488

Qinling-Dabie fold belt, China, 11, 392, 396–399
  age of metamorphism, 397
  compression (tectonic), 413
  exhumation, 415, 418, 419
  extension, 415
  geochronologic, 407
  geophysical constraint, 405–407
  I-type granitic rocks, 402–404
  magma genesis, 416
  paleomagnetic, 406–407
  paleotectonic, 411–417
  REE of granite, 403
  sedimentary basin, 404–405
  strike slip motion, 414–415
  thrusting, 413
  trace element, 402

quartz
  pseudomorph after coesite, 2, 3, 18, 20, 357, 368
  strength of quartz-rich rocks, 25

radiogenic isotopes
  Beni Bousera: Nd-Sr, 475, **476**; Sr, 475; Pb, 475, 476, **477**
  Dora-Maira, 163
  hydrothermal alteration, 477
  Ronda: Sr-Nd, 482; osmium, 482
Raman technique, 312, 445
rare earth elements (REE)
  clinopyroxene, Beni Bousera, **473**
  Kumdy-Kol, Kokchetav, 435
  peridotite, Beni Bousera, 472
  pyroxenite, 482, 493
  Zerendin series, Kokchetav, 435, 436, **436**, *438*
Raudhaugene, Norway, 306
retrogression texture, 165
rock chemistry
  Beni Bousera, lherzolite, 491
  Beni Bousera, pyroxenite, **472**
  Kumdy-Kol, metamorphics, *434*
  Western Gneiss Belt, 307, 309
  Zerendin series, **435**
Ronda peridotite massif, Spain, 22, 479, 483, 489
  age, 459
  diamond, 480, 493
  emplacement mechanisms, 496–500
  geology, 478
  graphite aggregates, 480
  graphite bearing, 480
  graphite octahedron, **481**
  petrology, 480–481
  tectonic setting, 479
  thermobarometry, 486–488
Rongcheng, Su-Lu, China, 370

sample descriptions
  Dora-Maira–Mont Viso, 201–203
San Chiaffredo unit, 210
sapphirine, 42
Satra Nappe, 266
Saxo Thuringian terrane, 19, 207
Scandinavian Caledonides, 244, 245, 271
Scotland, 323
Selje district, Norway, 301, 304, 306, 309, 331, 340
Sesia Lanzo zone, 14, 17, 146, 212, 235
Seve thrust complex, Sweden, 266, 280–281, 322
Sevier orogen, 149
Shima, Dabie, China, 371
Siberian lowland, 429
Sino-Korean craton, 10, 11, 164, 358, 392, 393, 395
Skaerfjorden area, Greenland, 322
  eclogite province, 322
Sneznik mountains, 18
Sognefjord, Norway, 5
South Dabie Terrane (SDT), 359, 383, 401, 402, 418
Sowie Gory, 19
Spitsbergen, 322

stability fields
  coesite-quartz, 24
  diamond-graphite, 24, 452
  jadeite-albite-quartz, 24
Stadlandet peninsula, Norway, 301, 327
  aerial photograph, **308**
stishovite, 35
strain
  coaxial constriction, 146
  continental rocks, 146
  Dora-Maira regime, 142
  Eoalpine, regional, 143
  rate, 169
Straumen, Norway, 312
stress
  deviatoric, 25, 339
  differential, 169
stretching lineations
  Dora-Maira, 139, 143
structures
  cross-section, Mont Viso–Dora-Maira, 185
  fabrics, Dora-Maira, 141–142
  Hongan blueschist, China, 401
  imprints, Dora-Maira, 148
  low angle normal fault, 152
  pre-UHPM, 144
  record of, Dora-Maira, **150**
  relic, 133; Dora-Maira, 139–141, **140–141**
  southern Dora-Maira, 137
  superimposed, 133
Su-Lu, China, 207, 316, 320, 356
  granulite facies, 369, 377
  pressure-temperature, **373**
subduction, 23, 146, 501
  Alpine, 182
  of continental crust, 457, 486
  European Mesozoic margin, 133
sulfur (sulphur) isotopes, 474
Sulu-Tjube, Kokchetav metamorphic mineral assemblage, 10
Sunnfjord, Norway, 8
Sunnmøre, Norway
  eclogite, 309
  metadolerite, 309
supracrustal continental rocks, 21
  thickness, 21
Susong group, Dabie, China, 359
Sweden, 322

Tan-Lu fault, 12, 13, 357, 384, 421
  strike-slip motion, 12, 401
tectonic models, 186, 283–287, 413–417, 499, 502
tectonic thinning, 171, 172
temperature of metamorphism
  Beni Bousera, 493–496
  graphitic pyroxenites, 496
  Ronda, 493–496
  Zerendin series, 498, *449*
Tengiz depression, 9
Tepla zone, 19

Tethyan
  ocean, 175, 183, 207, 210
  suture, 133
thermal boundary layer (TBL), 321
thermodynamic mixing models, 62–71, *64–65*
Tianshan, 207
Tibet, 22, 391
Tibetan ophiolite
  diamonds in podiform chromite, 484
  Dongiao massif, 484, **485**
  Luobusa massif, 484, **485**
Tongbai, China, 392, 396
  age of metamorphism, 397
Tongbai-Moziotan-Xiaotien fault, 11, 358
Torino, Italy, 13
Træna, Norway, 260
Tromsø, Norway
  eclogites, 269–271
  nappe complex, 322
Trondheim, Norway, 253, 301
Turgi Basin, Russia, 8

ultrahigh pressure metamorphism (UHPM)
  calculated *P-T* parameters, 23
  Caledonian, 281–287, 307
  collision, Bohemian massif, 17
  Dabie Mountains, 356, 382–383
  decompression, 143
  definition, 2, 4–21, 137–139
  emplacement, 502
  experimental studies on index minerals, 34–40
  experimental studies on mineral assemblage, 40–82
  formation of, 22–23
  geothermal gradients, 23
  melt generation, Kokchetav, 451
  melting of, 451
  metamorphic evolution, *216*, 452
  mineral assemblage, 40–82
  minerals, 2, 22
  model of formation, 334
  occurrences, **3**
  orogenic peridotites, 459
  P-T paths, 23, 148, 150, **150**, **198**, 452, 497
  peak metamorphic age, 22, 23, 201
  peak *P-T* conditions, 138, 236, 307, 447, 448
  petrogenetic grids, 40–82
  regional, 383
  retrogression, 143
  rocks, 2
  silicate phase *P-T* field, **36**
  size of blocks, 20–21
  strain during UHP, 146
  structural analysis, 133
  tectonic scenario, Dora-Maira, 150
  thermodynamic calculation, 40–82
  thermodynamic data, 55
  time of exhumation, 25
  ultramafic rocks, 456–503
Ulstein-Dimnöy, Norway, 330
Ultra-Dauphinois zone, 149

ultramafic rocks, 22
  mineral geothermobarometers, 486
uplift, 285
Ural mountains, Russia, 2, 207

Val di Susa, Italy, 207
Val Gilba, Italy, 113, 216
Val Maira, Italy, 207
Val Po, Italy
  Case Ramello, 217
Val Varaita, Italy, 113, 219
Variscasn orogeny, 17, 207
  continental collision, 18, 20
  thrust and strike slip faults, 17
Vartdalsfjorden district, Norway, 328, 329, 331
Venesca unit, Dora-Maira, 14, 15
  basement, 137
Vestlandet, Norway, 312, 323
Viso Mozzo, Italy, 191
  metamorphic temperature, 199

wagnerite, 114
Weihai, Su-Lu, 369
West Siberian basin, Russia, 8
Western Alps, 2, 34, 132, 159
  continental basement, Gran-Paradiso and
    Monte Rosa, 13
  description of, 161–164
  geologic map, **14, 134, 208**
  subduction, 22
Western Gneiss Region, Norway, 2, 5–8, 132, 300,
    341, 458
  age of UHPM, 8
  basal gneiss complex, 301
  creation and exhumation, 336
  eclogites, structure, 146

field relations, 300–303
geologic map **7, 246**
map, **302**
mineralogical evolution, 254
peak metamorphism, 7, 8
subduction, 22
Western Mediteranean, 22
white schist, 217
Wudang Mountains, 407
Wumiao, Dabie, China, 370

xenoliths
  eclogite, 457, 466, 478
  lherzolite, 457
  peridotite, 476
Xinxian, Henan, China, 357
Xinyang group, Henan, China, 396

Yakutian kimberlites, 98
Yangtze craton, 10, 395
  collision with Sino-Korean, 12, 13, 164, 357,
    392, 417–420
  margin, 12
  timing of collision, 406–407
yoderite, 41–42
Zerendin series, Kokchetav, 9, 10, 428, 429, 431
  diamond-bearing gneiss, 10, 430
  granitoid plutons, 430
Zermatt-Zass zone, Switzerland, 17, 21, 150, 161,
    163–164
zircon, 97
  coesite in, 104–105
  in diamonds, 104
  Kokchetav massif, 104–105
Zlote Mountains, 18

Printed in the United States
By Bookmasters